AF613902

Engineering Materials

Springer-Verlag Berlin Heidelberg GmbH

http://www.springer.de/engine/

Leonid Tushinsky · Iliya Kovensky ·
Alexandr Plokhov · Victor Sindeyev ·
Peter Reshedko

Coated Metal

Structure and Properties of Metal-Coating Compositions

With 264 Figures

Springer

Prof. Dr. Leonid I. Tushinsky
Materials Science Faculty
Novosibirsk State Technical University
Head of Novosibirsk Inter-University
Center for Problems of Modern
Materials Science
Corresponding Member of the Higher
School Academy of Sciences
Doctor Honoris Cause at the Silesian
Technical University, Poland and
Pridneprovsky Academy of Civil Engineering and Architecture, Ukraine
Post Office Box 99
630092 Novosibirsk, Russia

Prof. Iliya Kovensky
Head of Materials Science Faculty
Tyumen State Oil University
Minsk Str. 96, ap. 67
625048 Tyumen, Russia

Prof. Alexandr Plokhov
Materials Science Faculty
Novosibirsk State Technical University
Marx Avenue 20, ap. V-261
630092 Novosibirsk, Russia

Prof. Victor Sindeyev
Dean of the Mechanics and Technology Dept.
Novosibirsk State Technical University
Parkhomenko Str. 8, ap. 101
630100 Novosibirsk, Russia

Prof. Peter Reshedko
Electromechanics Faculty
Novosibirsk State Technical University
Marx Avenue 35, ap. 9
630073 Novosibirsk, Russia

Cataloging-in-Publication Data applied for
Die deutsche Bibliothek - CIP-Einheitsaufnahme
Coated metal: structure and properties of metal coating compositions / Leonid Tushinsky

(Engineering Materials)

DOI 10.1007/978-3-662-06276-0

http://www.springer.de

Originally published by Springer-Verlag Berlin Heidelberg New York in 2002
MyCopy version of the original edition 2002

Typesetting: Digital data supplied by authors
Translated from Russian by AUM Translations Services Ltd., Novosibirsk
Cover-Design: de'blik, Berlin
Printed on acid-free paper SPIN 10725644 62/3020 Rw 5 4 3 2 1 0
www.springer.com/mycopy

Introduction

Saving in metal, corrosion and wear control of machine parts are the problems of increasing priority. Conventional constructional materials are unable to provide reliability and durability of equipment under conditions of increased working speeds and loads, aggressive medium attacks and elevated temperatures. Solution to these problems involves change in properties of the surface layers of products, in the first instance, by means of depositing functional coatings on machine parts. Relative simplicity of the process and. practically unlimited possibilities of varying properties of coatings have resulted in their wide usage in machine building and transport industry, instrument engineering, radio electronics and other industries.

Coating provides means for enhancing constructional strength of steels and alloys[1]. It makes possible a combined method when one group of dislocation mechanisms is used for bulk hardening (grain refining, creation of stable subgranular dislocation structures), while the working surface is hardened through other mechanisms (dislocation density increase, separation of excess phases, etc). Combined hardening saves valuable alloying components and provides reduction in metal consumption of machines and constructions.

Coating allows manufacturing new composites that combine high durability (fatigue resistance, wear resistance) with sufficient reliability (fracture toughness); enhancement of operational stability of machine parts and tools as compared to stability attainable using conventional heat treatment methods; restoration of worn surfaces and, consequently, reduction in demand for spare parts.

However, wide application of coatings is hindered by poor development of special methods for evaluation of structure and properties of surface-hardened materials. This circumstance along with a lack of competent scientific analysis of experimental results not only makes the optimization of coating deposition regimes difficult, but may also compromise the undoubtedly promising practice.

Optimization of the technological processes of protective coating deposition depends on the improvement of old techniques and development of new methods for evaluation of the constructional strength of base metal-coating compositions. By now, investigation of coated metals has evolved into an independent experimental field with well-defined subject of study and specialized techniques.

The methods for evaluation of structure and properties should be considered as an integral part of continuous sequence: coating deposition – structure analysis – study of properties – refinement of technology – coating deposition, etc. Analysis

---

[1] Generalized parameter that determines material behavior under conditions maximally approximating field operation.

of the coating structure must be complemented and verified by the results obtained in studying mechanical, physical and special properties. This, in turn, allows refining technology and bringing it to the next stage of integral search. Then a new structure analysis is to be made, etc.

Studies of the combined hardening force to change the attitude to the classical strength–reliability–durability triad that forms a foundation for constructional strength of bulk-hardened metals. The quality of coatings is evaluated not only by their mechanical characteristics determining this triad but, first of all, by their special properties. For coated parts working, for instance, in frictional pair or under conditions of impact effect of abrasive particles the main criteria for durability are the rate and intensity of surface wear. When a product is used at high temperatures, it is essential to know the following special properties: the coefficient of gas permeability of the coating; coating–metal bonding strength over the required temperature range; the temperature coefficient of linear expansion; protective properties of the coating; its thermal conductivity, etc. Therefore, many papers deal with various methods for evaluation of special properties of coatings at different methodological levels.

However, the problem of the effect of coating on the constructional strength of a product is much more complicated than some authors tend to reckon, and it cannot be solved by studying the structure and properties of coatings alone. If, for instance, an increase in wear resistance of the surface layer of a shaft is accompanied by an abrupt decrease in fracture toughness of the base metal, coating deposition would be hardly expedient because of the increased probability of brittle failure of the part. Thus, not only the coating itself but also the base metal-coating composition as a whole should be studied.

The base metal-coating composition must meet an additional requirement of obligatory combination of high level of coating special properties with sufficient margin of strength, reliability and durability of the substrate. Base metal–coating system should be regarded as a structural entity. Therefore, this composition requires comprehensive research using up-to-date methods for evaluation of constructional strength, such as crack resistance and fatigue testing, as well as both static and dynamic tests.

In view of the above, we offer a classification of research techniques for structure and properties divided into two large sections: tests of coatings and tests of coated materials (see Fig. I). Each section comprises four groups of techniques. Structure studies and determination of coating-base metal bonding strength belong to both groups.

Undoubtedly, the presented classification does not pretend to be universal. It has its own drawbacks and does not rule out the possibility of an entirely different approach that can serve as a basis for systematization of methods for coatings and coated materials. At the same time, the suggested classification can assist in the unification of equipment, correct choice of proper testing techniques for particular tasks, encouragement of development and application of new methods in the research practice and production. Obviously, introduction of new ways of studying base metal-coating compositions will extend the classification.

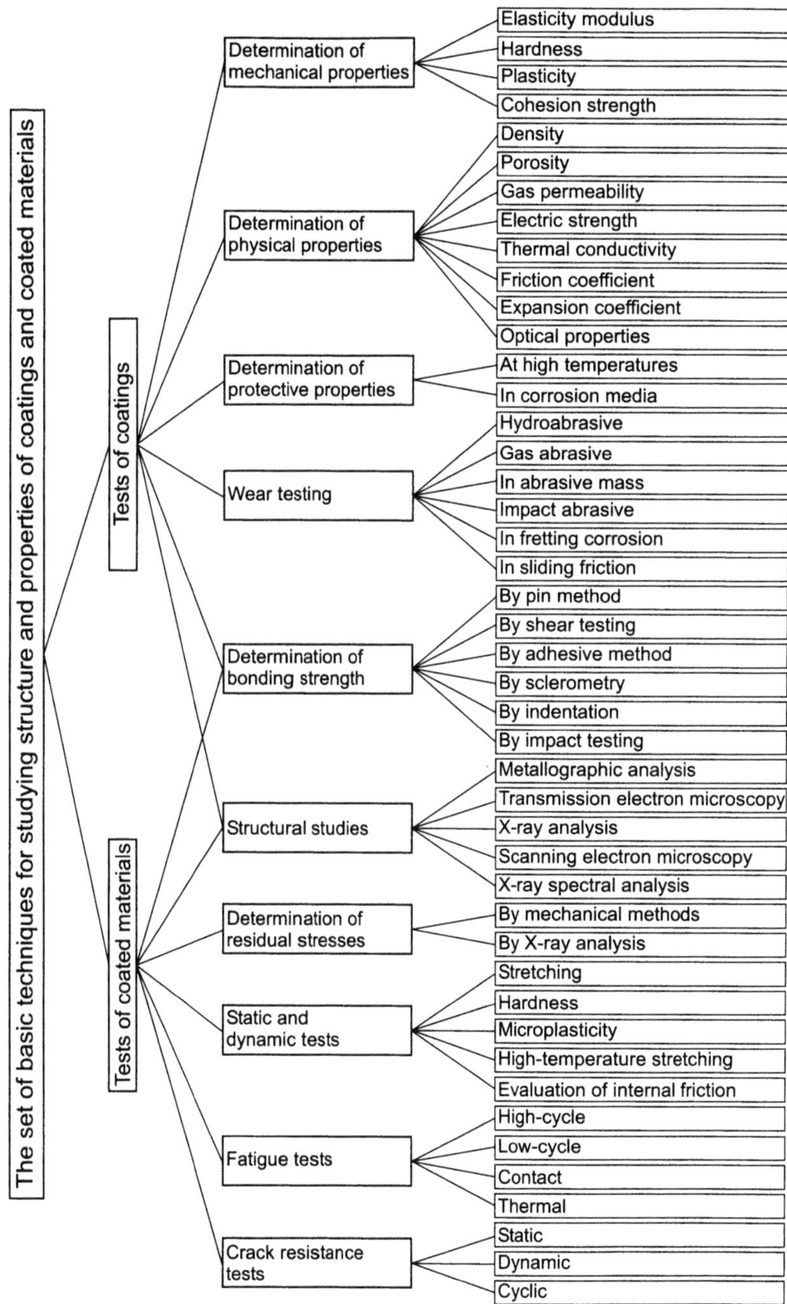

Fig. I. Classification of techniques for studying coatings and coated materials

Having definite scope, advantages, errors, applicability range, each of the techniques described in this book yields certain information on constructional strength of a coated product. However, none of these techniques, by itself, is capable of sufficiently complete investigation of the effects of external conditions in all their variety and interrelations. The chosen techniques must complement each other to model actual operational conditions of a coated part. Only their combined application may assist in correct solution of theoretical and experimental problems.

Comprehensive research of the base metal-coating composition is based on the interrelation of structure, properties and constructional strength (see Fig. II). When moving upwards from "simple" (mechanical and physical properties of coatings) to "complex" (wear resistance, protective properties and crack resistance of coated materials), a researcher may consider each previous step as a basis for transition to the next, higher step bringing him closer to constructional strength.

Since only a comprehensive research can provide the most reliable results determining the performance of products, this book describes classical, local and superfine methods, and also contains information regarding expediency of applying the methods in combinations with others. Along with analytical characteristics, procedures of preparing the specimens, and methodical aspects, descriptions of installations and instruments for conducting particular studies are also provided for each technique. These are mostly commercial devices often found in industrial and research scientific laboratories. At the end of each chapter an extensive bibliography including the most important and available sources (mostly in English) is provided to allow the reader to deepen and expand his knowledge of one or another method, if necessary.

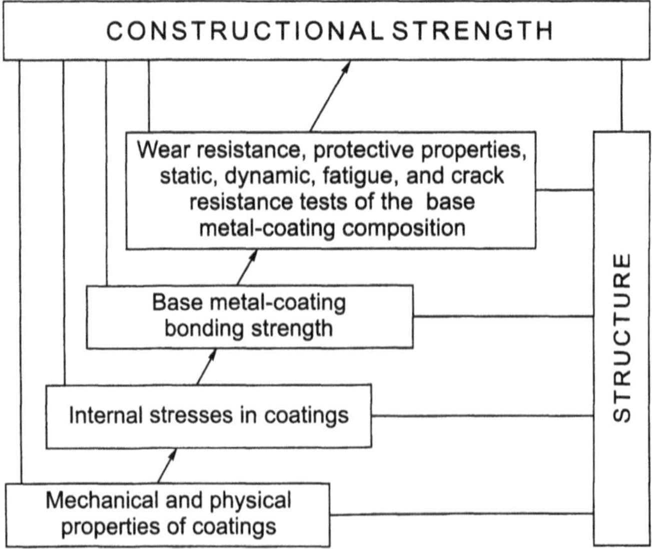

Fig. II. Scheme of complex studies of the base metal-coating composition

The experimental data included in the book to illustrate the capabilities of the covered methods is essentially based on the authors' own investigations. The authors express their heartfelt gratitude to their colleagues from Novosibirsk State Technical University and Tyumen State Oil University for valuable help in preparation of this book.

So far as the authors know, the present book is the first publication of this kind. It will be useful not only in scientific research, but also in daily tasks encountered by specialists in the field of protective coating deposition. The book could also be useful to students and post-graduate students working in this field.

Table of Contents

1 Structure

1.1 Formation of the Structure of Coatings

1.1.1 Peculiarities of the Structure Formation of Gas-Thermal Coatings

Formation of Boundaries

The structure of a coating is ultimately determined by the mechanism of its formation. For metallic powders three major processes contribute to the formation of structure:

- heating and melting (partial or complete), spherization;
- moving to the surface of the specimen due to interaction with the oxygen of the air, oxide film formation;
- dynamic contact with the surface of the base metal or an adjacent layer, deformation and spreading over the surface.

The latter process leads to formation of three new types of boundaries in addition to inter-granular and inter-phase boundaries common to compact materials. In the "base metal–coating" composition the boundaries between the deformed particles, between the layers, and between the base metal and the coating are also formed.

If the formation of the coating is accompanied by elevation of the base metal temperature (steel) up to 600–700 °C, a film of wustite (FeO) or a combination of iron oxides (FeO, Fe_2O_3, Fe_3O_4) covers the surface of the metal. For steels of complex composition the films of the oxides of the alloying elements can also be formed. A film of oxide at the "base metal–coating" interface significantly reduces the strength of the coating. However, if the oxide film is broken upon the collision of the particle with the base metal, strong inter-atomic bonds will form at the point of contact. Oxide films decorating the shapes of particles are also quite common between the particles inside the layers and at the inter-layer boundaries.

Powder particles often have nearly equilibrium, spherical shape. During the deposition of the coating the shape of microparticles changes due to imposed plastic deformation. The particles get stretched, become elongated and partially bent. From the point of view of spatial symmetry the system of surfaces becomes plane-oriented. In practice this is seen as the presence of equal-axes grains at the longitudinal section, and pronounced spatial orientation (oblongness) of particles at two mutually perpendicular transverse sections. Undeformed particles of regular shape can also be found occasionally.

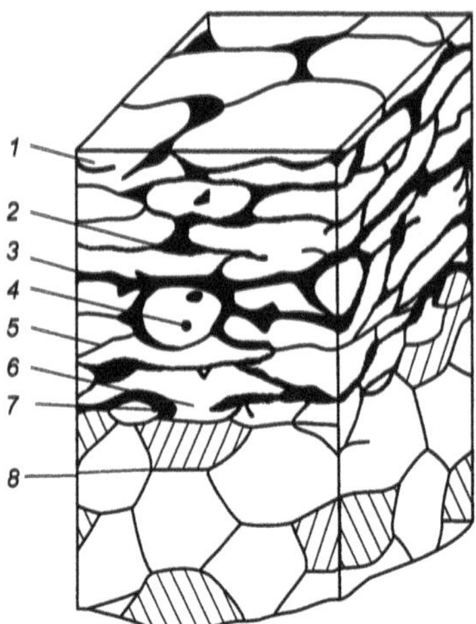

Fig. 1.1. A schematic structure of a gas-thermal metallic coating. *1* deformed particle; *2* pore at the boundary between the particles; *3* boundary between the layers; *4* undeformed particle; *5* oxide film; *6* area of welding between the coating and the base metal; *7* boundary between the coating and the base metal; *8* ferrite-pearlite structure of the base metal

Microstructures of the "base metal–coating" interface and the coating itself differ both qualitatively and quantitatively from the microstructure of the base metal. The following characteristic features of the coatings should be mentioned here: the coexistence of various combinations of nonequilibrium phases and even amorphous structures; the formation of highly branched pore structure with pores of different sizes and configurations; the presence of interconnected oxide films in the bulk of the coating [1.39, 1.58]. A simplified schematic structure of a coating is shown in Fig. 1.1.

Zverev et al. [1.103] have shown that formation of the boundaries between the particles depends on the type of inter-particle interactions during the deposition of a single layer. For example, when sputtering plastic nickel all the particles are in equal conditions, having been simultaneously activated by heating and plastic deformation by collision with the surface. Since identical composition and structure simplify formation of inter-particle bonds, no pores are formed at the inter-particle boundaries, the particles fit close to each other and are highly deformed, resulting in significant strength of thus formed coating.

Since the particles of the material to be deposited are heated to elevated temperatures in gas medium and become oxidized in the atmosphere of air, inter-particle boundaries are inevitably formed with the participation of oxides. However, metallic particles coalesce if the oxide film is broken. Sputtering in a controlled atmosphere preventing oxidation leads to complete coalescence of the particles over the entire surface of contact.

Formation of the inter-layer boundaries sometimes limits the efficiency of the detonation coatings. During the shot (formation of a single layer) together with particles having optimal speed there always are much slower particles which reach the surface later. They deposit on the already formed layer together with the products of acetylene combustion, e.g. pieces of soot, thus deteriorating inter-layer cohesion. Failure to observe the proper regime of sputtering leads to increase in porosity and the number of discontinuity flaws at the boundaries between the layers, and ultimately to peeling of the coating.

For plasma sputtering of coatings the formation of the inter-layer boundaries is determined by the difference in time spent by particles in the atmosphere. Time span of the particle deposition in a single layer is several orders of magnitude shorter than the interval between the deposition of adjacent layers, which depends on the size and shape of the specimen and the productivity of the setup. During the period of time between the deposition of two successive layers of coating the surface becomes contaminated by dust-like fractions of the deposited material or its oxides and by adsorption of gases from the atmosphere. The thickest contaminated layers due to gas adsorption are formed at the "base metal–coating" interface and the boundaries between the layers.

The work [1.51] gives an estimate of the characteristic dimensions of boundaries and inhomogeneities of the structure of plasma sputtered oxide coatings: thickness of the inter-layer boundaries about 1.0–10 μm, size of the deformed particles about 2.0–20 μm, thickness of the inter-particle boundaries up to 1 μm, extent of the inter-particle welding area about 0.1–0.5 μm, width of cracks between particles about 0.08–0.3 μm. Submicrostructure of the particles consists of grains and cracks with estimated width $(6–15)\cdot10^{-4}$ μm.

The bonding between the coating and the base metal is mainly determined by mechanical interaction and welding of coating particles with the metal. The prevailing type of bonding determines the structure of the "base metal–coating" interface and the transition zone, revealed by microscopic studies.

Interaction of the near-surface layers of the base metal with hot particles and high-temperature gas flows can lead to local phase transitions in the metal. Thus, studying plasma sputtering of molybdenum on polished steel surface having pearlite structure, Khasui [1.40] found sites with characteristic needle-like structure, which after checks for microhardness proved to be martensite. Apparently, in the thin layer near the "base metal–coating" interface, consisting of the pearlite grains, upon heating to temperature exceeding Ac_1 followed by rapid cooling a local phase transition with formation of tetragonal structure took place.

Effect of the Aggregate State of the Particles on the Formation of Structure for Plasma Sprayed Coatings

The presence in the gas flow of particles reaching the substrate in different aggregate states, and their participation in the formation of coatings are an important factor which determines the character and extent of structural and mechanical inhomogeneity of materials, produced using the gas-thermal deposition technique. An extensive treatise of this important topic can be found in the book of Kudinov et al. [1.51].

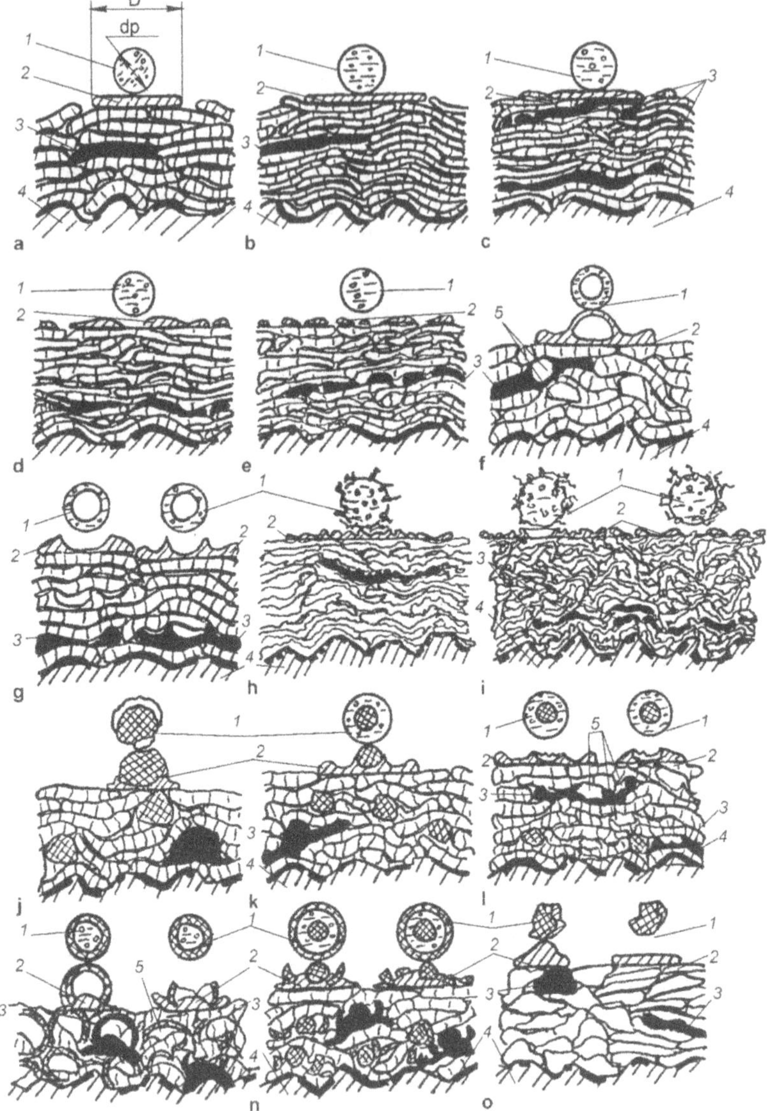

Fig. 1.2. Contributions from particles of the dispersed phase present in the flow in different aggregate states to formation of the structure of the sputtered material (Reproduced with permission from [1.51]). *1* particle in the flow prior to collision; *2* particle after the collision with already formed material; *3* contribution of the selected element to the structure of the sputtered material; *4* substrate; *5* pores

Fig. 1.2 [1.51] shows a scheme with contributions from particles in different aggregate states to structure formation of a plasma sprayed coating. If particles of one of the shown types prevail in the given flow, a specific type of structure is formed.

If the plasma sputtering flow consists mainly of completely molten particles, the structures shown in Fig. 1.2a–e,h,i are formed, depending on the speed of the particles, their temperature just prior to collision, degree of deformation and fragmentation during stacking in the layer. Increasing the degree of overheating of the material above the melting point leads to higher degree of particle deformation, resulting in structures with scales of smaller height and larger diameter. As a consequence of fragmentation of the overheated particles (Fig. 1.2c,d,e) upon collision with substrate or already deposited layer small grains of included material are embedded in the sputtered layer (Fig. 1.2c,d,e).

The structures shown in Fig. 1.2f,g are formed upon collision of particles having an inner hollow filled with gas with the substrate. Occurrence of such particles in the plasma sputtering flow leads to formation of large-scale pores in the formed material. For a flow consisting mainly of gas-filled particles the resulting coating has a high degree of porosity, determined predominantly by the pores of large size.

Because of the inhomogeneity of heating and cooling of the dispersed material in the plasma flow particles having a complex aggregate state can be formed in the sputtering flow (Fig. 1.2j–n). Fig. 1.2j shows a particle that was only partially molten in the flow without forming a closed envelope of melt. Similarly, Fig. 1.2k,l,m shows particles with solid cores completely covered with melt. Such particles fix at the surface due to the existing liquid phase, forming inclusions of hard grains in the sputtered material. Spontaneous chipping of these grains during formation of the layer gives rise to large-scale pores. A significant fraction of such particles in the sputtering flow or their overwhelming contribution to the material of the flow leads to formation of grainy-scaled and grainy types of structure.

Secondary solidification of the molten particles in the sputtering flow is sketched in Fig. 1.2m,n. In this case the particles fix at the surface due to liquid phase as well, but this time the latter is contained in a solid shell broken up upon collision of the particle with the substrate. Material formed from flows with significant contribution of such particles has complex and highly inhomogeneous structure. Particles having a liquid core and a solid shell tend to form material containing shreds of the shells, pores and scaly elements. The most inhomogeneous structure, however, is formed from particles, having a solid shell, a liquid layer and a solid core at the center. In this case, along with the elements mentioned above, the material also contains grainy inclusions (Fig. 1.2n).

Finally, it is also possible that absolutely solid particles take part in the formation of a plasma sprayed coating (Fig. 1.2o). This happens when molten particles immure solid, practically loose particles in the layer.

1.1.2 Electrocrystallization of Galvanic Coatings[1]

Formation of the Seeds of Crystallization

Formation of the electrodeposited coatings at the substrate starts with creation of the seeds of crystallization of the deposited metal on the substrate surface.

The seeds of the deposited metal formed at the cathode and intended to promote further growth of crystals should have specific shape and size lest they plainly dissolve back. For electrocrystallization of metals the seeds forming at the cathode usually consist of a small number of atoms (typically about ten atoms per seed) placed in the same plane next to each other (2D seeds, having the height of a monoatomic layer) or stacked over each other (3D seeds). Small 3D seeds mostly have a nearly semispherical shape, having minimal surface energy and approximating the equilibrium arrangement of atoms.

According to classical theory of nucleation the critical radius of a 3D seed capable of growing on a foreign substrate depends on overvoltage and is given by the Thompson equation:

$$r_3 = \frac{2\sigma V_m}{nF\eta}, \tag{1.1}$$

where σ is the specific inter-phase energy of the metal–electrolyte interface, V_m is the molar volume of metal, n is the charge of metal ions, and F is the Faraday constant.

The critical radius of a 2D circular seed is given by the following equation:

$$r_2 = \frac{\sigma_k S}{nF\eta}, \tag{1.2}$$

where σ_k is the specific edge energy of the seed having the height of a monoatomic layer, S is the area occupied by the seed.

The rates of formation of the 3D and 2D seeds are given by the following expressions:

$$\ln I_3 = k_1 - k/\eta^2; \tag{1.3}$$

$$\ln I_2 = k_1^{'} - k^{'}/\eta, \tag{1.4}$$

where $k_1, k, k_1^{'}, k^{'}$ are constants.

The presented expressions show that increasing overvoltage in the process of electrocrystallization of metals leads to increase in the rate of seed formation and to decrease in their critical size, thus favoring formation of dense and fine-grained coatings.

[1] This section is written by V.V. Povetkin

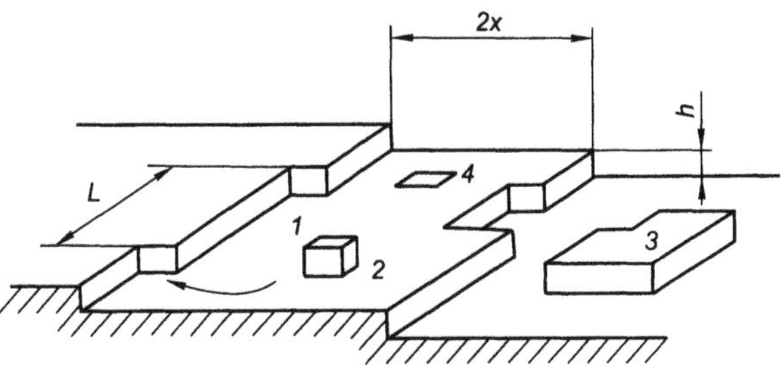

Fig. 1.3. The most important features of the crystal surface. *1* fracture; *2* atom at plane surface; *3* seed; *4* negative seed of dissolution

Growth of Crystals

Formed at the cathode seeds of supercritical sizes grow further by incorporating new particles into the crystal and completing the primarily formed atomic planes. The kinetics of the seed development generally determines the rate and character of the growth of the entire crystal surface.

Let us consider the general character of the growing surface [1.47]. Any metal surface contacting with electrolyte always has zones with specific geometric and energetic peculiarities. The first one to be mentioned here is a so-called step, or an edge of an uncompleted atomic layer or a stack of layers with each layer packed as a face of a monocrystal. An example of a step is the edge of a growing 3D or 2D seed. The direction of the step is typically parallel to one of the crystallographic directions.

The steps can be characterized by their height h, the average separation $2x_0$, and the average concentration of fractures or the average distance between the fractures l. The fractures (position "1" in Fig. 1.3) are the main points of growth of the crystal, i.e., the sites where the new particles are incorporated into the growing structure. A particular feature of the fracture is the fact that the configuration of the step and consequently the surface energy of the system do not change upon addition of a new particle to it. The fractures are thus called *the position of the repeated step.*

At each temperature the step has a certain equilibrium concentration of fractures determined by the minimum of the free energy $U+TS$ of the step. Since each fracture increases the entropy of the system, the concentration of fractures is governed by the entropic term in the expression for free energy.

At a certain *critical* temperature free energy of the step becomes equal to zero. At this point the steps completely merge, and the face of a growing crystal assumes the shape shown in Fig. 1.4. The entire surface of the crystal turns into "multilevel" system, in which an incomplete atomic layer carries several layers with gradually vanishing filling factor.

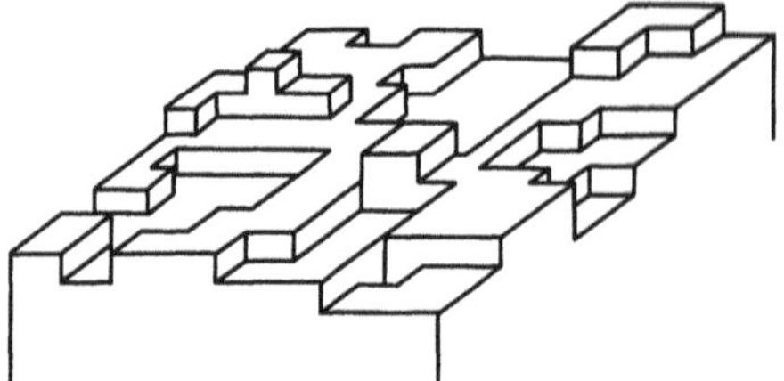

Fig. 1.4. "Multilevel" surface of the crystal face

In the process of electrocrystallization, especially in the conditions of impurity adsorption, such a "multilevel" surface is typically formed when the rate of nucleation is rather high, but the growth rate is insufficient.

The crystal grows differently at the "multilevel" and stepped surfaces. In the former case there is a high density of the points of growth scattered irregularly over the surface. New particles are added all over the surface, and only the overall movement of the metal-solution interface in the direction normal to the surface takes place. This is called normal growth of crystal.

In the case of stepped surface the new particles are added only at the fractures on the steps, leading to displacement of the steps across the face until the entire layer is completed. This is called lamellar growth of crystal.

There is no sharp distinction between the two types of growth: normal growth can be considered as the limiting case of the lamellar growth for complete merging of steps. And vice versa: if the step is not atomically smooth, normal growth *at the step* can take place.

The morphology of the formed surfaces provides the primary indication of the prevailing growth regime. Lamellar growth usually yields surfaces with clear crystal-like facing, and electron microscopy reveals parts of the surface with specific crystallographic orientations, as well as distinct steps at the faces.

In industrial applications galvanostegy is typically performed in the conditions, favoring the lamellar type of crystal growth. The crystals develop from 2D seeds that expand over the entire surface as layers of monoatomic or polyatomic height. In most cases the crystals grow through regular movement of the polyatomic layers of growth–"packets" with thickness of 0.01–1 μm depending on the regime of the electrolysis.

Formation of Continuos Coatings

Coalescence of the formed crystals usually starts from development of the intercrystal bridges with amorphous structure and smaller height as compared to crystallites. The bridges can be formed in different ways: via filling the space between the large crystals by numerous tiny seeds, developing narrow shapeless primary bridges, growing needle-like crystals connecting the forming crystals of the coating. The latter way of crystal interconnection is typical for refractory metals and their alloys, having high density of dislocations in the electrolytic coatings (10^{11}–10^{12} cm^{-2}).

Further development of thus formed continuos coating can proceed by several paths. If the crystals grow by the normal growth mechanism, the grains with their axis of maximum growth rate coinciding with the direction of ion supply to cathode (usually normal to the surface) will be the fastest to grow. These protruding crystals will also have the most favorable conditions for supply of discharging ions, thus having the best opportunity to develop across as well. As the coating is being formed, the number of grains in the cross-section parallel to the surface of the cathode gradually decreases, and the grains themselves develop a preferential orientation, that is, a texture. The coating will have anisotropic columnar structure and the most perfect texture.

For low current density and high electrolyte temperature (low cathode overvoltage) the crystals grow by the tangential mechanism. In these conditions the sites of active growth at the cathode are rather scarce, and the ions (atoms) are bound to diffuse across the growing crystal surface until they find a place to incorporate into the lattice. This lateral diffusion effectively leads to tangential supply of material to the points of growth, and the developing crystals have their long axes preferentially oriented along the surface of the cathode, thus forming lamellar structure of the coating.

Summing up, the process of formation of a continuos galvanic coating on indifferent substrates can be divided into four consecutive stages:

- development of isolated seeds of crystallization at the active sites of the substrate, that then expand into large crystals; the seeds grow into crystals by incorporating individual atoms or clusters of atoms that are supplied by the mechanism of surface diffusion;
- aggregation of individual crystals by the mechanism of intergrowth, or coalescence;
- coalescence of the crystalline aggregates and development of a continuous coating in the form of a net;
- gradual filling of the cells of the net by the depositing material to form a continuous layer – the galvanic coating.

1.2 Light Microscopy

Light microscopy (LM) is one of the primary methods to study the structure of metals and alloys. Although the applicability of the method is limited by its maximum magnification factor of x1500–x2000, restricting it to details of structure not smaller than approximately 0.2 μm, it is often adequate for study of galvanic coatings.

1.2.1 Metallographic Microscopes and Their Capabilities

Microstructure of the metallic material is analyzed using a metallographic microscope, employing in most cases light-field direct illumination. Light rays reflected from parts of the section surface perpendicular to the optical axis of the microscope are collected by the lens, and the rays reflected from irregularities of the surface are diffused. As a result the parts of the section perpendicular to the optical axis of the microscope appear lit, and the tilted portions of the surface look dark. This allows to figure out different elements of the structure, such as grain boundaries, that turn into grooves during preparation of the section, various inclusions, pores, cracks, etc.

Apart from light-field direct illumination, several other ways of illuminating the speciment are commonly used to improve image contrast:

- slanted illumination to study sections with developed relief of surface;
- dark-field illumination to get high-contrast images of the surface with clear grain structure;
- polarized illumination to study the structure of anisotropic materials;
- phase contrast technique, capable of detecting the differences in the heights of the relief as small as 5 nm and being particularly useful to study grain boundaries, twins, sliding lines and dispersed phases;
- interference contrast techniques, allowing not only to detect small changes in the microrelief of the surface, but also to estimate them quantitatively.

1.2.2 Characteristic Features of the Metallographic Analysis

Goals and Objectives of Light Microscopy

The main purpose of the metallographic analysis is to establish relations between technological parameters of the surface hardening process, physical, mechanical and performance properties of the coatings basing on qualitative and quantitative characteristics of its structure [1.7, 1.27, 1.79, 1.86, 1.101].

All methods of metallographic analysis fall into four main groups:

- determination of the oxide content by comparison with a reference scale [1.40, 1.89];
- phase qualitative analysis [1.9, 1.28, 1.37, 1.62];
- stereometric studies with computation of the number of particles per unit area of the section surface, determination of their sizes, number per unit volume, fractional composition of the phases [1.23, 1.89];
- computation of the grain composition of the powder [1.67, 1.68], determination of the structure of the powder particles [1.89].

Apart from establishing correlations between technology, structure and properties of the coatings, the results of metallographic studies help judge about the stability of the technological parameters of the coating deposition process, indirectly estimate cohesion of the coating with the base metal, study friction surfaces, detect cracks and discontinuity flaws, determine the mechanisms of adhesive and cohesive destruction of the coating, track changes in the structure of the coating under operating conditions, etc.

The structure of the coating is observed in the eyepiece of the microscope or analyzed from the picture.

Specimens

Three types of specimens are used for metallographic studies: specimens-witnesses, small items with deposited coating, and compact specimens cut from large parts. The place of cut is chosen so as to get a typical specimen for each given composition "base metal–coating". For inhomogeneous structure of the composition the specimens are cut from each characteristic zone of the part.

The most reliable information about the structure of the "base metal–coating" composition provides the study of transverse sections. Longitudinal and slanted sections are typically used only to confirm conclusions about the spatial structure of the composition made from the study of transverse sections. At the same time it is clearly impossible to anticipate all situations that a researcher can come across during the study of a given coating. When choosing the plane of section the following principle formulated by Saltykov [1.77] provides a useful guideline: "The aim of the researcher in each particular case should be to choose the most appropriate section position, so that the structure visualized by it be representative for the entire volume... to choose judiciously the point of view within the section surface, and to compute the average value of the structure parameter in such a way that it be applicable for the structure of the object as a whole".

Preparation of specimens for optical microscopy includes the following operations: cutting the specimen, its grinding, polishing, etching (if necessary). Typically transversal specimens with section surface normal to the coating surface are used. Reasonable height of a specimen is 2.0–2.5 cm with working surface area 1–2 cm^2. It is recommended to use softer cutting wheels for cutting hard materials and harder wheels for cutting plastic base metal, respectively. The size and abrasive of the cutting wheel in each particular case are chosen taking into account the hardness of the given composition. When cutting a specimen the instrument must always move in the direction from coating to the base metal to reduce the probability of peeling the coating due to tensile stresses at the point of the wheel exit onto the surface. Heating of specimens to high temperatures should be avoided, as additional stresses due to temperature gradients lead to cracking and peeling of the coating. Furthermore, excessive heating can change the structure of the base metal as well.

The specimens for optical analysis are typically cut using special cutting-off machines with diamond or el-boron wheels. In metallographic laboratories it is recommended to use abrasive floor machines.

A fast way to cut high quality specimens is to use abrasive-wire saw or spark cutting. In the former method an abrasive suspension is constantly fed to the area of friction between a metal wire and the specimen to cut. In the method of spark cutting an electric-spark discharge eroding the metal is created between a copper electrode and the specimen to cut. Both methods produce the surface of good finish without distorting the initial structure of the material.

To prevent breaking the corners and to increase the supporting area of the transversal section it is recommended to immerse the specimen in fusible alloys like Wood alloy or Roset alloy. Other options include using epoxy resin, organic glass, polystyrene, etc... Before immersing the specimens are placed in cylindrical mandrels with height 10–20 and diameter 30–40 mm.

Working on improving of the process of specimen preparation, Block et al. [1.11] suggested recommendations for preparation of specimens of thin vacuum coatings. Brindley and Leonhardt [1.14] recently suggested a new technique of metallographic studies using fluorescing epoxy resins.

Determination of the Oxide Content by Comparison with a Reference Scale

Technical and regulatory documentation for an item with a specific coating sometimes includes a three to five point reference scale. Visual inspection of the transversal section of the specimen and comparison it with the set of references provides a means to determine the degree of contamination of the coating with oxides [1.40].

For x400 magnification the references are presented as schematic structures characteristic for this particular metallic coating. The references have the shape of circles with diameter corresponding to the given actual area of the coating section (Fig. 1.5). When constructing reference scales all specific features of the structure are taken into consideration. In the situations when oxide content cannot be estimated by integer points fractional marks of 0.5, 1.5, 2.5, etc. are acceptable.

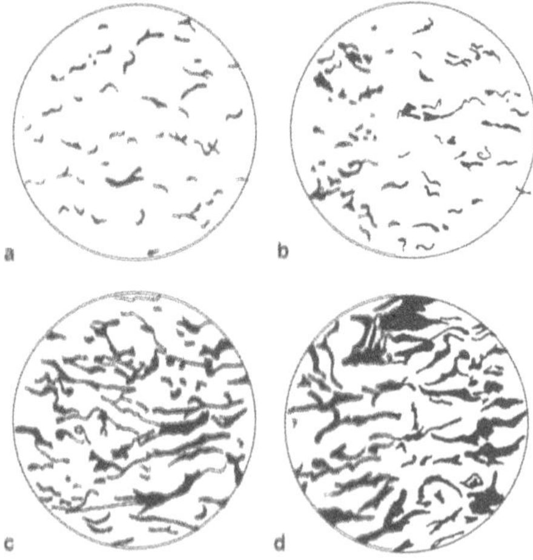

Fig. 1.5. Reference scale of oxide inclusions (x400): **a–d** correspond to 1 to 4 points, respectively

Miscrostructural analysis of not etched sections does not always allow to unequivocally discriminate oxides from elongated pores or stripes of plastic deformation of metallic particles. At the same time under dim illumination oxide inclusions tend to look lighter than pores. Another way to identify oxides is to deliberately defocus the picture. If moving the object stage of the microscope towards the lens impairs the contrast of the entire section surface, the dark elongated parts of the section should be considered oxide inclusions. A distinctive feature of the pores is the presence of optically sharp curved fragments relocating as the microscopic stage is moved.

Kreye [1.50] has demonstrated that new high capability systems of flame spraying of the third generation significantly reduce the fraction of oxides in the deposited iron alloys coatings.

Phase Qualitative Analysis

The method of metallographic analysis of the phase composition of the coating takes advantage of the differences in the reflectance of the particles constituting the coating. Different reflectance of the microphases leads to their different color contrast in the image. Metallographic phase analysis can be rather cumbersome and should generally be complemented with X-ray phase analysis and microhardness measurements.

The large number of phases forming in each given coating requires an individual approach to the analysis of each phase. This is especially true for coatings deposited with composite powders. Here we shall consider the procedure of phase metallographic study on the example of a common coating from exothermically reacting powders of nickel and aluminum, taking the color differences of the phases as the basis for identification [1.52].

According to equilibrium Ni–Al state diagram, the following phases can be present in the coating deposited from composite Ni–Al powders: solid solutions of Ni in Al, Ni in NiAl, Ni in Ni_3Al, Al in Ni, intermetallic compounds $NiAl_3$, Ni_2Al_3, NiAl, Ni_3Al. Since deposition is carried out in the atmosphere of air, the coating will probably also contain oxides of Al and Ni. High rates of deposition can lead to formation of non-equilibrium phases.

In the plane of section the phases Al, $NiAl_3$ (ε), Al–ε are colored white; Ni_2Al_3 (δ) – light blue; Ni_3Al (α_1), Ni (α) – light yellow; NiAl (β), Ni_2Al_3 – gray-blue; Al_2O_3 – dark gray. The solid solution of Ni in NiAl, being the phase of variable composition, changes its color from light blue to dark yellow depending on the concentration of dissolved nickel [1.52].

Analysis of not etched sections reveals the lamellarity of coatings, the presence of pores and oxides. However, in the analysis of a wide class of metallic coatings etching allows differentiation of the structural elements of the coating, estimation of the size and composition of the diffusion zone. Since the chemical activity of the coating materials and the substrate is quite different, it is sometimes necessary to use several etchants, because a single etchant fails to completely develop the structures of both the coating and the base metal simultaneously.

As an example Fig. 1.6 shows the structure of a common self-fluxing powder coating PG–SR4 with the following per cent composition: B – 3.4; C – 0.8; Si – 3.8; Cr – 17; Fe – 4.0; Ni – up to 100%.

The hot powder particle deforms upon collision with the surface of the base metal. Occurring repeatedly, this process leads to layer by layer formation of the coating (Fig. 1.6a) from highly anisotropic particles: the length and width of the particles is about an order of magnitude larger than their thickness in the direction, perpendicular to the surface. Successive packing of a large number of deformed particles inevitably leads to formation of pores (microhollows) (Fig. 1.6b), especially at the junctions. The pores are known to have mostly irregular shape and sizes up to 60 μm. The porosity is particularly high in the vicinity of the intact powder particles that were left undeformed during deposition and thus preserved their spherical shape (Fig. 1.6c,d). The quantity of such particles in the bulk of the coating varies approximately from 12 to 21%.

For fusing temperature 1040 °C (cooling in the air) the solid solution is a part of various eutectics or forms grains of irregular shape with size 2–5 μm (light phase in Fig. 1.6e) found between the phases of nickel boride Ni_3B. At higher temperatures (1100°C) and longer times (60 min) of fusing the solid solution forms dendrites. A thin layer of solid solution of elements in nickel is seen in the coating near the boundary with the base metal. The formation and growth of the first order axis dendrites from the layer is clearly visible (Fig. 1.6f).

The boride phase of nickel in the coating has the composition close to Ni_3B. This phase can form rounded grains or enter various eutectics. For example, regular petal-shaped structures built from interleaved layers of solid solution of elements in nickel and of nickel boride can be found in the fused coating (Fig. 1.6g).

Several types of eutectics can be formed in a self-fluxing alloy. Apart from the lobe-like structure shown in Fig. 1.6g eutectics with colonies resembling the lamellar structures in high-carbon steels (Fig. 1.6h) are found occasionally. The extent of a colony in the coating fused at 1040 °C can be as large as 120 μm with the width 10–20 μm. A detailed study at high magnifications shows that the colonies are in fact bunches of needles of Ni_2Si having diameter 0.22–0.58 μm separated by layers of solid solution with thickness 0.18–0.30 μm. A section of the colony (bunch) gives a picture of globules or short needles.

Crystals containing chromium carbides and bromides are very important structural elements of self-fluxing alloys, that determine a whole complex of application properties of the coating. Structure of the coating fused at 1040 °C is characterized by homogeneous distribution of small crystals over the entire cross-section of the coating. Together with the large number of small crystals having the size 0.35–1.00 μm larger crystals with the cross size up to 10 μm and inter-crystal distance 20–30 μm are also formed. Still larger crystals with the size up to 40 μm are found occasionally (Fig. 1.6i).

Summing up, after fusing and crystallization in the atmosphere of air the coating PG–SR4 has a multiphase structure. The pores and other defects disappear, being substituted by structural elements with the dimensions from fractions of μm (in eutectics) to tens of μm (grains of nickel and carbo-borides).

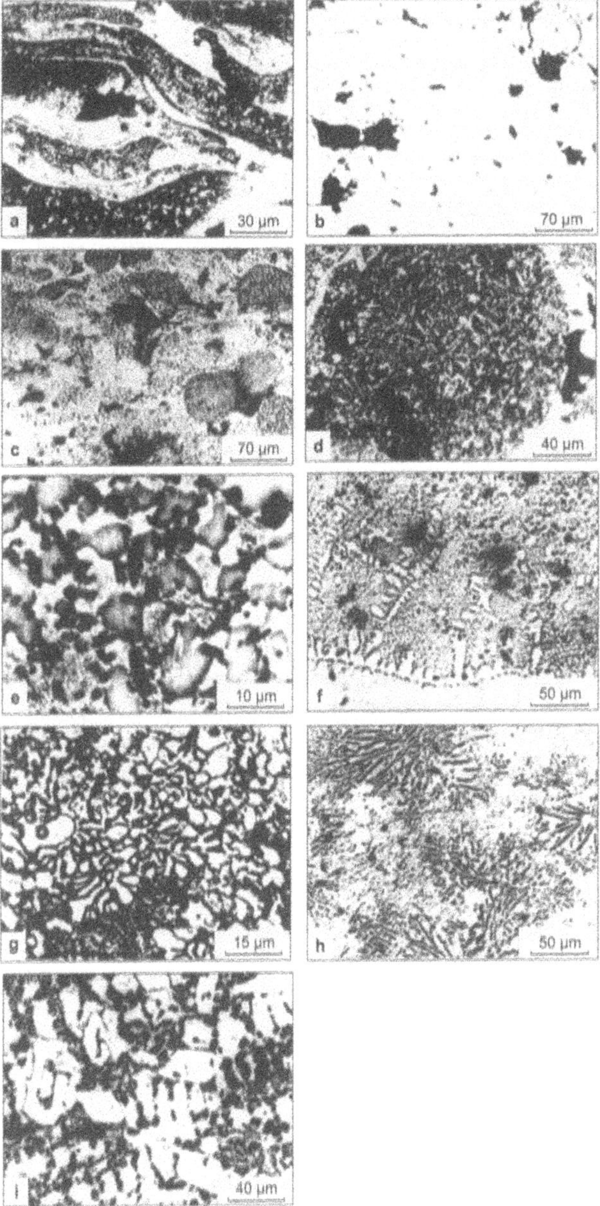

Fig. 1.6. Structure of the coating PG–SR4. **a** lamellar structure after deposition; **b** pores (not etched section); **c**, **d** etching revealed the shape of the coating particles; **e** grains of nickel (light phase); **f** dendrites; **g** petal-shaped eutectic Ni+Ni_3B; **h** needle-shaped eutectic Ni+Ni_2Si; **i** chromium carbides and borides [courtesy of Poteryaev Yu.P.]

Metallography of the Base Metal–Coating Interface

There are several problems associated with the study of the structure of the boundary and the transition zone between the coating and the base metal. Firstly, it requires separate etching of the coating material and the metal, which complicates preparation of high quality specimens. Secondly, if transition zone is formed during deposition of the coating, its dimensions are usually not large, thus making obtaining reliable information about the structure of the boundary layers difficult.

The character of the base metal–coating interface provides indication of the strength of their bonding. If the boundary is well defined, the dominating mechanism of the bonding of the coating to the base metal is mechanical interaction. In this case it is crucial that the coating abut tightly to the surface of the metal copying its relief. Discontinuities at the boundary point to the low quality of the deposition process. Smeared image of the boundary confirms formation of the transition zone and chemical interaction of the two materials.

Microstructural studies indicate that the boundary between many gas-thermal coatings and the steel substrate has irregular structure with large number of pores and discontinuity flaws (Fig. 1.7a). During shot-blasting cleaning of the surface before depositing the coating, the base metal is often mechanically cold-hardened at depths 20–100 μm. The distorted shape of the grains is retained after the coating has been deposited, which means that the metal of the substrate is not heated above its recrystallization point during the process of deposition.

Fusing causes structural changes in the coating, at the boundary and in the base metal (Fig. 1.7b). Fig. 1.7c, shows in details the microstructure of the steel–fused coating PG–SR4 composition. Fusing consisted of the following steps: heating in the oven to the temperature 1040 °C in the atmosphere of air, 10 min exposure interval, cooling in the air. After fusing the following elements of microstructure can be seen at the transversal section:

1- the fused coating, microhardness 8000–11000 MPa,
2- a layer of solid solution of elements in nickel (part of the coating), microhardness 3500–5500 MPa,
3- near-boundary layer of the base metal, microhardness 2000–3000 MPa,
4- the base itself, microhardness 1800 MPa.

Apparently, thermally activated diffusion of elements from the liquid bath of the coating plays the main role in the formation of the transition zone in Steel 20.

Cooling the composition Steel 20–coating PG–SR4 from 1040 °C in the air leads to formation of a continuos interlayer in the near-boundary zone of the base, as shown in Fig. 1.7c. For slow cooling (annealing) the transition zone has polyhedral ferrite-pearlite microstructure. Metallographic studies indicate that the volume fraction of pearlite and, hence, the carbon content in the near-boundary zone of steel increase upon fusing of the coating.

Summing up, alloying of the near-boundary volume of steel in the course of fusing affects both the quantitative phase composition of the metal upon quasi-equilibrium transformation (annealing) and the kinetics of the polymorphic $\gamma \rightarrow \alpha$ transition in iron.

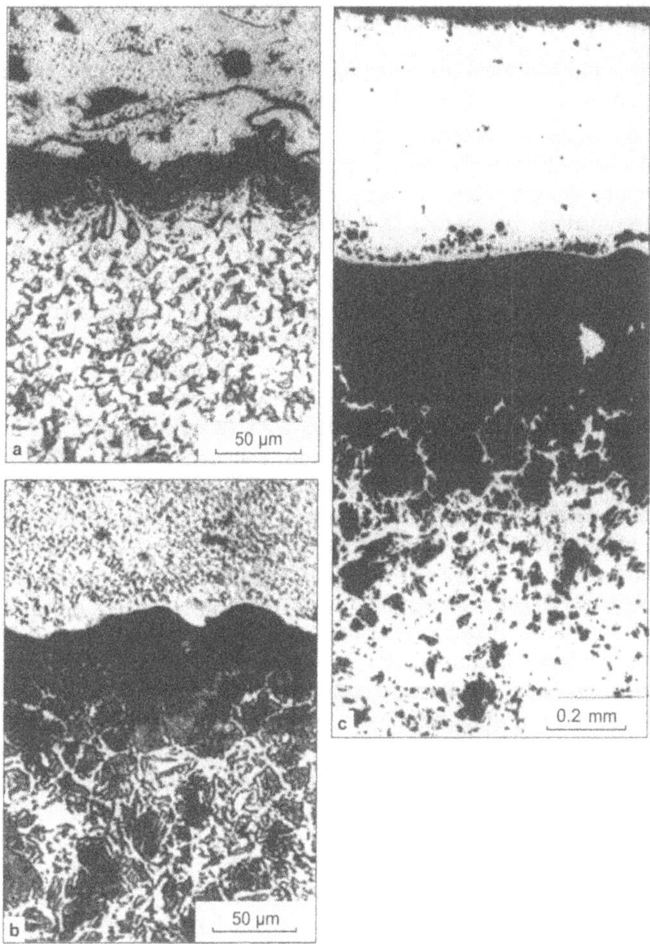

Fig. 1.7. Effect of fusing on the structure of the near-boundary zone in the composition Steel 20 – coating PG–SR4. **a** coating before fusing; **b** coating after fusing; **c** microstructure of the zone: *1* fused zone; *2* near-boundary zone of the coating; *3* surface layer of the base metal enriched with carbon; *4* base metal [courtesy of Poteryaev Yu.P.]

The structure of the continuos transition zone forming in steel during cooling in the air after fusing the coating can best be described as pseudoeutectic. Its formation is promoted by increased carbon content, reduced stability of austenite for boron content higher than 0.006%, and by formation of iron borides Fe_2B acting as centers of nucleation for crystallization of pearlite. Besides, during fusing a partial "dissolution" of the base metal and leveling of the coating–base metal boundary relief occurs.

High-Temperature Vacuum Metallography

Deposition of a coating is sometimes accompanied by heating the part above the point of $\alpha \rightarrow \gamma$ transition in steel. Thus direct studies of the structure of austentite and its decay products become necessary.

Since the existing processes of developing the structure of austentite using color vacuum etching are rather time-consuming (up to 30 min), such methods of thermal microscopy cannot be applied for studies of structural changes in steel upon bulk hardening by deformation.

A new technique of capturing rapid microstructural changes in steel and its coatings at high temperatures and under plastic deformations was suggested. The technique has significant advantages over conventional vacuum etching in the studies of undeformed austentite: it fixes the structure of austentite practically immediately, which is crucial for studies of dynamical processes; it sharply emphasizes weakly etching twin boundaries, creating color contrast of neighboring volumes; it reliably excludes the traces of moving boundaries of the austenite grains from the set of fixed boundaries; it is also quite visual.

Another important feature of the new technique is its applicability for analysis of the processes of structure formation in the course of bulk hardening of the substrate. The method allows to analyze the kinetics of not only the changes in the austenite structure under plastic deformation, but of the formation of the austenite decay products as well. And, finally, the method provides the possibility to study the structure of the deposited coating in details.

The studies are conveniently performed using the installations of the IMASH type with improved vacuum commutation system. A polished specimen is heated to desired temperature inside the vacuum chamber by electric current. The temperature is monitored by a thermocouple welded to the specimen. At the given moment a precisely dosed portion of air is let into the chamber. Under the action of oxygen the film of oxide covers the surface of the specimen. The thickness of the film at each particular spot is determined by the surface energy of the spot, which in turn is determined by crystallographic orientation of the surface and the concentration of defects. As a result the thickness of the film changes in a stepwise fashion upon transition from one grain to another. Adjusting the portion of air let into the chamber, the thickness of the oxide film can be kept within the range necessary for interference of the visible light. A stepwise change in the surface orientation then leads to stepwise change in the color of the oxidized surface.

After the primary oxide film has covered the surface of the specimen, further oxidation is limited by the rate of oxygen diffusion through the film and proceeds rather slowly. The rate of secondary oxidation of the surface is further reduced by accelerated cooling of the specimen, inherent in the technique. Because of this neither any following changes in the initial structure, nor polymorphic transformations leading to final structure of the substrate distort the resulting color relief picture.

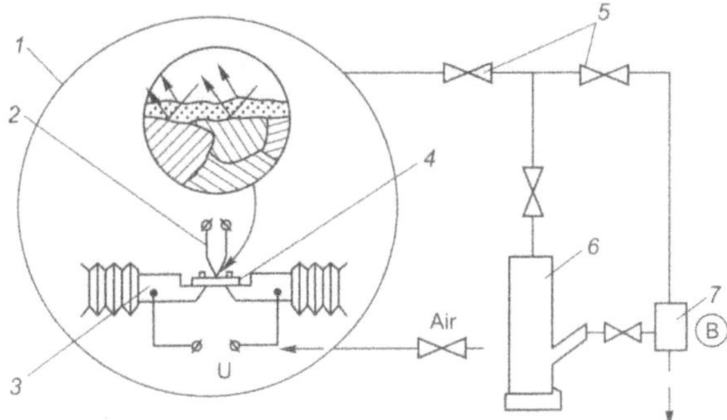

Fig. 1.8. Installation for developing the structure of steels and coatings using the high-temperature rapid color oxidation technique. *1* vacuum chamber; *2* thermocouple; *3* specimen holder; *4* specimen; *5* valves; *6* diffusion pump; fore pump; electric motor

The scheme of the installation for developing the structure of steels and coatings using the high-temperature rapid color oxidation technique is shown in Fig. 1.8. To capture structural changes occurring upon hot deformation, the air from the dosing cylinder must be let into the chamber at the moment of relieving the load. Interpretation of the produced color picture is complicated by a number of factors. The oxide film is developed against the background of the specimen relief produced by deformation. Initial boundaries of the grains (Fig. 1.9) can be seen for austenitization in vacuum, they become deeper and wider after the deformation. The grains themselves are also deformed – the lines of shear are clearly visible at their surfaces.

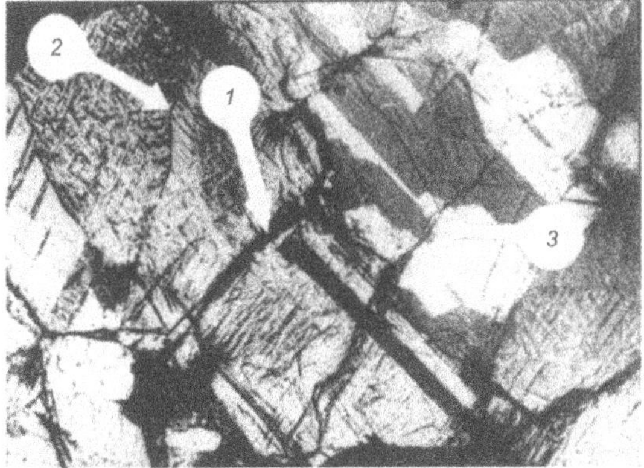

Fig. 1.9. Typical picture of oxidized surface of hot deformed steel (magnification x300). *1* initial boundary of the austenite grain; *2* boundary of the recrystallized grain; *3* trace of the initial austenite boundary

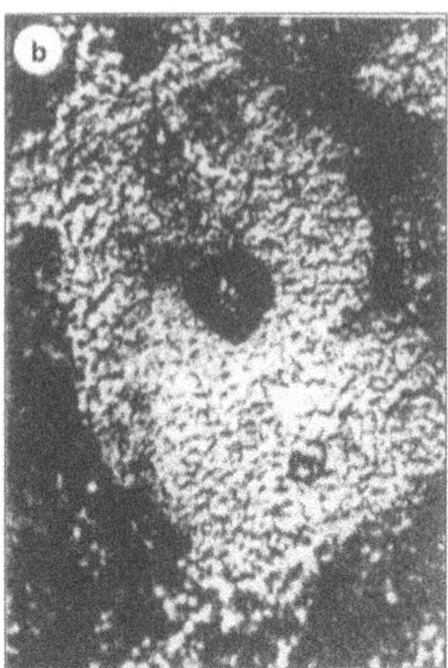

Fig. 1.10. Structure of the intermetallic coating PN85U15: **a** without etching (magnification X400), **b** after high-temperature vacuum etching (magnification x1000) [courtesy of Bataev V.A.]

The film of oxide copies the relief of the surface within the limits of each single grain and changes its thickness in a stepwise fashion upon transition from one grain to another. When inspecting such a surface with an optical microscope the interference of light rays in the film produces a characteristic color picture. The thickest film is formed at the edges of the grains because of their higher chemical activity. The boundaries of the grains formed by recrystallization typically are narrower than the initial boundaries. A trace of the initial boundary in the form of a groove can be found within the newly recrystallized grains.

Sometimes the technique of high-temperature vacuum etching helps visualize the fine structure of a coating in the situations where the conventional etching methods fail to give satisfactory results. As an illustration Fig. 1.10 shows the structure of the coating PN85U15 containing 15% of Al and 85% of Ni as obtained by the new technique. A block-like fine structure with the block size 1–3 μm is clearly seen.

Quantitative Stereometric Analysis

In this section we shall outline the procedure of quantitative stereometric analysis of the structure of coatings. To determine the number of particles per unit area of the section of the coating, a circle with diameter 79.8 mm (for magnification factor x100) lying completely at the coating surface is drawn on the picture of the surface or on the viewing screen of the microscope. The actual area of the surface confined within the circle equals 0.5 mm^2. All the particles that fell inside the circle are then counted. Larger magnification factors can be employed for study of fine coatings.

The actual number of particles n per 1 mm^2 of the coating surface area for x100 magnification inside the circle with diameter 79.8 mm is given by the following expression:

$$n = 2(z + K\omega), \tag{1.5}$$

where ω and z are the number of particles crossed by the delimiting circle and lying inside it, respectively, K is the correction factor depending on the number of particles inside the circle ($K = 0.46$ for $z = 50$).

If other magnification factors are used, the number of particles can be determined from the general expression:

$$n = 4(z + K\omega)/\pi D^2, \tag{1.6}$$

where D is the actual diameter of the delimiting circle in the plane of the section.

The size of the particles of coatings is conveniently estimated by counting the number of intersections between the boundaries of the particles and secant lines. Since any coating is a plane-oriented system, the orientation of the secants should correlate with the orientation axis of the coating. The number of intersections is maximal for a secant normal to the axis and minimal for a secant parallel to the axis of orientation. The secants are drawn on the image of the structure in the viewing screen of the microscope or on its picture. In case of direct observation the role of secants can also play eyepiece rulers of the microscope. At least 50 full particles must fall within the field of view of the microscope, and a secant must cross at least 10 of them. The actual length of each secant in the plane of the section is determined using a micrometer object. The number of intersections is counted for at least five typical parts of the section. The average length in mm of particles in each of three Cartesian directions d_L is then computed from the following equation:

$$d_L = L/N, \tag{1.7}$$

where L is the total length of all secants in the given direction in the plane of the section, mm; N is the total number of particles crossed by the secants.

The number of particles N_V in 1 mm^3 of the coating volume is found from the equation:

$$N_V = 0.7 N_x N_y N_z, \tag{1.8}$$

where the factor 0.7 accounts for nonequiaxiality of the particles, and N_x, N_y, N_z are the numbers of intersections between the boundaries of the particles and the secant lines in each of the three directions per 1 mm of the length of the secants in the plane of the section.

The per cent content of phases in the coating can be determined using a simple and efficient method of point quantitative analysis, which gives results with high precision and does not require sophisticated equipment. The method is based on the first rule of stereometric metallography: the fraction of a phase on the surface of a two-dimensional structure (section) is numerically equal to the fraction of random points that hit the phase.

Technically the method is conveniently realized with a special device called push integrator. The device is essentially a set of mechanical or electric counters that can cause displacement of the object stage of the microscope with the section under study. When the crosshair of the eyepiece points at a phase, a key of the counter for this phase is pressed. This increases the reading of the counter and moves the microscopic stage with the section to a new position. The key for the relevant counter is pressed, and the entire cycle is then repeated over and over again. After the test is completed the readings of the counters, giving the number of points for each phase, are recorded. When analyzing a coating it is recommended to discriminate not only the structural elements, but also pores and oxides at the boundaries of the particles. If a push integrator is not available, the points can be counted using special inserts with graticules for the eyepiece, counting the number of node points of the graticule inside each phase. The accuracy of analysis depends on the area of the section occupied by the phase and the number of points.

Phase content in this method is given be the following expression:

$$W = 100g/M, \tag{1.9}$$

where g is the number of points for the phase, and M is the total number of points.

All geometric measurements on the section have one common drawback – uncertainty caused by finite thickness of the boundaries separating the particles of a coating. An oxide film at the boundary can have thickness comparable to the average size of the particles, which causes systematic error in the results of quantitative microanalysis and complicates the measurement. Magnification factor of the microscope is another factor that influences the results of geometric measurements [1.77].

Calculation of the Grain Composition of Powder. Metallography of Particles

A specimen for determination of the grain composition is taken either from the packaging (a container, can, or bag) or directly from the flow with a special sampling probe. The specimen with mass 0.5–1.0 g is carefully mixed and packed and is sent for optical analysis. A microspecimen for the analysis is prepared in the following way: a small amount of the specimen powder is placed on the object plate of the microscope using a glass rod, one or two drops of dispersing immersion liquid (a 1–2% water solution of a surfactant, glycerol, castor, sun-flower seed, paraffin or cedarwood oil) is added to the powder, the mixture is evenly spread over the surface and covered with the cover glass. The grains in the prepared microspecimen should not overlap. Two parallel specimens are prepared for improved accuracy of analysis.

In optical estimation of the particle size distribution the longest chord in horizontal or vertical direction of a particle is taken as the particle size. Direct observation of the microscopic image of the powder specimen provides the most simple and efficient method for determination of the particle size, although pictures can also be used if necessary. In direct observation either the specimen is continuously moving, or different fields of view are consecutively studied. For continuous movement of the specimen in one direction maximal horizontal or vertical chords of all particles in the field of view, limited by a rectangle or circle with known dimensions, are measured. When studying separate fields of view the specimen is periodically moved by a step exceeding the diameter of the circle or the diagonal of the rectangle of the field of view. At least 625 particles should be measured. The range of measured maximal chords is divided into at least 6 intervals (classes) with average particle size for the class being a simple average of the lower and upper limits of the interval for the class. The grain composition (particle size distribution) of a powder is characterized by the ratio of the number of particles in each class to the total number of measured particles.

A specimen for metallographic analysis of powder particles is prepared in a slightly different way as compared to "massive" sections. In this case the specimen is an artificial pseudoalloy of an inert matrix and the dispersed phase–powder particles. A convenient material for the matrix is poreless, sufficiently strong and readily polishable dental plastic. The powder (one part), plastic and solvent (3 parts) are mixed and put into a mandrel. After complete curing of the plastic the specimen is taken out of the mandrel, and the required section is prepared and studied.

Metallography of Galvanic Coatings

As has already been mentioned, a general procedure of preparing a metallographic section includes cutting, grinding, polishing and etching of the specimen. For galvanic coatings the first two operations can sometimes be omitted, for example, when studying the microstructure of a coating in the plane parallel to the substrate (longitudinal sections). The substrate is initially prepared in the required shape of the specimen, and the low roughness of the produced surface of the coating allows skipping the operation of grinding. This becomes especially important for thin coatings, since grinding produces a deformed layer of quite large thickness, which can be unacceptable.

Longitudinal sections were used in [1.45] to study the processes of recrystallization during natural aging of electrodeposited tin coatings. Fig. 1.11a, shows the initial structure of the coating, which consists of irregular grains of different sizes. After aging for 3.5 months the coating consists of relatively equiaxial grains (Fig. 1.11b). Further aging leads to uniform coarsening of the grains, which acquire more regular facing (Fig. 1.11c). And finally, after 14 months of aging anomalous growth of separate grains is clearly seen (Fig. 1.11d). Apparently, the observed changes in the structure are caused by the processes of primary, collecting and secondary recrystallization.

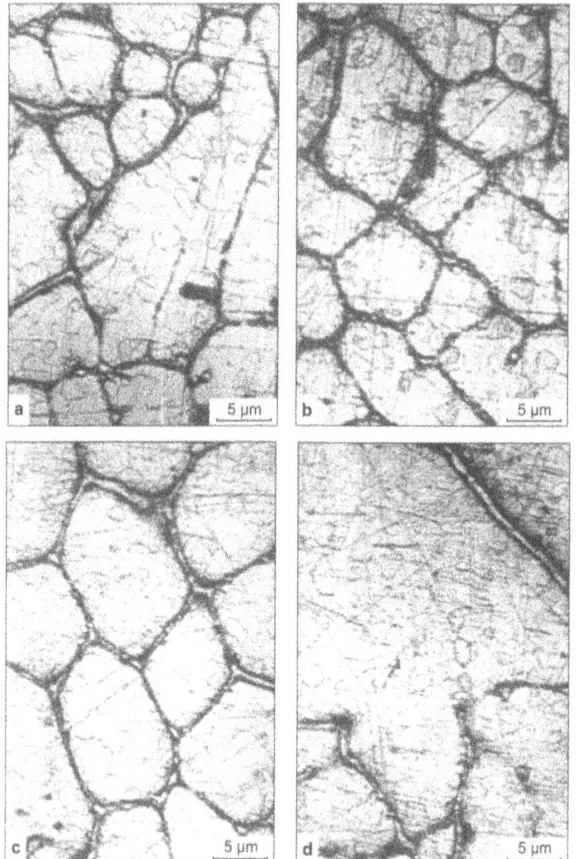

Fig. 1.11. Evolution of the structure of tin coatings deposited from sulfate electrolyte with aging. Aging time (months): 0 (**a**), 3.5 (**b**), 7 (**c**), 14 (**d**)

Transversal sections are used to study the profile of the microstructure of galvanic coatings and usually require all preparative operations. In [1.72] transversal sections were used to study the evolution of column-like structure of electrodeposited copper coatings upon annealing. Fig. 1.12a, shows the initial structure, typical for coatings deposited in a stationary electrolysis regime. Microstructure shown in Fig. 1.12b corresponds to onset of primary recrystallization. The figure shows that formation of relatively equiaxial recrystallized grains begins from the surface of the substrate and continues through the volume of the coating. Further annealing leads to coarsening of the recrystallized grains (Fig. 1.12c) and their anomalous growth (Fig. 1.12d). Similar to the previous example, these structural changes are identical with the stages of collective and secondary recrystallization in deformed metals.

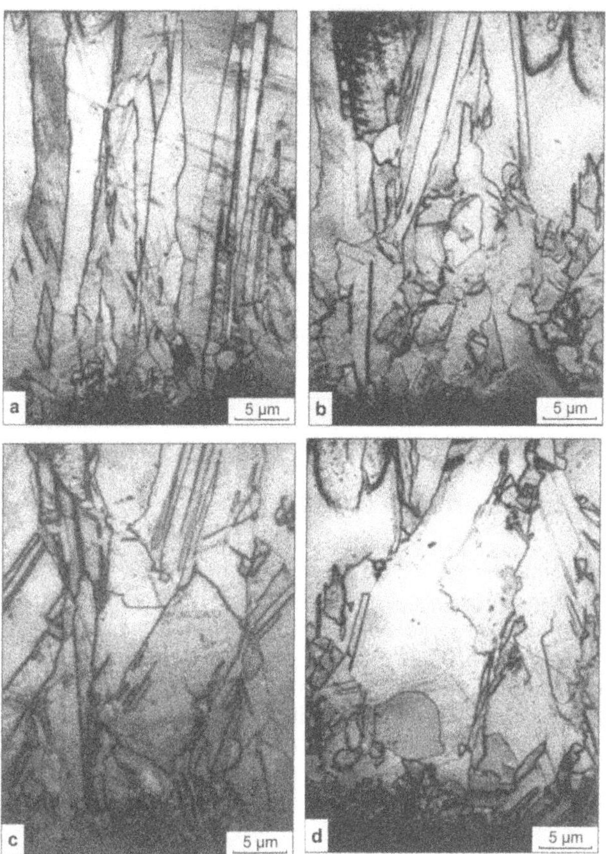

Fig. 1.12. Evolution of the structure of copper coatings deposited from sulfate electrolyte immediately after the deposition (**a**) and after annealing at 500 °C for 1 (**b**), 10 (**c**) and 100 (**d**) hours

Transversal sections are used to study the lamellar structure, texture, defects of the coating, the structure of the transition zone. However, the practically applicable thickness of a coating in this case is limited by the resolution of a light microscope and cannot be less than approximately 1 μm. This limitation is overcome in the technique of slanted section, covering the range of thickness δ' from hundreds of nm to hundreds of μm. Fig. 1.13 shows a slanted section of a specimen with a coating and the dependence of the "magnification factor" K on the cutting angle α of the section. After cutting the specimens with relatively planar surface they are ground with progressively finer sandpaper and then polished.

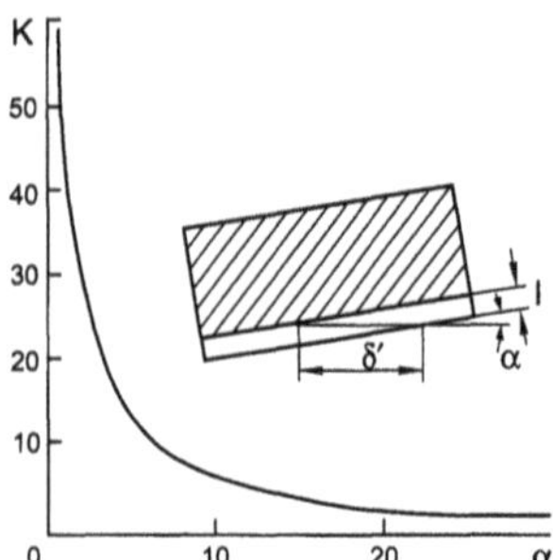

Fig. 1.13. Magnification of the visible thickness of a layer at a slauted section of a speciment

1.3 Transmission Electron Microscopy

Transmission electron microscopy (TEM) is widely used in the analysis of coatings, since a required specimen transparent for electron beam – a foil or replica – can be prepared for almost any coating. The method yields the structure of coatings in the submicroscopic range and gives an image with resolution of about 1 nm and a diffraction picture of the same portion of the surface in a single experiment. Today TEM holds one of the leading positions in the study of coatings.

1.3.1 Components of the Microscope and Its Working Principle

A transmission electron microscope consists of the following main elements: power supply, providing all systems of the microscope; vacuum station, creating the necessary vacuum level in the column of the microscope; column of the microscope, containing all electron-optical imaging components.

The optical scheme of the microscope is shown in Fig. 1.14. It consists of the system of object illumination (electron gun, condensing lens assembly with stigmator of the second condenser and electromagnetic deflection system) and the imaging system (objective lens with stigmator, projecting assembly with intermediate and projecting lenses).

Illumination system creates electron beam incident on the specimen. The beam of electrons ejected from a heated cathode is conditioned in the electron gun and then focused by the first and the second condenser. Upon passing through the specimen the beam is partially scattered. The scattered electrons are stopped by an aperture diaphragm. The electrons that were not scattered by the specimen pass through the diaphragm and are focused by the objective lens in the object plane of the intermediate lens, forming the primary magnified image of the object.

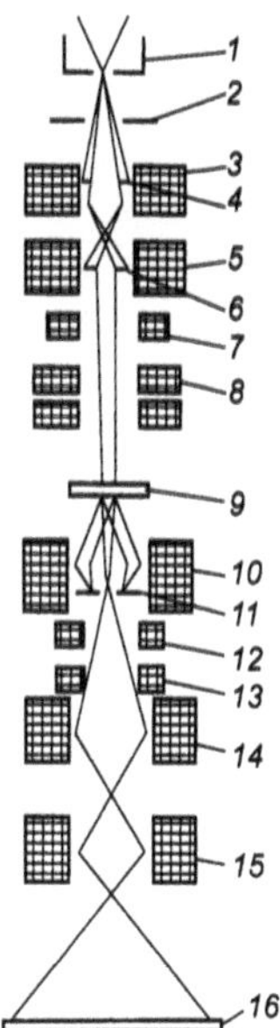

Fig. 1.14. Optical scheme of a transmission electron microscope. *1* cathode assembly; *2* anode; *3* first condenser; *4* diaphragm of the first condenser; *5* second condenser; *6* diaphragm of the second condenser; *7* stigmator of the second condenser; *8* deflecting system; *9* object; *10* objective lens; *11* aperture diaphragm; *12* stigmator of the objective lens; *13* stigmator of the intermediate lens; *14* intermediate lens; *15* projecting lens; *16* screen

The high resolution of the microscope is achieved by using a stigmator of electromagnetic type in the lens. The projecting assembly (intermediate and projecting lenses) creates the final magnified image on the fluorescing screen or a photographic plate. To obtain an electronogram the objective aperture diaphragm located behind the rear focal plane is moved out of the beam passage zone, and a selector diaphragm is placed in the image plane of the objective lens.

1.3.2 Specimens for Transmission Electron Microscopy

General Overview

Introduction of TEM had a significant impact on the development and formation of modern understanding of the fine structure of coatings. However, with rare exceptions, the pace of development and the level of electron microscopy studies lag behind the progress of other branches of metallography, e.g., metallography of steel. Such a situation can be explained by specificity of coatings as objects for electron microscopy study, which precludes direct transfer of conventional TEM techniques to this field.

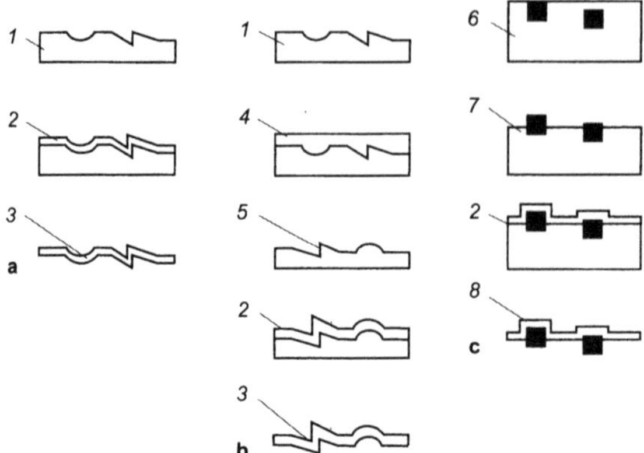

Fig. 1.15. Single step (**a**), two step (**b**) replicas, and replicas with extraction (**c**). *1* specimen; *2* carbon coating; *3* final carbon replica; *4* intermediate plastic film; *5* plastic replica (inverted); *6* polished specimen; *7* etched specimen; *8* final carbon replica with extracted particles

Galvanic coatings having thickness from 10–20 nm to 10–15 μm, as well as thicker gas-thermal coatings, often consist of several layers with complex phase composition. Due to differences in chemical and physical properties of the layers and the substrate etching often leads to formation of shorted pairs, thus complicating the development of microstructure. For a thin layer the layer can be completely removed before an etching relief develops on the surface. Cold-hardening during mechanical treatment of specimens distorts the microstructure of coatings, and the resulting hardened layer can be thicker than the coating itself. Another complicating factor is brittleness of thin foil specimens because of nonmetallic inclusions in the coating, the presence of embedded oxide phases and microcracks due to internal stresses.

A review of papers on TEM published during the last few years shows that the techniques of coating preparation and image interpretation are practically not discussed in the literature, the only exception being the work [1.52] which describes the preparation of replicas for study of protective coatings on steel. As for thin foils, the information about their preparation, including the techniques for separating the coating from its substrate, are scattered in numerous publications, and a summarizing review would be highly desirable. In the following sections we shall try to present the known data in a systematic way.

Replicas

There are three principally different ways of preparing replicas (surface prints). Single step replicas (Fig. 1.15a) are most commonly obtained by sputtering a layer of carbon, silica or other substances directly on the surface of the studied specimen. Sputtering is carried out in vacuum using standard equipment. Lacquer replicas (4% solution of collodion in amyl acetate) also give acceptable results. The replica film is then separated from the substrate either mechanically or, should any difficulties arise, by electrolytic etching of the specimen under the replica which has been dented in small squares with a thin needle.

If a single step replica cannot be separated from the specimen, as is the case with a very rough or uneven surface, e.g., with a sharp bend of a rather viscous material, two step replicas (Fig. 1.15b) are used. Here a film of carbon is sputtered on a previously prepared plastic print of the surface which is then dissolved.

The replicas of the third type, the replicas with extraction (Fig. 1.15c), are used in the study of two-phase structures. The surface of the specimen is etched in a special way, so that only the matrix is dissolved, leaving the inclusions of the second phase intact. Then a film of carbon or other material is sputtered at the etched surface, and the specimen is further etched so as to completely dissolve the matrix near the surface until the replica film with extracted particles separates from the surface. The particles in the replica retain the positions they occupied in the specimen.

Depending on the specific problem that is being solved, the replicas can be prepared in the following ways:

Replicas from a slanted section of a coating are used to determine the microstructure of different layers across the section, including the substrate.

The described above mechanical method of preparing a slanted section for light microscopy cannot be used for TEM specimens. An alternative technique of preparing a slanted section using electrochemical etching is shown in Fig. 1.16. When a specimen with a coating is periodically immersed in electrolyte, a slanted section is produced because of unequal etching efficiency with the height of soaking of the wetted surface with electrolyte. Electrolytes recommended for several common steel coatings are given in Table 1.1.

In order to get a carbon replica of the slanted section alone the specimen is screened before sputtering. The sputtering itself is carried out in two stages at different angles (15–30° for the first sputtering and 90° for the second one) to improve the contrast of the print and the mechanical strength of the replica. The replica is separated by electrochemical etching in the conditions when vigorous hydrogen production due to interaction of different layers and formation of insoluble reaction products are avoided, since the former process can destroy the replica, and the latter complicates the separation.

Replicas from a transversal section of a thin coating layer are used to study the variation of structure and phase composition of the coating with its thickness.

The procedure of preparing replicas from transversal sections includes preparation of a high quality transversal section as described above for light microscopy, sputtering of the replica and its separation following the guidelines for replicas from slanted sections.

Replicas from coatings in the plane parallel to the surface of substrate are used to study the effect of chemical inhomogeneity of the base metal, the presence of nonmetallic inclusions, grain orientation, etc. on the microstructure of coatings.

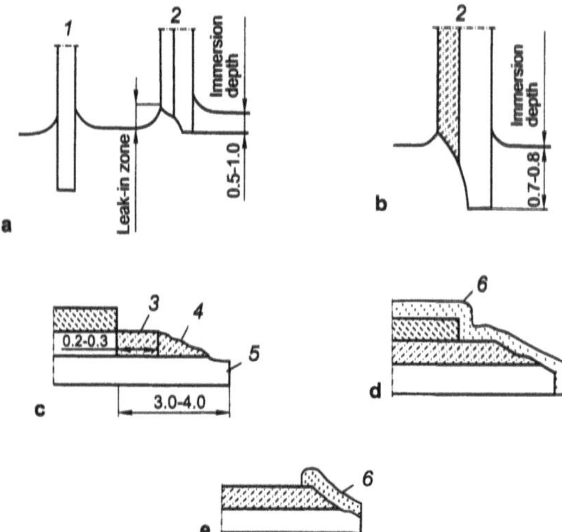

Fig. 1.16. Preparation of a slanted coating section on steel. **a** etching on first immersion; **b** etching on second immersion; **c** screening by metal foil; **d** sputtering of replica on slanted section; **e** prepared replica on slanted section after the screening foil has been removed and the edges cut.

In this case first the layers covering the sought phase are removed using chemical or electrochemical etching. After checking by one of the structural methods that the selective etching has uncovered the desired layer, a replica is sputtered and then separated, using the techniques described earlier.

Replicas with extraction. If the phase to be studied is buried in the bulk of the coating, first the upper covering layers are removed as has just been described. After the phase has been uncovered, the standard procedure of preparing a replica with extraction is performed. Just prior to electron microscopy studies the replica is treated with a 5–10% acid solution to dissolve the thin film of oxides and products of chemical interaction with electrolyte, which encapsulates the particles in the replica and seriously impedes the penetration of electron beam.

Replicas from the surface of coating exfoliation are used to study adhesion between a coating and its substrate. For highly bonded coatings exfoliation is forced artificially following the scheme of technological tests for bending. Weakly bonded coatings are separated mechanically. It is recommended to take the replicas from both the coating and the substrate side. The replicas are prepared following the standard procedure.

The above classification can be completed by carbon films deposited on the surface of a substrate prior to electrodeposition of galvanic coatings. Such replicas allow using microdiffraction and are thus extremely useful in studying the initial phases of electrocrystallization and the processes of formation of a continuous galvanic coating.

Table 1.1. Electrolytes used in preparation of replicas for electron microscopy studies of metallic coatings

Composition of electrolyte	Etching conditions	Purpose
5 ml of hydrochloric acid, 5 ml of chloric acid, 100 ml of water	$U = 1–2$ V	Slanted section of tin and zinc coatings
50 g of caustic soda, 85 g of metanitrobenzoic acid, 100 ml of water	Chemical etching at 80 °C	Selective etching of tin layers
200 g of chromic anhydride, 50 ml of orthophosphoric acid,	Chemical etching at 70–80 °C	Selective etching of zinc layers
1.2 ml of sulphuric acid, 500 ml of water		
4 ml of nitric acid, 100 ml of ethyl alcohol	Chemical etching	Etching of zinc on transversal sections
15 ml of chloric acid (60%), 100 ml of isopropyl alcohol	$U = 10–20$ V	Electropolishing of steel, zinc, nickel
20 ml of hydrochloric acid, 5 ml of isopropyl alcohol, 100 ml of water	$U = 1–2$ V	Separation of carbon replicas from tin coatings
2 ml of hydrochloric acid, 100 ml of methyl alcohol	$U = 4–6$ V	Separation of carbon replicas from zinc coatings
1 ml of hydrochloric acid, 100 ml of methyl alcohol	$U = 1–4$ V	Separation of carbon replicas with extraction of Fe–Sn phase from tin coatings
Concentrated sulphuric acid	Chemical etching	Etching of tin coatings for SAS (surface-active substance) extraction
1–2 ml of hydrochloric acid, 100 ml of water	$U = 0.5–1$ V	Extraction of SAS from tin into replica
100 ml of nitric acid, 100 ml of water	Chemical etching	Separation of thin chromium coatings from steel

Foils From Gas-Thermal Coatings

Electropolishing that is commonly used to prepare metallic specimens cannot be employed to study "base metal–gas-thermal coating" compositions for a number of reasons: structural inhomogeneity, poor electric conductivity of nonmetallic inorganic coatings, the presence of open-ended pores and microcracks.

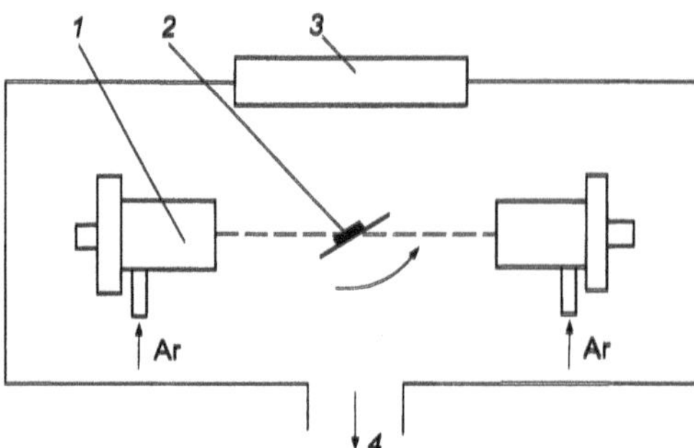

Fig. 1.17. Ion etching installation. *1* glowing discharge ion gun; *2* specimen holder; *3* vacuum chamber window; *4* vacuum pump outlet

A technique to produce very thin foils from a coating material, the so called ion etching, was suggested in [1.69]. In the method atoms are removed from a surface bombarded by ions with energy 3–6 keV, thus thinning the specimen with the rate 0.5–5 μm/hour.

An experimental installation for ion etching (Fig. 1.17) is implemented on the basis of a commercial vacuum station and has two glowing discharge ion guns. The specimen is oriented and vertically located with respect to ion beam by a specimen holder. Inert argon gas is supplied to each gun. Vacuum pumps reduce the residual pressure in the chamber of the station to 10^{-3}–10^{-2} Pa. To monitor the process of thinning and check for creation of openings in the foil an optical microscope is provided near the window of the vacuum chamber.

The presence of two ion guns and specimen rotation with the speed about 5 rpm ensure bombardment of both planes of the foil, even thinning and polishing, and negligible heating of the specimen. The processing of a specimen should begin with accelerating voltages about 5 kV and gun current about 100 mkA.

The specimens for ion etching are prepared in the form of disks with diameter 3 mm, thinned to 100–300 μm by grinding with a piece of fine sandpaper. Ion etching of the disks produces openings with edges transparent to electrons. The wedge-shaped openings accessible for electron microscopy analysis can be as wide as 100 μm.

The rate of thinning is quite low, and preparing a specimen can take rather long time, up to a week. On the other hand, ion etching produces a polished surface, giving the possibility of fine tuning the specimen. The thin portion can be produced in practically any part of the foil: at the top surface of the coating, in the center, near the base metal–coating interface (for foil specimen cut across the layers of the coating).

Ion etching is accompanied by several side effects that complicate identification of structure: rippling of the foil surface, appearance of black dots and rising of the specimen temperature. Rippling can be partially removed by optimizing the speed of specimen rotation. The black dots are the result of the increase in the number of defects in the foil material under ion bombardment.

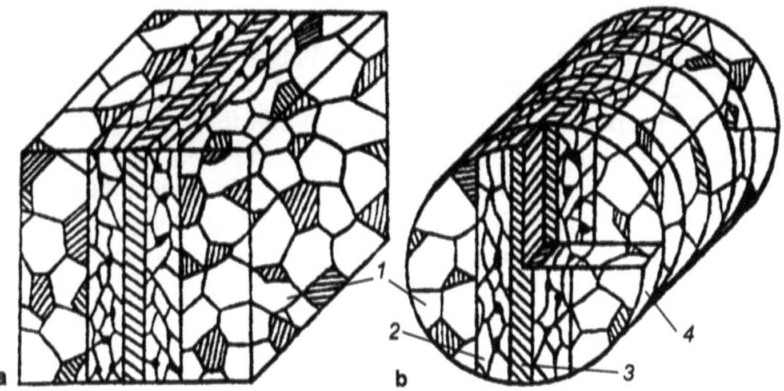

Fig. 1.18. Preparation of disk specimens for ion etching. **a** gluing; **b** cutting the disks; *1* base metal; *2* coating; *3* glue; *4* disk

The following procedure proved to be practical in the study of thin coatings. Two identical specimens are glued together in such a way that the surfaces of the coatings face each other (Fig. 1.18), and a cylinder is cut from this sandwich. The cylinder is then cut into discs, with each disk containing the basic metal, the coating and the glue, which are thinned by ion etching.

Electron microscopy analysis of thus produced foils reveals the fine structure of the entire transversal section of the coating and the substrate. Moreover, this technique allows a detailed study of the structure of the base metal–coating interface and judging about its changes during the deposition. Further improvements of the method can be found in the papers by Hafter and Ostreicher [1.33] and Michel et al. [1.59].

Ion thinning of foils can be used to study the fine structure of refractory materials produced by powder metallurgy (AlN, TiC, SiC, Si_3N_4), pored ceramics and coatings containing phases with different chemical properties.

Foils From Galvanic Coatings

To prepare a thin foil a galvanic coating must first be separated from the cathode substrate. Usually electrodeposited coatings are protected by a layer of acid-proof varnish (e.g., polystyrene) from the side of the growth surface, and the material of the substrate is completely chemically dissolved in a mixture of acids. Iron and low-carbon steel are readily dissolved in the mixture of orthophosphoric and sulphuric acids, copper and its alloys – in the mixture of sulphuric and chromium acids.

Coatings thicker than 0.1 mm are first mechanically or chemically thinned. In mechanical thinning the coating is ground with progressively finer sandpaper until a metal plate with thickness about 100 μm is obtained. Mechanical thinning does not change the structure of the coating.

Chemical thinning is appropriate when the material of a coating can be easily dissolved in a chemical reagent. In the process of thinning it is necessary to check the evenness of etching and avoid selective etching of separate phases, grain boundaries and other structural elements.

Several methods for preparation of specimens of thin foils from galvanic coatings for electron microscopy studies are commonly employed. By far the most widely used is the so called "window" technique. Here a thin plate prepared as described above is covered from both sides by varnish, leaving an unprotected "window" in the center. It is also recommended to protect with varnish the lower part of the current supply that would otherwise dissolve and contaminate the electrolyte. Then the plate is wired as anode and electropolished with a stainless steel cylinder serving as cathode. The composition of electrolytes and regimes of electropolishing for several common coatings are given in Table 1.2. For a properly chosen regime of electropolishing the surface of the coating becomes bright. If the surface becomes dented as if pricked with a pin, the voltage should be decreased. Etching of the specimen indicates that the voltage is too low and should be increased.

When a hole appears in the unprotected "window", the plate is taken out of electrolyte, washed, dried, and the thin edges of the hole are covered with varnish. Electropishing is continued until a large number of small holes is formed in the central part of the specimen. The specimen is then thoroughly washed, and the thin parts are cut out with a scalpel and placed on a mesh. Thus prepared objects can be studied with electron microscope.

If a coating has small area (about 10 x 20 mm^2) and is rather thin (2–20 μm) more convenient is another method of preparing the foils. In this case the disks with diameter 3 mm, equal to size of the chuck of the microscope specimen holder, are cut from the coating using a punch. Such a disk is then fixed between two plates with openings of varying sizes (2.0, 1.5, 1.0 mm) and electropolished. The polishing begins with plates having the openings of the largest size. After 1/3 of the coating layer has been removed, the plates are substituted for the ones with medium size openings, until only 1/3 of the initial thickness of the coating is left. The final finishing of the specimen is done with the plates with the smallest openings, until a through hole develops in the center of the disk.

A simple method of preparing thin foils from galvanic coatings at a given distance l' from substrate was suggested in [1.47]. A coating having thickness l is separated from its cathode substrate, covered by acid-proof varnish from the substrate side and electropolished from the other side to remove the layer with thickness l'', where $l' = (l - l'')/2$. The foil is then further thinned using the usual procedure of double-sided electropolishing. The method allows studying the structure of coatings from any depth, including the growth surface itself.

It should be noted here that, irrespective of the technique used for preparation of foils, the best results are obtained with fresh specimens immediately after preparation. If this is not practical, the foils should be stored in vacuum to prevent formation of oxide films on the surface.

Table 1.2. Electrolytes and regimes of electropolishing used in preparation of thin foils for electron microscopy studies of metallic coatings

Coating	Composition of electrolyte	Regime of polishing
Ag	60% thiourea, 24% CH_3COOH, 16% H_2SO_4	$T = 20–22$ °C, $i = 0.4 \div 0.5$ A/dm^2
Au	150 ml CH_3COOH, 6 ml H_2O, 30 g CrO_3	$T = 20–22$ °C, $i = 0.4 \div 0.5$ A/dm^2
Au–Co	25 ml C_2H_5OH, 25 ml glycerol, 50 ml HCl	No data
Bi, Bi–Sb	98% of saturated solution of NaCl in H_2O, 2 ml of 2% HCl	$T = 10$ °C, $U = 3 \div 5$ V, $i = 20$ A/dm^2, $\tau = 30$ s, current on periodically for 30 s intervals
Cd	65% H_3PO_4	T = 10 °C, vigorous stirring, $U = 3 \div 5$ V,
Co and its alloys	1. 40% HCl 2. Concentrated HCl 3. 23% $HClO_4$, 77% CH_3COOH	No data
Cr	20% $HClO_4$, 80% CH_3COOH	No data
Cu	1. 60–70% H_3PO_4 2. 66% H_3PO_4, 10% CH_3COOH 24% H_2O	$T = 10$ °C, $U = 5 \div 6$ V $T = -3$ °C, $U = 2.2$ V
Cu–Bi	66% H_3PO_4	$T = 10$ °C, stirring
Fe	490 ml H_3PO_4, 5 ml H_2SO_4, 100 g CrO_3	No data
Fe–Ni, Fe–Co, Fe–Mn	94% H_3PO_4, 6% H_2SO_4, 100 g H_2O	$T = 50$ °C, $U = 5$ V
Ni	1. 60% H_2SO_4 2. 80% CH_3COOH, 20% $HClO_4$	$T = 20$ °C, $i = 0.4 \div 0.5$ A/dm^2 No data
Ni–Mo, Ni–S	75% H_2SO_4	$T = 20 \div 40$ °C
Ni–Co, Ni–Fe, Ni–P	40% H_2SO_4 60% H_3PO_4	$T = 50$ °C, $U = 2$ V
Pd	50% CH_3OH, 33% HNO_3, 17% H_3PO_4	$T = 25$ °C, $U = 30$ V
Zn	37% H_3PO_4 in C_2H_5OH	$T = 10$ °C, stirring

1.3.3 Application of TEM

Both direct and indirect study of coatings with TEM provide qualitative and quantitative criteria to establish the correlation between the technological parameters of the sputtering process and physical and mechanical properties of coatings [1.25, 1.64], yield information about details of the structure of the base metal–coating interface [1.22, 1.29, 1.32, 1.80] and about the interaction of materials from different layers [1.2, 1.36]. A direct TEM study allows investigation of the coating structure [1.95]. Indirect studies help in finding out the cause for destruction of a coated part, classify the types of fractures, study the kinetics of crack propagation, determine the microrelief of the friction surface in the study of friction contact zone of wear-proof coatings, as well as the relief of the surfaces of the base metal and the coating after separating the latter.

TEM in the study of coatings was first employed for investigation of replicas. Today, when scanning electron microscopy has firmly established itself as the main technique for study of surfaces, the replicas have partially lost their importance. Nowadays they are mainly used in analysis of the surface morphology in the cases when resolution provided by scanning electron microscopy is not sufficient, and in study of microparticles that cannot be prepared as foils.

The work [1.46] provides an example of the judicious use of replicas in the study of the details of crystallite growth and formation of a continuous coating on the surface of a glass-carbon cathode during joint deposition of copper and cadmium. The coating was electrodeposited in galvanostatic regime with cathode current density 1 A/dm^2 from the electrolyte with mole composition $CuSO_4$ – 0.06, $CdSO_4$ – 0.04. Electrolysis time 5–100 s.

The results of TEM study show that as the time of electrolysis increases, the number of seeds at the cathode and their sizes also increase (Fig. 1.19). Some of the seeds gain crystallite shape in the process and coalesce, forming growth twins. Development of new seeds during electrolysis testifies to activation of initially passive sites at the surface of glass–carbon. With time isolated crystallites coalesce and form a continuos film of coating.

It was the electron microscopy analysis of replicas that provided the basis for elaboration of the model of coating development from initial stages of electrocrystallization to formation of continuous coating on the surface.

The method of replicas, however, suffers from a serious drawback – it does not provide information about the internal structure of material and the defects of the crystal structure. Because of this the method is being progressively replaced by the technique of direct transmission of electron beam through a thin foil of material.

The application of TEM has significantly boosted studies in the morphology and fine structure of galvanic coatings, including amorphous, that now attract substantial interest. An example of TEM studies of amorphous coatings can be found in [1.73] for thin foils of electrodeposited nickel–phosphorus coatings. Amorphous state in such alloys is reached by increasing the fraction of the non-metallic component.

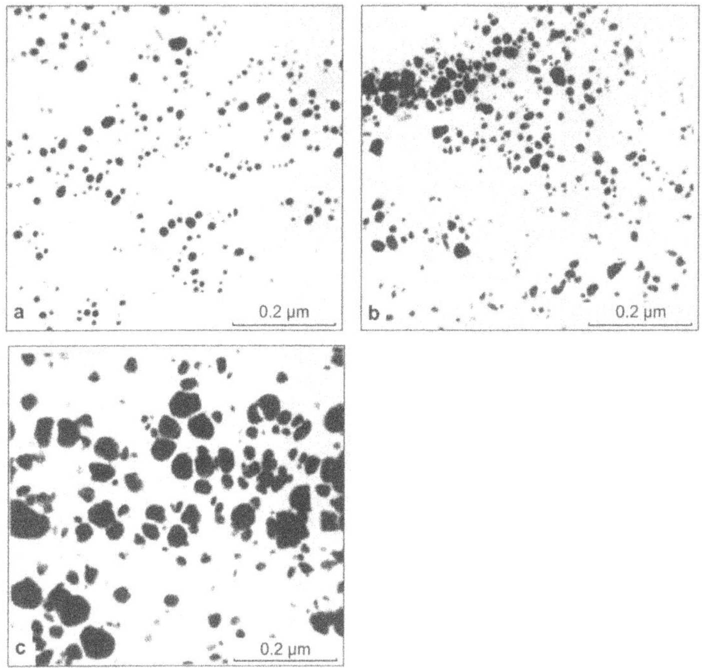

Fig. 1.19. Stages of growth of the Cu–Cd alloy coating. Electrolysis time: 5 (**a**), 10 (**b**), 40 (**c**) seconds.

The coatings with 1% of phosphorus have a pronounced coplanar dislocation structure with dislocations in parallel planes (Fig. 1.20a). As the doping degree increases the average grain size degreases (Fig. 1.20b), the grains acquire rounded shapes, the degree of dislocations in the alloy sharply increases, and the concentration of twins drops. Alloys with 11% of phosphorus form bright textureless coatings, consisting of nonfragmented spheroid grains (Fig. 1.20c).

Further enrichment of coating with phosphorus leads to formation of amorphous structure (Fig. 1.20d). Diffraction pattern now has two rings: an intense first one and a split second one with no reflexes from crystallite regions (Fig. 1.20e). Such a pattern is typical for amorphous metallic materials, and electron microscopy images show spotty contrast with characteristic size about 4 nm. The character of the contrast does not change when changing from light field to adjusted dark field images, which is typical for absorption nature of the contrast. Apparently, amorphous coatings have regions with density lower than the density of matrix. Similar defects in amorphous films of various origin are sometimes referred to as "superlattice". The universality of the "superlattice" for amorphous state of material and its interpretation as density fluctuations is the most important conclusion for practical studies, which will doubtlessly proliferate as amorphous galvanic coatings become more and more common.

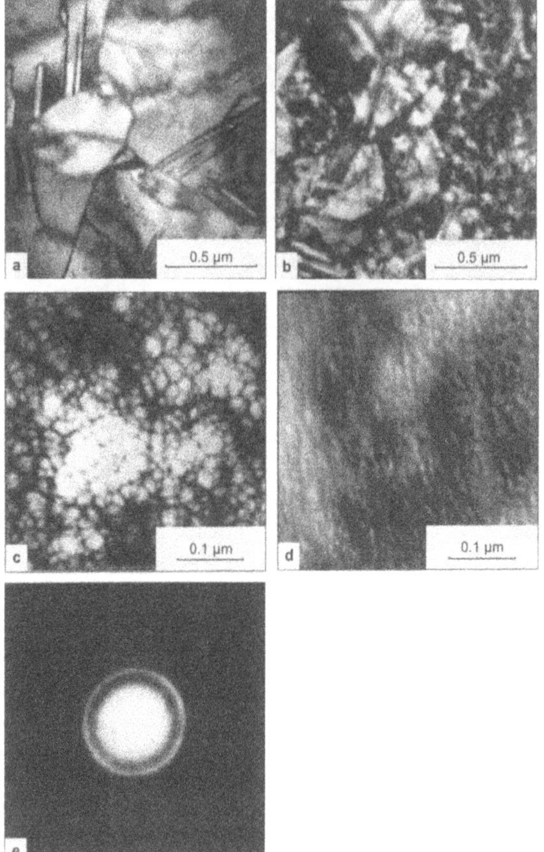

Fig. 1.20. Electron microscopy images of the structure of Ni–P galvanic coating with phosphorus content 1 (**a**), 3 (**b**), 11 (**c**), 16 (**d**) per cent and electron-diffraction pattern of the specimen *d* (**e**)

1.3.4 Prospects of Electron Microscopy Studies

Further developments in the field of electron microscopy studies of coatings will probably come from the improvements of the transmission electron microscopy technique in the following directions:

- use of high voltage electron microscopes (accelerating voltage 1000 kV and higher), capable of imaging thicker objects (up to about 1 µm) and producing diffraction pictures from smaller regions, thus making the procedure of specimen preparation much simpler;

- improvement of spatial resolution to 0.1 nm, allowing application of direct imaging techniques and thus providing insights into atomic structure of grain boundaries and subboundaries, phase interfaces, various structural imperfections of the crystal lattice;
- introduction of analytical methods, using add-on devices to electron microscope to gain not only information about chemical composition of microregions, but other quantitative parameters of the coating as well (electronic structure, atom location in the crystal lattice of complex substances, degree of atom dislocation around a defect, etc.)

1.4 Scanning Electron Microscopy and X-ray Spectrum Microanalysis

1.4.1 General Overview

The principal difference between scanning electron microscopy (STM) and TEM is the way the image of a specimen is produced. In STM the picture of an object is formed by scanning its surface with an electron probe with diameter 5–10 nm, and information is delivered mostly by reflected and secondary electrons. The method of imaging determines the main methodological possibilities of the STM technique.

Firstly, a direct study of the structure of coating surface in a wide range of magnification factors from x10 to approximately x30000 with rather high resolution (about 10 nm) becomes possible. This makes the tedious and difficult procedure of preparing special specimens transparent to electron beam, like replicas and foils, unnecessary.

A typical specimen for a modern scanning microscope has a diameter about 20 mm and thickness about 10 mm, although much smaller specimens can be readily studied as well. The procedure of specimen preparation is very straightforward, and as a rule just cleaning the surface of the specimen with appropriate solvents in an ultrasonic bath is sufficient.

Secondly, STM provides focus depth, exceeding hundreds of times the depth attainable in TEM, thus producing stereoscopic pictures of the structure with clear spatial pattern of all its elements.

And finally, STM has a number of analytical possibilities, which greatly extend its application area and provide diverse and often unique information. With STM the studies of a specimen by observing the appearance of “daylight”, studies of crystallographic and dislocation structure in the channeled electron mode, tracking of structural changes in a specimen in the course of experiment (deformation, wear, corrosion, heating, ion etching), analysis of local chemical composition become possible. The latter forms the basis for microprobe studies that will be treated in more details later.

1.4.2 Components of the Microscope and Its Working Principle

In a scanning electron microscope (Fig. 1.21) the primary electron beam produced by electron gun is accelerated in high voltage (up to 50 kV) electric field and focussed into the probe with a system of electromagnetic condenser lenses. The scanning generator synchronizes the position of the electron probe moving across a specimen with the position of the imaging points at the displays of the observation and photoregistration CRTs, maintaining direct correspondence between the specimen and its image. Interaction of the primary electron beam with a specimen leads to emission of secondary and reflected electrons and characteristic X-radiation. Secondary and reflected electrons are caught by the collector, and after conversion and amplification are used to modulate the intensity of the beam of the imaging CRT, i.e., to produce a raster image of the object.

Secondary electrons are emitted from the near-surface layers of the specimen with maximum thickness for metals not more than 5 nm. The actual value of the emission coefficient depends on the local potential barrier for electrons created by the specimen surface at each particular point. Change in the number of emitted secondary electrons as a function of the surface topology produces a half-tone image on the imaging screen of the microscope and creates an illusion of stereoscopic picture.

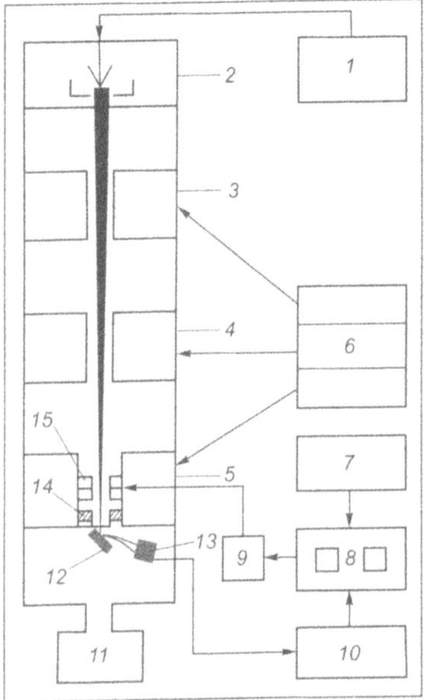

Fig. 1.21. Principal scheme of a scanning electron microscope. *1* high voltage power supply; *2* electron gun; *3–5* condenser lenses; *6* lens power supply; *7* scanning generator; *8* observation and photoregistration screens; *9* magnification control unit; *10* video amplifier; *11* vacuum system; *12* specimen; *13* collector system; *14* stigmator; *15* scanning coils

Reflected electrons originate at depth several times larger than thickness of the layer imaged by secondary electrons. The value of the reflection coefficient depends on the atomic numbers of the excited elements, and the higher the average atomic number of elements constituting this particular portion of the specimen, the larger the number of reflected electrons. Specimen parts with higher emission of reflected electrons look brighter on the screen, and the image of the specimen has a pronounced black and white character without half tones and illusion of stereoscopicity. However, since contrast formation is determined by variations in the chemical composition of the specimen, the image provides a sort of map of the distribution of elements on the surface. Furthermore, the distribution of induced current at the specimen yields an image in absorbed electrons, which complements the reflected electrons image and also serves to obtain contrast as a function of the atomic number of elements.

The mode of electron detection used in each particular situation depends on the purpose of the study and produces specimen images in different signals.

The characteristics of STM place it in an intermediate position between light microscopy and TEM (see Table 1.3). However the three methods should not be considered as competitors and rather complement each other, implying combined application.

1.4.3 Scanning Microscopy of Base Metal–Gas-Thermal Coating Composition

Scanning electron microscopy is widely used in the studies of materials with gas-thermal coatings, providing among others the following possibilities:

- analysis of fractures in coated parts [1.16, 1.61, 1.84]
- study of the mechanisms of deformation and cracks formation [1.15, 1.34]
- analysis of structure in transversal and longitudinal sections of coatings [1.31, 1.81]
- determination of the microstructure of the transition zone in the vicinity of the coating–base metal interface [1.8, 1.41]
- quantification of the coating porosity [1.91]
- study of the surface structure of coatings and base metals after treatment [1.65, 1.76]
- analysis of changes in surface relief caused by wear [1.26, 1.60, 1.63, 1.83, 1.92] and corrosion [1.85].

Preparation of specimens with coatings for STM studies usually causes no problems and is performed following guidelines for preparation of metal specimens. Special attention should be paid to preservation of specimen relief during mechanical treatment, coating exfoliation and chipping must be avoided. Specimens with nonconducting coatings prior to STM study should be covered with a thin layer of carbon or metal to provide drain path for electrons.

Table 1.3. Comparative characteristics of light microscope (LM), scanning electron microscope (STM) and transmission electron microscope (TEM)

Characteristics	LM	STM	TEM
Resolution			
Routine	5 μm	0.2 μm	10 nm
Available to operator of high qualification	0.2 μm	10 nm	1 nm
Requiring special measures	0.1 μm	0.5 nm	0.2 nm
Focus depth	Small	Large	Moderate
Modes of operation			
Transmission	+	+	+
Reflection	+	+	–
Diffraction	+	+	+
Other	Some	Numerous	None
Specimen			
Preparation	Usually straightforward	Straightforward	Tedious, artefacts possible
Range and type	Various, original or its replica	Various, original or its replica	Only thin original or replica
Maximum allowable specimen thickness	Massive	Moderate	Very thin
Environment	Arbitrary	Usually vacuum	Vacuum
Usable space	Small	Large	Small
Field of view	Rather large	Rather large	Limited
Signal	Only picture	Allows processing	Only picture
Cost	Low	High	High

1.4.4 Scanning Microscopy of Electrodeposited Coatings

The work [1.46] provides an example of application of fractography to study the main structural features of highly textured Fe–Ni, Fe–Co and Bi–Sb alloys. The texture is created by partially ordered anisotropic grains and subgrains with sizes in the direction perpendicular to the coating surface exceeding the dimensions in the plane parallel to the growth surface of the coating. Transversal fractures of the coatings show column-like aggregates of elongated crystallites, oriented with their long axes normal to the plane of the substrate (Fig. 1.22). The height of the aggregates is several times larger than their width (diameter) which equals the average grain size. The width remains nearly constant across the entire thickness of the coating, which implies that the texture is formed by autoepitaxial growth of each successive grain on the surface of the previous one.

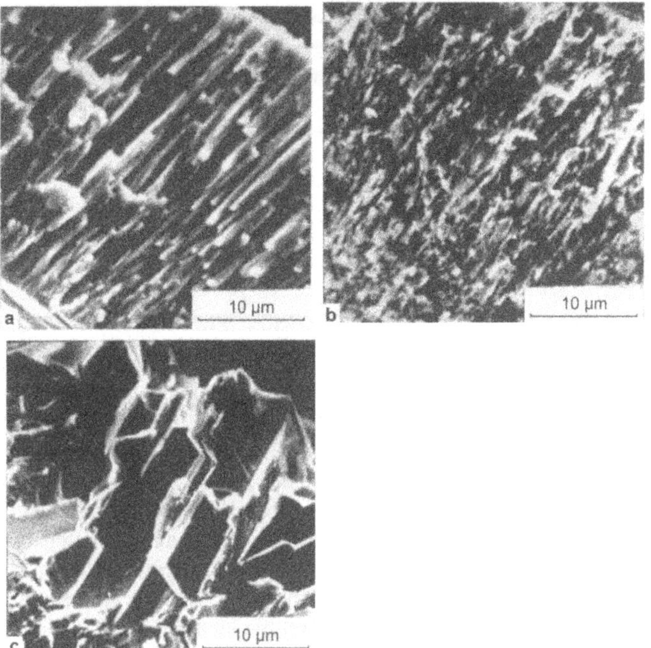

Fig. 1.22. Fractures of the coatings from alloys: Fe–Ni (**a**); Fe–Co (**b**); Bi–Sb (**c**)

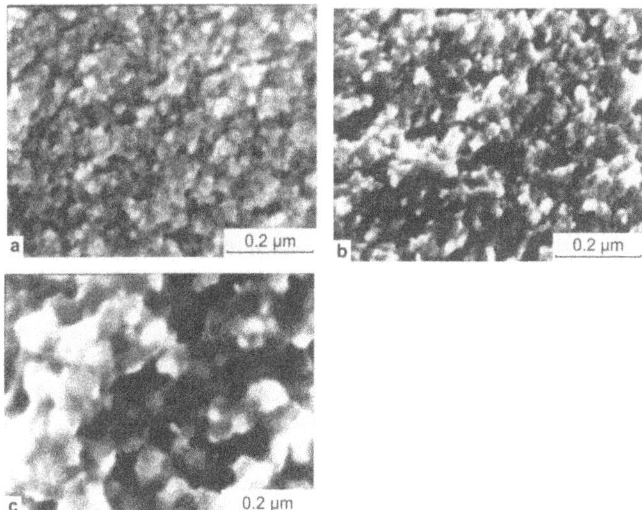

Fig. 1.23. Fractures of copper coating deposited from ethylene diamine electrolyte immediately after electrodeposition (**a**) and after annealing for 1 hour at 300 °C (**b**) and 500 °C (**c**)

Table 1.4. Effect of annealing on certain characteristics of copper coatings

State of coating	Rate of corrosion [g·m^2/hour]	Plasticity [%]	Average pore diameter [nm]
After electrolysis	0.6	3.0	20
After annealing (1 hour)			
at 300 °C	0.8	9.5	30
at 500 °C	1.8	2.2	75

The results [1.46] of fractographical studies of copper and nickel coatings (Fig. 1.23) pointed at an increase of coating porosity in the course of annealing. The latter was termed secondary porosity to distinguish it from primary porosity developing during electrocrystallization itself. It was shown that low temperature annealing does not result in appreciable secondary porosity. The pores only slightly increase in diameter due to migration of hydrogen to collectors and annihilation of vacancies at the walls of internal hollows in the coating. The coating retains satisfactory corrosion resistance with nearly threefold gain in plasticity due to relaxation of internal stresses (Table 1.4) as compared to the state immediately after the deposition. High temperature annealing, on the contrary, leads to abrupt increase in pore sizes and their volume fraction with corresponding deterioration of strength, plasticity and protective properties of the coating. The analysis of fractures allowed to prescribe the regime of thermal treatment, guaranteeing an optimal combination of performance and mechanical properties of the coating.

1.4.5 Prospects of STM

The examples we have provided in the preceding section illustrate the power of qualitative STM analysis. However, the quantitative possibilities of STM have not yet been completely explored and are currently limited to a rather narrow set of characteristics like average grain size, porosity, etc. One of the ways of increasing the informativity of quantitative STM studies of structure and revealing the features inaccessible for visual inspection would be the use of coherent optic techniques, namely Fraunhofer diffraction patterns from polycrystalline specimens.

Another promising direction of developing the STM technique would be the use of stereoscopic images that can be readily obtained on any scanning microscope with tilting specimen stage. The pictures can be used for both visual inspection and for quantitative measurements in three dimensional space.

An obvious extension to STM methodology that would lead to qualitative breakthrough in the study of galvanic coatings is a wider application of computers for automatic image analysis, increased accuracy of quantitative measurements and improved speed of data processing.

And the last but not the least, STM can be used for microanalysis of chemical composition of the coatings. The technique of X-ray spectrum microanalysis is covered in the following sections.

1.4.6 Construction of Microprobe. X-ray Spectrum Microanalysis Technique

The method of X-ray spectrum microanalysis (XSM) utilizes the characteristic X-radiation from chemical elements contained in microvolume exited by incident electron beam. Spatial resolution of the technique in the analysis of massive specimens is practically limited only by probe diameter (about 1 μm). XSM is capable of determining all elements from boron ($Z = 5$) to uranium ($Z = 92$). Overlapping of characteristic lines of elements can cause certain difficulties in determination of a number of elements in the presence of others, and special methodical approaches for overcoming this problem have been devised. The minimal sensed concentration of an element, or its threshold of detectability, is about hundredth of per cent for elements with $Z > 11$, for light elements (boron, carbon, nitrogen, oxygen) the sensitivity is much lower (1–5%).

A principal scheme of an X-ray spectrum microanalyzer is shown in Fig. 1.24. A focussed beam of electrons is generated by electron optical system consisting of an electron gun and a system of lenses and diaphragms. Wavelength and intensity of characteristic X-ray lines excited in the specimen are measured by spectrometer system consisting of crystal analyzers and detectors. A light microscope is used to select the part of the specimen for analysis and monitor the object in the process of analysis.

The spectrometer is tuned to desired element by adjusting the position of the crystal analyzer, which reflects only the characteristic X-rays incident at angles satisfying the Bragg condition. The reflected radiation is collected by detector, amplified and registered by a ratemeter or another registering device. Modern microprobes, along with Bragg spectrometers, often have energy dispersion spectrometer systems as well, which allow simultaneous observation of all elements at once. The relative simplicity of a characteristic X-ray spectrum having few lines makes qualitative element analysis fast and reliable. Addition of a device for continuous feeding of a specimen under the probe and a recording potentiometer allows to obtain the distribution of the determined element in the direction of specimen feeding.

Scanning of the electron probe in XSM yields an image of the specimen surface in characteristic radiation of the analyzed element. A map of the qualitative distribution of the given element on the surface is produced on the screen of CRT.

The basis of quantitative XSM analysis is the proportionality of the emitted characteristic radiation intensity to the concentration of elements in the analyzed microvolume. The content of an element in the specimen is determined from comparison of intensities of characteristic lines from the specimen and from a reference containing 100% of the measured element. The references are placed in the specimen chamber on a special microscopic stage together with the object. It should be noted here that computation of element content in a multicomponent system is a rather demanding task requiring great care to take all necessary corrections into consideration.

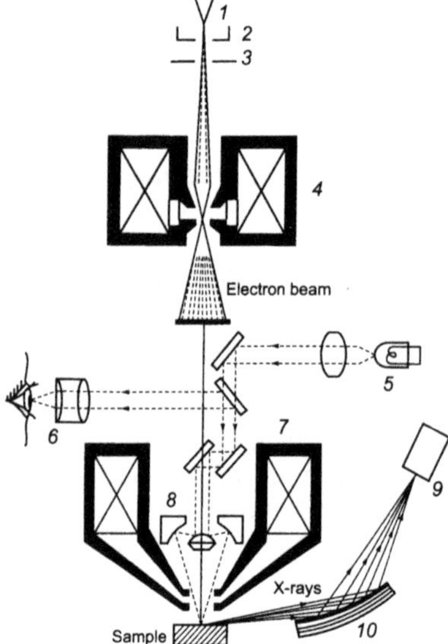

Fig. 1.24. X-ray spectrum microanalyzer. *1* electron gun filament; *2* Venelt cylinder; *3* anode; *4* condenser lens; *5* light source; *6* eyepiece; *7* objective lens; *8* reflective objective; *9* radiation detector; *10* crystal analyzer

Since scanning can also yield a surface image in secondary electrons, a microprobe as a rule is at the same time a scanning electron microscope.

1.4.7 Applications of XSM

XSM is mostly used to determine local chemical composition of inclusions, phases, grains and to study the distribution of a given element between components of the structure or layers of a coating [1.35, 1.66, 1.82, 1.100]. The latter is the main application area of XSM in electrodeposition, since during electrocrystallization of alloys the elements can be distributed unevenly both by thickness and by surface of the coating, which has negative impact on performance and technological properties of the coating.

An instructive example is provided by electrodeposition of Sn–Pb alloy in the production of PCBs. The alloy is deposited with a twofold purpose of protecting the conductors during etching and improving the soldering-ability of lead holes. However, alloy deposition from a standard electrolyte gave deviations from average composition as large as 26–28%, which rendered impossible mechanized soldering and alloy reflow. An XSM study of alloy composition homogeneity with thickness of the coating (30–100 μm) and along hole axes helped choose the regimes of electrodeposition giving minimal deviation of composition, not exceeding 6–7%.

Studies of chemical heterogeneity of electrodeposited alloys of metals from Fe subgroup (Fe–Ni, Fe–Co, Fe–Mn) [1.46] demonstrated that first layers, accessible to analysis, are enriched with the more electronegative component, and its content in alloy gradually decreases with increasing coating thickness. For formation of massive coatings with thickness more than 30 μm, when current density is maximum for a stationary regime, the components of the alloys also tend to distribute unevenly with respect to structural elements. Coatings forming near the regions of maximum current consist of large chemically nonuniform spherolites with boundaries enriched by 3–8% by the more electronegative (doping) component of the alloy as compared to their cores.

Laminarity and concentration gradient with coating thickness are also rather common for alloys, but usually these can be easily removed by diffusion annealing. Unfortunately, XSM results on the degree of coating laminarity by chemical composition and on the character of component distribution cannot be considered entirely conclusive. The characteristic size of chemical heterogeneity is often comparable to diameter of the probe, and only averaged information about component distribution is thus obtained. More reliable results are expected from layer-by-layer studies of coatings using Auger and photoelectron spectroscopy, which seem very promising in this respect.

The already cited work [1.46] also provides an example of applying the XSM technique to estimate the degree of hydrogen pickup by galvanic coatings. The content of hydrogen in electrodeposited Co coatings was determined indirectly by comparing the intensities of the CoK_α lines from a reference (99.99% Co) and the specimen. Such a comparison led to conclusion that maximum concentration of hydrogen (about 5%) is reached in the initial layers of the coating. As the thickness of the coating increases to about 50 μm, the concentration of hydrogen gradually decreases and then remains practically unchanged. It should be noted, however, that interpretation of experimental data in this case is complicated by the presence of impurity elements other than hydrogen in the coating.

As a concluding remark it is worth mentioning that in contrast to STM, the surface of the specimens subject to XSM must be perfectly flat without any marks or relief, since the latter are a major source of errors in the determination of concentration. Preparation of the specimen generally does not cause any serious difficulties, although it does require certain thoroughness. Special care should be taken to avoid embedding of abrasive into the surface of prepared specimen or formation of corrosion products during etching.

1.4.8 Combined Use of STM and XSM

Although X-ray microanalyzers and scanning electron microscopes were developed independently, a clear tendency today is to combine the two in a single instrument providing the functionality of both STM and XSM. Electron-optical systems of such hybrid microprobe analyzers are capable of forming electron beam in a wide range of diameters (from 5–20 nm to 0.2–2.0 μm) and allow easy switching from STM to XSM mode of operation simply by switching to another registration system.

Combined study of galvanic coatings with both techniques is still in its initial stage, but there is no doubt that the combination of STM and XSM will very soon attract the attention it deserves of experimentalists working in the field of galvanic coatings.

1.5 X-ray Structure Analysis

1.5.1 Fundamentals of X-ray Structure Analysis

X-ray structure analysis (XSA) is basically the study of structure using diffraction patterns from X-rays scattered by the analyzed object. Since the wavelengths of X-rays (of the order of ≈1 Å) are comparable with interatomic distances in crystallic structures, the crystals are natural diffraction gratings for X-rays. The phenomenon of diffraction is illustrated in Fig. 1.25. A beam of X-rays S_0 incident on the crystal at a grazing angle θ is scattered by atoms of the crystal. The scattering can be considered as a sort of "reflection" from atomic planes. However, in contrast to visible light X-rays are "reflected" selectively, only when the wavelength λ, interplanar spacing d_{hkl} and diffraction angle θ satisfy the Wolf-Bragg law of diffraction:

$$n\lambda = 2d_{hkl}\sin\theta, \tag{1.10}$$

where $n = 1, 2, 3, \ldots$ is diffraction order.

The "reflections", or to be more precise the diffraction maxima, are registered on a photographic plate or by any other suitable registration method. For a known wavelength λ determined by the material of anode of the used X-ray tube, each value of d_{hkl} corresponds to certain angle θ. The Wolf-Bragg law allows to determine d_{hkl} for the entire set of atomic planes forming the lattice by measuring diffraction angles θ.

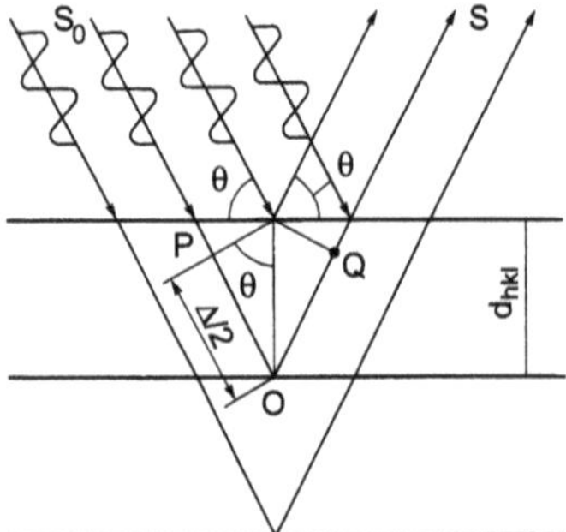

Fig. 1.25. Diffraction of X-rays on crystal lattice

Diffraction maxima corresponding to "reflections" from different atomic planes generally have different intensity depending on distribution of atoms in the lattice. The intensity is measured either from the degree of blackening on photographic plate or from the height of peaks on diffractogram. Together with angle position of the diffraction line and its shape the intensity of a line is one of the most important characteristics in XSA.

1.5.2 Instrumentation and Some Methodical Features of XSA

XSA instruments can be roughly divided into two classes, X-ray chamber and diffractometer type, according to the method of registering the diffraction pattern. In the former the pattern is fixed on a photographic film, while the latter have a goniometer to record the angle distribution of radiation intensity using a counter of quanta, followed by data output for external processing. Of the two the most widely used are commercial diffractometers.

A diffractometer employs the Bragg-Brentano focussing scheme (Fig. 1.26). The rays exiting the window of X-ray tube are passed through slits limiting the primary X-ray beam, reflected from a flat specimen and focussed on the entrance slit of the quanta counter. A goniometer handles changes and measurements of the angles between the incident beam and specimen surface and between the incident beam and the counter. Angular velocity of the counter is exactly two times higher than the speed of specimen rotation, and the rays reflected from a specimen always find the counter. Scanning the angle between the incident beam and the surface of a specimen, angular positions of the specimen giving diffraction maxima are sequentially found.

The most common X-ray source for XSA is an X-ray tube with anode covered by a layer of metal: Cr, Fe, Co, Ni, Cu, Mo or Ag. The characteristic K_α line of the metal is used, and the K_β component is suppressed with either filters or monochromators.

An object for a typical X-ray diffractometer is a flat specimen with diameter (or side of the square) 5–20 mm or more. Such a specimen can be easily prepared from virtually any coating. An additional advantage as compared to other specimen types is that polishing and etching are not necessary.

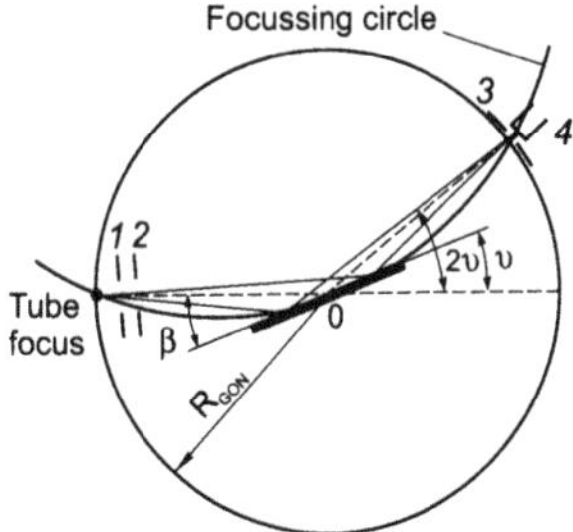

Fig. 1.26. Bragg-Brentano scheme of X-ray photography. *1–2* slits limiting the primary beam; *3* entrance slit of the counter; *4* counter of quanta

1.5.3 Application of XSA to Development of New Coatings and Analysis of Their Performance in Real Conditions

The main task of XSA in material science is identification of a material using its crystal structure characteristics, which is very important for modern materials such as multicomponent and multiphase coatings [1.21, 1.30, 1.75, 1.78]. XSA is also widely employed to reveal various imperfections of crystal structure in real substances [1.74, 1.102] and to analyze atomic structures of partially ordered and noncrystalline objects like amorphous coatings [1.44, 1.94].

The application of XSA in the analysis of crystal structure comes in two directions:

- development and quality control of new coatings with prescribed properties;
- analysis of coating performance in real conditions (under oxidation, in a friction pair, etc.)

There is extensive literature on application of XSA in the development of new coatings. Marel et al. [1.56] have reported the structure and composition of AlN–Si_3N_4 coatings produced by physical deposition from vapor in the temperature range 1000–1320 °C on a graphite substrate with protective layer of SiC. The results of comparative studies of TiN coatings deposited in vacuum using different methods have been presented in [1.42]. Ural et al. [1.90] have studied the structure of thermobarrier coatings on the basis of ZrO_2–20%Y_2O_3, deposited using the technique of electron bombardment evaporation with temperature of the base metal 1050 °C. Londa et al. [1.54] have analyzed the process of formation of the ($TiAl_3$+Al) coating on a low-alloyed steel substrate using a combined technique of sputtering followed by treatment with an infrared laser. In their review on nondestructive control of the base metal–coating composition Crostack et al. [1.20] referred to the method of X-ray structure analysis as to one of the primary and most informative techniques. Heat resistance and oxidation mechanisms of corrosion-resistant coatings have been studied in details in works [1.3, 1.5, 1.18, 1.55]. The mechanisms of wear have been addressed in [1.1, 1.88].

1.5.4 Application of XSA to Analysis of Galvanic Coatings

XSA is still the most widely used and universal method in the investigation of galvanic coatings. XSA is used for study of crystal structure, qualitative and quantitative phase analysis, determination of size, distribution and location of individual crystallites in polycrystalline structures, analysis of type, concentration and distribution of lattice defects, analysis of atomic structure of partially ordered and noncrystalline coatings.

Classical approaches of XSA have already been extensively treated in a number of reviews and monographs published in the sixties and seventies. For example, a presentation of X-ray studies of lattice defects can be found in the work [1.70], an overview of textures – in the work [1.43]. In the following sections we will restrict ourselves only to original methods developed in the last years.

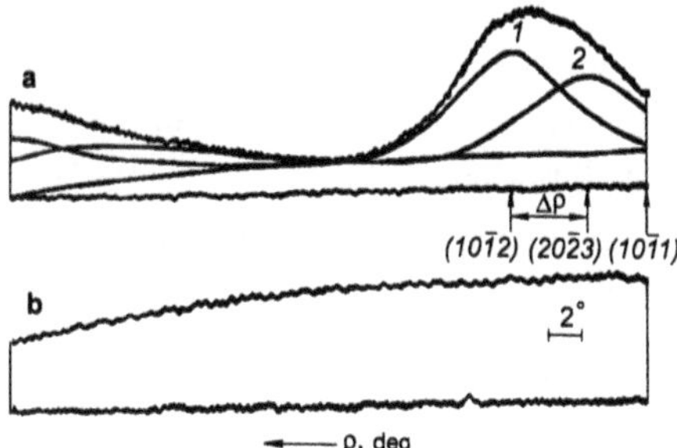

Fig. 1.27. Curve of texture interference record $(10\bar{1}1)$ of electrodeposited cadmium (**a**) and reference (**b**) (adopted from [1.93]). *1*, *2* actual profiles of texture maxima $(10\bar{1}2)$ and $(20\bar{2}3)$, respectively

Studies of Texture

Orientation of crystallites is very common for the structure of a coating, and XSA is probably best suited to qualitative and quantitative study textures [1.6, 1.10, 1.46, 1.97].

The degree of structure perfection can be characterized by two parameters: the fraction of crystallites having a given orientation, and the dispersion of the texture, i.e., the spread of orientations of the crystallites from this fraction about the principal axis direction of the texture (for normal distribution). To determine the parameters polar figures for the specimen and a non-textured reference are first constructed from X-ray diffractometry data. The parameters are then calculated from orientational density in the vicinity of the texture axis direction. The higher the ratio of the first parameter to the second one, the more perfect is the structure.

Vorobiev and Girin [1.93] suggested a technique for quantitative characterization of a multicomponent axial texture directly from the curve of texture record for the specimen, taken using the Schultz scheme. The technique yields the actual profiles of texture maxima, taking into account their relative intensities and angular distance between the maxima (Fig. 1.27). The principles of the technique can be formulated as follows.

For a coating having a double axial texture with axes $\langle h_1k_1l_1\rangle$ and $\langle h_2k_2l_2\rangle$ normal to the surface of the specimen the ratio of the intensities of the lines $\{h_1k_1l_1\}$, $\{h_2k_2l_2\}$ from the specimen and the reference (i.e., $\sigma_{\{h_1k_1l_1\}} = I_{1\{h_1k_1l\}}/I_{ref\{h_1k_1l_1\}}$ and $\sigma_{\{h_2k_2l_2\}} = I_{2\{h_2k_2l_2\}}/I_{ref\{h_2k_2l_2\}}$) in a diffractogram taken in the Bragg-Brentano scheme will be the larger, the larger is the number of crystallites oriented with their reflecting plane $\{hkl\}$. If we denote the intensities of texture maxima corresponding to axes $\langle h_1k_1l_1\rangle$ and $\langle h_2k_2l_2\rangle$ as I_1 and I_2, respectively, their ratio will have a well defined value: $I_1 = nI_2$, where $n = \sigma_{\{h_1k_1l_1\}}/\sigma_{\{h_2k_2l_2\}}$. The angular distance $\Delta\rho$ between the two texture maxima corresponds to angle φ between crystal planes $\{h_1k_1l_1\}$ and $\{h_2k_2l_2\}$. The actual profiles of the texture maxima can then be easily constructed using the well known Rechinger method of separating $K_{\alpha_1\alpha_2}$ doublet (Fig. 1.27), which in turn allows calculation of quantitative characteristics of the structure: the fraction of unordered crystallites, the fraction of textured crystallites, and the average scattering angle of the axial texture.

The described technique of estimating the degree of texturing of a galvanic coating allowed to establish a correlation between the wear resistance of galvanic chromium coatings and the quantitative parameters of their texture.

The presence of a sharp axial texture often complicates X-ray analysis of the structure of electrodeposited coatings. If the axis of such a texture $\langle h_1k_1l_1\rangle$ is oriented perpendicular to the surface of the specimen, all diffraction maxima in a diffractogram taken in the Bragg-Brentano scheme, except for $\{nh_1nk_1nl_1\}$, will probably be masked by background. To register the maxima $\{h_2k_2l_2\}$ and $\{nh_2nk_2nl_2\}$, required for determination of fine structure elements, Vorobiev and Girin [1.93] employed a goniometer add-on device, capable of working in transmission and reflection modes and yielding a complete pole figure.

In the reflection mode the specimen can be tilted around the horizontal axis. The intensity distribution curve corresponding to a given section of the projection sphere is obtained by slow rotation of the specimen in its own plane and synchronous relocation of the registering device.

The specimen is rotated around the horizontal axis of the goniometer until the normals to the reflecting planes $\{h_2k_2l_2\}$ and $\{nh_2nk_2nl_2\}$ of an axially textured specimen coincide with the bisectrix of the angle between the incident and reflected X-ray beams (the angles between the incident and reflected rays with specimen rotation axis were equal, with the axis lying in the plane of the specimen). The technique allowed registration of diffraction maxima, required for determination of mosaic block size and microdistortions in electrodeposited chromium coatings, which could not be obtained with the conventional method of registering a diffractogram. It was also demonstrated that "X-ray amorphism" of the coatings reported in some earlier works had been artefactual.

Study of Amorphous Coatings

If the conclusion about the "X-ray amorphism" of chromium coatings was a consequence of underestimation of the degree of texturing of the coating, similar inferences in the case of electrodeposited manganese coatings are a result of the fineness of the coating structure. Electrodeposition of manganese from chloride and sulfate water solutions at high current densities results in formation of a coating with the structure of α-Mn, with degree of dispersion rapidly increasing with increasing current density, especially in the presence of sulfur and selenium impurities. Conventional diffractograms of such coatings usually have one or two smeared maxima, that are practically lost in the background.

As a result a commonly accepted point of view about "nonphanerocrystalline" structure of manganese coatings have emerged. However, computer simulations of the dependence of the diffractogram line intensity distribution on the degree of α-Mn crystallite dispersion have demonstrated that the experimentally observed diffractogram practically coincides with the theoretically calculated one for the size of α-Mn crystallites equal to 2.0 nm. Hence, it can be concluded that "X-ray amorphous" manganese in this particular case is highly dispersed α-Mn.

Such a complex approach, combining experimental and computational methods, has great potential in the studies of amorphous and ultradisperse coatings since due to unique combination of properties these start to become the subject of extensive studies and application. A complex study would first of all include X-ray structure analysis and other diffraction techniques – electron and neutron diffraction, as well as atomic radial distribution function analysis. Diffraction patterns for X-rays, electrons and neutrons for amorphous materials as compared to patterns for crystalline bodies have no sharp reflexes and demonstrate only smeared intensity maxima. Radial distribution function contains information about the character of the short range ordering of the structure, which in the absence of long range ordering of atoms is the main distinguishing feature of the amorphous state.

As an example of structure characterization of amorphous alloys we shall refer to nickel-phosphorus galvanic coatings studied in [1.73]. The coatings with varying content of phosphorus were obtained by electrodeposition from nickel-plating electrolytes with varying concentration of sodium hypophosphite (electrolyte composition, g/l: $NiSO_4$ – 200, $NiCl_2$ – 30; regime of electrolysis: i_c = 10 A/dm^2, T = 80 °C, pH 1.0). The diffractograms were taken using Co–K_α radiation.

Fig. 1.28 shows the results of XSA for electrodeposited alloys of different composition. The coatings with up to 11% of phosphorus are supersaturated solid solutions of phosphorus in nickel (equilibrium solubility of phosphorus in nickel does not exceed 0.5%) and crystallize in a face-centered cubic lattice. The diffractograms have clear reflexes (111) and (200). As the concentration of phosphorus in alloy increases, the first diffraction maximum (111) becomes more and more diffuse, and the intensity of the second one (200) decreases.

Diffractograms of specimens with more than 11% of phosphorus show a single diffuse peak (halo), corresponding to the most intense line (111) of nickel. With further increase of phosphorus concentration up to 21% the width of the halo substantially increases. Microdiffraction patterns of such specimens from electron microscope (Fig. 1.20e) consist of two rings: an intense first one and a split second one, with no reflexes from crystalline regions. The character of diffraction patterns obtained by the two techniques unambiguously point to the amorphous structure of coatings with more than 11% of phosphorus.

It is interesting to follow the transition of amorphous nickel-phosphorus coatings into crystalline upon heating, since thermal stability is of utmost importance for practical application of amorphous alloys. The process of such a crystallization often has a complex nature and proceeds in several stages with formation of a number of metastable phases. Furthermore, the differences in heating conditions have a pronounced effect on the mechanism of phase-structure transformations and the morphology of the newly forming phases.

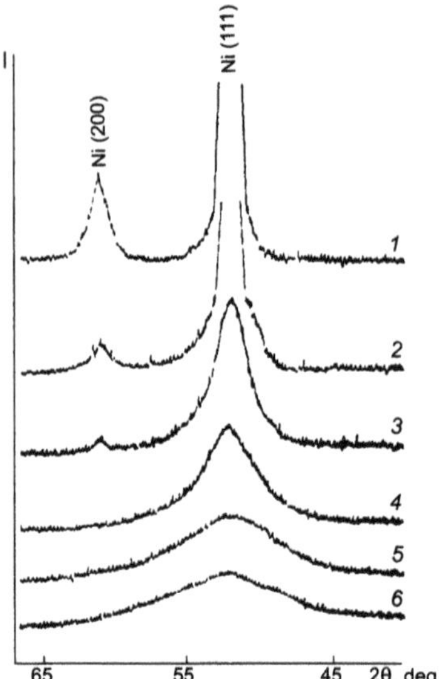

Fig. 1.28. Fragments of diffractograms for electrodeposited Ni–P alloy. Phosphorus content: 1% (*1*); 3% (*2*); 8% (*3*); 11% (*4*); 16% (*5*); 21% (*6*)

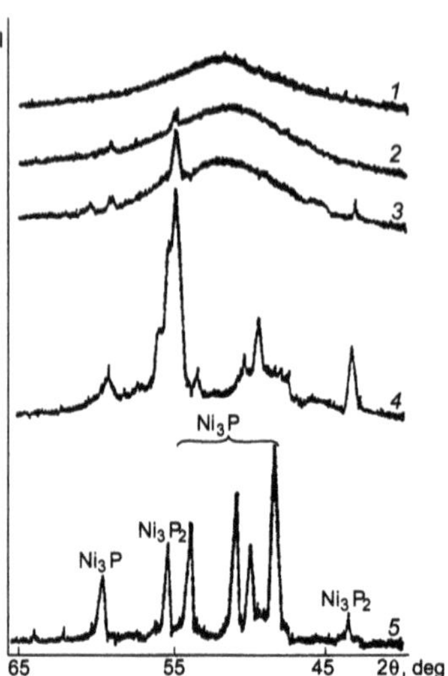

Fig. 1.29. Fragments of diffractograms for electrodeposited Ni–P alloy immediately after the deposition (*1*) and after annealing at 270 °C (*2–4*) and 500 °C (*5*). Annealing time – 10 (*5*); 15 (*2*); 30 min (*3*); 1 hour (*4*)

Fig. 1.29 shows the results of XSA for alloys with 21% of phosphorus. It can be seen that annealing the initial amorphous alloy at 270 °C leads to crystallization of an intermediate metastable phase Ni_5P_2. After annealing for 3 hours the amorphous coatings are completely transformed into crystalline. Extending annealing time to 10 hours leads to onset of formation of a stable phase Ni_3P along with the metastable Ni_5P_2. The broadening of the diffraction lines indicates that the formed crystals are highly dispersed. Nevertheless, even 25 hours of annealing at 270 °C is not sufficient to completely convert the alloy into its equilibrium state. A plausible explanation can be that the high activation energy of the process of Ni_3P crystallization requires higher annealing temperatures. And indeed, increasing the temperature of annealing to 400 °C leads to further interaction of the initially formed nickel phosphide with nickel $Ni_5P_2 + Ni = 2Ni_3P$ to form the stable phosphide Ni_3P, characterized by a body-centered tetragonal lattice with parameters $a = 0.892$ and $c = 0.440$ nm.

High and Low Temperature X-ray Investigations of Coatings

First documented attempts on high temperature X-ray studies of galvanic coatings have already appeared in the literature. Although this technique is still used very occasionally, it is the most reliable method for investigation of polymorphic and phase transitions and structure of phases that are stable only at elevated temperatures. High temperature X-ray studies on a DRON–3,0 diffractometer (Russia) are possible using a commercially available add-on device UVD–2000 with the following technical specifications:

Specimen size	15 x 12 x 2 mm^2
Operating temperature range [°C]	
in vacuum	20–2000
in inert gas	20–1800
in air	20–1200
Operating environment	vacuum 10^{-4} mm mercury, inert gas, air
Accuracy of temperature stabilization [°C]	±4

A specimen is heated in the working chamber using a resistive electrical furnace. The desired temperature is set manually and then is measured and stabilized automatically.

The capabilities of the high temperature X-ray investigations were demonstrated in [1.71], where the nature of the cubic phase of electrodeposited cobalt was studied. Cobalt coatings deposited from highly acidic solutions (pH < 2) have a biphase structure with close-packed face-centered and face-centered cubic lattices. Until recently there was no agreement about the nature of the metastable cubic phase.

Some researchers argued that in electrodeposited cobalt along with α-Co, having a close-packed face-centered lattice, a high-temperature β-Co modification with face-centered cubic lattice can also be formed. The others suggested that the cubic phase of cobalt coatings is cobalt hydride, forming when hydrogen produced at cathode is incorporated into the structure of the coating.

The diffractograms were taken at a DRON–3,0 diffractometer. The specimens were annealed in an UVD–2000 device in a stepwise fashion with step size 30–50 °C, the pauses in heating were used to record the diffraction lines. From diffractograms the integral line intensities I of the (101) line of the close-packed face-centered phase and the (200) line of the face-centered cubic phase of cobalt were determined and plotted as a function of temperature in the form $\ln(I/I_0) = f(T)$, where I_0 denotes the intensity of the corresponding line at room temperature. These two specific lines were chosen as indicators of structure and phase transformations because they do not overlap with reflexes from the other phase.

Fig. 1.30 shows the experimental temperature dependencies of the intensities of the (101) and (200) lines of the close-packed face-centered and the face-centered cubic phases of a cobalt coating, respectively. Logarithm of a line's intensity is known to change linearly with temperature. The experimental plots show a linear dependency of $\ln(I/I_0)$ on temperature up to about 250–280 °C. Cobalt and hydrogen are known to form two types of unstable hydrides, CoH and CoH_2, that readily decompose upon heating to 45 °C and 150 °C, respectively. Since, other conditions being equal and force factor being taken into account, the intensity of a line is proportional to the volume fraction of the given phase in a mixture, decomposition of hydrides would have led to an increase in the volume fraction of the α-Co phase and vanishing of the face-centered cubic phase volume fraction. The $\ln(I/I_0)=f(T)$ plot would have departured from the linear dependencies of the intensities of the two lines, which is clearly not the case in the indicated temperature range. Consequently, it can be concluded that the cubic phase observed in cobalt coatings at room temperature is not a hydride modification of the metal.

Increasing annealing temperature from 280 °C to the point of phase transition from close-packed face-centered to face-centered cubic lattices (420 °C) leads to some increase in the integral intensity of the (101) line and decrease in the intensity of the (200) line as compared to linear $\ln(I/I_0)=f(T)$ dependency, since a partial decomposition of the β–Co phase now takes place. In the course of the phase transition the intensity of the (101) line of α–Co abruptly drops, completely vanishing when the specimen is heated to 480 °C. Concurrently an increase in the integral intensity of the (200) line of the face-centered cubic phase is observed. Only face-centered cubic phase of Co is seen for annealing temperatures above 480 °C.

The following slow cooling is accompanied by a reverse transition from face-centered cubic to close-packed face centered structures and a complete phase recrystallization. Below 420 °C and down to room temperature only lines of the stable α-Co phase show up in diffractograms. Such coatings have inferior hardness, corrosion resistance and coercive force as compared to initial metastable modifications.

The remark about the still limited use of high temperature X-ray studies of galvanic coatings applies equally to low temperature investigations as well. At the same time, the latter can prove really efficient in the studies of coatings in the situations, when intense afterelectrolysis processes of lattice reorganization take place at room temperature.

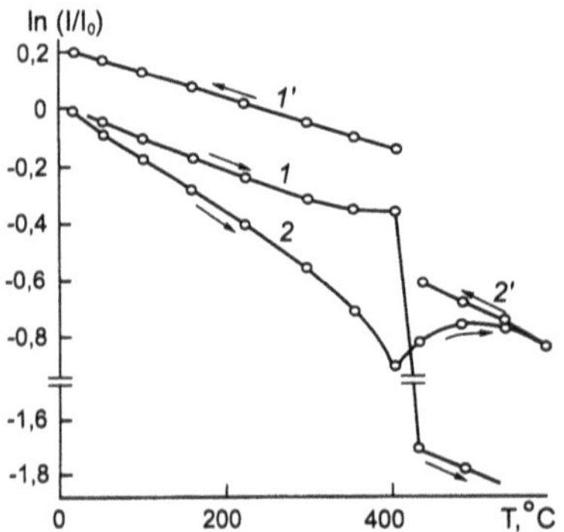

Fig. 1.30. Temperature dependencies of the intensities of the (101) line of the close-packed face-centered phase (*1*, *1'*) and (200) line of the face-centered cubic phase (*2*, *2'*) of a cobalt coating, deposited from electrolyte with pH 1.5, during heating (*1*, *2*) and following cooling (*1'*, *2'*) (marked by arrows)

If this is the case, a specimen sometimes is cooled below room temperature using a heat-conducting block with upper part contacting dry ice and controlling the temperature with a thermocouple, which can be rather troublesome. A more elegant solution is provided by a low temperature add-on device URNT–180 (Russia) designed for use with X-ray diffractomer DRON–3,0.

The specifications of the device are as follows:

Specimen size	Ø12 mm, thickness 1 mm
Operating temperature range [°C]	-180–+120
Temperature setting precision, [°C]	±3
Accuracy of specimen temperature stabilization, [°C]	±0.3

A specimen is cooled by evaporating liquid nitrogen. The rate of evaporation and the temperature of nitrogen gas flow is controlled by a dedicated control unit. The desired temperature is set manually and then is measured and stabilized automatically.

Application of Synchrotron Radiation

The rate of afterelectrolysis processes can be so high (e.g., due to migration of interstitial atoms) that certain irreversible changes have ample time to occur in the interval between the end of electrodeposition and the start of X-ray mapping. Low temperature X-ray diffractography cannot help in these situations, and the radical solution is XSA study of galvanic coatings directly in the course of electrolysis. However, attenuation of the X-ray flux in the bulk of electrolyte causes serious methodical difficulties. The first X-ray diffraction study in the cell was reported by Polukarov [1.70] who used a thin film (100–200 μm) of diluted 0.002 n silver nitrate solution covering the electrode, but this method has not become popular since the real concentrated electrolytes cannot be used, and the electrolysis itself cannot be performed correctly. Synchrotron radiation opened new prospects in the studies of the structure of galvanic coatings immediately during electrodeposition.

The unique properties of synchrotron radiation allow a significant reduction of the contribution from background scattering by solution and a rapid registration of a series of diffractograms. A special diffractometer with an electrochemical cell of specific design performs monochromatization and conditioning of the synchrotron radiation beam and registration of the diffracted radiation (Fig. 1.31) [1.12]. The body of the cell is made from teflon and holds a foil cathode, circular anodes located symmetrically with respect to cathode, and a capillary. In such a geometry of electrodes the coating is deposited simultaneously on both sides of the foil, thus improving the sought diffraction of synchrotron radiation. Cell windows are covered with 10 μm teflon film. The beam of synchrotron radiation enters through the entry window and is scattered on the foil covered with a coating. The diffracted radiation exits through the other window having conical shape to increase the range of registration angles.

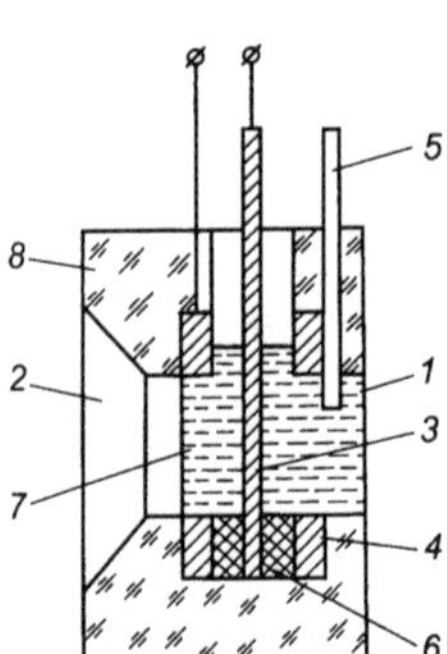

Fig. 1.31. Electrochemical cell for X-ray diffraction studies using synchrotron radiation (adopted from [1.12]). *1* entry window; *2* exit window; *3* cathode; *4* anode; *5* capillary; *6* teflon insulator; *7* electrolyte; *8* cell body

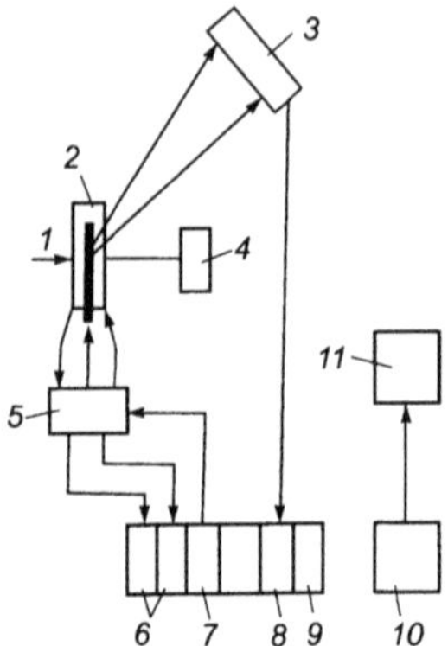

Fig. 1.32. Scheme of taking a diffractogram during electrolysis using synchrotron radiation (adopted from [1.12]). *1* beam of synchrotron radiation; *2* electrochemical cell; *3* detector; *4* trap; *5* potentiostate; *6* analog-to-digital converter; *7* digital-to-analog converter; *8* digitizer; *9* buffer memory; *10* computer; *11* display

The regime of electrolysis is controlled by a dedicated computer via a potentiostate interfaced through a digital-to-analogue converter (Fig. 1.32). The computer also monitors the regime through two analog-to-digital converters, one of which measures potential at the electrode, and the other registers cell current. Diffractograms are taken during electrolysis at preset time intervals, digitized and stored in buffer memory. The system is capable of taking a diffractogram in 1 ms.

The results of first experiments using synchrotron radiation are rather encouraging [1.12]. In their studies of tin deposition on a copper substrate the authors managed to register reflexes from tin already at initial stages of the process, when thickness of the coating did not exceed 50 nm. The dynamics of the transformation of tin lattice substructure from nonequilibruim-distorted lattice at the early stages to the equilibrium structure of a thick tin layer in the course of deposition was also observed. The study of the peculiarities of nickel structure transformation during cathode hydrogen pick-up and the results on the initial stages of the nickel hydride formation and its effect on the structure of nickel matrix provide another example of successful application of synchrotron radiation in XSA investigations. There is no doubt that further improvements of instrumentation and methods aimed at increasing the sensitivity of synchrotron radiation diffractometers towards structural changes of the surface layer will further extend the capabilities of XSA of galvanic coatings.

1.6 Electron Spectroscopy Techniques

1.6.1 Potential of Electron Spectroscopy Techniques

Of all emissive methods of surface analysis the most popular for investigation of coatings have become the techniques based on the Auger effect and photoionization – the so called Auger electron spectroscopy (AES) and photoelectron spectroscopy, commonly referred to as electron spectroscopy for chemical analysis (ESCA). The unique analytical characteristics of these methods (see Table 1.5) have made them an indispensable tool for investigation of phenomena related to oxidation, corrosion (including intercrystalline corrosion), passivation of metal surface, strength of bonding between the coating and its substrate, and many others [1.4, 1.13, 1.19, 1.24, 1.53, 1.57, 1.99].

1.6.2 Fundamentals of AES and ESCA Techniques

The method of AES is based on energy analysis of secondary Auger electrons. A high energy electron from the primary beam incident on a specimen knocks out an electron from one of inner atom shells, e.g., the *K* shell, and puts it in a high energy outer shell. Thus produced excited atom can return to its ground state by filling the created electron vacancy with an electron from a higher lying shell, say, the *L* shell. Such a transition is accompanied by releasing the energy determined by the energy difference between the two shells $E_L–E_K$, which can take the form of emitting either a photon, thus producing the characteristic X-ray radiation, or another electron from the *L* shell, a so called Auger electron. Since Auger electrons are emitted only from discrete levels, their energy is characteristic of the element, providing its energy signature. Hence, energy analysis of Auger electrons can be used to determine the element composition of specimens.

Table 1.5. Analytical characteristics of AES and ESCA

Characteristic	AES	ESCA
Sensitivity of analysis [%]	10^{-1}–10^{-2}	1–10^{-2}
Relative accuracy of analysis [%]	5–10	5–10
Sensitivity to elements	$Z > 2$	$Z > 2$
Depth resolution, [nm]	0.3–3	0.3–3
Surface (linear) resolution	0.5–5·10^3 nm	2–10 mm
Can provide information about chemical bonding	+	++
Can be used for nondestructive analysis	+	++

"+" possible in principle;
"++" best suited for.

Because primary electrons have rather high energy, Auger electrons can come from rather deep layers of the specimen. However, most of these "deep" electrons lose their energy in inelastic collisions before they leave the surface of the specimen. Typically only Auger electrons from the near surface layer with thickness 0.3–3 nm contribute to observed Auger spectra, which makes the AES technique particularly valuable for studies of thin near surface layers.

The method of ESCA is based on energy analysis of electrons, emitted from inner atom shells under X-irradiation. ESCA can provide information about the chemical state of the atom of a given element, since the energy of emitted electron is equal to difference between the energy of incident photon and the chemical bond energy. Although penetrating power of X-rays is rather high, in the energy range currently used in photoelectron spectrometers the depth resolution determined by the free path of emitted electrons is approximately equal to resolution of AES.

1.6.3 Instrumentation and Experimental Techniques

A basic Auger spectrometer (Fig. 1.33) consists of a working chamber containing electron gun, specimen holder and energy analyzer, and a registration system [1.17]. Since the yield of Auger electrons is very sensitive to the state of the surface, the presence of adsorbed atoms and films of adsorbed gases, a rather high vacuum (10^{-6}–10^{-8} Pa) is required inside the chamber and is commonly provided by oil-free pumps. Furthermore, modern spectrometers have provisions for cleaning the surface to be studied directly inside the chamber using either high temperature desorption or ion bombardment.

The operating principle of an Auger spectrometer is determined by the type of energy analyzer it is equipped with. Most common are spectrometers with retarding field and with cylindrical mirror analyzers. The latter principle was used in a scanning Auger spectrometer, in a sectioning spectrometer and other instruments.

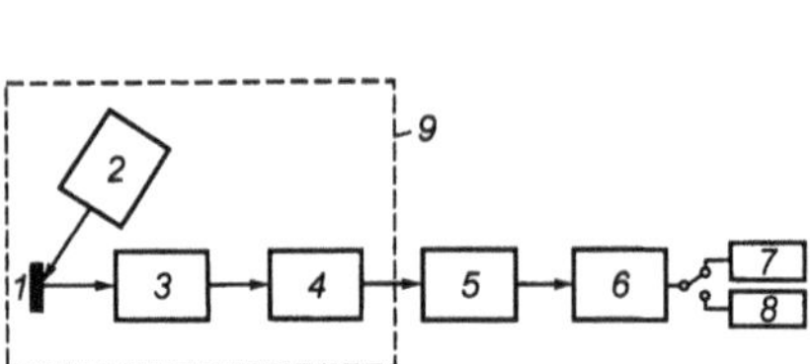

Fig. 1.33. Functional diagram of an Auger spectrometer (adopted from [1.17]). *1* specimen; *2* electron gun; *3* energy analyzer; *4* secondary electron multiplier; *5* preamplifier; *6* phase-sensitive detector; *7* oscilloscope; *8* plotter; *9* vacuum chamber

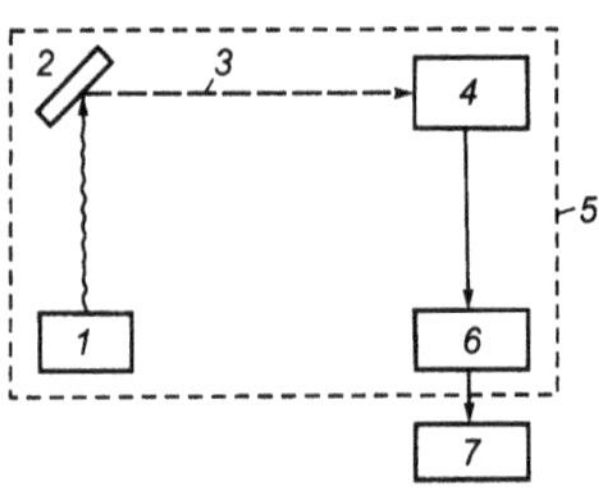

Fig. 1.34. Functional diagram of a photoelectron spectrometer (adopted from [1.17]). *1* source of electromagnetic radiation; *2* specimen; *3* photoelectrons; *4* energy analyzer; *5* vacuum chamber and magnetic shield; *6* collector of photoelectrons; *7* control system

A photoelectron spectrometer (Fig. 1.34) consists of a source of electromagnetic radiation, an energy analyzer, a collector of photoelectrons and a control system [1.17]. A common source of radiation is an X-ray tube with magnesium, aluminum, sodium or yttrium anode. Photoelectron spectra are usually obtained using electrostatic energy analyzers with deflecting or retarding field. The emitted electrons are collected by a channel electron multiplier or a microchannel plate, which allows detection of very low currents down to counting single electrons. The signal from electron collector is amplified and goes to an amplitude discriminator and then to a multichannel analyzer. Each channel of the analyzer measures the current of electrons of specific energy (channel step 0.1 eV, time of signal accumulation about 1 s). The spectrum is then processed by a computer and can be presented in either analog or digital form.

Many photoelectron and Auger spectrometers come with various add-on devices such as heating and cooling units, the units for chipping, cutting, breaking the specimens directly in the working chamber.

The techniques of AES and ESCA naturally complement each other and are often combined in one spectrometer (Fig. 1.35) or with other techniques of surface study, e.g., in the spectrometers of the ESCALAB family. As a rule, such spectrometers are equipped with a computer, a chamber for specimen processing and locking, a glove box with inert gas atmosphere, a specimen manipulator and other facilities that greatly extend the analytical capability of the spectrometer. However, the following point should be kept in mind when studying a specimen on a combined spectrometer.

ESCA is an inherently nondestructive method of analysis, since X-ray radiation creates practically no radiation defects in the surface of material. This is generally not true for AES, where the surface is probed by a beam of electrons. The combination of ESCA with AES and especially with ion etching (for surface cleaning or layer analysis) requires certain caution in interpretation of results and in using the same specimen for other studies, since ion or electron bombardment can significantly change the structure and properties of the surface up to its amorphization.

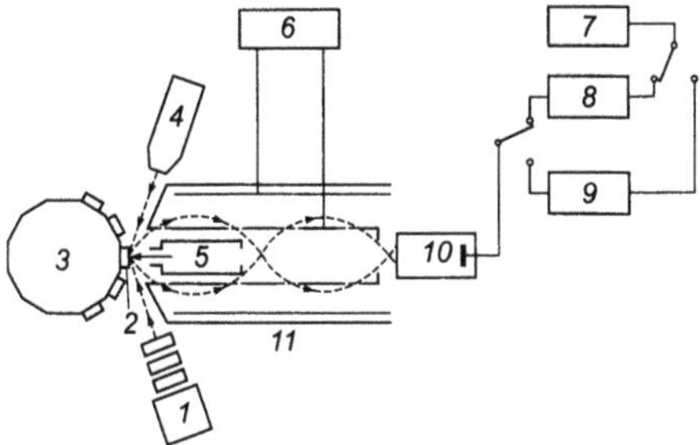

Fig. 1.35. Functional scheme of the ESCA/AES spectrometer PH–548 (adopted from [1.17]). *1* ion gun; *2* specimen; *3* carousel specimen holder; *4* X-ray tube; *5* electron gun; *6* control system; *7* plotter; *8* registering device for Auger electrons; *9* ESCA analyzer; *10* electron multiplier; *11* energy analyzer

1.6.4 Application of AES and ESCA

An example of applying the AES technique to investigation of coatings is provided by work [1.74], in which the authors studied galvanic Cu–Bi coatings deposited from trilonate electrolyte at different cathode potentials. Fig. 1.36 shows typical Auger spectra for alloy with 35% of bismuth taken at a JAMP–10 spectrometer. Argon ion etching with energy 3 keV and absorbed current 4 mkA was used to study the near surface layers. Content of elements in the layers was determined using the method of element sensitivity coefficients, and the profiles of copper, bismuth and oxygen distribution by thickness were obtained.

As the spectra clearly demonstrate, the surface of the untreated coating is heavily contaminated with impurity elements – carbon, oxygen and nitrogen, while the content of the main components of the alloy near the surface is rather low. After etching the surface with argon ions the intensities of the lines from nonmetallic impurities become significantly lower. The profiles of element concentration (Fig. 1.37) indicate that concentration of copper increases when moving from the surface into the bulk of the coating, while concentrations of bismuth and oxygen decrease. The content of elements stabilizes at depth about 20 nm, which implies that the surface is covered with a film of oxide having thickness about 20 nm.

Another work of the authors [1.46] considers the mechanism of improving the adhesion of galvanic coatings with thermal treatment by the following scheme: hardening of the base metal–electrodeposition–annealing. Using the industrial bismuth, copper and nickel coatings as an example, it was demonstrated that improvements in bonding between the base and the coating with annealing are explained by excess concentration of hardening vacancies that intensify the diffusion processes.

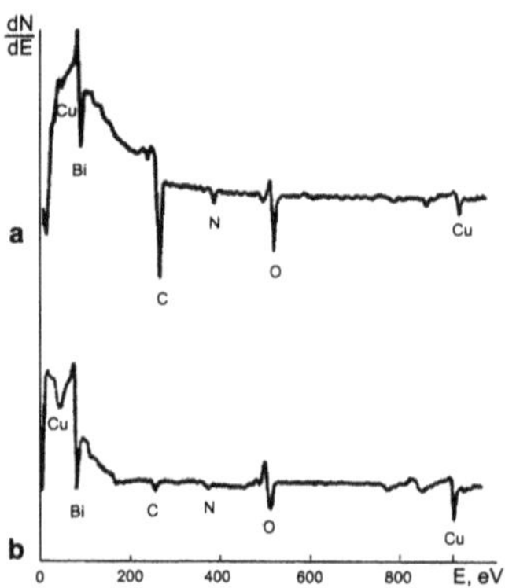

Fig. 1.36. Auger spectra from electrodeposited Cu–Bi coating immediately after the deposition (**a**) and after etching the surface with argon ions for 8 minutes with the rate of 5 nm/min (**b**)

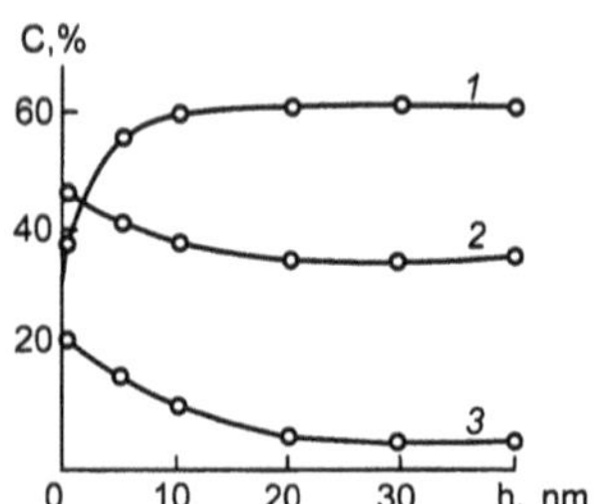

Fig. 1.37. Profiles of concentration distribution by thickness of the Cu–Bi coating for copper (*1*), bismuth (*2*) and oxygen (*3*)

An AES layer analysis of the coating and the base reveals an extended zone of mutual penetration of atoms of contacting materials as compared to coatings deposited using a conventional technology.

The relatively good surface resolution of the AES technique, determined by the diameter of the primary electron beam (from about 5 μm down to as low as about 50 nm) has led to development of scanning Auger electron spectroscopy, which uses an electron beam for probing the spatial distribution of elements in the near surface layers. Scanning AES yields information about the surface topography, and combining it with ion etching as described above renders such a study comprehensive. A variation of the scanning approach in electrodeposition is identification of foreign inclusions which cause peeling, bulging, lowered adhesion and discoloration of galvanic coatings.

It has already been mentioned that sensitivity of element analysis using AES is superior to the sensitivity provided by ESCA. The same is also true for analysis of spatial distribution of components, since even a focussed beam of X-rays still illuminates a rather large part of the surface, with the size about 2–10 mm (as compared to 50 nm–5 μm for AES).

However, ESCA as compared to AES provides more reliable information about the chemical state of atoms, allowing discrimination between an isolated atom and an atom bonded in a compound. The energy of photoelectrons is unambiguously determined by the valence state of an atom, and the latter shows up in spectra as a characteristic shift of electron lines. Variations of bond energy can be easily accounted for from elementary models, and determination of chemical states of molecular compounds from chemical shifts in their ESCA spectra is today a relatively straightforward task.

Interpretation of chemical shifts in AES spectra, on the contrary, is rather tedious because of the two stage character of the Auger effect. For this reason the authors using AES in their studies could only approximately suggest the chemical substances constituting the film at the surface of galvanic coatings.

An example of original application of ESCA to solve a problem of long standing in coating science can be found in work [1.87]. The authors clearly demonstrated that during electrodeposition of a coating on an alumina oxide substrate, a metallic coating can form directly on the surface of anode oxide, without prior mandatory complete filling of micropores. When conducting the experiment, the photoelectron exit angle (glancing angle) relative to the surface of the specimen was decreased down to 5°, which significantly improved the surface sensitivity of ESCA and facilitated the discrimination between surface effects and phenomena in the bulk of the specimen.

And finally, it is worth mentioning, that although ESCA does not provide the sensitivity of analysis attainable with AES, it has an important advantage of simplicity and accuracy of normalizing spectrum line intensities in determination of substance content of the surface under study.

1.7 Resonance Techniques

1.7.1 Nuclear Gamma Resonance (Mossbauer Spectroscopy)

The method of nuclear gamma resonance (NGR) is based on the effect of resonance absorption of γ-quanta by nuclei in solids without energy losses for nuclear recoil, discovered by Mossbauer. The unprecedented energy resolution of the technique, providing the accuracy of energy determination as high as 10^{-8}–10^{-9} eV, makes NGR an efficient tool for investigation of various hyperfine interactions in solids. Most important results to date have been obtained in the study of magnetic structure, electron density distribution, phase composition, phase transitions and processes of ordering [1.17, 1.98].

Principles of the Method

A diagram of a typical Mossbauer spectrometer is shown in Fig. 1.38. A vibrator moves a radioactive source of γ-quanta back and forth with speed v relative to absorbing specimen, creating a Doppler shift of emission line relative to absorption line. A detector measures the flux of γ-quanta transmitted through the specimen. The dependence of the radiation flux transmitted through the absorber on the Doppler shift value, obtained by changing the speed of source movement, yields the Mossbauer absorption spectrum.

The source in a Mossbauer spectrometer can move relative to absorber either with constant speed or with constant acceleration (speed ramp). When operating a spectrometer in the constant speed mode the spectrum is obtained point by point by successive measurement of the transmitted flux for each value of source speed. This significantly increases time required to record a spectrum, but allows a finer scanning step and improved accuracy of analysis in the study of singlet lines or fragments of complex spectra. In the constant acceleration mode the speed of source relative to absorber changes from its maximum positive to its maximum negative value in each measurement cycle, which allows one shot detection of the entire absorption spectrum in the chosen range of speeds.

The choice of spectrometer and its mode of operation is ultimately determined by the problem to be solved. A brief description of some γ-spectrometers and experimental procedures is given at the end of the section.

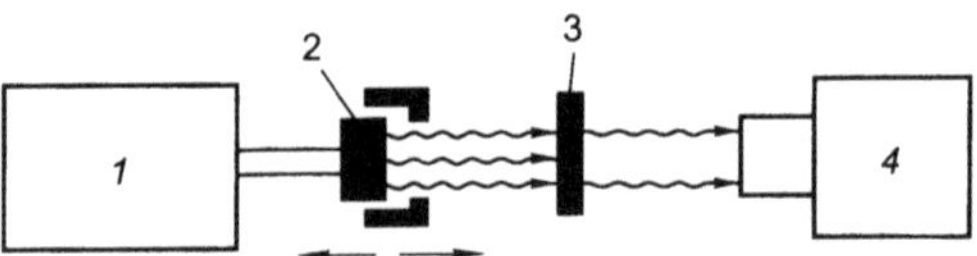

Fig. 1.38. Diagram of a Mossbauer spectrometer. *1* vibrator; *2* source of γ-quanta; *3* absorbing specimen; *4* detector

Table 1.6. Common Mossbauer isotopes and their characteristics

Isotope	Energy of emitted γ-quanta [keV]	Mean life-time of nu-clear excited state [ns]	Width of ab-sorption line [x 10^8 eV]	Recoil energy [x 10^3 eV]	Resonance absorption cross-section [x10^{19} cm^2]
^{57}Fe	14.4	98	0.46	1.95	23.6
^{61}Ni	67.4	5.3	8.6	40.0	7.21
^{67}Zn	93.0	9400	0.005	69.3	1.18
^{119}Sn	23.8	18	2.5	2.56	13.3
^{121}Sb	37.1	3.5	13	6.12	11.8
^{125}Te	35.6	1.4	33	5.44	2.86
^{181}Ta	6.25	6800	0.007	11.6	17.0
^{182}W	100.1	1.4	33	29.6	2.4
^{195}Pt	98.9	0.17	270	26.9	61.0
^{197}Au	77.3	1.8	25	16.3	0.44

A Mossbauer spectroscopy experiment is only possible if one of the isotopes of the studied element has an appropriate nuclear transition. The Mossbauer effect has been found for several tens of elements, but the overwhelming majority of all practical studies have been done with ^{57}Fe (more than a half of all reported studies), ^{119}Sn and ^{197}Au. A list of several most common Mossbauer isotopes with some of their relevant parameters is given in Table 1.6.

Direct studies of metals (and their alloys) having no Mossbauer nuclides is impossible, and of those having only short-lived nuclides is technically difficult. However, such systems can be studied indirectly by introducing isotopes acting as sources of Mossbauer radiation. In the case of galvanic coatings this operation is rather straightforward, since the isotopes can be introduced into the coating electrochemically, immediately in the process of electrodeposition.

Parameters of Mossbauer Spectra and Their Interpretation

The following main parameters can be extracted from a Mossbauer spectrum: isomeric shift, number of lines (hyperfine structure) in the spectrum, width and height of individual absorption lines (Fig. 1.39). All of them bear information about the details of crystal and magnetic structures of electrodeposited coatings.

Isomeric shift of line. Experimentally the value of isomeric shift δ is determined as the difference between the position of the absorption line center of gravity and the zero value of speed on the axis of source speed (Fig. 1.39a). Isomeric shift is observed in situations when the source and the absorber are chemically different, and is sometimes referred to as the "chemical shift". The absolute value of the shift is determined mostly by charge density of electrons and is thus very sensitive to valence state of the atom. The latter circumstance allows using isomeric shifts for identification of chemical states of atoms in compounds and alloys.

Although systematic studies of isomeric shifts allowed to accumulate substantial experience in their interpretation, the latter can still cause serious difficulties in certain situation, as is the case for alloys of transition metals, for which there is no rigorous electron theory. As of today the correlation between the isomeric shift and chemical properties has not yet been established to such a degree as to allow using solely NGR without turning to complementary techniques.

Hyperfine structure of spectrum (number of lines). The hyperfine structure of a Mossbauer spectrum reflects the effects of magnetic dipole and nuclear quadrupole interactions, which cause splitting of nuclear energy levels.

In magnetically ordered systems hyperfine splitting is characterized by effective magnetic field H_{eff} experienced by the resonance nuclei. The value of H_{eff} can be determined from the difference in the positions of the centers of gravity of the outermost peaks in the split line (Fig. 1.39b). The splitting of energy levels and selection rules for magnetodipole transitions lead in the case of most commonly used Mossbauer isotope ^{57}Fe to a six-line absorption spectrum (Fig. 1.39). The relative intensities of the sextet lines depend on the average orientation of specimen magnetic moments with respect to γ-ray direction and in the absence of anisotropy obey the 3:2:1:1:2:3 proportion. The presence of preferential magnetic force lines directions changes this distribution of intensities, which can be used for detection of magnetic anisotropy. Thus, texturing of galvanic coatings leads to redistribution of the relative intensities of the sextet lines, which reflects the change in the orientation of the easy magnetization axis. Magnetic anisotropy was also found in Mossbauer studies of electrodeposited amorphous iron–phosphorus alloys during annealing.

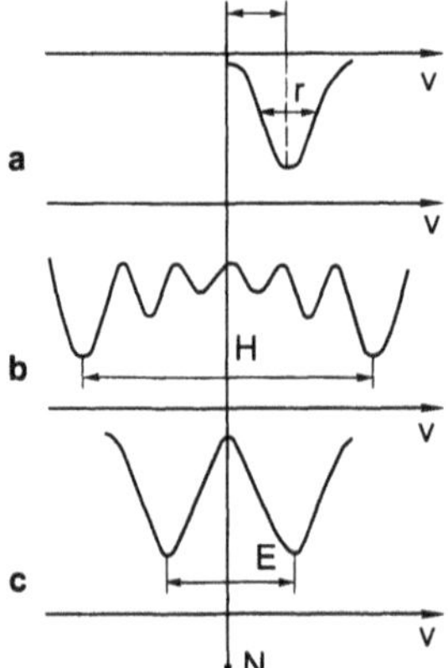

Fig. 1.39. Typical shapes of Mossbauer absorption spectra. **a** singlet line with isomeric shift; **b** hyperfine magnetic splitting; **c** spectral doublet

The effective magnetic field is very sensitive to the presence of impurity atoms in the local environment of the resonance atoms. The concentration dependencies of the average effective magnetic field, especially for iron alloys, have been carefully studied and are usually linear, although a more complex behavior has also been reported for a number of alloys.

The notion of hyperfine structure turned out to be particularly valuable in the study of phase transitions, when magnetic structure of the phases is significantly different, especially for transitions from ordered to paramagnetic state. For example, in the case of ^{57}Fe the spectrum changes from magnetically split sextet to a single line or doublet.

Wertheim et al. [1.96] studied the formation of β-phase of nickel hydride during electrochemical introduction of hydrogen into nickel foil. Prior to studies the foil was doped with radioactive ^{57}Co using diffusion assisted electrodeposition. For low concentrations of hydrogen in nickel, when the coating has the structure of a solid solution with face centered cubic lattice, the Mossbauer spectra display a characteristic six-line spectrum, which corresponds to ^{57}Fe distributed in nickel. As the concentration of hydrogen increases, nonmagnetic hydrides are formed. Starting from the alloy composition $NiH_{0.7}$ the magnetic moment of the foreign atom of iron changes, and the spectrum transforms into the single line of ^{57}Fe in $NiH_{0.7}$.

Another important parameter of the hyperfine structure is the quadrupole splitting ΔE. In nonmagnetic materials, when there is no magnetic splitting, quadrupole interaction leads to splitting of the absorption line into doublet with separation between the two lines equal to ΔE (Fig. 1.39c). Quadrupole splitting is sensitive to the point symmetry of the immediate environment of the Mossbauer atom and is nonzero if the local environment has symmetry lower than cubic.

This property of quadrupole splitting makes the parameter ΔE suitable for studies of point structural defects, since even in lattices with cubic symmetry the presence of cationic vacancies connected with the iron ion leads to breaking of the symmetry and transformation of the spectrum. The appearance of quadrupole splitting in a NGR spectrum can also indicate a phase transition accompanied by transformation of the cubic syngony into a structure with lower symmetry.

Quite often NGR studies are carried out for materials which demonstrate the effects of both magnetic dipole and electric quadrupole interactions. Such a combined interaction changes the relative positions of the six line of the hyperfine structure, substantially complicating the spectrum.

Line width and height. The width r of a Mossbauer line having Lorentzian shape is measured at half its height (see Fig. 1.39a). The lowest attainable value of the line width is its natural width determined by the lifetime of the nuclear level. Experimentally measured widths are usually several times larger. There are several sources of broadening of a Mossbauer line. For electrodeposited substitution and interstitial alloys the main contribution comes from concentration broadening, caused by differences in conditions of resonance γ-quantum absorption for nuclei having different local environments. When the resonant atoms in a lattice can be found in several different configurations of impurity atoms, the resulting spectrum is a superposition of spectra corresponding to the same nuclear transition, but in each of the local environments. Such a complex spectrum will generally have broader and possibly even distorted lines.

Table 1.7. Dependence of the short range ordering parameter α_0 on the composition of Fe–Ni and Fe–Co alloys

Ni content [atomic %]	α_0	Co content [atomic %]	α_0	Ni content [atomic %]	α_0	Co content [atomic %]	α_0
5	0	4	0	18	0.34	25	0.32
8	0.18	11	0.16	25	0.40	32	0.42
13	0.30	18	0.23	30	0.40	40	0.42

The intensities of lines in a complex Mossbauer spectrum are proportional to probabilities of different configurations of impurity atoms in the local environment of a resonant atom and thus provide quantitative information about the immediate environment. Together with the data on the influence of local environment on the effective magnetic field these characteristics allow investigation of processes accompanying agglomeration of impurity atoms and phase separation.

The methodic aspects of the described studies can be illustrated using the investigation of structural details of electrodeposited Fe–Ni and Fe–Co coatings as an example [1.46]. The alloys with atomic content of cobalt and nickel listed in Table 1.7 were deposited on copper cathodes from sulphate electrolytes with total concentration of salts of the deposited metals equal to 1.26 M. Resonance absorption spectra were taken on a YaGR–4 NGR spectrometer (source – ^{57}Co in chromium) in the constant acceleration (the entire sextet) and constant speed (the outermost peaks, corresponding to maximum positive and negative values of speed) modes, respectively. The spectra were interpreted using a specialized computer program. All lines were assumed to have Lorentzian shapes. The types of sites occupied by iron atoms were identified using the available data on the influence of transition metal ions on the effective magnetic field of iron in alloys (Table 1.8) in the model of combined first two coordination spheres, since impurity atoms in farther coordination spheres make negligible contribution to effective magnetic field on the ^{57}Fe nucleus.

Table 1.8 shows that a single Ni atom in the first two coordination spheres increases the effective field at an iron nucleus by 9.4 kOe, and an atom of Co – by 13.3 kOe. The quality of fitting the experimental data by the model was estimated using the χ^2 criterion, which in all cases was better than $3\cdot10^{-5}$.

Table 1.8. Contribution of the nearest atoms of 3D-transition metals in solid solutions to effective magnetic field and isomeric shift of iron

Metal	ΔH [kOe]	$\delta\cdot10^2$ [mm/s]	Metal	ΔH [kOe]	$\delta\cdot10^2$ [mm/s]
Ti	-19.1	-1.3	Mn	-23.0	-1.6
V	-24.3	-2.3	Co	+13.3	–
Cr	-26.9	-2.0	Ni	+9.4	–

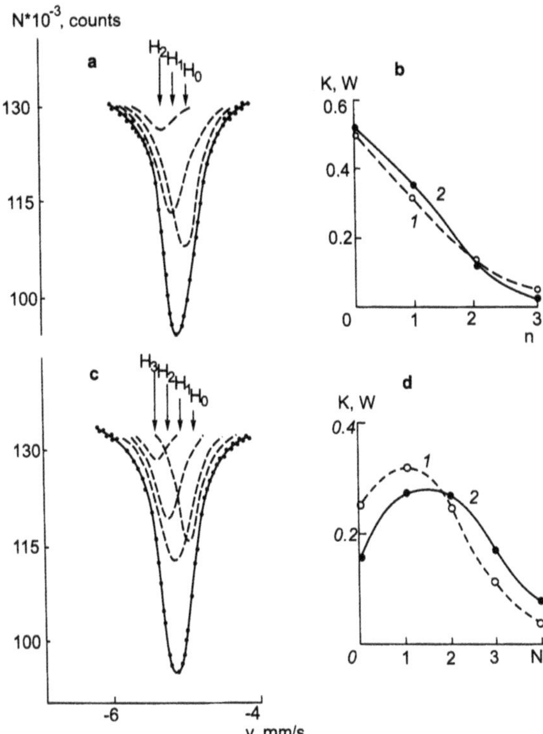

Fig. 1.40. The outermost peaks of the γ-resonance spectra of alloys Fe–4%Co (**a**) and Fe–11% Co (**c**) and the results of their interpretation (**b, d**). *1* the fraction of iron atoms *K* having the varying number *n* of Co atoms as their nearest neighbors, obtained from experimental spectra; *2* theoretically calculated fraction of iron atoms *W* (see text for details)

γ-Resonance spectra for all alloys have a shape typical for magnetic splitting with broadened outermost peaks. Fig. 1.40a shows one of the outermost peaks for alloy with composition Fe–4%Co. An analysis shows that it is a superposition of several primary spectra, corresponding to iron atoms in different local environments with the following values of average effective magnetic fields: $H_{eff\,0} = 332$ kOe – no Co atoms as nearest neighbors for iron atoms; $H_{eff\,1} = 345$ kOe – one, and $H_{eff\,2} = 358$ kOe – two Co atoms in the first two coordination spheres of an iron atom. The areas under the spectra corresponding to iron atoms in different environments give the fractions of iron atoms, having one, two, etc. Co atoms as their nearest neighbors.

Fig. 1.40b shows the fractions $K(n)$ of iron atoms having no Co atom as neighbors ($n = 0$), and having one ($n = 1$), two ($n = 2$) and three ($n = 3$) Co atoms near an iron atom for alloy Fe–4%Co, obtained from experimental spectrum of

Fig. 1.40a. Also shown are the theoretically predicted fractions $W(n)$ of iron atoms found in different local environment of Co atoms, obtained from binomial distribution assuming a completely disordered solid solution:

$$W(n) = \frac{14!}{n!(14-n)!} C^n (1-C)^{14-n}, \tag{1.11}$$

where n is the number of neighbor Co atoms in the first two coordination spheres, C is Co concentration in the alloy (atomic per cent).

The theoretical (W) and experimental (K) values are in a reasonably good agreement. The coincidence of the fractions of iron atoms with n Co atoms as nearest neighbors, obtained from the areas of the components of the decomposed spectrum, with theoretical values for completely disordered alloy confirms, that in the electrodeposited Fe–4%Co alloy Co atoms in the solid solution are distributed practically randomly. Similar results were also obtained for the Fe–5%Ni alloy.

Increasing the concentration of the alloying constituent in alloys leads to transformation of their Mossbauer spectra, causing redistribution of the intensities of components corresponding to iron atoms in different local environments of cobalt or nickel atoms. Since the changes of γ-resonance spectra are identical for all alloys, only the spectrum for the Fe–11%Co alloy is shown in Fig. 1.40c. Fig. 1.40d shows the corresponding experimental and theoretical fractions of iron atoms found in different local environments of Co atoms. The departure of the experimentally obtained values (K) from theoretically predicted (W), in particular an increase of the intensity of the peak reflecting the number of iron atoms in purely iron atom environment ($n = 0$), points at the formation of regions enriched with Co, i.e., at the formation of Co atoms agglomerates in solid solution.

The degree of deviation of a solid solution from the completely disordered state is quantitatively characterized by the parameter of short range ordering α_0:

$$\alpha_0 = \frac{[K(n=0) - W(n=0)]}{W(n=0)} \tag{1.12}$$

As Table 1.7 shows, the value of α_0 increases with increasing the content of the alloying constituent, reaching a plateau for alloys with 25–30 atomic per cent of the constituent. A paramagnetic doublet, reflecting the formation of a principally new nickel atom local environment for iron atoms, appears in the center of the sextet on the Mossbauer spectrum of the Fe–25%Ni alloy (Fig. 1.41).

Such a behavior of the short range ordering parameter is explained by increased concentration of the alloying elements in enriched regions, the composition of which ultimately reaches the composition of a separate phase. However, a conclusive decision about the character of the nickel or cobalt atom aggregates seen in Mossbauer spectra, whether they are a new intermetallic phase or clusters of atoms, requires application of diffraction techniques.

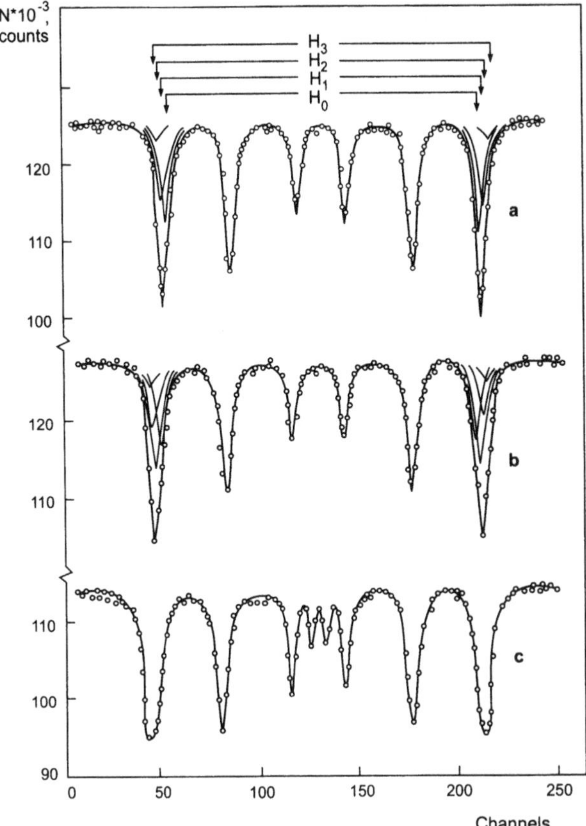

Fig. 1.41. Mossbauer spectra of electrodeposited Fe–Ni alloys for Ni content 5 (**a**), 13 (**b**) and 25 atomic percent (**c**), respectively

Comparison of the Mossbauer data with XSA results shows, that for alloys with nickel or cobalt content up to 25–30% the concentration inhomogeneities take the form of clusters with body-centered cubic lattice identical to the matrix lattice, while for the Fe–25%Ni alloy they are fine inclusions of an intermetallic phase γ-(Fe, Ni) having face-centered cubic lattice with parameter $a = 0.3573$ nm. In the alloy with composition Fe–32%Co, the spectra of which do not have the paramagnetic doublet, the inhomogeneities presumably form a superstructure based on the Fe_3Co formulation.

The illustrated identification of structural features has an important practical outcome, since the degree of concentration heterogeneity of alloys has a pronounced effect on many physical and mechanical properties of the coatings and provides means of regulating them in rather wide limits [1.46].

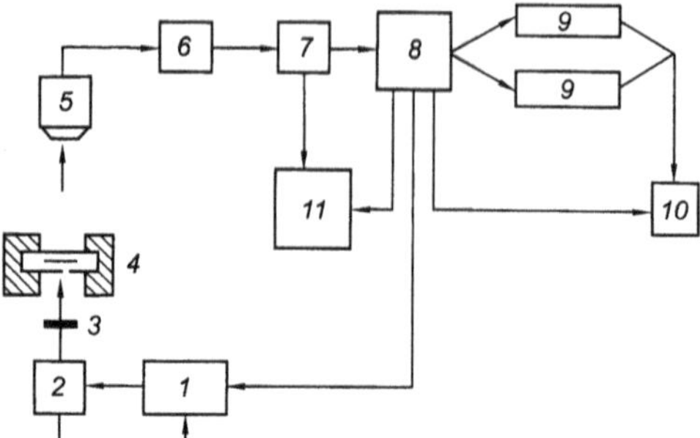

Fig. 1.42. Functional scheme of the NGR spectrometer YaGRS–4. *1* low frequency generator; *2* electrodynamic vibrator; *3* source of γ-quanta; *4* specimen; *5* detector; *6* amplifier; *7* discriminator; *8* control unit; *9* scaling units; *10* digital recorder; *11* multichannel analyzer

Experimental Details

The specimens for Mossbauer spectroscopy usually have the form of thin foils (20–30 μm for iron), and the procedure of their preparation from galvanic coatings is relatively simple. As a rule the substrate of the coating is chemically dissolved, sometimes followed by electropolishing, as was done in the already cited work on the study of Fe–Ni and Fe–Co alloys. Thin foils allow operating the spectrometer in the transmission mode, when the detector is placed after the specimen (Fig. 1.42). This scheme of conducting a Mossbauer experiment is most commonly employed in modern commercial NGR spectrometers, for example, in the YaGRS–4 spectrometer (Russia).

The main units of the spectrometer perform the following functions: high-voltage power supply (not shown in the figure) provides the required voltage for a PMT – the detector of γ-rays; the vibrator assembly houses the radioactive source, a collimator to form a parallel beam of γ-rays of the specimen, and the detector; control panel (also not shown) provides access to operating parameters of the spectrometer and allows to set the required working conditions; digital recorder is used to output the experimental data on the paper tape when operating spectrometer in the constant speed mode; multichannel pulse analyzer provides data output to display in the constant acceleration mode.

Electrodynamic vibrator moves the source of γ-quanta. The speed of movement is controlled by the signal from the low-frequency generator, driving the translation coil of the vibrator. γ-quanta transmitted through the specimen are registered by the proportional counter of the γ-ray detector. The signal from the output of the detector is amplified and fed to pulse discriminator. Control unit handles setting the mode of operation, choosing source speed and registration time, accumulation and discrimination of pulses. Depending on the mode of operation the data is output either to digital recorder via scaling units, or to multichannel analyzer directly from the discriminator.

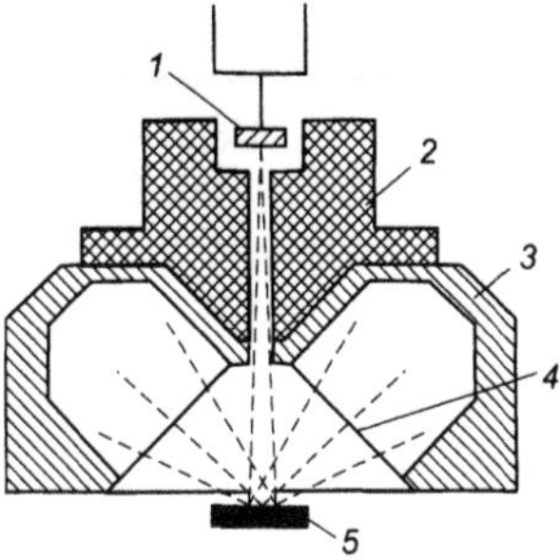

Fig. 1.43. Back reflection chamber of a Mossbauer spectrometer. *1* source of γ-quanta; *2* collimator; *3* body of the counter; *4* window; *5* specimen

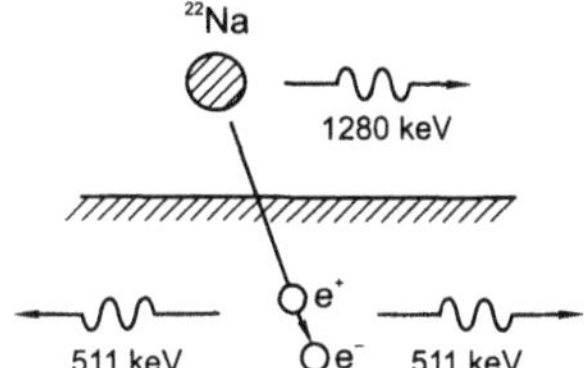

Fig. 1.44. Scheme of electron-positron annihilation

If for some reasons thin foils cannot be obtained, a spectrum from a massive specimen can be taken in the reflection mode, when a detector registers γ-quanta emitted after the resonance absorption in the specimen and back reflection. To compensate for the loss of intensity, about an order of magnitude higher source activity and a spectrometer of special construction is required in this case. In one of the designs γ-radiation is registered by a proportional counter with cone-shaped window made from 100 μm thick polyvinyl film covered with a 200–300 nm thick layer of aluminum mounted inside the back reflection chamber (Fig. 1.43).

The construction of modern Mossbauer spectrometers allows using various add-on devices to conduct γ-resonance studies in a wide range of experimental conditions. The specimen can be put into a heater or a cryostat, in the gap of an electromagnet or inside a superconducting solenoid, etc...

1.7.2 Positron Annihilation

General Overview

The technique of positron annihilation takes advantage of the propensity of positrons and electrons of a substance to annihilate upon interaction. A suitable radioactive isotope is usually used as a source of positrons (Table 1.9), the most common being ^{22}Na, having the longest half-life period and the highest decay efficiency [1.17].

A positron is emitted in the act of radioactive decay of an atom of a radioactive isotope and is accompanied by a γ-quantum, in the case of ^{22}Na having energy 1280 keV. After entering into metal the positron is slowed down to thermal speeds by internal electric field (so called thermalization) and annihilates in the collision with one of the conduction electrons, emitting two γ-quanta with energy 511 keV (Fig. 1.44). Average lifetime of positron before annihilation is 100–300 ps.

Table 1.9. Sources of fast positrons

Isotope	Half-life period	Decay efficiency [%]	Isotope	Half-life period	Decay efficiency [%]
^{22}Na	2.6 yr	89	^{58}Co	71.3 day	15
^{55}Co	18.2 hr	60	^{64}Cu	12.7 hr	19
^{57}Ni	36.0 hr	50	^{90}Nb	14.7 hr	54

In a practical study the following characteristics of electron-positron annihilation are usually registered: positron lifetime τ, angular distribution of annihilation γ-quanta, and Doppler broadening of the annihilation line Δ.

Instrumentation

The lifetime of positrons τ is experimentally determined as the time interval between the moments of emitting the prompt 1280 keV quantum and the annihilation γ-quanta. One of the possible realizations of this scheme is a system of fast-slow coincidence, shown in Fig. 1.45 [1.38]. In the slow channel one of the detectors of a single-channel analyzer registers the 1280 keV quantum, and the other detector – the 511 keV quantum, while the fast channel supplies stable time markers. The signals are fed to the input of a multichannel analyzer. A typical spectrum of positron lifetimes in metal is the ratio of the count rate in each channel of the multichannel analyzer to the total number of channels, expressed in the units of time [1.38].

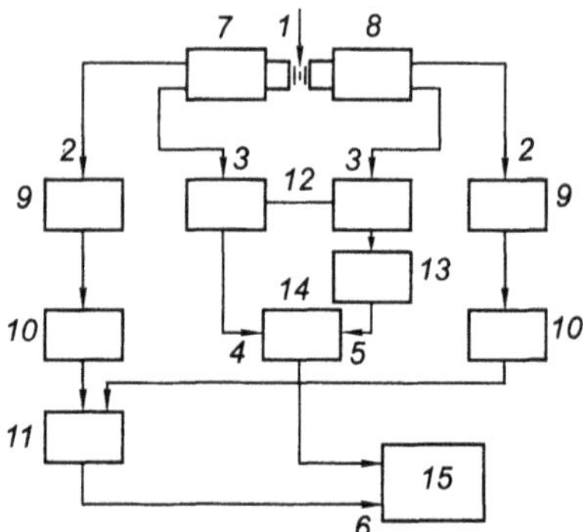

Fig. 1.45. Scheme of experimental setup for measuring positron lifetime (adopted from [1.38]). *1* source of positrons and specimen (sandwich); *2* slow; *3* fast; *4* start; *5* stop; *6* control input; *7*, *8* detectors of quanta with energy 1280 keV and 511 keV, respectively; *9* main amplifier; *10* single-channel analyzer; *11* coincidence unit; *12* fast discriminator; *13* delay; *14* time to pulse height converter; *15* multichannel analyzer

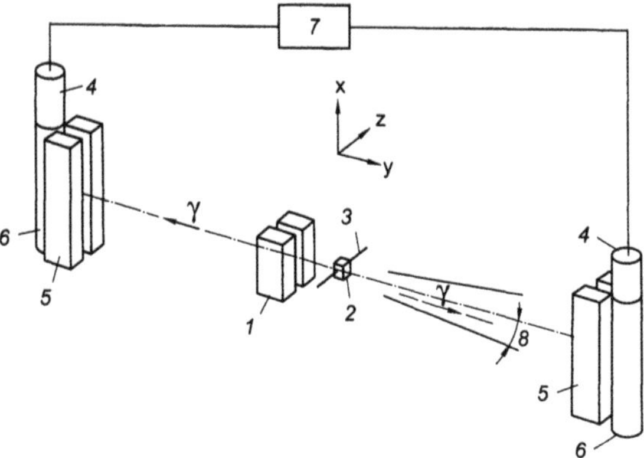

Fig. 1.46. Principle of measuring angular correlation (adopted from [1.38]). *1* slit collimator; *2* specimen; *3* source of positrons; *4* photomultiplier tube; *5* detector collimator; *6* scintillator; *7* coincidence unit

The angular distribution of annihilation γ-quanta is obtained by measuring the deviation of the angle θ between the two quanta from 2π. A possible scheme of measuring the deviation is the so-called long slit geometry scheme, shown in Fig. 1.46. Here two detectors are put on both sides of the positron source and specimen. The pulses from the detectors come to coincidence unit and then to counter. The small value of the deviation angle requires high angular accuracy of measurement, which is achieved by increasing the distance between the detectors to 2–5 m. One of the detectors moves in the horizontal plane and registers the angles θ. Angular distribution of the annihilation quanta is then the dependence of the coincidence count rate from the angle between the two quanta $I(\theta)$.

The Doppler broadening of the annihilation line can be measured by an energy resolving spectrometer. Half width of the broadened lines is usually of the order of about 3 keV, typical energy resolution of the spectrometer is about 1.2–1.5 keV.

Applications

The technique of positron annihilation is most effectively employed in the studies of the details of formation and distribution of crystal lattice defects in electrodeposited coatings, since trapping of positrons by the defects changes their characteristic lifetimes, angular distribution and Doppler broadening.

Positrons are trapped and localized by the following defects of the lattice: vacancies, divacancies, vacancy clouds, pores and dislocations. The positrons are generally not sensitive to isolated interstitial and impurity atoms, packing defects and microscopic defects with the size larger than 1 μm. Positrons can be trapped by the grain boundaries, but this interaction becomes important only for grain sizes smaller than 1 μm.

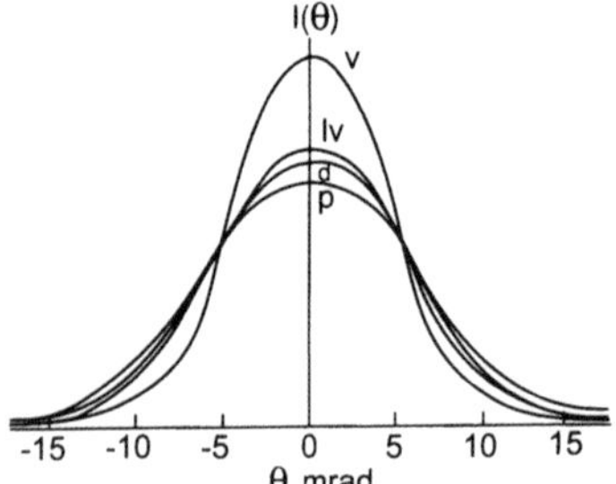

Fig. 1.47. Typical curves of angular correlation for annihilation of positrons in defect-free regions of the lattice (*p*), dislocations (*d*), isolated vacancies (*1v*) and pores (*v*), respectively

The existing theory of positron annihilation in real crystals allows to deduce the internal electron structure of defects, their type and concentration from the observed changes in annihilation characteristics. For example, the lower density of conduction electrons in a defect results in a longer average positron lifetime τ and a higher peak value H_c on the curve of angular correlation. For both τ and H_c the following relation holds true $F_v > F_{1v} > F_d > F_p$, where F_p, F_d, F_{1v}, and F_v are the characteristic parameters F for positron annihilation in perfect regions of the lattice, dislocations, vacancies and pores, respectively.

Fig. 1.47 shows typical curves of angular correlation for different regions of the lattice. If positrons annihilate on several different types of lattice defects, the behavior of the annihilation parameters is more complex, although the model of trapping still allows to distinguish the defects and separate their contributions using computer simulations.

The method of positron annihilation has made only its first steps in the investigation of galvanic coatings, but the results are very encouraging. Interpretation of the experimental annihilation characteristics allowed to distinguish three types of defects that are formed during electrodeposition and have an effect on positron trapping: vacancies, dislocations, and three-dimensional defects comprising not less than three vacancies. The latter are most plausibly hydrogen segregations at grain boundaries. Annihilation parameters change in the course of natural aging and annealing of the specimens, reflecting the transformations of crystal defects. Thus, aging is accompanied by growth of the three-dimensional defects and decrease of the density of dislocations and concentration of defects, while heating above 200 °C results in migration of vacancies and segregation of vacancy clusters at grain boundaries.

An example of successful application of the method of positron annihilation to study coatings is provided by the work [1.46]. The authors determined the concentration of vacancies in nickel coatings deposited from industrial electrolyte for bright nickel-plating (composition, g/l: $NiSO_4{\cdot}7H_2O$ – 250, $NiCl{\cdot}6H_2O$ – 50, H_3BO_3 – 25, saccharin – 1, butindiol – 0.15; regime: T – 40 °C, i_c – 1 A/dm^2, pH – 2).

Angular distributions of annihilation quanta were measured on a setup with point-linear geometry providing angular resolution about 1 mrad. The used source of positrons ^{22}NaCl (β^+, γ) had activity of 6 mkCurie.

The defects of nickel coatings were annealed in vacuum (residual pressure 0.133 *P*) by gradually raising the temperature of the specimen in 25 °C steps and allowing 20 min of exposure at each temperature. After annealing the *f*-parameter of angular distribution (the ratio of the sum of count rates at angles $\theta = (\pm 1.5;\ 1.0;\ 0.5)$ mrad to the sum of count rates at angles $\theta = (\pm 6.0;\ 6.5;\ 7.0;\ 8.0;\ 8.5)$ mrad) was measured at room temperature. A reference specimen was also annealed at 800 °C for 1 hour.

The change of the *f*-parameter in the course of annealing can be related with both the change of the defect type and the change of the integral concentration of defects – centers of annihilation. Therefore, the change of the defect type in the course of annealing was monitored using the *R*-parameter, which does not depend on the concentration of defects and is determined solely by their type. The parameter is defined as follows. Angular distribution of the annihilation quanta was measured at room temperature for a specimen annealed at the temperature, at which an extremum at the curve $f=F(T)$ is observed. Then

$$R = \left(H'_d - H'_p\right) / \left(H''_d - H''_p\right), \tag{1.13}$$

where H'_d and H'_p are the sums of the count rates at angles ±1.5 mrad for normalized on equal area spectra for annealed experimental and reference specimens, respectively, H''_d and H''_p are the sums of the count rates at angles ±9 mrad for normalized on equal area spectra for annealed of experimental and reference specimens, respectively.

Some representative experimental results are shown in Fig. 1.48. The value of *f*-parameter increases upon heating the specimen from 20 °C to 100 °C. Comparison of the values of *R*-parameter at the two temperatures shows that it is the change of the type of lattice defects trapping positrons that is responsible for the growth of *f*-parameter. Simultaneously a sharp decrease of the lattice parameter and a certain decrease in the level of microdistortions is also observed.

The effects observed for low temperature annealing are explained by interaction of defects with impurity atoms having low migration energy. In nickel coating the role of such impurity plays atomic hydrogen having migration energy as low as 0.38 eV. Despite the tendency of diffusion-mobility hydrogen to egress, its residual concentration in electrodeposited coatings remains rather high. As an interstitial element, hydrogen is located in the interstitial space of the nickel lattice and is naturally captured by defects having effective negative charge, in particular by vacancies. Since the experimental specimen is actually a supersaturated solid solution of hydrogen in nickel, the lattice parameter of the coating is substantially higher than its equilibrium value.

Low temperature annealing leads to thermal release of hydrogen from the vacancies. The released hydrogen finds its way to sinks, which accounts for the decrease of the lattice parameter and increase of the *f*-parameter value (Fig. 1.48).

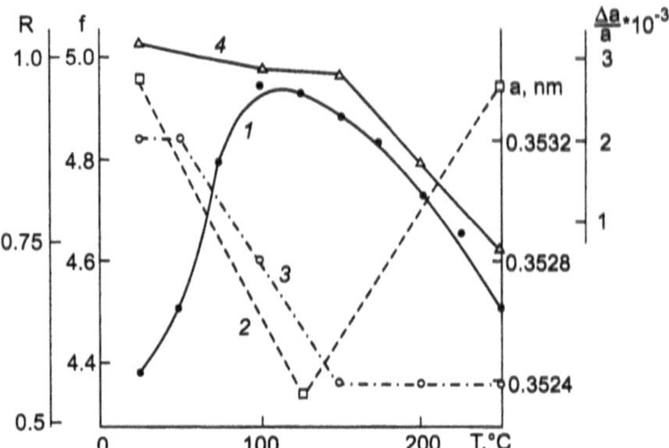

Fig. 1.48. Dependence of the *f*-parameter (*1*), *R*-parameter (*2*), lattice parameter *a* (*3*), and the level of microdistortions $\Delta a/a$ (*4*) on the temperature of annealing

The vacancies now can act as traps for positrons and the dominating annihilation centers. This inference is confirmed by the data presented in the already cited work [1.46], where an increase of the first component in the positron lifetime spectrum for electrodeposited nickel to values typical for annihilation of positrons on vacancies was clearly demonstrated. Comparison of the values of *R*-parameter at room temperature and at 100 °C (see Fig. 1.48) leads to conclusion, that vacancy as an annihilation center replaces a more extended type of defect, which is responsible for positron trapping in the initial specimen. Nickel in the given experimental conditions is known to crystallize with a characteristic substructure, and the extended defects can be, for example, grain boundaries or subboundaries.

The value of *f*-parameter at 100 °C, which, as has just been demonstrated, reflects annihilation of positrons trapped by vacancies, can be used to estimate the atomic concentration of the vacancies. In the framework of the positron trapping model the concentration of vacancies is given by the following expression:

$$C_v = (1/\mu\tau_0)(f_d - f_p)/f_p, \tag{1.14}$$

where $\mu = 2.2\cdot 10^{15}$ s^{-1} is the rate of positron trapping by nickel monovacancies; $\tau_0 = 110$ ps is positron lifetime in defect-free nickel; f_d and f_p are the values of *f*-parameter for a specimen annealed at 100 °C and for a control specimen, respectively.

The value $C_v = 10^{-4}$ obtained for a specimen annealed at 100 °C implies ultra-high concentration of vacancies in the initial specimen, which is probably 1–2 orders of magnitude higher that the value found experimentally. And indeed, for activation energy of 0.7 eV the onset of migration of vacancies to sinks is about 70 °C, and part of vacancies in the experimental conditions becomes lost for calculations.

Increasing annealing temperature from 120 °C to 250 °C intensifies egress of hydrogen and vacancies to sinks. The lattice parameter stabilizes and approaches its equilibrium value, the level of microdistrortions substantially decreases. Migration of vacancies assists in redistribution of dislocations and promotes perfection of the substructure and formation of well-defined polygonal boundaries [1.74]. The relative fraction of positrons that annihilate in vacancies decreases as compared to the fraction of positrons annihilating at subboundaries, which is manifested in decreased value of *f*-parameter and increased value of *R*-parameter (Fig. 1.48).

The presented results outline some of the possible applications of the positron annihilation technique to investigation of galvanic coatings and illustrate the large potential of the method, which undoubtedly has wide prospects.

The method is applicable to study polycrystalline specimens with or without texture. The specimens can be round, square or can have arbitrary shape with area 25–100 mm^2. Desirable specimen thickness is determined by the depth M of positron penetration in the material and should be (3–5)M. Positrons resulting from decay of the ^{22}Na isotope have penetration depth equal to 28 and 13 μm for nickel and iron, respectively [1.38]. The procedure of specimen preparation is straightforward and includes removal of corroded layers from the surface by electropolishing and following cleaning with alcohol. The main shortcomings of the technique are the complexity of equipment and long acquisition times, typically 10–20 hours, depending on activity of the source and the desired statistical accuracy [1.48].

References

1.1 Ahn HS, Kwon OK (1993) Wear behaviour of plasma-sprayed partially stabilized zirconia on a steel substrate. Wear 162–164:162–164

1.2 Alimoussa A, Casanove MJ, Roucau C, Villard C, Schwerdtfeger M, Diciocio L, Moriceau H, Villegier JC (1992) TEM characterization of high Tc superconductive multilayered thin films. Mater Sci and Eng B 14:353–356

1.3 Ansart F, Ganda H, Saporte R, Thaverse JP (1995) Study of the oxidation of aluminum nitride coatings at high temperature. Thin Solid Films 260:38–46

1.4 Asami Katsuhiko, Moriya Takeshi, Aihara Tomoyasu, Masumoto Tsuyoshi (1996) Surface characterization of sputter-deposited Cu–Ta alloys by Auger electron spectroscopy. Sci Repts Res Inst Tohoku Univ A 42:225–229

1.5 Atic M, Zarzycki J, R'Kha C (1994) Protection of ferritic stainless steel against oxidation by zirconia coatings. J Mater Sci 13:266–269

1.6 Attar F, Johannesson T (1995) Adhesion and X-ray elastic constant evaluation of CrN coating. Thin Solid Films 258:205–212

1.7 Barbezat G, Nicoll AR, Sickinger A (1993) Abrasion, erosion and scuffing resistance of carbide and oxide ceramic thermal sprayed coatings for different applications. Wear 162–164:529–537

1.8 Bartlett Andrew H, Mashio Roberto Dal (1995) Failure mechanisms of a zirconia–8wt% yttria thermal barrier coating. J Amer Ceram Soc 78:1018–1024

1.9 Benarioua Y, Bouaouadja N, Wendler B (1996) Carbide formation in titanium coatings deposited on carbon steel. J Mater Sci Lett 15:1067–1069

1.10 Benhenda S, Benjemaa N, Bourir M (1994) Effect of pulse plating parameters on electrical contact behavior of nickel coatings. Trans Compon, Packag and Manuf Technol A 17:303–308
1.11 Block G, Hirschfeld D, Lichtenauer IM, Pohl A (1992) Preparation of VPS and PVD coatings for metallographic and TEM investigations. Microchem 2:279–282
1.12 Boldyrev VV, Lyakhov BN, Tolochko BP (1989) Diffractometry with synchrotron radiation (in Russian). Nauka, Novosibirsk
1.13 Bredendiek–Kamper S, Jenett H, Bubert H (1991) AES investigation of thermally sprayed Al_2O_3 coatings on steel. Fresenius' J Anal Chem 341:346–348
1.14 Brindley WJ, Leonhardt TA (1990) Metallographic techniques for evaluation of HSBF thermal barrier coatings. Mater Charact 24:93–101
1.15 Castro R, Chard G, Smith Ronald W, Rollett Anthony D, Stanek Paul W (1992) Toughness of dense $MoSi_2$ and $MoSi_2$/tantalum composites produced by low pressure plasma deposition. Scr Met and Mater 26:207–212
1.16 Chen HC, Pfender E (1996) Microstructure of plasma-sprayed Ni–Al alloy coating on mild steel. Thin Solid Films 280:188–198
1.17 Cherepin VT, Vasiliev MA (1992) Methods and instrumentation for surface analysis of materials: A reference book (in Russian). Nauchnaya Mysl, Kiev
1.18 Chevelier S, Bonnet G, Colson JC, Larpin JP (1998) Influence of a reactive element oxide coating on the high temperature oxidation of chromia-formed alloys. J Chim Phys et Phys–Chim Biol 95:2083–2101
1.19 Chou TC, Nieh TG, Tsui TY, Phar GM, Oliver WC (1992) Mechanical properties and microstructures of metal / ceramic microlaminates Part I Nb/$MoSi_2$. J Mater Res 7:2765–2773
1.20 Crostack HA, Jahnel W, Meyer EH, Selvadurai U (1996) Developments of nondestructive testing of surface coatings. Weld World 37:114–120
1.21 Curtis CL, Gawne DT, Priestnall M (1994) The processing and electrical properties of plasma-sprayed yttria-zirconia. J Mater Sci 29:3102–3106
1.22 Dehm G, Soheu C, Ruhle M, Raj R (1998) Growth and structure of internal Cu/Al_2O_3 and Cu/Ti/Al_2O_3 interfaces. Acta Mater 46:759–772
1.23 Deuis RL, Yeliup JM, Subramanian C (1997) Aluminum composite coatings produced by plasma transferred arc surfacing technique. Mater Sci and Technol 13:511–512
1.24 Dietrich S, Schneegans M, Moske M, Samwer K (1996) Investigation on metallurgical properties for VLSI applications. Thin Solid Films 275:159–163
1.25 Edris H, McCarthey DG, Sturgeon AJ (1997) Microstructural characterization of high velocity oxy-fuel sprayed coating of Inconel 625. J Mater Sci 32:863–872
1.26 Erdemir A, Switala M, Wei R, Wilbur P (1991) A tribological investigation of the graphite-to-diamond-like behavior of amorphous carbon films ion beam deposited on ceramic substrates. Surface and Coating Technology 50:17–23
1.27 Gard D, Dyer PN (1993) Erosive wear behavior of chemical vapor deposited multilayer tungsten carbide coating. Wear 162–164:552–557
1.28 Goswami J, Raghunathan L, Devi A, Shivashankar SA, Chandrasekaran S (1996) Chemical vapor deposition of thin copper films using a new metalorganic precursor. J Mater Sci Lett 15: 573–575
1.29 Guilemany JM, Nutting J, Dong Z (1997) Coating-substrate bonding after HVOF thermally spraying WC–Co onto a Ti–6%V alloy. J Mater Sci Lett 16:1043–1044

1.30 Guo Zhongcheng, Yang Xianwan, Liu Hong-Kang (1997) Microstructure of electrodeposited Ni–W–P–SiC composite coatings. Trans Nonferrous Metals Soc China 7:22–26

1.31 Guo Zhongcheng, Yang Xianwan, Liu Hong-Kang, Wang Zhiuin, Wang Min (1996) Characteristics of Re–Ni–B–Al_2O_3 coating material. Trans Nonferrous Metals Soc China 6:33–36

1.32 Habazaki H, Shimizu K, Skeldon P, Thompson GE, Wood GC (1996) The composition of the alloy/film interface during anodic oxidation of Al–W alloys. J Electrochem Soc 143:2465–2470

1.33 Hafter J, Ostreicher K, Sung C (1992) Improved method for the preparation of TiN-coated WC–Co-based samples for cross-sectional AEM investigation. Microsc Res and Techn 23: 329–342

1.34 Hainsworth SV, Bartlett T, Page TF (1993) The nanoindentation response of systems with thin hard carbon coatings. Thin Solid Films 236:214–218

1.35 Harmsworth PD, Stevens R (1992) Phase composition and properties of plasma-sprayed zirconia thermal barrier coatings. J Mater Sci 27:611–615

1.36 Hartung F, Ewert JC, Dzirk J, Schmitz G (1998) Interreaction of Cu/Au thin films observed by high angle hollow cone darkfield electron microscopy. Scr Mater 39:79–85

1.37 Hu C, Baker TN (1997) A new aluminum silicon carbide formed in laser processing. J Mater Sci 32:5049–5051

1.38 Hunger GI (1985) Selected methods of investigation in physical metallurgy (in Russian). Metallurgia, Moscow

1.39 Jacobson BE (1980) Methods of characterization of coating microstructures. Thin Solid Films 73:331–345

1.40 Khasui A (1975) Deposition techniques (in Russian). Mashinostroenie, Moscow

1.41 Kiiski AA, Ruuskannen PR, Rubin JB (1996) Wear resistant coatings produced by shock-wave compaction of powders. Met and Mater Trans A 27:2297–2304

1.42 Kim Soo-Hyun, Chung Deuk-Seok, Park Ki-Chul, Kim Ki-Bum, Miu Seok-Hong (1999) A comparative study of film properties of chemical vapor deposited TiN film as diffusion barriers for Cu metallization. J Electrochem Soc 146:1455–1460

1.43 Kochergin SM, Leontiev AV (1974) Formation of textures in the course of electrocrystallization of metals (in Russian). Metallurgia, Moscow

1.44 Kolawa E, Sun X, Reid JS, Chen JS, Nickolet MA, Ruis R (1993) Amorphous $W_{40}Re_{40}B_{20}$ diffusion barriers for <Si>/Al and <Si>/Cu metallizations. Thin Solid Films 236:301–305

1.45 Kovenski IM, Povetkin VV (1989) Afterelectrolysis phenomena in metallic coatings (in Russian). Zaschita Metallov 25:367–371

1.46 Kovenski IM, Povetkin VV (1994) Methods of investigation of electrodeposited coatings (in Russian). Nauka, Moscow

1.47 Kovenski IM, Povetkin VV (1999) Physical metallurgy of coatings (in Russian). Joint Venture "Internet Engineering", Moscow

1.48 Kovenski IM, Kuznetzov PV, Povetkin VV, Makhmudov NA (1991) Study of point defects in electrodeposited coatings using positron annihilation technique (in Russian). Zaschita Metallov 27:1369–1371

1.49 Kovenski IM, Povetkin VV, Venediktov NL (1992) Modification of protective properties of coatings by annealing (in Russian). Zaschita Metallov 28:338–341

1.50 Kreye H (1998) A comparison of HVOF systems-behavior of materials and coating properties. Galvanotekhnika 89:3006–3013

1.51 Kudinov VV, Pekshev PYu, Belaschenko VE, Solonenko OP, Safiulin AP (1990) Plasma-sprayed coatings (in Russian). Nauka, Moskow

1.52 Kulik AYa, Borisov YuS, Mnukhin AS (1985) Gas-thermal deposition of composite powders (in Russian). Mashinostroenie, Leningrad

1.53 Li XN, Jin S, Dong C, Zhang Z, Gong ZX, Ma TC (1997) Transmission electron microscopic studies of ternary FeNi–silicide layers prepared by metal vapor vacuum arc ion implantation. Thin Solid Films 304:196–200

1.54 Londa Yrene, Shinya Masanoby, Takemoto Mikio (1994) Coatings of aluminide intermetallic compounds on steel utilizing a hybrid techigue of spraying and IR–laser fusion. Mater and Manuf Processes 9:495–505

1.55 Lou Hanyi, Tang Youjung, Sun Xiaofen, Guan Hengrong (1996) Oxidation behavior of sputtered microcrystalline coating of superalloy K17F at high temperature. Mater Sci and Eng A 207:121–128

1.56 Marel A, Marti P, Henry F, Armas B, Bonnet R, Loubradou M (1997) Nanostructure and local chemical composition of AlN–Si_2N_4 layers grown by LPCVP. Thin Solid Films 304:256–266

1.57 Masanobu Izaki, Takashi Omi (1996) Structural characterization of martensitic iron-carbon alloy films electrodeposited from an iron (II) sulfate solution. Metand Mater Trans A 27:485–486

1.58 McPherson R (1981) The relationship between the mechanism of formation, microstructure and properties of plasma-sprayed coatings. Thin Solid Films 83:297–310

1.59 Michel H, Remy M, Czerwiee T, Angun K (1991) Structural study of titanium-nitride coating interfaces related to plasma diagnostics. Mater Sci and Eng A 139:276–283

1.60 Moonir–Vaghefi SM, Saatchi A, Hedjazi J (1997) Tribological behavior of electroless Ni–P–MoS_2 composite coatings. Z MetallK 88:498–501

1.61 Mora-Marquer JG, Lira-Olivares J (1987) A study of crack initiation and propagation in Ni–Cr thermally sprayed coatings using acoustic emission techniques. Thin Solid Films. 153:243–252

1.62 Murakami Kenji, Okamoto Taira, Matsumoto Miroshi, Miyamoto Yoshinari, Irisana Tsuyoshi (1993) Structure and mechanical properties of a thermal-sprayed nickel–20wt% chromium alloy. Mater Sci and Eng A 160:181–187

1.63 Nicoletto G, Tucci A, Esposito L (1993) Comparative dry wear behavior of hard coatings. Wear 162–164:925–929

1.64 Ohmori Akira, Zhou Zhan, Inoue Katsunori (1992) Penetration behavior into connected porosities of plasma sprayed $A1_2O_3$ coatings with liquid Mn_+. Trans JWRI 21:302–303

1.65 Olivas JD, Mireles C, Acosta E, Barrera EV (1997) Surface characterization of plasma spray metal deposition procedures using X-ray photoelectron spectroscopy. Thin Solid Films 299:143–148

1.66 Pauleau Y, Gouy–Pailler Ph (1992) Characterization of tungsten - carbon layers deposited on stainless steel by reactive magnetron sputtering. J Mater Res 7:2070–2079

1.67 Pawlowski Lech (1991) Applications and properties of thermally sprayed oxide ceramics. Powder Met Int 23:357–362

1.68 Pawlowski Lech (1991) The properties of plasma sprayed aluminum–aluminum oxide cermets. Surface and Coating Technology 48:219–224

1.69 Pilyankevich AN, Britun VF (1981) Preparing the specimens of refractory compounds for transmission studies with electron microscope (in Russian). Poroshkovaya Metallurgia 11:97–101

1.70 Polukarov YuM (1968) Formation of crystal lattice defects in metals (in Russian). VINITI, Moscow

1.71 Povetkin VV, Kovenski IM (1986) Structure formation in electrodeposited cobalt coatings (in Russian). Electrokhimia 22:1171–1175

1.72 Povetkin VV, Kovenski IM (1989) The structure of electrodeposited coatings (in Russian). Metallurgia, Moscow

1.73 Povetkin VV, Ustinovschikov YuI (1985) Electron microscopy studies of the structure of electrodeposited nickel-phosphorus coatings (in Russian). Izv. Academii Nauk USSR. Metally 13:187–189

1.74 Povetkin VV, Kovenski IM, Ustinovschikov YuI (1992) Structure and properties of electrodeposited alloys (in Russian). Nauka, Moscow

1.75 Rudajevova A (1993) Thermal diffusivity of plasma-sprayed coatings of ZrO_2 with 8wt % Y_2O_3 and ZrO_2 with 25 wt % CeO_2. Thin Solid Films 223:248–252

1.76 Saitoh Hidetoshi, Kyuno Takunari, Hosoda Ichiro, Urao Ryochi (1996) Surface morphology of polycrystalline diamond films etched by Ar^+ beam bombardment. J Mater Sci 31:603–606

1.77 Saltykov SA (1976) Stereometric metallography: Stereology of metallic materials (in Russian). Metallurgia, Moscow

1.78 Salvato M, Attanasio C, Coccorese C, Maribato L, Prischepa SL (1994) Superconducting and structural properties of BSCCO thin films by molecular beam epitaxy. Cryogenics 34:859–862

1.79 Sampath S, Gansert R, Herman H (1995) Plasma-spray forming ceramics and layered composites. JOM: J Miner, Metals and Mater Soc 47:30–33

1.80 Shao Beiling, Liu Ansheng, Wang Xiaohua, Li Yonghong, Zou Linying, Wang Jing, Zhang Jianguo, Shi Changyong, Zhou Yiry (1996) Microstructure of interface between HVOF sprayed WC–Co coating and spring steel substrate. Trans Nonferrous Metals Soc China 6:76–81

1.81 Shirkhanzadeh M (1995) Fabrication and characterization of alkoxyl-derived nanophase TiO_2 coatings. Nanostruct Mater 5:33–40

1.82 Singh Rajiv K, Qian F, Damodaran R, Moudgil S (1996) Laser deposition of hydroxyapatite coatings. Mater and Manuf Processes 11:481–490

1.83 Su YL, Lin JS (1993) Friction and wear behavior of a number of ceramic-coated steels matched as sliding pairs to various surface-treated steels. Wear 166:27–35

1.84 Tawancy HM, Sridhar N, Abbas NM, Rickerby D (1998) Failure mechanism of a thermal barrier coating system on a nickel-base superalloy. J Mater Sci 33:681–686

1.85 Tjong SC, Ku JS, Wu CS (1994) Corrosion behavior of laser consolidation chromium and molybdenum plasma- sprayed coatings on Fe–28Mn–7Al–1C alloy. Sci Met et Mater 31:836–839

1.86 Tomlinson WJ, Bransden AS (1994) Sliding wear of laser alloyed coating on Al–12%Si. J Mater Sci Lett 13:1086–1088

1.87 Trapeznikov VA, Shabanova IN (1988) X-ray electron spectroscopy of ultrathin surface layers of condensed systems (in Russian). Nauka, Moscow

1.88 Tushinski LI, Plokhov AV (1986) Study of structure and physico-chemical properties of coatings (in Russian). Nauka, Novosibirsk

1.89 Tushinski LI, Plokhov AV, Stolbov AA, Sindeev VI (1996) Structural strength of the base metal-coating composition (in Russian). Nauka, Novosibirsk

1.90 Ural O, Mitchell TE, Heuzer AH (1994) Microstructures of Y_2O_3–stabilized ZrO_2 electron beam-physical vapor deposition coatings on Ni–base superalloys. J Amer Ceram Soc 77:984–992

1.91 Usmani Saifl, Sampath Sanjay (1996) Erosion studies on duplex and graded ceramic overlay coatings. JOM: J Miner Metals and Mater Soc 48:51–54

1.92 Vancoille E, Celis JP, Roos JR (1993) Dry sliding wear of TiN-based ternary PVD coatings. Wear 165:41–49

1.93 Vorobiev GM, Girin OB (1979) Quantitative evaluation of multicomponent axial texture of electrodeposited metals (in Russian). Zavodskaya Laboratoria 45:546–547

1.94 Wang M, Schmidt K, Reichelt K, Jiang X (1992) The properties of W–C:H films deposited by reactive pf sputtering. J Mater Res 7:1465–1472

1.95 Wang Yuansheng (1998) Grain preferential orientation of Ag/Ni multilayered film. Trans Nonferrous Metals Soc China 8:472–476

1.96 Wertheim G, Hausmann A, Zander B (1977) Electronic structure of point defects (in Russian). Atomizdat, Moscow

1.97 Wheeler NS (1995) Microstructural characterization of cobalt–tungsten coated graphite fibers. J Res Nat Inst Stand and Technol 100:641–659

1.98 Woodruff DP, Delchar TA (1986) Modern techniques of surface science. Cambridge University Press, Cambridge

1.99 Wu Kee-Rong, Bayer Raymond G, Engel Peter A, Sun DC (1998) Wear of physical vapor deposition TIN coatings sliding against Cr-steel and WC counterbodies. Trans ASME J Tribol 120:482–488

1.100 Xiang Xinghua, Zhu Jingchuan, Yin Zhongda (1997) Microstructure of particles NiCoCrAlY in gradient plasma sprayed coating ZrO_2/NiCoCrAlY. Trans Nonferrous Metais Soc China 7:106–110

1.101 Yushchenko K, Borisov Yu, Murashov A, Kolomytsev M, Lugscheider E, Iokiel P, Feldhege M (1995) Thermal sprayed aluminium-silicon alloy coatings. Mater and Manuf Processes 10: 837-841

1.102 Zhang N, Wang Y, (1992) Dislocations and hardness of hard coatings. Thin Solid Films 214:4–5

1.103 Zverev AI, Sharivker SYu, Astakhov EA (1979) Detonation sputtering of coatings (in Russian). Sudostroenie, Leningrad

2 Mechanical Properties of Coatings

2.1 Elasticity

2.1.1 General Overview

Elasticity is the ability of a coating material to resist changes in its volume or shape under mechanical stress due to increase in internal energy. The main characteristic of this property is the elasticity modulus (Young's modulus), which describes the resistance of a coating material to deformation. Numerically it represents the ratio of applied stress increment to the ensuing increment of elastic deformation. In other words, elasticity is a measure of the stiffness of the coating material.

Resistance to elastic deformation is not only used as a measure of strength and hardness of material. This value is also required to calculate residual stresses, one of the main characteristics of a coating. Other common elastic properties, such as shear modulus and the Poisson coefficient, are rarely determined for coatings.

Below the point corresponding to the limit of proportionality, the deformation in medium-carbon and high-carbon steels is proportional to the applied stress, and the slope of the linear part of the bending curve characterizes the bending elasticity modulus. Further loading of bulk metal specimens leads to deviation from Hooke's law (i.e. the metal flows). The behavior of metal coatings of similar chemical composition is quite different (Fig. 2.1, curve 2).

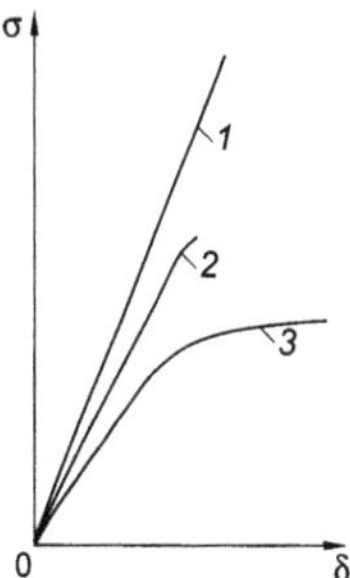

Fig. 2.1. Bending stress (σ)–deformation (δ) diagram for several types of coatings. *1* nonmetallic inorganic brittle coating; *2* high-strength metal coating; *3* metal coating of medium and low-strength

Bending of metal coatings does not cause even a slight plastic deformation, and destruction occurs abruptly and without any signs of flow. The destruction surface is perpendicular to tensile stress. For low-strength metal coatings (Fig. 2.1, curve 3) elastic deformation is observed only for rather small stresses, and residual deformation develops with increasing the load. Typical brittle failure is characteristic for non-metallic inorganic coatings (Fig. 2.1, curve 1) [2.76].

Elastic modulus (E) of a coating depends on loading scheme and is 5–10% higher for compression than for tension. The elastic modulus for coatings based on iron changes in the range (0.5–1.0)·10^5 MPa, i.e. differs by factor of 2. The Young modulus of pure iron is a constant determined only by interatomic interaction forces and equal to 2.0·10^5 MPa. In spite of significant differences in chemical composition, the elastic moduli of structural, tool, and special steels differ not more than by 10%.

Young modulus for metals practically does not depend on their structure and thermal treatment. Alloying and plastic deformation also have no pronounced effect on the elastic modulus. Heating of the material show leads to decrease of E, and there is a direct correlation between the temperature coefficients of Young modulus and of linear extension. The latter is caused by an increase of interatomic distances in the lattice with increasing temperature, which leads to reduction of interatomic interaction forces.

The situation for gas-thermal, plasma, detonation and other coatings is quite different. Here Young modulus depends not only on the temperature at which the test is conducted, but also on the method of deposition, technological regimes, porosity, etc. Therefore the Young modulus should be determined experimentally for each particular coating. The attempts to develop quantitative correlation between porosity and elastic modulus did not succeed.

All methods for determining the elastic modulus can be divided into two groups: static and dynamic. In the static methods the sought characteristics are calculated as proportionality coefficients at a static loading. Dynamic methods are based on the relation between the speed of elastic wave propagation in a material and the elastic modulus.

2.1.2 Determination of Elastic Modulus for Bending

In this method the elastic modulus is determined by static loading of the specimen under test resting freely on two supports (Fig. 2.2) within the proportionality limits of a bending curve.

Bending tests have the following advantages as compared to tension. The suggested loading scheme is much less demanding in terms of applied stresses, which allows to carry out detailed study of brittle materials such as coatings often are. Furthermore, no heads are required for specimens used for bending tests. This is important since fabrication of heads for specimens made of nonmetallic inorganic materials is rather difficult.

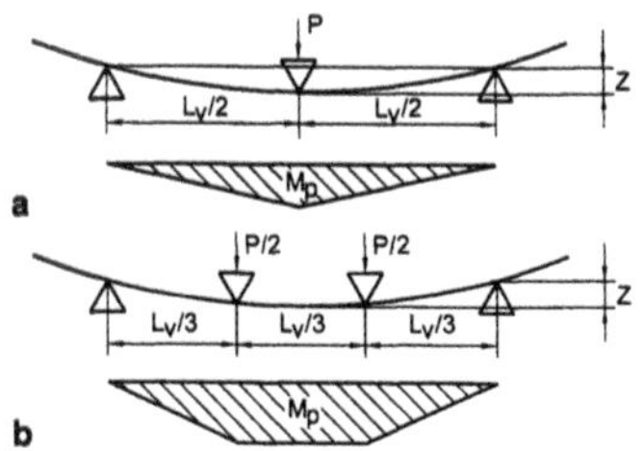

Fig. 2.2. Three-point and four-point schemes of bending loading: P load; L_v distance between the supports; Z bending flexure; M_P distribution diagram of the bending moment

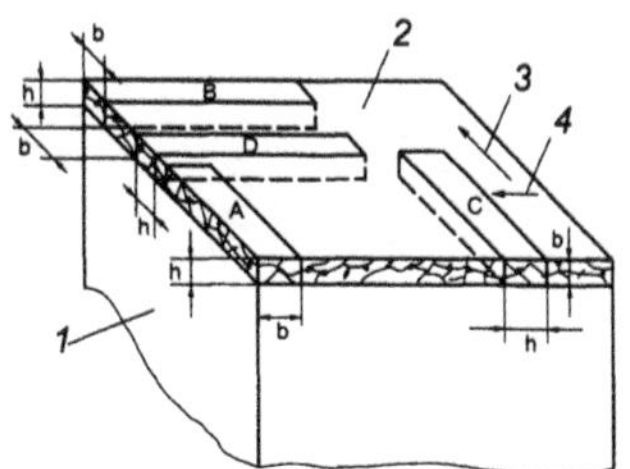

Fig. 2.3. Cutting the specimen taking into account the directions of the coating anisotropy axes. *1* metal substrate; *2* coating; *3* longitudinal direction of the part with deposited coating; *4* transverse direction of the part with deposited coating

A specimen for bending test usually has the shape of rectangular parallelepiped with the length (L) 80 mm, width (b) 5±0.5 mm, and height (h) 3±0.2 mm. A proportionally scaled specimen can also be used if required.

The coatings are usually characterized by pronounced anisotropy, and the results of the test would typically strongly depend on the arrangement of layers and their orientation within the material. It is recommended to cut the specimen taking into account the directions of the anisotropy axes, so that one of the anisotropy axes coincide with the long axis of the specimen (Fig. 2.3). The elastic modulus should be determined for at least two configurations, for example, A and C, or B and D.

An especially difficult problem is the fabrication of sample bars from ceramic coatings. Kudinov and Ivanov recommend the following fabrication method [2.50]. The coating with thickness 3.5–4.0 mm is deposited on a cylindrical mandrel of aluminum alloy having diameter 80 mm and length 90 mm. The coating is then cut by skive along the generatrix of the cylinder down to the metal substrate and is mechanically separated from the mandrel (Fig. 2.4).

A convenient framework for mechanical testing of ceramic coatings in general, and for estimation of their elastic characteristics in particular, is provided by the certification strategy of modern ceramics developed by Gogotsi and Zavada [2.30]. Numerous works of Gogotsi and Zavada with co-authors published during the last 15 years have crystallized into an elegant system of certification of weakly deforming materials [2.29]. All certification tests are carried out on a specimen in the form of a bar with dimensions 3.5×5.0×50.0 mm. Young modulus for such a specimen may be determined by static or dynamic methods. For the first case the distance between the supports is equal to 20 and 40 mm for three-point and four-point bending test, respectively. The size of the basic specimen practically coincides with the standards proposed by the USA and Germany [2.30].

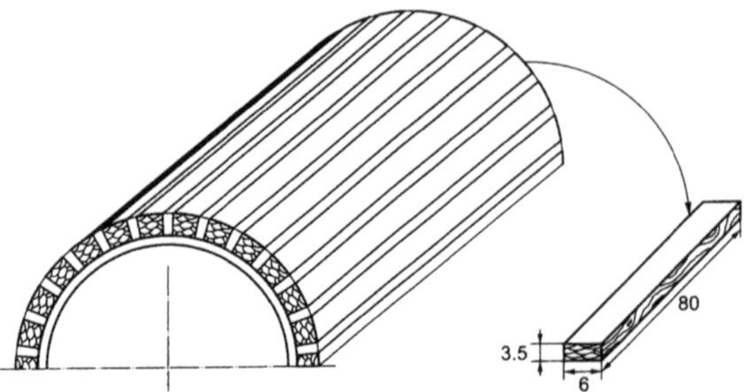

Fig. 2.4. Cutting ceramic specimens from a coating deposited on a cylindrical mandrel

At least five specimens having identical structure and axis of anisotropy are used for the tests. The specimens should not have broken corners and edges or large surface pores.

After separation from the substrate the bars of the coating material are ground in the mandrel to the required dimensions. Lateral surfaces should be flat and parallel to each other. All lateral planes of specimens are finished to roughness $Ra \leq 0.63$–0.32 μm.

Transverse dimensions of the specimens are measured at three points in two mutually perpendicular directions. Measurement error should not exceed 0.01 mm. To improve the accuracy it is recommended to normalize the specimens (see Sect. 3.5). Any tearing machine having special bending gadget, providing uniform relative displacement of loading edge(s) and supports, and allowing to measure loads to ±1% and bending flexure to ±2% may be employed for tests.

Commercially available tearing devices allow to determine elastic modulus of coatings with accuracy down to 1 kPa. The devices are intended for determination of elastic and strength characteristics of plastics and ceramics at various loading schemes including the bending loading. The wide range of testing loads (from 0.02 to 50 kN) and velocities of the active grip displacement (from 0.5 to 1000 mm/min), as well as high accuracy of measuring loads and deformations (bending flexure), make such devices a preferable choice for determination of coating strength and elastic characteristics.

A scheme of a typical test installation is shown in Fig. 2.5. The specimen 6 is loaded with two grips 5 and 7. The passive upper grip 5 is suspended at the elastic element of the dynamometer 4. The active lower grip 7 is mounted on the moving cross-arm 8 of the loading mechanism 9. The stress applied to the specimen by the moving cross-arm via the grip is measured by the elastic element of the dynamometer and indicated by arrow 2 on the load scale 3. The plots of deformation versus stress and stress versus time are automatically recorded by the registering device 1.

To provide the required loading scheme the test installation must be equipped with mechanism having loading edge(s) and supports of standard shapes and radii. The loading mechanism should provide the option of adjusting the positions of the supports and loading edges.

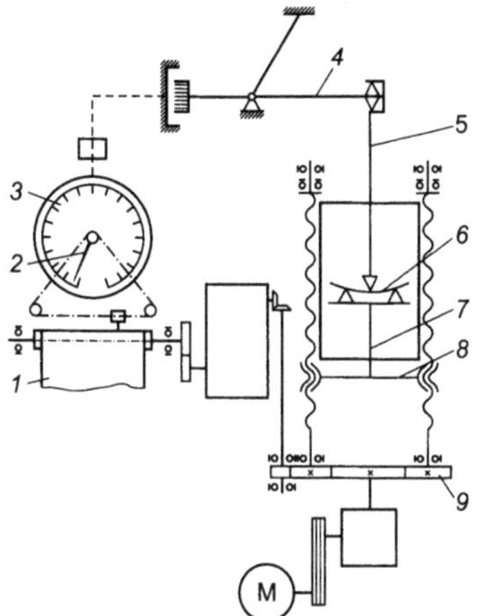

Fig. 2.5. Diagram of installation for mechanical testing of plastics and ceramics. *1* registering device; *2* indicator arrow; *3* load scale; *4* elastic element of the dynamometer; *5*, *7* upper and lower grips; *6* specimen; *8* moving cross-arm; *9* loading mechanism

In addition to conventional equipment special facilities and methods have also been developed to determine mechanical characteristics of thin foils and coatings [2.33, 2.59].

A long-term project on development and fabrication of instrumentation for certification of mechanical properties of brittle materials is been carried out by Gogotsi at the Institute for Problems of Strength (Ukraine). The created specialized installations can also be used for experimental study of coatings.

In many cases the researchers gave preference to development of stand-alone units consisting of a loading block providing the load for a specimen, self-contained measuring system comprising stress, bending flexure and acoustical emission sensors with necessary electronic transducers [2.30, 2.70]. A setup for study of deformation diagrams for ceramics in the wide range of temperatures (20–1500 °C) was developed on the basis of a tensile machine. High temperature blocks with cassette devices were designed for batch testing of multiple specimens. The blocks are capable of simultaneous heating of up to 10 specimens to the required temperature (1500 °C) in air or neutral environment and feeding them consequently to the loading supports.

A specimen is placed on the supports on its wide side in such a way that equal distances are left from the supports to each end of the specimen. The distance between the supports should be 75% of the total length of the specimen. The load is applied until the relative deformation of edge fibers (ε) reaches 0.2%. Bending stress for three-point loading scheme should be applied at the middle point of the specimen normal or parallel to the layers of the coating. In the four-point loading scheme the specimen is loaded by two edges pressing in the middle third part of the space between the supports. Bending flexure is estimated by the displacement of the movable part of the loading device as shown in Fig. 2.2. The relative deformation of the edge fibers for three- and four-point loading schemes can be calculated from the following equations, respectively:

$$\varepsilon = \frac{6Zh}{L_V^2}; \tag{2.1}$$

$$\varepsilon = \frac{Zh}{0.185 L_V^2}. \tag{2.2}$$

Next, two values of the load values together with respective bending flexures are determined from the diagram within the proportionality range of the bending curve. For three- and four-point schemes the bending elastic modulus (E_{bend}) in MPa can be calculated from the following equations:

$$E_{bend} = \frac{L_V^3 \left(P_2 - P_1\right)}{4\, b\, h^3 \left(Z_2 - Z_1\right)}; \tag{2.3}$$

$$E_{bend} = \frac{0.185\, L_V^3 \left(P_2 - P_1\right)}{b\, h^3 \left(Z_2 - Z_1\right)}, \tag{2.4}$$

where L_v is the distance between the supports in mm; P_1, P_2 are the loads in N; b is the width of the specimen in mm; h is the thickness of the specimen in mm; Z_1, Z_2 are the bending flexures corresponding to loads P_1 and P_2 in mm.

The four-point loading scheme, commonly referred to as "pure bending", usually gives more reliable results since in this case the maximum bending moment is created over a sufficiently long part of the specimen, as opposed to localized bending moment at a single cross-section for three-point loading.

For many ceramic coatings the measurements of elastic characteristics by static methods is rather complicated [2.69]. Under loading they experience brittle failure already at small deformations, which can be measured only by high-sensitive indicators or tension stations with high amplification coefficients. However, amplification of a slowly changing weak signal without its transformation leads to significant errors. Application of static methods for estimation of elastic constants of ceramic materials and coatings in the temperature range over 1500 °C is even more troublesome. There is no pronounced limit of the proportionality range on the deformation diagrams, deformation recording systems are cumbersome and complex, each temperature range requires an appreciable number of specimens to be tested due to statistical scatter of their physical and mechanical properties.

Therefore elastic characteristics in such complex conditions are commonly measured by dynamic methods, which allow to determine temperature dependencies of Young modulus when studying the influence of the regimes of deposition and subsequent treatment for the same specimens without their destruction. A further advantage of dynamical methods is the possibility of simultaneous determination of the shear modulus (G). Experimental setups for determination of E and G from eigenfrequencies of the specimen are described in the review [2.86].

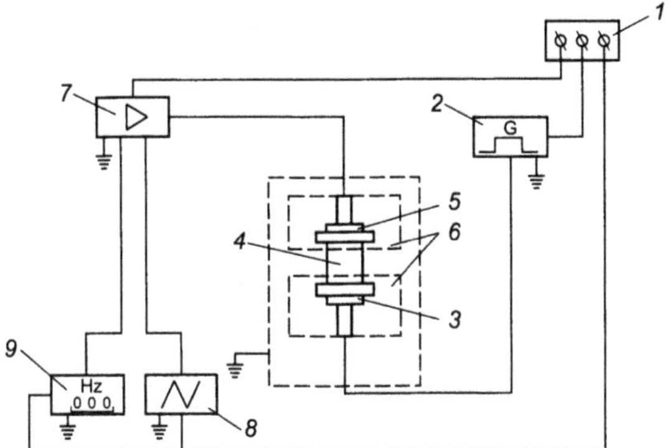

Fig. 2.6. Device for dynamical determination of Young modulus. *1* stable power supply; *2* ultrasonic generator; *3* sensor (piezoelectric element); *4* specimen; *5* receiver (piezoelectric element); *6* shield; *7* amplifier; *8* oscilloscope; *9* frequency meter

2.1.3 Dynamic Method for Determination of Elastic Modulus

The method is based on a physical law relating the eigenfrequencies of the specimen to the elastic modulus of the material under test. The eigenfrequencies are usually determined by exciting longitudinal ultrasonic oscillations of the specimen.

Barvinok argues that dynamic methods can yield elastic characteristics of coatings with higher accuracy since amplification of an alternating signal from a sensor does not pose serious problems [2.4].

A specimen for dynamic test has the shape of a 60 mm long bar with rectangular cross-section with dimensions 8±0.2 mm (width) and 6±0.2 mm (thickness). The end sides of the specimen must be parallel to within 0.2 mm as determined by a clock-type indicator. When cutting the specimen the directions of anisotropy axes should be taken into account. The density of the coating is determined separately to within 0.01 g/cm^3 on a witness specimen. The test is performed on the specimens from one batch, prepared using identical technological regimes and all having the same sizes and roughness parameters.

The testing device consists of a specimen holder, a generator of longitudinal oscillations with fine frequency control in the range from 20 to 100 kHz, and a device for determination of the resonance frequency with the error not exceeding ±0.2%. A scheme of the device is shown in Fig. 2.6.

The specimen is placed between the two piezoelectric elements (sensor and receiver), and the frequency of the generator is slowly increased until the lowest resonant frequency, corresponding to eigenfrequency of the specimen, is reached.

The resonant frequency is measured with accuracy better than 0.025%. The dynamic elastic modulus (E_g) in MPa is then calculated from the following equation:

$$E_g = 4L^2 \rho f^2; \tag{2.5}$$

where L is the length of the specimen in mm; ρ is the density of the coating in g/cm^3; f is the measured eigenfrequency in Hz.

However, although the dynamical methods offer a number of advantages, their application is hindered by the difficulties of separating the deposited coating from its metal substrate, and by errors introduced by this procedure. These problems were to a significant extent solved by Barvinok, who in his work [2.4] on the study of plasma composite multi-layer coatings suggested improved techniques, in which he was able to overcome the above-mentioned difficulties.

The techniques allow to determine the elastic characteristics of each sequentially deposited layer without separating it from the preceding layers or the metal substrate. Furthermore, they allow reliable measurement of Young and Hook moduli and Poisson coefficient for the specimen as a whole. An additional advantage of the techniques is the possibility to determine elastic characteristics on a standard equipment in a factory laboratory. The essence of the techniques is as follows: thin layers of coating are sequentially either deposited on the plate or removed from it, each time measuring the eigenfrequency of the entire system. Knowing the eigenfrequences, mass and geometric parameters of the plate, and using the obtained data, the distribution of the elastic modulus across the plate can be readily determined [2.4].

2.1.4 Other Methods for Determination of Elastic Characteristics

A new approach to determination of elastic characteristics of strengthening coatings was suggested in [2.23]. In the new method Young modulus and Poisson coefficient are determined from the results of static stretching tests of a single standard specimen with symmetrical double-sided coating. In the process of stretching the longitudinal deformation of the base metal substrate, and the longitudinal and transverse deformation of the coating are measured. The elastic characteristics of gas plasma coatings could then be calculated from suggested formulas.

The authors [2.56] used this method to study Young modulus and Poisson coefficient of plasma coatings utilized in nuclear power industry. In particular, it was found that compositions with dysprosium oxide have the best overall combination of mechanical properties for thin coatings, while compositions with yttrium oxide compounds are preferable for thicker coatings.

In their review paper Bulichev and Alekhin [2.9] have demonstrated that testing of coatings by continuos indentation allows determination of Young modulus with sufficient accuracy. Taking into account the simplicity of micro-hardness measurements as compared to those considered above, the indentation method for measuring E will most likely be applied in practice for non-destructive control of coatings [2.80].

Another approach to estimating the properties of coatings was given by Attar and Johannesson in [2.3], where x-ray diffraction technique was used to determine the elastic modulus of CrN coating having thickness 4.4 μm and hardness 2120 MPa.

2.1.5 Results and Discussion

For brittle materials the dynamic E_d and static E_s elastic moduli have the same value, while for relatively brittle materials $E_s < E_d$. As it has already been mentioned, Young modulus for coatings significantly differs from this characteristic of the compact material and is determined by the structure of the coating [2.15, 2.39, 2.49, 2.77, 2.83]. In some cases due to structure differences the two elastic moduli become practically incomparable. For a film in amorphous state the value of Young modulus is 20–40% lower than for a crystalline specimen. The coatings having clearly pronounces anisotropy or texture also have anomalous properties [2.34].

Using dynamic method, Kudinov and Ivanov [2.50] attempted to establish correlation between Young modulus and the porosity of plasma coatings. Open porosity was varied from 5 to 30% in 24 deposition regimes for 6 materials. The authors came to the conclusion that there is no direct influence of porosity on E, and the scatter of the values is substantial. This result implies that Young modulus for materials with different porosity should be determined by experimental methods only [2.50].

Dekhtyar et al. [2.20] have shown that thermal treatment of plasma coatings increases the elastic modulus by 30–50%. The elastic moduli of the coating immediately after plasma sputtering and after thermocycling from 1000 °C down to room temperature were studied by Vaidya et al. [2.77]. The measured Young moduli for plasma aluminum, nickel and molybdenum metallic coatings turned out to be approximately three times lower than for their compact counterparts [2.52].

The already cited work [2.20] also provides interesting data on layer-by-layer determination of the elastic modulus for plasma metallic coatings. The results of test conducted using original home-built equipment showed, that for layer by layer deposition of nichrome and tungsten the values of the elastic modulus are constant across the thickness of each layer and only slightly (1–8%) change for different layers of the same material. The factors that determine the temperature state of the particles of the deposited coating have stronger effect on elastic characteristics of plasma coatings than factors determining the temperature state of the base metal [2.20].

Eaton and Novak [2.25] developed a model correlating deposition parameters with hardness and elastic modulus of a coating. The authors studied 33 regimes of plasma sputtering of ZrO_2 coatings with 20% of Y_2O_3 for controlled change of deposition technology. Mechanical characteristics were measured in the four-point loading scheme. Variations of elastic modulus in the range from 20.1 to 47.6 GPa were observed. The model was found to be applicable for 72% of tests.

2.2 Hardness

2.2.1 General Overview

Hardness is the resistance of a material to local deformation or fracture upon impressing into its surface a harder body called indentor. Hardness reflects a large complex of mechanical properties. Tests for hardness are used for controlling the quality of new coatings [2.21, 2.28, 2.61], finding surface changes of the base metal [2.76], estimating structural heterogeneity across the cross-section of a coating [2.48], investigating wear and behavior of the coating [2.31, 2.75], determining bond strength between the coating and its base metal [2.36], revealing microfragility of coating, etc.

Hardness determination is the most widely used technique of all existing methods of studying mechanical properties of materials. Simplicity and accessibility of the method make it suitable for determination of the hardness of finished and intermediate products. After the testing, the products found acceptable may be used in an ordinary way since the method itself is non-destructive, and the prints left on the surfaces of the parts during the test can be easily removed. Numerical values yielded by the tests are used to control thermal treatment and choose the most efficient deposition regime [2.5, 2.19, 2.43, 2.46, 2.73]. However, hardness is an ill-defined concept, which explains the wide spectrum of methods and instruments for its determination. Conventional character of all measuring methods for determining hardness led to conventionality of hardness scales corresponding to the methods. The units of hardness are also conventional. The lack of theoretical correlations between hardness and physical properties of materials does not allow a definition of hardness units in terms of the basic units of International System (SI).

Hardness measurements have the following important advantages: they do not cause serious damage to specimen under test; do not require fabrication of special specimens; just a small surface area of the specimen must be prepared by a grinding wheel or sandpaper to carry out the test.

All hardness determination methods can be divided into two groups, static and dynamic, according to the way the load is applied to the specimen. Of the two the static methods, in which the load is increased smoothly and gradually, are used more frequently. The three commonly used variations of the static methods are static indentation of a hardened steel ball, a diamond cone, or a diamond pyramid, referred to as Brinell, Rockwell, and Vickers methods, respectively. Each of them is capable of measuring hardness within a certain range.

2.2.2 Determination of Hardness by Brinell

Hardness by Brinell (*HB*) is mainly determined for soft base metals, such as annealed or normalized steel, bronze, brass, etc. When determining the HB hardness, a small steel ball of diameter D is pressed into the tested specimen by the load P, applied during a certain time (Fig. 2.7).

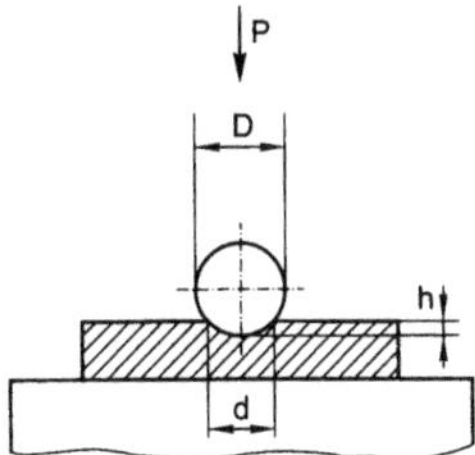

Fig. 2.7. Scheme of hardness measurement by Brinell

Hardness number by Brinell is determined as the ratio of the load P (N) to surface area of the spherical print (mm^2) according to the equation:

$$HB = \frac{2P}{\pi D\left(D - \sqrt{D^2 - d^2}\right)}, \tag{2.6}$$

where D and d are the ball diameter and the diameter of the print in mm, respectively.

For improved accuracy of hardness measurement the balls of the largest standard diameter, 10 mm, should be used. Using a large ball when measuring hardness of cast iron or coarse-grained materials yields more accurate average hardness numbers of composite structures consisting of several components.

The diameter of the indenter ball and the applied load are chosen for each particular specimen, taking into account its character, shape and sizes. Thin specimens, especially with strongly curved surfaces, and thin-walled hollow specimens should be tested at low loads.

The base surface of the specimen under test should rest tightly on the working surface of the hardness meter. The load should be applied strictly normal to the tested surface and should gradually build up to the given value. Failure to observe these conditions can lead to significant errors.

When determining the hardness number HB, the diameter of the print should be measured in two orthogonal directions, taking the average of the two measurements as the print size. The difference in the measured diameter of the print should not exceed 2% of the smaller value.

The diameter of the indenter ball D, the load P, exposition time, and minimum allowable thickness of the tested specimen can be found in special standard tables.

The diameters of the resulting prints d in the Brinell method should lie in the range $0.2D < d < 0.6D$. If this requirement is not met, the test is considered invalid.

The known dependence between the hardness by Brinell HB and the depth of the print h allows determination of h and the area of the print F:

$$h = \frac{P}{\pi D\, HB}, \tag{2.7}$$

$$F = \pi Dh. \tag{2.8}$$

The measurements should not produce any deformation at the opposite side of the specimen under test.

Hardness number by Brinell depends on prior hot or cold mechanical treatment of metals by pressure, thermal treatment, and on the temperature at which the test is being conducted. The hardness number by Brinell for steels correlates with their breaking point σ_B:

$$\sigma_B \approx 0.35\ HB. \tag{2.9}$$

There is also a correlation between the hardness number by Brinell and the yield point of a non-hardened metal σ_Y:

$$\sigma_Y \approx 0.383\ HB. \tag{2.10}$$

Sizable dimensions of the prints allow measuring hardness with high accuracy. Therefore the Brinell method may be used in the studies of plastic deformations and deformation processes, when the size of the plastic area is commensurable with the size of the resulting print.

Brinell method has been implemented both in stationary instruments with mechanical loading and in portable hardness meters used to measure hardness of bulky parts directly at the site.

2.2.3 Measurement of Hardness by Rockwell

In the Rockwell method a standard diamond cone (scales A, C, D) or a steel spherical point (scales B, E, F, G, H, K) is impressed into the surface of the specimen or part. First the preliminary load P_0 is applied, and then the main load P_1 is added. The hardness is determined from the impression depth of the indenter under the load P after removing the main load (P_1) (Fig. 2.8):

$$P = P_0 + P_1. \tag{2.11}$$

The surface of the specimen should be flat because when testing cylindrical samples the indentor penetrates deeper as compared to flat specimens of the same hardness, which leads to underestimation of the hardness value.

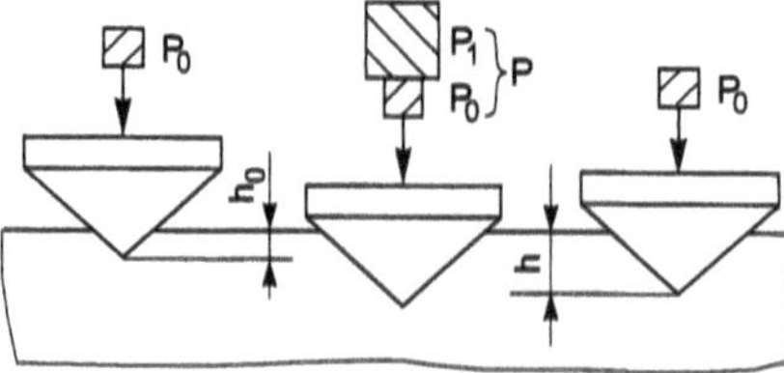

Fig. 2.8. Scheme of hardness measurement by Rockwell

Table 2.1. Loads and measuring ranges for different scales of hardness by Rockwell

Scale of hardness	Symbol for units	Preliminary load P_0 [H / kgf]	Main load P_1 [H / kgf]	Total load P [H / kgf]	Working range [units]
A	HRA	98.07 / 10	490.3 / 50	588.4 / 60	20–88
B	HRB	98.07 / 10	882.6 / 90	980.7 /100	20–100
C	HRC	98.07 / 10	1373 /140	1471 /150	20–70
D	HRD	98.07 / 10	882.6 / 90	980.7 /100	40–70
E	HRE	98.07 / 10	882.6 / 90	980.7 /100	70–100
F	HRF	98.07 / 10	490.3 / 50	588.4 / 60	60–100
G	HRG	98.07 / 10	1373 /140	1471 /150	30–94
H	HRH	98.07 / 10	490.3 / 50	588.4 / 60	80–100
K	HRK	98.07 / 10	1373 /140	1471 /150	40–100

According to the State Standard # 9012, the thickness of the specimen or coating should exceed the indentation depth at least by a factor of 10.

The minimum allowable thickness of the specimen or coating can be found from Figs. 2.9, 2.10.

The instruments for measuring hardness should provide loads indicated in Table 2.1.

The value of hardness is determined using scales corresponding to set loads. Most common are scales A and C. In this case a diamond cone with vertex angle 120° and vertex radius of curvature 0.2 mm is used as indenter. For the B scale a steel ball 1.588 mm in diameter is used instead of the cone. The Rockwell method for measuring hardness is rather fast because the value of the hardness number is read directly from the scale of the hardness meter.

Hardness by Rockwell is determined in conventional units. One unit of hardness corresponds to axial displacement of the indenter by 0.002 mm. The residual deformation is determined after removing the main load. The hardness number is read from the scale of a clock-type indicator, which automatically shows in mm the remainder from the subtraction of the difference between the two impression depths corresponding to two loads from a certain constant given in mm.

The hardness number by Rockwell for scales A and C is calculated from the following expression:

$$HRA = HRC = 100 - \frac{h - h_0}{0.002}, \tag{2.12}$$

where h_0 is the penetration depth of the indentor pressed into the specimen under test by the preliminary load P_0; h is the penetration depth for total load P, measured after removing the main load P_1 leaving the preliminary load P_0.

The hardness number for scale B is calculated from the following expression:

$$HRB = 130 - \frac{h - h_0}{0.002}. \tag{2.13}$$

The distance between the two neighboring prints or from the center of any print to the edges of the specimen should be at least 3 mm.

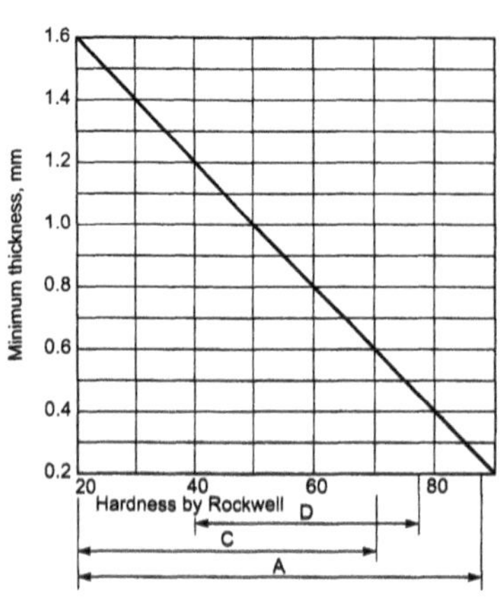

Fig. 2.9. Minimum allowable thickness of specimen (coating) depending on the scale and expected hardness for measurements using scales A, C, and D

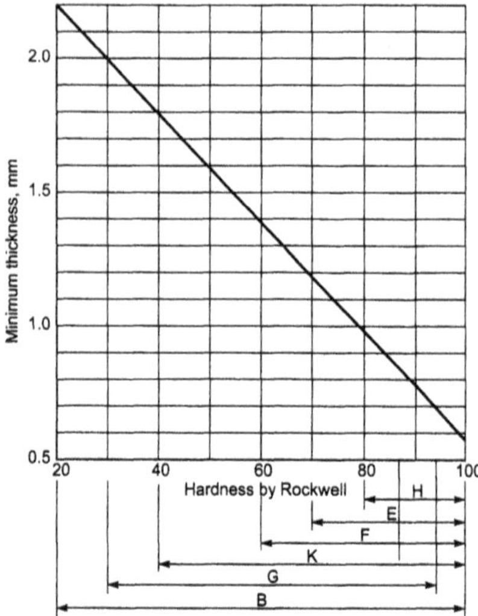

Fig. 2.10. Minimum allowable thickness of specimen (coating) depending on the scale and expected hardness for measurements using scales B,E,F,G,H, and K

The Rockwell method of measuring hardness has the following advantages over the Brinell method: ability to test parts of high hardness; simplicity in determining hardness number by reading its value directly from the scale of indicator without any calculations or tables; low surface failure rate of the method; high efficiency of measurements; possibility to operate the installation by untrained personnel.

The disadvantages of the Rockwell method as compared to the Brinell method are as follows: lack of unified hardness scale; arbitrary choice of scales; significant influence of spurious factors on the measured hardness number; impossibility of repetitive verification of the results.

The Rockwell method tests may also be used for estimation of bonding strength of thin coatings with the metal substrate [2.36].

2.2.4 Measurement of Hardness by Vickers

The Vickers method of measuring hardness is similar to measurements by Brinell, but the indenter ball is substituted by a diamond pyramid.

When measuring hardness by Vickers, a regular quadrangular pyramid is impressed into the specimen under test by a load applied for a certain period of time.

After removing the load the diagonal of the print left on the specimen surface is measured (Fig. 2.11).

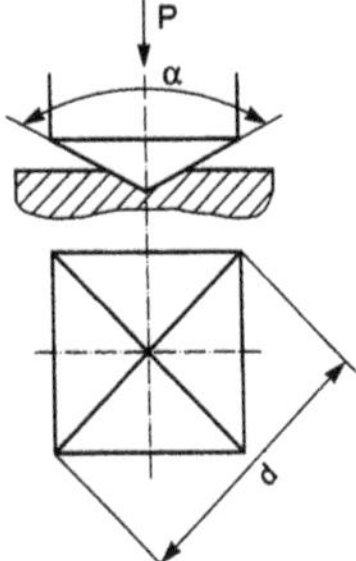

Fig. 2.11. Scheme of hardness measurement by Vickers

The hardness number by Vickers (State Standard #2999) is calculated from the following expression:

$$HV = \frac{2P\sin(\alpha/2)}{d^2}, \tag{2.14}$$

where P is the load in N; $\alpha = 136°$ is the angle between the two opposing faces of the pyramid at the vertex; d is the arithmetic average of the lengths of diagonals of the print in mm.

The standard loads for impressing the diamond pyramid are 50, 100, 200, 300, 500, and 1000 N. Lower loads can also be used if a proper instrument is available.

The minimum allowable thickness of the specimen or coating under test is 1.2 times the diagonal of the print for steels and 1.5 times the diagonal for non-ferrous metals. The measurement can be considered valid only if there are no signs of deformation on the other side of the specimen.

To improve the accuracy of the measurements it is recommended to use as high loads as practical. When measuring the hardness of a coating, the less is its thickness, the lower should be the load. If the thickness of the specimen is not known, several measurements at different loads, e.g., $P = 100$; 200; 300 and 500 N, should be performed. If the metal substrate has no effect on the results of measurement, equal or close values of hardness will be obtained. If the hardness decreases or increases with increasing the load, the latter should be reduced until two measurements at close values of the load yield equal or close results.

The minimum allowable thickness of the specimen or coating of a nonferrous metal can be found from the nomogram (Fig. 2.12).

The special studies have demonstrated that the presence and character of coating cracks should also be taken into account. When the indentor hits the crack or its vicinity the measured hardness value decreases approximately by 40%.

The analysis of the results of measurement in different conditions has shown that the variation coefficient, calculated for the coatings of different hardness, increases with increasing hardness. Thus to keep the same statistical accuracy of the measurement it is necessary to proportionally increase the number of indentations for coatings with higher hardness.

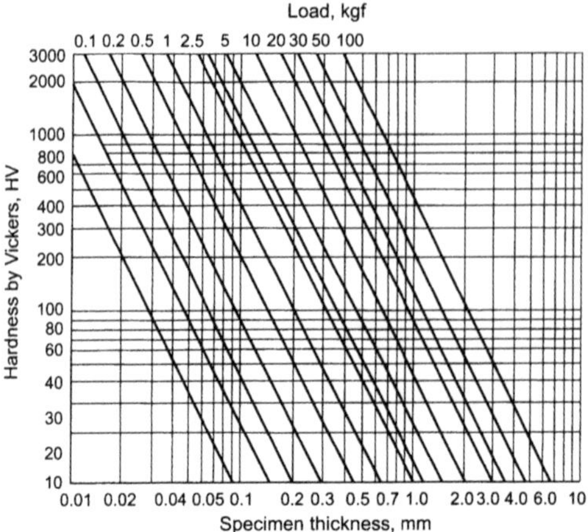

Fig. 2.12. Minimum allowable thickness of a specimen or coating

The Vickers method has important advantages: independence of the measured value of hardness on the load; ability to test the hardest materials; negligible surface damage; possibility to determine the hardness for comparatively thin layers.

The disadvantages of the method are as follows: fragility of the diamond pyramid and inability to test coarse-grained materials due to distortion of print shapes.

The Vickers method is capable to measure hardness *HV* in the range 8–1000 for steel specimens up to 200 mm high.

2.2.5 Measurement of Microhardness

The method of microhardness is widely used in the study of local properties of bulk strengthened materials and various coatings. It allows accurate and fast quality control of materials, in particular very small parts, and provides means to determine the properties of thin surface layers and coatings [2.40, 2.63, 2.82]. The method can be easily combined with studies of the microscopic structure of materials [2.37, 2.87], allows extensive physical-chemical investigations of structural transformations, and is sometimes indispensable [2.16, 2.17, 2.60, 2.66].

In the method of microhardness the hardness of microvolumes of coatings can be determined for loads of up to 4.9 N. The main purpose of the method is determination of the resistance of separate particles and structural components to plastic and elastic deformation, and finding anisotropy of hardness for various regions of the coating. According to the State Standard #9450, two test methods are distinguished: using reduced and not reduced prints. In the former method, the print parameters are determined after removing the load, while in the latter impressing of a diamond point into the studied area is accompanied by simultaneous measurement of the print depth. The second method is used when additional character-

istics of material are required: elastic recovery, relaxation, and elasticity modulus. This method characterizes the resistance of material to elastic deformation at indentation of a diamond point during a given period of time. After removing the load and the indenter point the print parameters are measured, from which microhardness can be determined using special formulas and tables.

The testing procedure requires small parts with coating or specially prepared specimens (witnesses). It is expedient to have two sets of specimens: one set for the tests with loading along the coating layers (to prepare transverse sections), and another set for the tests with loading normal to the layers (to prepare longitudinal sections). The roughness of the local area for the print should be not worse than $Ra = 0.32$ μm. The sections are prepared as the samples for metallographic studies. When preparing the sections special care should be taken to exclude possible changes in coating microhardness due to heating and hardening. The tested surface must be dry and clean. To fix the required alignment of indentor's axes relative to the coating surface and eliminate shift, flexure and turning of the section, it is recommended to fix the section at the object stage with the help of plasticine into which the specimen is impressed by a laboratory hand press.

Diamond points of various shapes and sizes are used for measurements (Fig. 2.13). The most common is the point in the shape of tetrahedral pyramid with a square base (the Vickers pyramid). The point with the shape of trihedral pyramid (the Khrushchev-Berkovich pyramid) is used for very hard materials ($HV > 1000$), because for equal lateral surface and height, the trihedral pyramid has more perfect sharpening then the tetrahedral pyramid. Thin coatings are tested by the point with a working part in the form of tetrahedral pyramid with a rhombic base (the Knoop pyramid). A bicylindric point is used for testing the microhardness of very thin (less than 3 μm) coatings.

Prior to testing the microhardness meter is mounted on an absorber placed on a massive base fixed to the wall or special foundation to dampen occasional oscillations of the rod with the indentor. Bumps and device vibration drastically increase the measurement error, and the sensitivity to external influence on impressing of the diamond point increases with reducing the load on the indentor.

When preparing the test, it should be taken into account that the accuracy of microhardness measurement is effected by the following objective and subjective factors: incorrect choice of the print area (the pyramid is indented close to a pore, edge or boundary of the layer); increased roughness of the tested surface; incorrect ratio between the coating thickness and the print depth (hardness of the base metal effects the apparent coating microhardness); the applied load is incorrect (due to friction losses in the loading mechanism, the actual load experienced by the indenter may be decreased, while fast sinking of indenter will apparently increase it); inaccurate measurement of the print diagonals (due to insufficient magnification of objective lens, poor lighting of the microscope view field, or incorrect reading from the scale of micrometer).

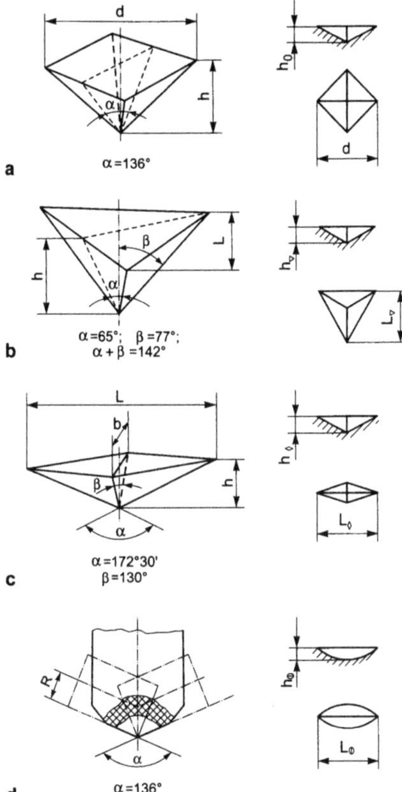

Fig. 2.13. The shapes of diamond points: **a** tetrahedral pyramid with a square base (the Vickers pyramid); **b** trihedral pyramid with a base in the shape of an equilateral triangle (the Khrushchev-Berkovich pyramid); **c** tetrahedral pyramid with a rhombic base (the Knoop pyramid); **d** bicylindric point

When choosing the load for the study of longitudinal sections, the ratio of assumed coating thickness and the print depth should be considered. If the thickness of the studied coating is not known, the hardness should be measured several times with a successive increase in load. If the material of the substrate has no effect on the results of microhardness measurements, the obtained values will be close. If microhardness decreases or increases with increasing the load, the latter should be decreased until equal or close results will be obtained for two close values of the load. This procedure of choosing the load is appropriate for low-porous, homogeneous and isotropic coatings. As for gas-flame, plasma and detonation coatings, it is expedient to measure the coating thickness by one of the existing methods. Then, considering the ratio between the coating thickness and the print depth and diagonal, the optimal load is chosen [2.26].

The following relationship must be satisfied for microhardness measurements on longitudinal sections:

$$h = \frac{d}{7} \leq \frac{\delta}{10}, \tag{2.15}$$

where δ is the coating thickness; h is the depth of indentation; d is the print diagonal.

Violation of this condition leads to artifactual correlation between the coating microhardness and thickness and the load at the indenter.

On the one hand, the requirement to keep the proper ratio between the thickness of a thin coating and the print depth means that very small loads which can be delivered only by microprocessor controlled equipment should be used. On the other hand, this implies that a scanning electronic microscope should be used for analysis of a small print (≤3 μm). The only reasonable alternative to the conventional approach is to take the load–print depth dependence. However, elastic and plastic deformations are superimposed on each other, and their ratio depends on the material and type of loading. Therefore, direct measurement of thin (up to 5 μm) coating hardness is possible only using expensive and complex equipment. With the standard methods it is difficult and sometimes impossible to keep the ratio between the coating thickness and depth of indenter impression. Besides, it is incorrect to compare results obtained for coatings with different thickness or those deposited on different materials due to plastic deformation of the base metal and punching of the coating. Thus, it is necessary to correct the results of the measurement for the coating thickness and substrate hardness.

Voevodin et al. [2.81] have determined that the error can be corrected using an expression, which considers the ratio between the area S of the entire print and the areas of the coating Sc and substrate $S_s = S - S_c$:

$$H^c = \frac{SH_s - S_c H_c}{S - S_c}, \tag{2.16}$$

where H^c is the corrected value of hardness; H_s and H_c are the substrate and coating hardness, respectively.

The authors argue that their method allows correction of the results of microhardness measurements on hard coating, when the depth of indenter impression exceeds the thickness of the coating. The required condition is that the hardness of the base metal be less than the hardness of the coating.

After the load on indenter has been chosen, the specimen is installed on the object stage in such a way that the surface under study is normal to the direction of load application and indenter axis. The region where the print is to be impressed should be free of fractures. Layer boundaries, pores and inclusions must be at least twice the length of the diagonal of expected print away from the site of impression. The distances between the centers of adjacent prints should be 3 times larger than the print diagonal. At transverse sections, the distance from the center of the print to the outer edge of the coating should be not less than the double length of diagonal. The specimen is loaded gradually at a constant rate, which has no effect on the print sizes. For anisotropic coatings, one of the print diagonals should coincide with the lamination orientation.

The indentor is gradually sunk during 15 s and kept under load for 5–10 s. Too rapid loading causes the impact effect of indentor on the coating surface, which can lead to underestimation of microhardness due to coating breaking above a pore.

After removing the load the print is placed in the view field of a microscope. The necessary number of prints is determined by the required accuracy, but taking into account the specificity of the coating structure, at least 30 prints should be prepared and studied.

When testing brittle coatings, if chips or cracks are found on the print surface or adjacent regions the print is discarded. In this case the correct values of microhardness can be obtained by gradually decreasing the applied load until no chips and cracks are produced (if it is not necessary to define the coefficient of brittleness).

When measuring the microhardness by the method of reduced print, the microhardness number is determined by dividing the value of the normal load applied to the diamond point (in kilogram-forces) by the area of the print lateral surface (in square millimeters).

1. For a tetrahedral pyramid with a square base, the microhardness number *HV* is found from the following expression:

$$HV = \frac{1.854P}{d^2}. \tag{2.17}$$

2. For a trihedral pyramid with the base in the shape of an equilateral triangle, the microhardness number *HV* is found from the following expression:

$$HV = \frac{1.570P}{l^2}. \tag{2.18}$$

3. For a tetrahedral pyramid with a rhombic bottom, the microhardness number *HV* is found from the following expression:

$$HV = \frac{12.873P}{l^2}. \tag{2.19}$$

4. For the bicylindric point, the microhardness number *HV* is found from the following expression:

$$HV = \frac{4.168P}{l^3}. \tag{2.20}$$

In Eqs. 2.17–2.20 P is the normal load applied to the diamond point, N; d is the arithmetic average length of the square print diagonals, mm; l is the print size, mm (Fig. 2.13). When measuring microhardness by the method of unreduced print, the microhardness number is determined by dividing the value of the normal load applied to the diamond point (in kilogram-forces) by the area of the print lateral surface (in square millimeters), corresponding to its measured depth.

Depending on the shape of the diamond point (Fig. 2.13), the microhardness numbers are defined by the following expressions.

1. For a tetrahedral pyramid with a square bottom:

$$HV_h = \frac{0.03784P}{h^2}. \tag{2.21}$$

2. For a trihedral pyramid with the base in the shape of an equilateral triangle:

$$HV_h = \frac{0.03797P}{h^2}. \tag{2.22}$$

3. For a tetrahedral pyramid with a rhombic bottom:

$$HV_h = \frac{0.01385P}{h^2}. \tag{2.23}$$

4. For a bicylindric point:

$$HV_h = \frac{0.07292P}{h^{3/2}}. \tag{2.24}$$

In Eqs. 2.21–2.24 P is the normal load applied to the diamond point, kgf; h is the print depth, mm.

The microhardness number is calculated from Eqs. 2.17–2.24 or taken from standard tables. The microhardness values by Vickers and Knoop almost coincide for loads in the range 1–5 N. However, the Knoop print depth is only about 1/30 of the length of the print large diagonal, which is a distinct advantage of the Knoop method as compared to the Vickers method. Since the indentation depth of the pyramid is low, the method is used for study of very thin layers of material. A disadvantage of the Knoop indenter is the lower symmetry of the prints. An example of combined use of several methods for measuring microhardness and comparison of the results has been provided by Bull et al. [2.7].

Another elastic characteristic of coatings, often encountered in the studies of non-metallic inorganic coatings, the coefficient of brittleness K_f, can be determined if at least a hundred prints are available:

$$K_f = \frac{N_{cr}}{N} \cdot 100, \tag{2.25}$$

where N_{cr} and N are the numbers of prints with defects (cracks, chips on the print surface or surrounding area) and the total number of prints for the given load, respectively.

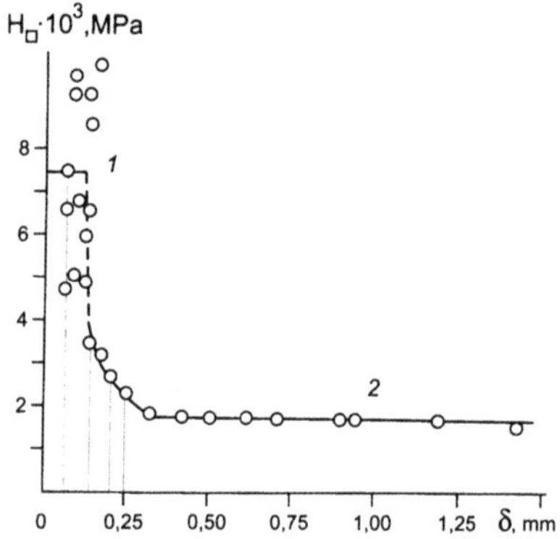

Fig. 2.14. Dependence of the microhardness of ZrO_2 plasma coating (*1*) and its base metal (*2*) versus the distance from the surface [courtesy of Bataev A.A.]

Even for identical coatings deposited using similar technologies the values of microhardness from different authors differ by 20–50 %. According to Byakova, the scatter of reported values for the metal-based coatings with a body-centered cubic lattice is as high as 7–10 GPa, reaching 20–37 GPa for brittle ceramic coatings [2.11].

To improve the reliability of the results, it is necessary to indicate the load on the indenter, the type of indenter, the roughness and procedure of surface preparation before indentation.

The values of microhardness obtained on the transverse sections of substrates with coatings give differentiated information about the properties of surface layers [2.46], and moreover, about the coating thickness, estimated by a drastic drop of the measured microhardness at the coating-substrate interface (Fig. 2.14).

A somewhat different picture may be observed when the coating is fused with its base metal, and profound structural transformations take place in the transition zone [2.5]. For this case, a peak of microhardness precedes the abrupt drop (Fig. 2.15).

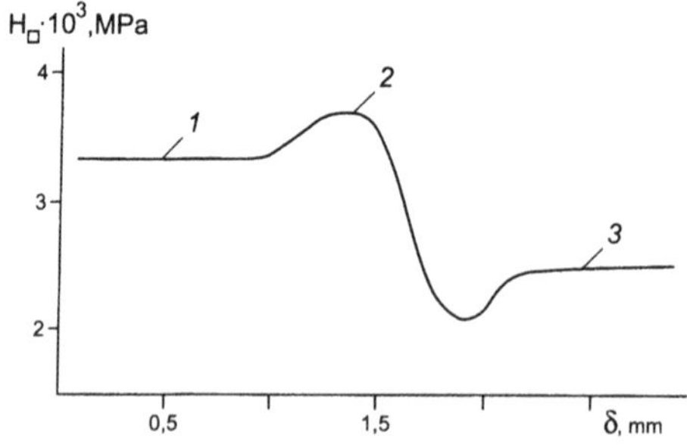

Fig. 2.15. Characteristic zones of a fused coating. *1* coating; *2* fused zone; *3* metal substrate

Interesting conclusions can be gleaned from the results of several hundreds measurements of microhardness for the fused and non-fused self-fluxing coatings. The average microhardness for non-fused coating and its substrate is 9.5 GPa and 6.6 GPa, respectively, with the distributions close to normal. The dispersion for non-fused coatings is somewhat smaller than for the fused ones. A plausible explanation is that the time of interaction with the flame of the burner is considerably longer than the time of deposition, and fusing is accompanied by intense formation of carbide and boride phases with very high microhardness and soft bonding, which leads to increased value of dispersion.

In the cases when composition and properties of the coating change across its thickness, it is expedient to perform measurements by indenter pressing into the flat surface of a slanted section of the coating.

When measuring the microhardness at slanted sections, it should be remembered that the remaining thickness of the coating layer decreases as the site of test is moved from the coating surface to the coating-substrate interface. Hence, for a given load, the error in microhardness measurement will increase with approaching to the interface. More accurate values of microhardness in this zone can be obtained if the load applied to the indenter is changed according to the change in the thickness of the remaining coating layer.

2.2.6 Peculiarities of Testing the Coating by Continuous Indentation

Because of its simplicity and ease of use, the method of reduced print is widely used in research and industrial laboratories. However, the works of the last decade indicate that the method of unreduced print is also a promising, objective and efficient test for microhardness [2.32, 2.53, 2.85]. In his detailed review, Bulychev has illustrated the recent advances and prospects of the tests by continuous indentation of a pyramid [2.8]. Previously, together with Alekhin, he considered theoretical aspects of this problem and demonstrated that the methods of testing materials with constant registration of the indentation diagram had made a qualitative leap in their development.

A new automated measuring complex “MTI-3M”, developed for measuring microhardness of materials and coatings, has been designed for the studies of micromechanical properties with continuos registration of the indentation diagram on the coordinates “load–print depth” [2.8].

Thus obtained indentation diagram allows to estimate the peculiarities of microdeformation in a wide range of applied loads and print depths, to determine rheological characteristics of surface layers of the tested material (microcreep, relaxation), to study the process of crack formation and its connection with contact interaction, to estimate integrally micromechanical properties of the coating (microhardness, elasticity, plasticity), and to determine the effect of the base metal on the microhardness of the deposited material.

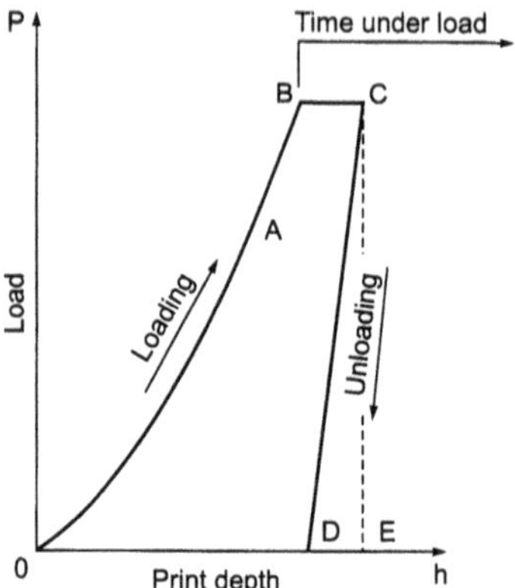

Fig. 2.16. A typical indentation diagram

A typical indentation diagram is shown in Fig. 2.16. The loading region 0AB at the diagram is a functional dependence of the print depth h versus applied load P, and is commonly approximated by a power dependence:

$$P = ah^n, \tag{2.26}$$

where a and n are the two constants, determined from the indentation diagram.

This region yields the value of unreduced microhardness according to relation:

$$HV = A\frac{P}{h^2}, \tag{2.27}$$

where the constant factor A depends on the geometrical shape of the indenter.

The region BC of the diagram describes rheological properties of the tested material, in particular, it characterizes the microcreep phenomenon. The region CD, determined by relaxation of energy accumulated during viscoelastic deformation caused by indentor penetration, provides information on plastic properties of the material.

Analysis of the diagram areas, proportional to works done at loading and unloading stages, allows estimation of the deformational a_{def} and relaxation a_{rel} capabilities of the tested material. The total work A is composed from work of residual deformation A_{rf} (area OBCD) and the work of the aftereffect forces A_{rel} (area CED).

$$a_{def} = \frac{A_{rf}}{A}, \tag{2.28}$$

$$a_{rel} = \frac{A_{rel}}{A}. \tag{2.29}$$

The automated measuring complex MTI-3M consists of the following units:

- microhardness meter,
- microprocessor controller,
- videoterminal device,
- alphanumerical printer.

The complex provides a standard interface for data output to the printer and data upload to computer for storage and further processing [2.8].

2.2.7 Microhardness as a Characteristic of the Coating Strength Properties

Systematic investigations of possible applications of the microhardness techniques to the strengthened layers and special measurements with coating indentation are carried out by Byakova and Gorbach in the National Technical University of Ukraine [2.12].

After generalization of a large body of experimental results on a wide spectrum of carbide, nitride, boride, and silicide coatings produced by thermodiffusion saturation, vacuum deposition and chemical precipitation, Byakova and Gorbach came to the following conclusions [2.12].

The large scatter in the reported values of coating microhardness H_μ cannot be explained only by differences in their chemical composition as it was assumed previously. The scatter for brittle high-strength coatings is caused by material destruction in the near-contact area, which is usually neglected. Besides, it was shown that a decrease in H_μ observed with increasing P is a consequence of the increasing contribution of plastic deformation, which, similar to microhardness, reaches stable values at loads $P > P_{cr}$. The upper limit of the critical load P_{cr} is determined by the nature of the tested material and depends on the level of residual stresses in the coating.

It then follows that a more accurate and reliable procedure of estimating the microhardness should address the following peculiarities of the test:

- preliminary quantitative measurements of H_μ in a wide range of permissible loads, whose upper limit is determined by the thickness of tested layer;
- measurement of the H_μ vs. P dependence;
- determination of the upper limit of the critical load P_{cr};
- final quantitative measurements of H_μ at loads $P \geq P_{cr}$.

The authors justify this procedure of measuring microhardness by the observation that microhardness, which can be determined experimentally for both viscous and brittle materials, in fact characterizes several different deformation processes. For viscous metallic materials microhardness demonstrates the resistance of the material to plastic deformation. For brittle materials, however, microhardness reflects their resistance to two types of deformation, plastic and elastic. Their relative contribution in the region of low loads changes, and in the general case in standard tests it should be considered as unknown [2.12].

To determine the true hardness of thin coatings, Burnett and Rickerby [2.10] suggested a theoretical model, which considers the contributions from both coating and base metal to the measured results. The experimentally obtained values of hardness are determined by the values of coating and substrate hardness. As an example the hardness of the coating from titanium nitride on tool steel substrate was determined using the suggested model.

The effect of residual compression stresses on microhardness was tested using TiN coatings, deposited on a hard alloy substrate [2.65]. A direct correlation between the two properties was clearly demonstrated. However, the distribution of deformations across the depth of a coating does not always correlate with its hardness and internal stresses.

To investigate the micromechanical properties of high-strength TiN, HfN and ZrN coatings deposited on the WC–Co substrate, and the properties of the substrate itself, the relationship between microhardness by Vickers and by Knoop and the applied load was also studied [2.41]. Using the Khrushchev–Berkovich pyramid, Chou et al. succeeded in determination of both the microhardness and Young modulus of multilayer films [2.16, 2.17].

Zhang and Wang have analyzed the possibility of determining the microhardness for the coatings from TiN, TiC, TaC, etc., as a function of the density of dislocation [2.87]. Wang et al. [2.82] have compared the estimated microhardness for high-speed steel and for TiN coating, calculated using a finite-element method, with experimentally determined microhardness by Vickers, measured at loads of 4–200 g. The model calculations assumed that properties of the coating and the base metal were isotropic. A good correspondence of experimental results and data calculated by the suggested model was obtained.

Unfortunately, many publications give the values of microhardness of thin coatings, obtained without strict adherence to requirements of standards. Underestimation of hardness increases with decreasing thickness of the coating, increasing the load, reducing the hardness of the substrate hardness or its elasticity modulus. The error can also be aggravated by poor adhesion of the coating to its substrate.

2.2.8 Measuring Microhardness by Scratching

Scratching is widely used as a measure of microhardness [2.7, 2.63]. Usually a groove is scratched on the tested surface, and its width is measured. The microhardness is defined as a ratio of the load applied to the diamond pyramid to the specific area of the pyramid contact with the tested surface. Successively changing the load on the indenter, the groove is made 5, 10, 20 μm wide. The load corresponding to the given width of the groove is called hardness by Martens. The commonly used methods are by an edge of a tetrahedral pyramid, and by an edge or facet of a trihedral pyramid. Hard materials should be tested by a trihedral pyramid. Coating adhesion can also be measured by scratching.

The microhardness for a tetrahedral and a trihedral pyramids is determined from the following equations:

$$H_{\square P} = 3.708\ P\ /\ b^2; \tag{2.30}$$

$$H_{\nabla P} = 3.138\ P\ /\ b^2; \tag{2.31}$$

where $H_{\square P}$; $H_{\nabla P}$ are the sought microhardnesses; P is the load in N; b is the groove width in mm.

When measuring microhardness by scratching, the groove length should be at least 0.25 mm. The width of the groove can be measured with an error up to 0.001 mm.

To study the effect of nitrogen content in TiN coatings on the critical normal load and coating adhesion, the authors [2.79] used scratching by a diamond indenter with radius of curvature 0.2 mm. Coatings with the thickness of 2.1 ± 0.1 µm were deposited on stainless steel with HV hardness HV of 1.6 GPa and high-speed steel (with 7 mass % of W) by the method of reactive ion plating.

The scratching was performed using a specially constructed device at constantly increasing load. The critical normal load was determined by the method of acoustic emission. During the test, the tangential force of scratching was also measured, and the friction coefficient was calculated [2.78].

It was determined that the critical normal load decreases dramatically with an increasing nitrogen content in the coating. Three different regions may be distinguished as the content of nitrogen is gradually increased. The first region corresponds to pure titanium. Here significant plastic deformations typical for soft coatings occur. Then comes the transition second region, and finally the third region, characterized by formation of cracks normal to scratching direction typical for hard coatings. This dependence on the content of nitrogen should be taken into account when choosing nitrogen pressure in the deposition of TiN coatings [2.79].

Scratching a slanted section by a diamond indenter can provide important information about the hardness of coatings and surface layers. Together with hardness estimation by the groove width, this method can be used for determination of material brittleness and its adhesion to the substrate, as well as other characteristics [2.7].

2.2.9 Hardness Meters, Microhardness Meters

The devices for hardness measurements are commercially available from several producers. The leading ones are Wolpert, Reicherter (Germany), Wilson (USA), Meier (Austria), and Akashi (Japan). Similar, but less universal, devices are also available from several Russian companies. The universal hardness meters have provisions for measurement by two and sometimes three methods, Rockwell, Brinell and Vickers, on a single installation. Besides, sometimes loads and indenters untraditional for Russia are used in these universal devices.

Since the mid eighties, hardness studies with ultra-low loads become increasingly popular [2.2, 2.32, 2.71]. Thus, a method of microhardness measurements at ultra-low load of 0.005–2 cN using the ultra-microhardometer supplied by Paor (Austria) was described in [2.51]. The ultra-microhardometer is mounted on the coordinate table of a scanning electron microscope. Recommendations on optimal positioning of the specimen and measurements of the dimensions of the print left by indenter were illustrated on the example of TiN, HfN and NbN coatings. Microhardness measurements using the ultra-microhardometer are useful for coatings having thickness from one to several microns.

An Australian company CSIRO Div. of Applied Physics has developed the UMIS–200 device, which allows to make precise measurements of ultra-microhardness for films with thickness as low as 0.6 μm [2.2].

2.2.10 Fracture Toughness of Coatings

Analysis of the coating structure demonstrates that the coating always has microcracks forming during the deposition process [2.55, 2.58] and the following thermal treatment [2.18, 2.35, 2.42, 2.47, 2.67, 2.74]. A developing crack can ultimately lead to complete destruction of a construction element [2.6, 2.22, 2.24].

Special tests on fracture toughness are an important part of the process of selecting materials for parts that will work in the conditions posing the danger of destruction. The testing technique is based upon the modern mechanics of destruction (for details see Chap. 8).

It is well known that a crack works as a concentrator of stresses, with the maximal stress accumulating in the corner of the crack. For a static load, the local stresses at the crack corner are proportional to the coefficient of stress intensity K defined as

$$K = \sigma\sqrt{\alpha\pi l_{cr}}, \tag{2.32}$$

where σ is the nominal stress in a solid without considering the crack; α is the coefficient depending on the shape of the body and the geometry of the crack; and l_{cr} is the crack length.

The intensity coefficient K may be interpreted as a stress before the crack at the distance $\pi/2$ (mm) from its corner.

Spontaneous development of the crack and spontaneous fracture begins when the value of K builds up to a certain critical value K_C in the course of slow growth of the crack:

$$K_C = \sigma_C\sqrt{\alpha\pi l_C}, \tag{2.33}$$

where σ_C is the critical value of stress in the body; l_C is the critical length of the crack. The intensity coefficient K_C does not remain constant for material, but depends on the thickness of the specimen under test (Fig. 2.17).

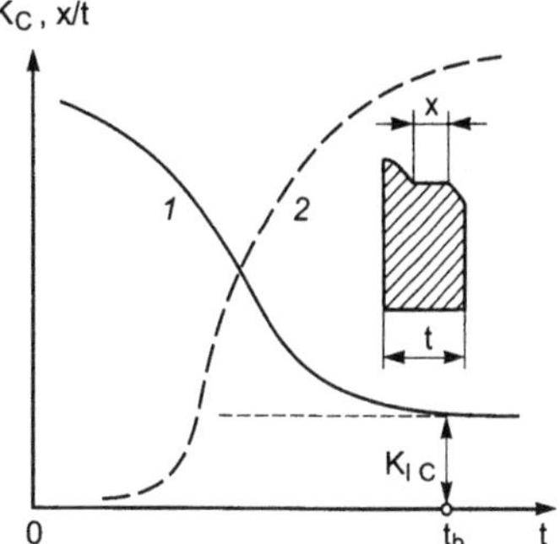

Fig. 2.17. Intensity coefficient K_C (*1*) and fraction of brittle fracture *x/t* (*2*) as a function of the thickness of the specimen with a crack. K_{Ic} is coefficient of intensity in the conditions of planar deformation; t_B is specimen thickness at which a planar deformation state is reached; *x* is thickness of the brittle fracture zone. Cross-section of the specimen is hatched

In a thin specimen subjected to a load normal to the crack plane, a planar stressed state having no stress along the crack is developed. Such a specimen upon destruction experiences bulk plastic deformation, and a considerable part of the resulting fracture is oblique.

As the thickness of the specimen is increased, when the coefficient K_C takes its minimal value K_{Ic}, a transition from the planar stressed state to a planar deformed state takes place at the corner of the crack, which becomes apparent in the absence of transverse deformation.

In this case pure brittle failure occurs with no signs of plastic deformation, and the fracture is produced without slanting. The thickness of the specimen at which the planar deformation state is reached should satisfy the condition

$$t_B \geq 2.5\left({K_{Ic}}/{\sigma_{0.2}} \right)^2, \tag{2.34}$$

where $\sigma_{0.2}$ is the conventional yield strength of the material.

The fracture toughness, measured in the conditions of planar deformation state, does not depend on the shape and size of the specimen, nominal stress and crack sizes, and is determined by the structure of material, ambient temperature and the rate of deformation.

Many countries have worked out standards regulating the procedure of determination of the fracture toughness. The tests are performed on the specimens of four types having different shapes and sizes. A sharp notch is made on the surface of the specimen, from the corner of which a fatigue crack develops. The specimen with the fatigue crack is stretched or bent to the point of destruction. During the test, the applied load P and the crack displacement V are registered, and the fracture diagram for the specimen is plotted.

From the fracture diagram the nominal load on the specimen P_Q is determined. From the values of P_Q and the average crack length the stress intensity coefficient K_Q is calculated. If the value of K_Q satisfies the relation characterizing unstable development of the crack in the conditions of planar deformation state:

$$t_B \geq 2.5\left(K_Q / \sigma_{0.2}\right)^2, \tag{2.35}$$

then the value of K_Q is taken as the true value of the critical coefficient of stress intensity K_{Ic}. If this relation is not satisfied, the nominal value of intensity coefficient K_Q is equal to the critical coefficient of intensity K_C for a specimen of the given thickness under the maximum load. To find the stress intensity coefficient K_{Ic} for this material, a specimen of a higher thickness is required.

The main difficulty with the determination of the crack resistance parameter K_{Ic} is the strict requirements on the dimensions of the specimen for which the destruction occurs in the conditions of the planar deformation state. Testing of moderate and low strength materials requires a substantial increase in the size of the specimen and application of powerful testing equipment.

A large body of available results for testing of massive specimens demonstrates that the parameter K_{Ic} most accurately characterizes the construction strength of the material. Unfortunately, determination of fracture toughness for coatings by the standard method is impossible because too large specimens should be tested. At the same time, several method for determination of K_{Ic} which do not require an initial crack in the specimen have been suggested, in particular, the method of micro-indentation, yielding the K_{Ic} of a thin surface layer of the part or its coating. Evans et al. have treated theoretical and applied aspects of this problem in their numerous papers.

The local load of the specimen in the course of testing is performed on Vickers hardness meter or on a microhardness meter equipped with a Vickers pyramid. To exclude the influence of the material of the substrate on the accuracy of the test results, the thickness of the coating must exceed the indentation depth and the diagonal of the print at least by a factor of 10 and 1.5, respectively.

Upon impressing the Vickers indenter into a brittle material the cracks develop around the indentation site. The cracks can be divided into three classes: radial, surface cracks developing from the corners and sides of the print, and lateral cracks propagating parallel to the surface. The simplest method is determination of K_{Ic} on the stage of spreading of a radial crack (Fig. 2.18).

In this case the expression for determination of the stress intensity coefficient (crack resistance) (in MPa·m$^{1/2}$) has the following form:

$$K_{Ic} = 0.074\, P / l_{cr}^{3/2}, \tag{2.36}$$

where P is the load on the indentor, N; l_{cr} is the length of the radial crack, m. The expression was obtained with the assumption that the length of the radial crack exceeds the diagonal of the imprint at least 2.5 times.

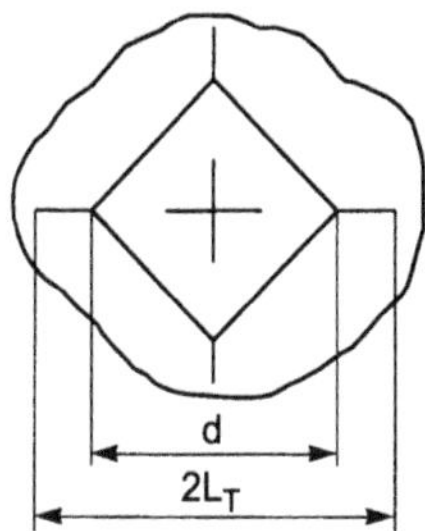

Fig. 2.18. A print from the Vickers indenter with radial cracks in the coating

The checks carried out by standard methods for determination of K_{Ic} confirmed the validity of the suggested approach. For brittle materials the ratio $2l_{cr} / d = 2.5$ can be satisfied for moderate indenter loads, below 4.905 N, which can be provided by a microhardness meter. More plastic materials require higher indentation loads which can be produced by a Vickers hardness meter, and the diagonal length of the print and the crack length is measured with a microscope. To get a sharper image of the crack, the surface of the specimen is ground before impressing the indenter, and is slightly etched in 5% alcohol solution of nitric acid after the printing.

Developing this method, Hainsworth et al. [2.32] used a scanning microscope to study deformation and formation of cracks in the indentation zone of a pyramid under very low loads. The results were analyzed for coatings having thickness 0.1–1.0 μm, taking into account the elastic and plastic properties of the coating and the base metal.

Sglabo and Dal Maschio have extended the application of the mechanics of destruction on measurements of adhesion by indentation of a Vickers pyramid at the interface between the coating and the substrate [2.72].

2.2.12 Plotting the Construction Strength Diagrams

Construction strength is a generally accepted measure of the reliability of parts in the given operational conditions. Construction strength is estimated from a complex of different mechanical properties determined under static and dynamic testing: breaking and yield points, fatigue strength, fracture toughness, impact elasticity, etc.

For metallic coating, the most convenient representation of the construction strength diagram can be given on the coordinates fracture toughness K_{Ic} – yield point σ_Y. Such diagrams have been successfully used for estimation of the construction strength of materials which are similar to coating in their physical-chemical properties. Besides, among the entire spectrum of mechanical properties, it is these characteristics of metallic coating that can be easily determined using readily available techniques.

Indeed, determination of the crack resistance coefficient K_{Ic} by the microindentation technique does not cause any difficulties and became one of the most common in the estimation of mechanic characteristics. As for the yield point, testing of many coating for tension poses certain difficulties because of small thickness of the coating and the complexity of the procedure of separating the coating from the substrate. However, the existing correlations between the strength and other characteristics that can be easily and reliably determined allow indirect estimation of the yield point. One of these equations relates the yield point σ_Y to the hardness by Vickers HV as follows:

$$HV = \sigma_Y \left[0.28 + 0.60 \left(\frac{3}{3-\lambda} \right) \ln \frac{3}{\lambda + 3\mu - \lambda\mu} \right], \qquad (2.37)$$

where $\mu = (1+\nu)\sigma_y/E$; $\lambda = 6(1-2\nu)\sigma_y/E$; E is the elastic modulus; ν is the Poisson coefficient.

This relationship is valid for materials having radial shift of substance from the point of indentation without build up on the surface and is applicable to high-strength coatings. Using Eq. 2.37, the dependence of HV versus σ_Y for different values of ν can be plotted in coordinates $HV/E - \sigma_y/E$ (Fig. 2.19a). Careful examination shows that the elastic moduli for commonly used metal and alloy coatings lie in the interval 196–224 GPa, and the Poisson coefficients – in the range 0.15–0.25, covering a rather narrow locus on the plot. Thus, the plot allows to easily find σ_Y from the known value of HV with an accuracy about 10%.

After the intensity coefficient K_{Ic} and the yield point σ_y have been found by the described techniques, the construction strength diagram is plotted.

The experimental diagrams of construction strength have been plotted for electrodeposited Fe, Fe–C, Fe–Mo, Fe–W, and Fe–V coatings subjected to different thermal treatment. As an example, a construction strength diagram of a Fe–Mo alloy is shown in Fig. 2.19. The diagram helps in choosing the regime of thermal treatment that would provide the desired combination of strength and crack resistance. It is worth noting that annealing at 400 °C dramatically reduces the yield point σ_y and increases the stress intensity coefficient K_{Ic}.

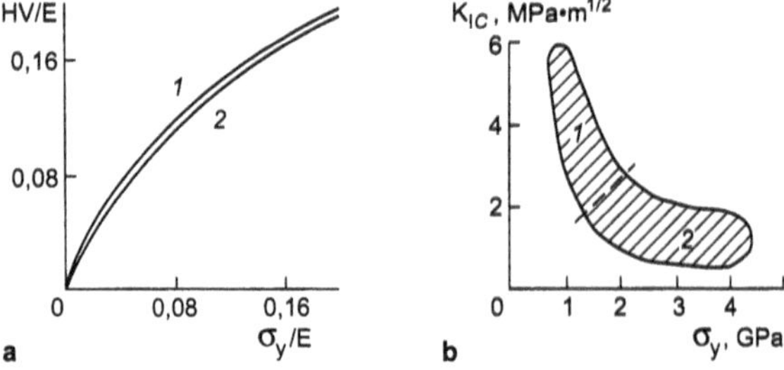

Fig. 2.19. Relationship between HV/E и σ_y/E for the Poisson coefficient ν in the range from 0.15 (*1*) to 0.25 (*2*) (**a**), and the diagram of construction strength of electrodeposited Fe–Mo coating annealed at temperatures above 400 °C (*1*) and below 400 °C (*2*) (**b**)

Plotting of construction strength diagrams opens new possibilities in estimation of mechanical and working properties of coatings. Only the first steps have been made in this direction, but the potential of such an approach is rather promising.

2.3 Bending and Tensile Strengths

2.3.1 General Overview

Bending and tensile strengths are the stresses corresponding to maximum loads that a coating may withstand in tests for bending and tension, respectively. It may seem that the strength of a coating is an auxilary characteristic, because coatings do not bear their own mechanical stresses. However, the strength of a coating is an indirect indication of its wear resistance [2.45], which is important, since the tests of wear resistance are usually long and tedious. Besides, the knowledge of coating strength helps estimate the role of coating in the strength of the entire metal–coating composition, which often experiences considerable mechanical loads [2.68].

The knowledge of the ultimate strength is also required to calculate the intrinsic stresses from the results of testing and to estimate the probability of the coating fracture.

As opposed to massive specimen and plastic metallic coatings, for thin and brittle ceramic coatings the surface-to-volume ratio becomes an important factor. Surface defects, acting like stress concentrators, become more and more important. The role of defects increases for thinner coatings.

The defects in gas-thermal coatings are formed due to buildup of thermal stresses as they reach a level exceeding the breaking point of the coating material. Formation of defects takes place predominantly in the zones with structural irregularities having the weakest binding between the particles.

The residual deformation in metal coatings is most probably caused not by plastic flow as for compact metals, but rather by generation and growth of microcracks.

The breaking points of metallic coatings are in the range 15–200 MPa, being 5–10 times lower than the corresponding values for the compact metal, which can be explained by their structural peculiarities. The main reasons for lower ultimate strengths of coatings were first formulated by Kudinov and Ivanov [2.50], and later this list was extended by Barvinok [2.4]:

1. The total area of zones between the particles where strong bond have been formed is only a fraction of the total contact area.
2. The strength of a welding spot is lower than for a compact material due to macro and micro defects of the binding.
3. A lower density of the coating (80–95%) as compared to the density of compact material.
4. High intrinsic stresses developed in the material in the process of deposition.
5. Absence of recrystallization processes at the boundaries of the welded particles.

6. The presence of pores, oxide inclusions, cracks and other defects which can work as concentrators of stresses.

Information on the techniques for determination of the ultimate strengths of thin coating is scarce and scattered, which complicates their study and practical applications. The most widely used are the methods for determination of the coating breaking ultimate strength by three-point bending and single-axis tension [2.14, 2.42, 2.62]. The compressive strength and the yield point are determined less frequently [2.57].

The former method is applied mainly for testing of brittle inductile coatings. The latter one can be used for any coating. The ultimate strength for brittle non-metallic coatings characterizes their fracture resistance, while for plastic metallic coatings it provides a measure of plastic deformation resistance.

The breaking strengths of a coating for different schemes of loading (tension, bending, compression) are different. For example, the compressive strength is several times higher than the tensile strength.

The reported results of mechanic testing have large scatter of experimental values, since the tests require microsamples which are difficult to prepare, may have inhomogeneous structure, microcracks, local porosity, etc. Therefore the tests are usually performed with at least 6–10 specimens.

2.3.2 Determination of Bending Strength

In this method the limiting value of the fracture load applied to the middle of a prismatic specimen resting freely on supports is determined, and then the required characteristics are calculated.

The standard methods of mechanical testing of materials are of little use for estimation of coating strength. Their small thickness and difficulty with mechanical treatment pose substantial problems in the preparation of the specimens.

The specimens with the length up to 80 mm, width up to 8 mm and thickness of 2–5 mm are prepared by either cutting (see Fig. 2.4 above) or depositing on a removable mandrel. The following device is recommended for preparing a batch of 24 identical specimens (Fig. 2.20). The device consists of a roll fixed in a chuck of a lather or other spinning device, and two disks with cartridges for 24 graphite bars which serve as substrates for deposition of coating that can be easily removed afterwards. Murakami et al. [2.62] deposited a plasma coating on a cooled rotating steel plate. The base metal can be coated by copper, graphite or asbestos. Then the coating was separated from the substrate and cut into flat specimens.

Special care should be taken in treating the planes of brittle ceramic coatings, since the state of their surface significantly effect the results of testing.

Since the ultimate strengths for sufficiently plastic metallic coatings are calculated from the results of measurements of the actual width and thickness of the specimen at the place of fracture, the cross-section of the specimen can vary in a wide range (±20%), and the requirements on the accuracy of preparation of metallic coatings and the roughness of their surface are lower than for ceramic coatings. Only deep defects, as cracks, chips, and burrs must be avoided.

The recommended dimensions of the specimen are 3.5x5.0x50 mm (a reference specimen for certification of modern ceramics by mechanical properties) [2.30]. The specimen is loaded in three- or four- point bending scheme [2.67] with the distance between the supports equal to 40 mm, which simplifies the construction of the loading device that can be placed inside a heater and is thus convenient for determination of the temperature dependence of the ultimate strengths.

Specimens with two times smaller cross-sections loaded in the three-point bending scheme with the distance between the supports equal to 20 mm are recommended for testing of very strong ceramics or when only small specimens are available.

Micromachines for studies of mechanical properties (as well as for determination of elastic properties) should provide the following functionality: registration of small loads and deformations; reliable installation and precise centering of specimen in holders; low rate of loading.

The loading scheme for three-point bending is shown in Fig. 2.2. In this case the bending strength ($\sigma_b{}^n$) in MPa can be calculated from the following expression that takes into account the horizontal component of the bending torque:

$$\sigma_b^n = \frac{3PL_V}{2bh^2}\left(1+\frac{4Z^2}{L_V^2}\right), \tag{2.38}$$

where P is the maximum load, N; L_V is distance between the supports; b is the width of the specimen, mm; h is the thickness of the specimen, mm; Z is the bending flexure in the middle of the interval between the supports in the moment of destruction, mm.

2.3.3 Determination of Tensile Strength

In this method a coating deposited on a special specimen is torn, and the tensile strength is calculated taking into account the size of the specimen and the thickness of the coating.

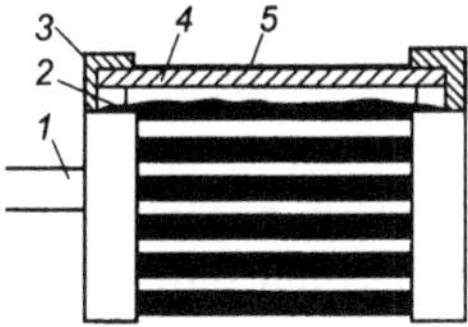

Fig. 2.20. Device for depositing specimens. *1* roll; *2* disk; *3* cartridge; *4* graphite bar; *5* coating

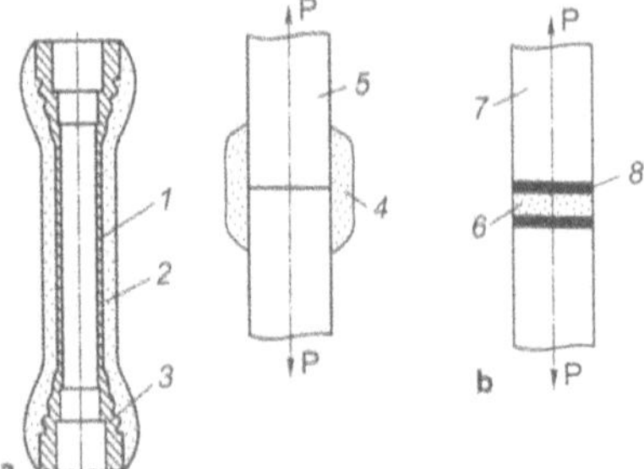

Fig. 2.21. Specimens for tension tests in longitudinal (**a**) and transverse (**b**) directions. *1*, *4*, *6* coating; *2* thickening on the ends of the tube; *3* tube; *5*, *7* mandrels; *8* glue

In the scheme of single-axis stretching, several types of specimens are used for estimation of strength in the longitudinal and transverse directions (Fig. 2.21). The tensile strength in the longitudinal direction and the relative extension for different metallic coatings are estimated on tubular specimens. A steel tube (Fig. 2.21a), having thickenings at the ends and internal thread, is covered with a coating. To localize the place of fracture, a 0.05 mm thick notch is cut in the central part of the coating.

After the tube is removed, the coating specimen is fixed into the holders of a tearing machine. To facilitate removing of the tube, it is recommended to put a thin layer of sodium chloride on the surface before deposition, which can later be dissolved in water.

In another method of testing the strength in the longitudinal direction, the coating is deposited on cylindrical surface of two mandrels tightly pressed to each other.

The tensile strength in the transverse direction is determined by the glue technique. The coating, separated from the base metal, is glued at both sides to two mandrels (Fig. 2.21b). The cohesion strength can be determined if the destruction takes place at the coating rather than at the glue. In the latter case the only meaningful conclusion is that coating strength is higher than the glue strength.

In Japan a standard specimen for testing the coating consists of two bushes A and B (Fig. 2.22). Khasui and Morigaki [2.44] propose to thread the cylindrical parts which would increase the binding between the coating and the base metal. The assembled specimen is fixed with a finger which is a part of the A bush and slides into the matching opening at the end surface of the B bush. To prevent inadvertent disassembling of the parts the bushes are fastened together with a screw which is removed after the deposition of the coating.

After assembling the specimen is ground along the generatrix of the cylinder followed by shot blasting. The quality of the specimen is determined by the gap between the two bushes. If after cold hardening by shot blasting the line separating the two bushes disappears, the specimen is considered suitable for tests. These precautions prevent accumulation of residual stresses in the zone of destruction of the coating.

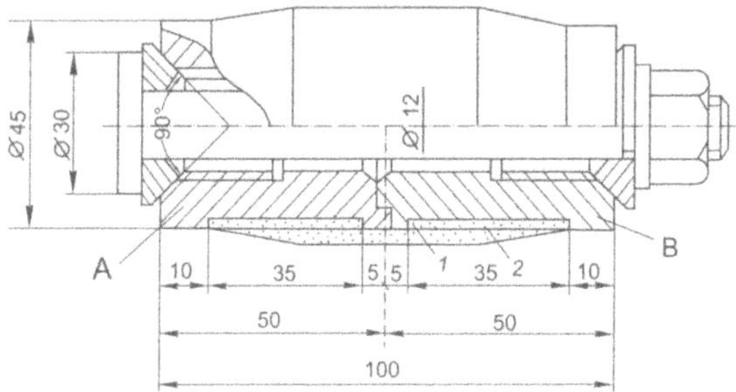

Fig. 2.22. Specimen for tension tests of coatings (adopted from [2.44]). *1* coating; *2* thread

To test the tensile strength of the coating, the screw is removed, and the specimen is put into the tearing machine and is loaded until destruction along the annular deposited layer.

The tensile strength (σ_b^p) in MPa can then be defined from the equation:

$$\sigma_b^p = \frac{4P}{\pi\left(D^2 - d^2\right)}, \tag{2.39}$$

where P is the maximal load, N; D is the diameter of the central part of the cylindrical specimen after depositing the coating (including coating thickness), mm; d is the diameter of the central part of the cylindrical specimen prior to deposition, mm.

The coating has higher strength when the stretching force is directed parallel to the deposited surface. If the stretching force is directed normal to the surface, the tensile strength is only 1/5 to 1/10 of the strength of the first case. The tensile strength for forces applied at right angle, depends mainly on the time interval between the deposition of successive layers, on the deposition productivity of the apparatus, and on the quality of the substrate preparation. The smaller the interval between depositions, the higher the productivity of the apparatus and the roughness of the substrate, the higher the tensile strength of the coating.

In his comprehensive review [2.34] Hardwick has considered the methods of mechanical testing of thin foils under single-axis stretching. This scheme of loading gives easy for interpretation results and provides opportunities for study of the destruction process. However, the procedure of preparing the specimens for single-axis tensile testing of foils is rather complex, and a strong "edge" should be taken into account.

A device for stretching tests of thin film coatings was developed by Iljinsky and Lyakh [2.38] on the basis of analytical scales with provisions for automatic recording of the tension diagram (Fig. 2.23).

The loads up to 10 N are produced by an electromagnet and measured with the accuracy of ±10%. The deformation is registered by an inductive sensor. The specimen is fixed in a holder and carefully centered. The tension diagram is first recorded by an X–Y recorder on coordinates "electric current–displacement" and then transformed into the coordinates "load–extension".

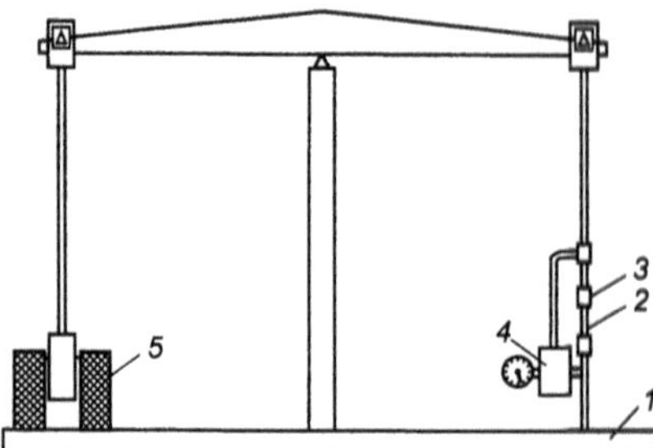

Fig. 2.23. Device for tension tests of microspecimens (adopted from [2.38]). *1* scales; *2* specimen; *3* holder; *4* displacement sensor; *5* loading electromagnet

Several researchers have attempted to apply the method of acoustic emission for study of the kinetics of coating destruction [2.84]. The method provides an opportunity of localization and identification in space and time of continuousness failure for bending and extension. The main experimental parameters are the signal amplitude which reflects an increase of the linear size of the defect, and signal intensity, i.e., the number of elementary acts of redistribution of stress fields per unit time.

2.3.4 Results and Discussion

Analyzing the mechanical properties of plasma-sprayed coatings (nichrome, Ni+16%BN; nichrome+25%NB; Al+25%BN) Barvinok [2.4] demonstrated that the bending strength, depending on technological regimes, can vary in the range from 18 up to 230 MPa. The key factors that determine the value of σ_B are the arc current, granulation of the powder, and the distance of sputtering. A variation of arc current from 300 up to 500 A changes the ultimate strength by a factor of 2.1. A change in powder granulation from 200–315 down to 40–70 μm or a change in the sputtering distance from 200 mm down to 80 mm increases σ_B by a factor of 1.6.

The complex character of the effect of technological regimes on the coating strength is caused by several factors inherent in plasma spraying [2.4], such as the different energy states of the particles, their velocity and degree of spheroidization, the fraction distribution of the particles. The strength of the coating also depends on the temperature, which is settled in the deposited layer, and on the time for which the layer remains at elevated temperature.

The temperature and duration of plasma spraying determine the probability of return processes, recrystallization, and thermoplastic deformation.

In general, all technological factors increasing the enthalpy of the plasma jet and thus the degree of melting of the sputtered particles produce a stronger metallic contact between particles and yield a stronger coating.

Fedorenko et al. [2.27] have generalized the results of studies on the influence of structure-phase transformations in detonation-sprayed coatings on their strength for the case of testing in three- and four-point loading schemes. The strength of the coatings with thickness up to 0.25 mm is high enough, and the character of material fracture is almost identical to the fracture of a compact material. The structural and phase heterogeneity, high level of intrinsic stress for a thicker coating (>0.25 mm) reduces the plastic properties of the coating, rendering it more brittle as compared to the compact counterpart.

The authors of [2.1] correlate the structural parameters of a multilayer coating with its strength. The laminate microfilm composites Al/Al_2O_3 were obtained by deposition of aluminum in vacuum with periodic leaking of oxygen into the vacuum chamber. The overall thickness of the composite was 25 μm, the thickness of Al_2O_3 layers was 5 nm, and aluminum layers - from 50 to 500 nm. The tests were performed at room temperature, at 577 °K and 683 °K. The crucial factor determining the strength of the composite was the distance between the layers of Al_2O_3. The highest yield point σ_y of 450 MPa is reached for the composite with the distance between the layers λ equal to 50 nm. The dependence of of σ_y on λ is described by Petch's equation.

A model developed by Eaton and Novak [2.25] (using $80\%ZrO_2$ and $20\%Y_2O_3$ plasma-sprayed coating as an example) assumes that the measured strength characteristics of specimens are the functions of mechanical interaction of particles only. The metallographic analysis indicates that this may really be the case, because the boundaries between the particles look rather sharp. Besides, relatively low levels of strength and nonlinear behavior of the loading diagram imply the absence of welding. The determined range of the ultimate strength is from 14.7 up to 53.8 MPa.

The Weibull coefficient, reflecting the scatter of experimental data and the distribution of cracks of different sizes, has also been suggested as a measure of the coating strength [2.63]. The strength of the following plasma-sprayed coatings was thus tested: Al_2O_3, Al_2O_3–2.5%TiO_2, ZrO_2–5% CaO, ZrO_2–6% Y_2O_3. The studies show that Weibull coefficient provides a rather accurate measure of the quality of coatings.

2.4 Impact Strength

2.4.1 General Overview

Impact strength characterizes the resistance of the coating material to dynamic impact and is determined by the energy required for fracture upon a momentary (shorter than 10^{-3} s) shock loading. Impact strength can be used for testing specimens both having a cut (concentrator) or without the cut in the temperature range from -100 up to +1000 °C. The measured values are the total work required for coating fracture, and the impact strength itself, i.e., the total work divided by cross-section of the specimen at the concentrator. The principle of testing is destruction of a specimen freely resting on supports by dynamic loading. The obtained characteristics allow estimation of fragility and viscosity of the coating under dynamic impact for the given testing conditions.

2.4.2 Specimens

The size and shape of the specimens used for tests are shown in Fig. 2.24. The specimens in the form of rectangular bars from the coating material are prepared by one of the techniques described earlier. When cutting the specimens anisotropy of the coating should be taken into account (see Fig. 2.3).

Nonmetallic inorganic coatings can be tested on specimens without stress concentrators. If required, the stress concentrator (a notch) is cut simultaneously on several specimens bundled together by an emery grinder or a skive after finishing the narrow lateral sides. Good results can be obtained if the notch is cut by a single-tooth mill. In this case the tooth profile should be controlled, because it determines the shape and size of the concentrator on a specimen. However, milling is of little use for specimens of hard and brittle coatings.

Taking into account the factors like alignment of the direction of impact relative to positions of the supports, coating anisotropy and the position of the concentrator notch, it is recommended to prepare two sets of specimens (Fig. 2.25). Each set should include at least three specimens for a given technology of coating deposition and the direction of the cut.

2.4.3 Equipment

The scheme of testing for shock bending is shown in Fig. 2.26. The specimen is fixed in the way of a pendulum having weight G, which is lifted at the height H_1 (angle α). The falling pendulum crashes the specimen and raises to the height H_2 (angle β).

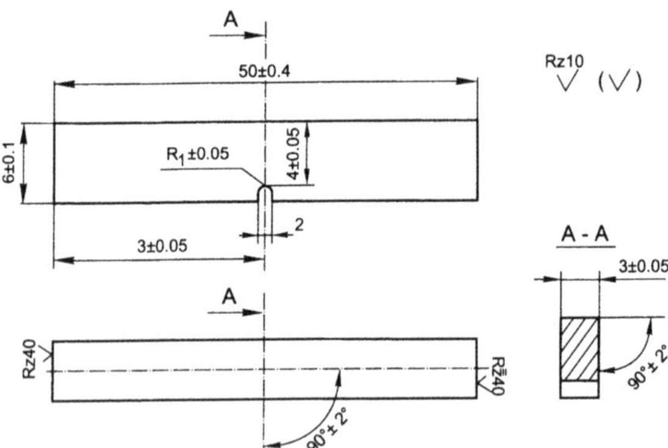

Fig. 2.24. Specimen for measuring impact strength of a coating

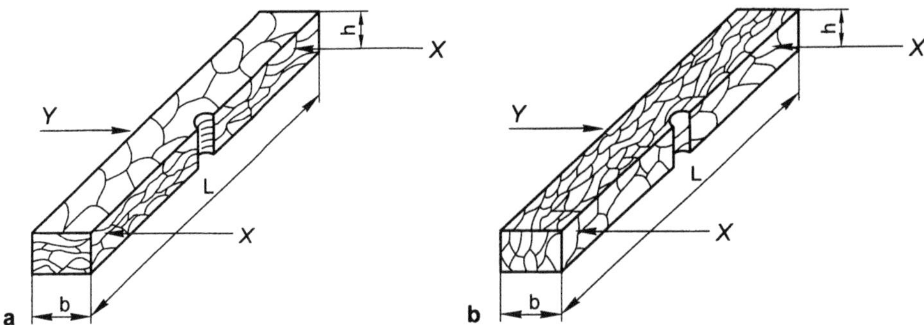

Fig. 2.25. Variants of cutting and mounting specimens taking into account anisotropy of a coating. Cutting: **a** longitudinal; **b** transverse. Directions: *X* reaction at supports; *Y* the impact

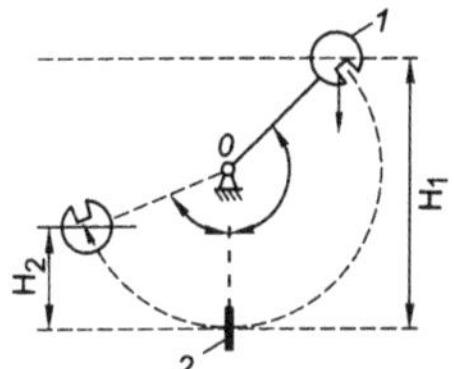

Fig. 2.26. Scheme of testing for shock bending

The height H_2 is smaller than H_1 because a part of the pendulum energy was spent on destruction of the specimen. The work *K* for destruction is equal to:

$$K = GH_1 - GH_2 = G(H_1 - H_2). \tag{2.40}$$

Taking into account the characteristics of the studied object (small cross-section and mostly brittle character of coating destruction), a pendulum hammer with low energy content of the type used in the investigation of dynamical properties of plastics is recommended for testing. A removable hammer with the energy content 2–3 times higher than the expected fracture energy can also be used. As a useful option the pendulum hammer can also be equipped with a thermal chamber, thus allowing estimation of impact strength in the temperature range from -90 to +300°C.

Landa et al. [2.54] have designed an original installation for measuring the impact strength of ceramic specimens, capable of testing the materials with low fracture energy at impact velocities 0.1–4.0 m/s and temperatures up to 1600 °C. The impact is delivered by a free falling load.

In tests with commercial pendulum hammers the specimen can be cooled or heated to the desired temperature by appropriate add-on devices (furnace, thermostat, cooling chamber, etc.). A commonly used cooling agent is a mixture of liquid nitrogen (or solid carbon dioxide) with alcohol, petroleum-ether, or benzine.

2.4.4 Preparation for Testing

Since the impact strength of a porous material is sensitive to the atmosphere humidity, the dried specimens must be kept into dry closed dessicators, plastic bags or other packaging that can maintain a constant humidity. The transverse dimensions of the specimen are measured in three points and averaged. A micrometer with the tip shaped to the profile of the concentrator is used for measuring the width at the position of the concentrator. A hammer providing the required ratio of the expected fracture energy to the maximum energy content is chosen and installed. The temperature of testing is fixed at the level depending on the operational conditions of the coated part.

2.4.5 Testing

The heated or cooled specimen is quickly transferred on the supports. The pendulum is released by turning the stopper lever. The hammer then delivers the impact to the side opposite to the concentrator notch; the impact line must pass through the concentrator and between the supports, normal or parallel to the coating layers. The value of the work for destruction is read from the scale, and the impact strength of the coating KC in J/cm^2 is calculated from the equation:

$$KC = K/S_0, \tag{2.41}$$

where K is the work for destruction, J; S_0 is the specimen net cross-section under the concentrator notch, cm^2.

Only results obtained for specimens fabricated using identical technology, with identical profiles of the concentrator notch, and tested at the same temperature can be compared.

2.4.6 Results

A characteristic feature of impact strength testing of high-strength coatings is low values of the impact strength and absence of any explicit dependence of the impact strength on the testing temperature (Table 2.2). Tests performed at room temperature demonstrated that KC for plasma-sprayed coatings PN85U15 with per cent composition Al – 15; Ni – 85, and PG–SP4 with per cent composition B – 3.0; C – 0.8; Si – 3.9, Cr – 16.0, Ni – the remainder is a factor of 45–80 lower than the impact strength of Steel 45 (0.45% C).

The study of fracture surfaces for coatings PN85U15 and PG–SP4 did not reveal any variations in surface pattern when changing temperature from +50 to –60 °C. In all studied specimens, the signs of both brittle and ductile fractures were observed.

Table 2.2. Impact strength of plasma-sprayed coatings, MJ/m^2 [courtesy of Bataev A.A.]

	Temperature of the test [°C]								
Material	-196	-60	-40	-20	0	+20	+50	+100	+500
Steel 45	–	0.18	0.24	0.36	0.44	0.63	0.60	–	–
PN85U15	0.016	–	–	–	–	0.014	–	0.019	0.015
PG–SP4	0.008	–	–	–	–	0.008	–	0.008	0.012

Castro et al. [2.13] have studied the impact strength of $MoSi_2$ and $MoSi_2$–Ta coatings. It was found that for these samples after the dynamic impact the cracks propagate through the interface between the layers.

References

2.1 Alpas AT, Embury JD, Hardwick DA, Springer RW (1990) The mechanical properties of laminated microscale composites of Al/Al_2O_3. J Mater Sci 3:1603–1609

2.2 Anast M, Bell JM, Bell TJ, Ben-Nissan J (1992) Precision ultra-microhardness measurements of sol-gel-derived zirconia thin films. J Mater Sci Lett 11:1483–1485

2.3 Attar F, Johannesson T (1995) Adhesion and X-ray elastic constant evaluation of CrN coatings. Thin Solid Films 258:205–212

2.4 Barvinok VA (1990) Stressed state control and properties of plasma-sprayed coatings (in Russian). Mashinostroenie, Moscow

2.5 Benarioua Y, Bouaouadja N, Wendler B (1996) Carbide formation in titanium coatings deposited on carbon steel. J Mater Sci Lett 15:1067–1069

2.6 Bianco R, Rapp RA, Smialek JL (1993) Chromium and reactive element modified aluminide diffusion coatings on superalloys: Environmental testing. J Electrochem Soc 140:1191–1203

2.7 Bull SJ, Rickerby DS, Gent JT (1991) Quality assurance assessment of thin films. Surface Eng 7:143–145

2.8 Bulychev SI (1992) Progress and prospects of material testing by continuous impressing of indentor (in Russian). Zavodskaya Laboratoria 3:29–36

2.9 Bulychev SI, Alekhin VP (1990) Testing of materials by continuous indentation (in Russian). Mashinostroenie, Moscow

2.10 Burnett PJ, Rickerby DS (1987) Assessment of coating hardness. Surface Eng 3:69–76

2.11 Byakova AV (1995) Peculiarities of microhardness determination in the estimation of structural strength of coatings (in Russian). Problemy Prochnosti 9:44–54

2.12 Byakova AV, Gorbach VG (1994) Strength to deformation and estimation of brittle strength of coatings with initial field of residual stresses (in Russian). Problemy Prochnosti 1:51–57

2.13 Castro R, Chard G, Smith Ronald W, Rollett Anthony D, Stanek Paul W (1992) Toughness of dense $MoSi_2$ and $MoSi_2$/tantalum composites produced by low-pressure plasma deposition. Scr Met and Mater 26:207–212

2.14 Chen HC, Pfender E (1996) Microstructure of plasma-sprayed Ni–Al alloy coating on mild steel. Thin Solid Films 280:188–198

2.15 Chou TC, Nich TG, McAdams SD, Phar GM (1991) Microstructures and mechanical properties of thin films of aluminum oxide. Scr Met and Mater 25:2203–2208

2.16 Chou TC, Nich TG, McAdams SD, Phar GM, Oliver WC (1992) Mechanical properties and microstructures of metal/ceramic microlaminates. Part 2. Mo/Al_2O_3 system. J Mater Res 7:2774–2784

2.17 Chou TC, Nich TG, Tsui TY, Phar GM, Oliver WC (1992) Mechanical properties and microstructures of metal/ceramic microlaminates. Part 1. Nb/$MoSi_2$ system. J Mater Res 7:2765–2773

2.18 Crostack HA, Jahnel W, Meyer EH, Selvadurai U (1996) Developments of nondestructive testing of surface coatings. Weld World 37:114–120

2.19 Czerwinski F (1998) Diffusion annealing of Fe–Ni alloy coatings on steel substrates. J Mater Sci 33:3831–3837

2.20 Dekhtyar LI, Loskutov VS, Zilberman BV (1982) Determination of elastic moduli in heterogeneous materials (in Russian). Fizika I Khimia Obrabotki Materialov 11:11–14

2.21 Deuis RL, Yeliup JM, Subramanian C (1997) Aluminum composite coatings produced by plasma transferred arc surfacing technique. Mater Sci and Technol 13:511–512

2.22 Diao DF, Sawaki Y (1995) Fracture mechanisms of ceramic coatings during wear. Thin Solid Films 270:362–366

2.23 Dolgov NA, Lyashenko BA, Veremchuk BS, Dmitriev YuV (1995) On the determination of elastic characteristics of protective coatings (in Russian). Problemy Prochnosti 7:48–51

2.24 Du HL, Datta PK, Lewis DB, Burnell-Gray JS (1996) Enhancement of oxidation/sulphidation resistance of Ti and Ti–6Al–4V alloy by HfV coating. Mater Sci and Eng A 205:199–208

2.25 Eaton H, Novak R (1986) A study of the effects of variations in parameters on the strength and modulus of plasma- sprayed zirconia. Surface and Coating Technology 3:257–267

2.26 Engel PA, Chitsaz AR, Hsue EY (1992) Interpretation of superficial hardness for multilayer platings. Thin Solid Films 207:144–152

2.27 Fedorenko VK, Milman YuV, Kadyrov VK, Ivaschenko RK, Ivaschenko OV (1987) Physical-mechanical properties and destruction of powder detonation coatings (in Russian). Poroshkovaya Metallurgia 11:88–94

2.28 Friesen T, Haupt J, Gissler W, Bata A, Bata PB (1991) Ultrahard coatings from Ti–BN multilayers and by co-sputtering. Surface and Coating Technology 48:169–174

2.29 Gogotsi GA (1991) Deformation behavior of ceramics. J Eur Ceram Soc 1:87–92

2.30 Gogotsi GA, Zavada VP (1994) Certification of mechanical properties of modern ceramics (in Russian). Problemy Prochnosti 1:68–75

2.31 Guo Zhongcheng, Yang Xianwan, Liu Hong-Kang, Wang Zhiuin, Wang Min (1996) Characteristics of Re–Ni–B–Al_2O_3 coating material. Trans Nonferrous Metals Soc China 6:33–36

2.32 Hainsworth SV, Bartlett T, Page TF (1993) The nanoindentation response of systems with thin hard carbon coatings. Thin Solid Films 236:214–218

2.33 Hankock P, Chien HH, Nicholls JR, Stephenson DJ (1990) In-site measurements of the mechanical properties of aluminide coatings. Surface and Coating Technology 43–44:359–370

2.34 Hardwick DA (1987) The mechanical properties of thin films: a review. Thin Solid Films 154:109–124

2.35 Harmsworth PD, Stevens R (1992) Phase composition and properties of plasma-sprayed zirconia thermal barrier coatings. J Mater Sci 27:611–615

2.36 Heinke W, Leyland A, Matthews A, Berg G, Friedrich C, Broszert E (1995) Evaluation of PVD nitride coatings using impact, scratch and Rockwell-C adhesion tests. Thin Solid Films 270:431–438

2.37 Huang Chi-Tung, Dun Jeng-Gong (1997) Microhardness of (Ti, Al)N films deposited by reactive RF magnetron sputtering. J Mater Sci Lett 16:59–61

2.38 Ilinsky AI, Lyakh GB (1978) Methods of mechanical testing for films and foils (Review) (in Russian). Zavodskaya Laboratoria 12:1507–1511

2.39 Jankowski A (1995) Metallic multilayers at the nanoscale. Nanostruct Mater 6:179–190

2.40 Jehn HA, Thiergarten F, Ebersbach E, Fabian D (1991) Characterization of PVD (Ti,Cr)N hard coatings. Surface and Coating Technology 50:45–52

2.41 Jindal PC, Quinto DJ (1988) Load dependence of microhardness of hard coatings. Surface and Coating Technology 3–4:683–694

2.42 Katayama Sakae, Hashimura Masayuki (1997) Effects of microcracks in CVD coating layers on cemented carbide and cermet substrates on residual stress and transverse rupture strength. Trans ASME J Manuf Sci and Eng 119:50-54

2.43 Kerkush IR, Melnik PI (1997) Properties improvement of plasma-sprayed iron coatings by chromium plating. Poroshkovaya Metallurgia 40:230

2.44 Khasui A, Morigaki O (1985) Hard Facing and Sputtering (in Russian). Mashinostroenie, Moscow

2.45 Kiiski AA, Ruuskanen PR, Rubin JB (1996) Wear resistant coatings produced by shock-wave compaction of powders. Met and Mater Trans A 27:2297–2304

2.46 Kobayashi A, Yamahiji K, Kitamura T (1991) Effect of heat treatment on high hardness zirconia sprayed coating by means of gas tunnel type plasma spraying. Trans JWRI 20:47–52

2.47 Kolawa E, Sun X, Reid JS, Chen JS, Nickolet MA, Ruis R (1993) Amorphous $W_{40}Re_{40}B_{20}$ diffusion barriers for <Si>/Al and <Si>/Cu metallizations. Thin Solid Films 236:301–305

2.48 Kovensky IM, Povetkin VV (1999) Physical metallurgy of coatings (in Russian). Joint Venture "Internet Engineering", Moscow

2.49 Krus D (Jr), Hoffman RW (1993) Finite element studies of tensile testing on thin film multilayers. Thin Solid Films 236:225–229

2.50 Kudinov VV, Ivanov VM (1981) Plasma deposition of high-melting coatings (in Russian). Mashinostroenie, Moscow

2.51 Kuhnemann S, Kopacz U, Jehn H (1987) Erfahrungen beim Einsatz eines Ultramikrohartetesters zur Prufung von dünnen Hartstoffschichten (in German). Prakt Metallogr 24:382–390

2.52 Kuroda S, Clyne TW (1991) The quenching stress in thermally sprayed coatings. Thin Solid Films 1:49–66

2.53 La Fontaine WR, Paszkiet CA, Korhonen MA, Li Che-Yu (1991) Residual stress measurements of thin aluminum metallizations by continuos indentation and X-ray stress measurement techniques. J Mater Res 6:2084–2090

2.54 Landa NI, Shpindler SS, Kolechkin YuK (1988) Installation for determination of impact viscosity of ceramics (in Russian). Zavodskaya Laboratoria 9:1138–1140

2.55 Lawrynowicz DE, Wolfenstine J, Lavernia EJ (1995) Reactive synthesis and characterization of $MoSi_2$/SiC using low-pressure plasma deposition and 100% methane. Scr Met and Mater 32:689–693

2.56 Lyashenko BA, Veremchuk BS, Dolgov NA, Ivanov VM (1996) Investigation of strength and strain properties of compounds with plasma deposited coatings (in Russian). Problemy Prochnosti 6:57–60

2.57 Ma Dejun, Xu Kewei, He Jiawen (1997) Numerical simulation for measuring yield strength of thin metal film by nanoindentation method. Trans Nonferrous Metals Soc China 7:66–68

2.58 Mahajan S, Wen JG, Ito W, Yoshida Y, Kubota N, Liu CJ, Morishita T (1994) Growth and superconductivity of C-axis in plane aligned $YBa_2Cu_3O_7$ films fabricated by the self-template method. Appl Phys Lett 65:3129–3131

2.59 Maksimovich GT (1974) Micromechanical studies of properties of metals and alloys (in Russian). Nauchnaya Mysl, Kiev

2.60 Mehrotra PK, Quinto DT (1986) High-temperature microhardness profiles of hard CVD coatings. High Temp–High Pressures 18:199–210

2.61 Monaghan DP, Teer DG, Laing KC, Logan PA (1994) The state-of-the-art in thin protective coatings. Surface Technol Int June:2–5

2.62 Murakami Kenji, Okamoto Taira, Matsumoto Miroshi, Miyamoto Yoshinari, Irisana Tsuyoshi (1993) Structure and mechanical properties of thermal-sprayed nickel–20 wt% chromium alloy. Mater Sci and Eng A 160:181–187

2.63 Müller D, Fromm E (1995) Mechanical properties and adhesion strength of TiN and Al coatings on HSS, steel, aluminum and copper characterized by four testing methods. Thin Solid Films 270:411–416

2.64 Ostojic P, Berndt CC (1988) The variability in strength of thermally sprayed coatings. Surface and Coating Technology 34:43–50

2.65 Perry AJ, Jagner M, Wolner PF, Sproul WD (1990) Aspects of residual stress measurements in TiN prepared by reactive sputtering. Surface and Coating Technology 1–3:234–244

2.66 Pisarenko GS, Lyashenko BA, Kozub YN (1997) Methods of high-temperature mechanical testing of inorganic heat-resistant coatings. In: Pisarenko GS (ed) Heat-resistant coatings for protection of constructional materials (in Russian). Nauka, Leningrad, pp 50–60

2.67 Quinn G (1989) Flexure testing of advanced ceramics. Br Ceram Trans J 3:94–95

2.68 Sampath S, Gansert R, Herman H (1995) Plasma-spray forming ceramics and layered composites. JOM: J Miner, Metals and Mater Soc 47:30–33

2.69 Sampath S, Tiwari R, Gredmundsson B, Herman H (1991) Microstructure and properties of plasma-sprayed consolidated two-phase nickel aluminides. Scr Met and Mater 25:1425–1430

2.70 Saunders SR, Vetters HR (1997) Standardization of test methods for the mechanical properties of thin coatings. Thin Solid Films 299:82–87

2.71 Seino Y, Hida N, Nagai S (1992) Mechanical properties of diamond thin films prepared by chemical vapor deposition. J Mater Sci Lett 11:515–517

2.72 Sglavo VM, Dal Maschio R (1990) Adhesion testing of plasma-sprayed ceramic coatings on metals by an indentation technique. Eur Appl-Res Repts Nucl Sci and Technol Sec 7:1487–1494

2.73 Smagorinski M, Tsantrizos P, Grenier S, Entazarian M, Ajersch F (1996) The thermal plasma near-netshape spray forming of Al composites JOM: J Miner, Metals and Mater Soc 48:56–59

2.74 Takeda Koichi, Ito Mitihisa, Takeuchi Sunao, Sudo Katui, Koda Masamichi, Kazama Koichi (1993) Erosion resistant coating by low-pressure plasma spraying. ISJI International 33:976–981

2.75 Tamura Motonori, Fukuda Kanao (1993) Properties and tribological behavior of Ti(C,N) coatings deposited by reactive ion plating. ISIJ International 33:949–956

2.76 Tushinsky LI, Sindeev VI, Plokhov AV (1996) The structure and mechanical properties of modified surfaces of engineering materials (in Russian). Novosibirsk State University Press, Novosibirsk

2.77 Vaidya R, Zurek AK, Castro R, Wolfenden A, Hosman SL, Subramanian KN (1994) Alumina coated Ti–25Al–10Nb–3V–1Mo for improved oxidation resistance. J Adv Mater 26:16–22

2.78 Vally JA (1986) A review of adhesion test methods for thin hard coatings. J Vac Sci and Technol A 4:3007–3014

2.79 Vally JA, Molarius JM, Korhonen AS (1987) The effect of nitrogen content on the critical normal force in scratch testing of TiN films. Thin Solid Films 154:351–360

2.80 Vancoille E, Celis JP, Roos JR (1993) Mechanical properties of heat-treated and worn PVD TiN, (Ti,Al)N, (Ti,Nb)N and Ti(C,N) coatings as measured by nanoindentation. Thin Solid Films 224:168–176

2.81 Voevodin AA, Spassky SE, Erokhin AL (1991) Determination of thin coating microhardness with consideration of its thickness and hardness of the substrate (in Russian). Zavodskaya Laboratoria 10:45–46

2.82 Wang HF, Yang X, Bangert H, Torzicky P, Wen L (1992) Two-dimensional finite element method simulation of Vickers indentation of hardness measurements on TiN–coated steel. Thin Solid Films 214:68–73

2.83 Wang M, Schmidt K, Reichelt K, Jiang X, Hübsch H, Dimigen H (1992) The properties of W–C films deposited by reactive rf sputtering. J Mater Res 7:1465–1472

2.84 Wen LS, Zhang HT, Zhou XK, Guan K, Liao B, Cao S (1987) Acoustic emission research on the fracture behavior of plasma-sprayed Ni–Al coatings during bend testing. Surface and Coating Technology 30:115–123

2.85 Whitehead AJ, Page TF (1992) Nanoindentation studies of thin film coated systems. Thin Solid Films 220:277–283

2.86 Zaitsev GG, Barabanov AF, Kuteinikov AF (1980) Determination of elastic constants of graphitized carbon materials by dynamic methods in the temperature range of 20–3000 °C (Review) (in Russian). Zavodskaya Laboratoria 1:81–86

2.87 Zhang N, Wang Y (1992) Dislocations and hardness of hard coatings. Thin Solid Films 214:4–5

3 Physical Properties of Coatings

3.1 Density

3.1.1 General Overview

Several kinds of porosity are distinguished as applied to porous materials depending on what quantity is taken and measured as the "volume of a solid body". The true density is defined as the ratio of the body mass m to the volume of its compact nonporous matrix V_r:

$$\rho_{true} = \frac{m}{V_r}. \tag{3.1}$$

This true porosity is a fundamental feature of a solid body; however, its real values can be estimated only for crystalline bodies of known composition invoking reliable X-ray structure data for calculation of crystalline lattice parameters provided that its defects can be disregarded. Therefore the true density is often called "X-ray" or "theoretical" density [3.80].

Calculation of the true density from X-ray data is impossible for most of porous materials with amorphous and mixed structure, while other density measurement techniques give values approaching ρ_{true} to only some extent. Usually the density obtained through the use of pore-filling picnometric substances is taken as the true one. Since the experimental values of this density in many respects depend on properties of a picnometric substance, in particular, on its infiltration ability, so this kind of density is called a picnometric density.

The density of a porous body, calculated from its total volume including the solid skeleton and pores, is denoted the apparent density:

$$\rho_{app} = \frac{m}{V}. \tag{3.2}$$

The apparent density of coatings is about 60–90% of the analogous compact material. This is caused by development of pores, formation of oxides, phase transitions and other physical-chemical phenomena accompanying deposition of a coating [3.36, 3.61, 3.72].

These picnometric and apparent densities are required for calculation of the total porosity and for analysis of chemical and phase transitions occurring during deposition of coatings and operation of products [3.46, 3.55].

3.1.2 Measurement of the Apparent Density by the Hydrostatic Weighing Method

The hydrostatic weighing method consists in detachment the coating from the base metal with its subsequent weighing first in air, then in a liquid that wets the coating but does not interact with it. The apparent density is calculated from the weighing results.

Once the coating is mechanically detached from the base metal, a test piece cut out of it must be polished, its sharp edges are smoothed down and the burrs are removed. The specimen mass should be 5 g or more, and at least three replicate specimens are manufactured.

The specimens must be dried up in a drying cabinet at the temperature of 110–135 °C. The specimen mass is considered to be constant if the subsequent weighing procedure carried out in an hour after completion of the drying process gives the difference less than 0.1% as compared to previous weighing. If the standard air humidity cannot be provided in the laboratory, the specimens must be put into a desiccator and taken out in turn immediately before weighing.

An analytical beam balance is recommended for weighing. A balance beam or quadrant scales can be used for weighing by the equalizing moment method.

The weighing device should be equipped with a special fitting for hydrostatic weighing. The following schemes can be used for either balance beam or quadrant scales (Fig. 3.1). In quadrant scales one of the pans is fixed on a shortened hanger 7, while a thin suspension string 3 fabricated of stainless steel is attached to it. The shortened hanger, suspension string, and pan are zeroed.

Weighing is also possible without shortening of the hanger. A support for the cylinder 3 (Fig. 3.2a) is put in such a way as to provide free vertical motion of a scale pan. When the cylinder is in the down position (Fig. 3.2b), a hook is attached to one of the pans from the underside, while a suspension string freely passing through the holes in the scale base and in the table is tied to the hook. The hook and suspension string are counterbalanced by a weight placed on another pan.

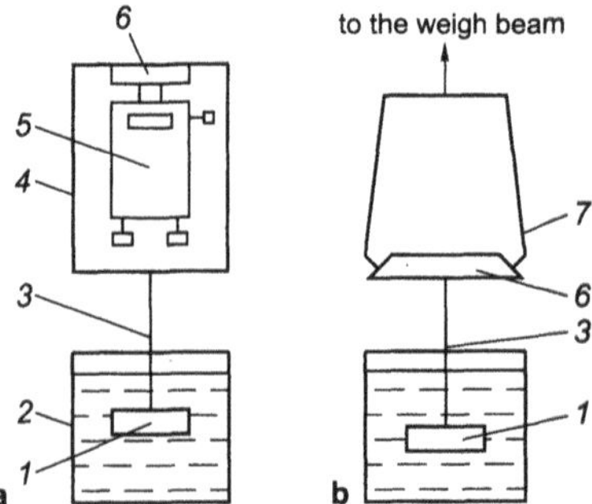

Fig. 3.1. Hydrostatic weighing setup with a shortened hanger: **a** using quadrant scales; **b** using balance-beam scales. *1* coating; *2* cylinder for hydrostatic weighing; *3* suspension string; *4* suspension frame; *5* quadrant scales; *6* pan; *7* shortened hanger

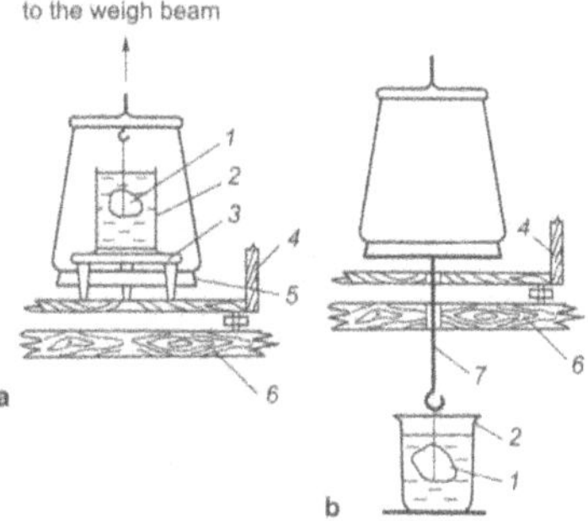

Fig. 3.2. Hydrostatic weighing setup without shortening of a hanger: **a** cylinder with the liquid is positioned above the pan; **b** cylinder with the liquid is positioned below the pan. *1* coating; *2* cylinder with the liquid; *3* support; *4* scale base; *5* pan; *6* cover of the table; *7* hook

In the course of routine laboratory measurements of density a special hydrostatic balance would be appropriate for use. Featuring a simple design, this balance provides easy and quick weighing of a specimen in the working liquid.

Testing involves the following main operations: weighing of a specimen in air, sealing of the surface pores in its coating, weighing of the specimen with sealed pores in water.

Specimens are taken from the desiccator and weighted in air one after the other. Then surface pores of the coating are sealed through the use of paraffin treatment or application of a thin layer of varnish or petrolatum. After prior drying and heating up to 50–60 °C, the specimens suspended on a fine string are submerged into molten paraffin so that the entire surface would be coated with a thin layer of paraffin after cooling. Once a specimen is submerged into varnish and the excess of varnish is removed, it must be kept in a drying cabinet at the temperature of 50–60 °C for 10–15 min. Thereafter a thin layer of petrolatum is applied on the surface and rubbed into surface pores.

Specimens with closed surface pores are attached to the suspension string and weighed in water; in so doing no air bubbles evolved from a specimen are permissible.

The apparent density of the coating (ρ_{app}) in g/cm is calculated from the formula[1]:

$$\rho_{app} = \frac{\rho_w m_1}{m_2 - m_3}, \tag{3.3}$$

where m_1 is the mass of a specimen weighed in air, g; m_2 is the mass of a specimen with closed surface pores, weighed in air, g; m_3 – the mass of a specimen with

[1] The mass of material applied to sealing surface pores is not taken into account due to its small quantity.

closed surface pores, weighed in water, g; ρ_w – water density at the testing temperature, g/cm.

3.1.3 Determination of Picnometric Density of a Coating Material

The essence of picnometric method is that a previously fragmented coating is put into a picnometer (a vessel of a certain volume and proper shape) and weighed along with it. Then the remaining volume is filled with a picnometric liquid and the system is weighed again; thereafter the density is calculated from the obtained results.

Pieces of coating mechanically detached from the base metal with the total weight of 3–5 g are ground to the least particle size of 20 μm. Each test requires at least three weighed portions. Retort-shaped spherical glass picnometers, which are generally used for routine measurements, may be closed either with a blind ground-in stopper or with a ground-in stopper with a capillary hole. The best results are obtained for picnometers with the capacity of 25–100 ml. However, it is worth noting that a sufficient accuracy of measurements can be achieved at smaller volumes too.

The picnometric liquids used in these tests should meet the following requirements: no chemical reactions with the coating; good wetting; small coefficient of dynamic viscosity; low values of surface tension and vapor pressure. The recommended liquids are xylene, toluene, kerosene, and ethyl alcohol.

The previously crushed and screened powder made of the coating material must be dried out at the temperature of 110–135 °C for two hours and kept in a desiccator.

A dry and clean picnometer with a mass determined in advance is filled with a picnometric liquid up to a mark, following which it is plugged by a stopper and weighed. Next the liquid is poured out and the picnometer is accurately rinsed and dried anew. After the picnometer is cooled down to room temperature, the powder from the desiccator is poured into the picnometer. Small particles of the coating settled on the device neck are removed by a cotton wad. Furthermore, the picnometer with a weighed portion inside is plugged by the stopper and weighed again. Then the picnometer is filled with the liquid up to the half of its volume. Air is removed from the powder by periodic shaking (but not by turning over) of the vessel. The liquid is added to the mark only once the open pores become soaked with liquid, i.e. after the air has come out completely.

Small coating particles adhered to the stopper and the neck of the picnometer are rinsed down by the picnometric liquid. After the air is entirely released, the level of the picnometric liquid is added up to the mark using a pipette.

The excess of liquid is taken away by thin strips of filter paper. Then the picnometer is plugged by the stopper and weighed again.

The picnometric density of the coating material (ρ_{pic}) in g/cm^3 is calculated from the formula:

$$\rho_{pic} = \frac{(m_3 - m_1)}{(m_2 - m_1) - (m_4 - m_3)}, \tag{3.4}$$

$$\rho_l = \frac{m_2 - m_1}{V_{pic}}, \tag{3.5}$$

where m_1 is the mass of the empty picnometer, g; m_2 is the mass of the picnometer with the liquid, g; m_3 is the mass of the picnometer with a powder, g; m_4 is the mass of the picnometer with the powder and picnometric liquid, g; ρ_l is the density of the picnometric at the test temperature, g/cm^3; V_{pic} is the picnometer volume, cm^3.

If the tested powder has many closed pores or the open pores are insufficiently filled with liquid, then picnometric procedure gives a systematic error.

3.1.4 Results and Discussion

The density and porosity of coatings are closely related, but most of the researchers consider the porosity the main characteristic of material. However, this point of view was disputed in some papers. For instance, a profound effect of the density on the other properties of ceramic coatings is emphasized in paper [3.95]. The coatings with the content of 6–8% Y_2O_3–ZrO_2 was deposited by the plasma method on different substrates using various kinds of powder and two types of plasma burners with variable technological parameters. The coating density was 85–91% of the theoretical level. It has been demonstrated that the hardness, endurance in a gas-abrasive medium, cohesion strength, and the thermal stability of the coatings are proportional to their density.

The technological parameters of deposition process influence the coating properties indirectly through its density. There was an attempt to calculate the density of gas-thermal coatings on the base of the random function theory [3.105], which provided a distribution of the average density over the coating thickness. The calculated data with accounting for the particle density variations from one layer to another are close to the experimental results and may be used for density estimation.

According to Kudinov, the apparent density of a coating produced by the plasma deposition of high-melting metals (W, Mo, Nb) depends on the thickness (h) of a deposited layer [3.56]. Experimental curves $\rho_{app}=f(h)$ for plasma-produced coatings are described by an empirical equation of the following kind:

$$\rho_{app} = \rho_{th} + (\rho_0 - \rho_{th})e^{-bh^2}, \tag{3.6}$$

where ρ_{app} is the density of a sprayed layer with the thickness h; ρ_0 is the coating density at $h \to 0$; b is a coefficient describing the rate of density reduction with growing thickness; ρ_{th} is the density of a very thick coating.

The value of ρ_{th} is a horizontal asymptote for the curve $\rho_{app} = f(h)$ at $h \to \infty$. In reality the level of ρ_{th} is achieved at $h = 2.5$–3.0 mm. The value of ρ_{th} can be calculated or determined from extrapolation [3.53]. The coefficient b can be calculated by the least-squares method as the slope of the line plotted on coordinates $\ln\left[\frac{(\rho_0 - \rho_{th})}{(\rho_{app} - \rho_{th})}\right]$ and h^2.

The values ρ_{th} and b are convenient indexes of density distribution over the coating thickness for different technological regimes.

Using several plasma and ceramic coatings as an example, it has been demonstrated that all the factors increasing the heating of sprayed particles provide a higher density of coating and reduce the rate of density decrease with growing thickness [3.53]. The key factor is the arc current growth. A shorter spraying distance also leads to a higher density of a spayed layer [3.59, 3.84].

The densities of plasma and detonation-produced coatings were compared in the paper [3.111]. The following powders were deposited: Cr_2O_3, WC–12%Co, Al_2O_3, TiC–20%Ni, etc. It has been shown that detonation-produced coatings have a higher density than plasma-sprayed ones. The effect of thermal treatment (annealing at 1100 °C) on the density of nickel aluminide after plasma deposition was considered by Sampath et al. [3.90]. A sprayed coating had the density of 6.29 g/cm^3. The annealing caused grain coarsening and an increase in density up to 6.41 g/cm^3.

Not only the methods and parameters of spraying influence the coating density, but also the granular composition of a powder. Curtis et al. [3.26] came to the conclusion that a finer powder in plasma spraying of ZrO_2–8%Y_2O_3 creates a higher coating density.

The coating density is often crucial for working properties, in particularly, for corrosion resistance. The authors of [3.50] in their recent work confirmed the conclusion reached in the study [3.95]. They prove that the key feature that creates a higher resistance to point corrosion in TiN–films is a high density.

3.2 Porosity

3.2.1 Concept of Material Porosity

The porosity is the ratio of the entire pore volume to the total volume of a porous body.

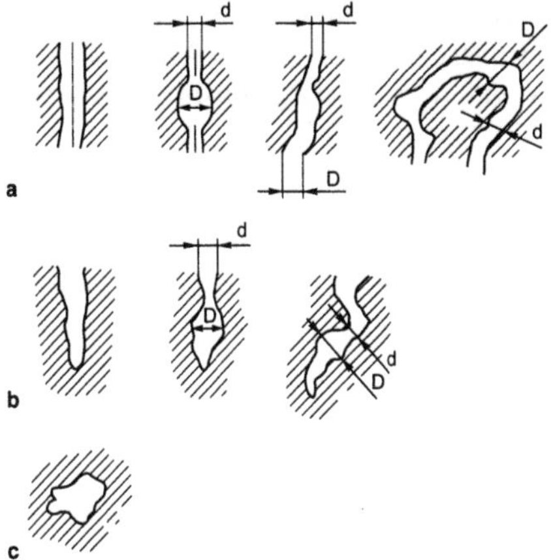

Fig. 3.3. Pore classifications (according to [3.80]): **a** open through pores; **b** open blind holes; **c** internal closed pores. *D*, *d* maximal and minimal pore diameters

The structure of porous materials is determined by its size and internal geometry. Regular porous structures are distinguished which consist of regularly interstratified elements in the form of separate pores joined by channels. For irregular porous materials, the size of pores (or their ensembles), their arrangement, and interconnection are random. Most of the porous materials, including the coatings, are classified into the second class of porous bodies. Such structures are characterized by a set of pores of every sort and kind, which may differ in size, shape, orientation, spatial location, and interconnection type (Fig. 3.3).

The peculiarities in geometry and space arrangement of pores are a basis for geometric modeling of porous structures [3.80].

A specific porous material can be described using the concepts of the maximal and average size of the pores, the pore size distribution and the pore sinuosity coefficient [3.23, 3.107]. The maximal pore size is an important characteristic of the material because it determines the maximum size of particles that might pass through it. The average size of the pores is usually given for the comparison of powder materials. The pore size distribution provides information about the number or volume of pores of a specific size and data about their size range – this is a more complete characteristic than the maximum or average size of pores. The sinuosity coefficient (γ) is determined as a ratio of the pore length (H_p) to the thickness of porous material (H):

$$\gamma = \frac{H_p}{H}. \tag{3.7}$$

The theory proved that for a porous medium consisting of identical spherical particles, the sinuosity coefficient varies from 1.205 to 1 with a porosity variation from 0.259 to 0.476 [3.107].

The parameters determining conductivity (the electric and thermal conductivity), elasticity (elasticity modulus), and resistance to plastic deformation (yield point, hardness) decrease monotonically with the increasing porosity; in a general way, they are adequately described by the equation [3.48].

$$\beta = \beta_0 (1 - P)^m , \tag{3.8}$$

where β is the numerical characteristic of a certain property of a porous body; β_0 is the same characteristic for structurally identical material but without pores; P is porosity, fractions of unity; m is the power degree, higher than one.

3.2.2 Research Methods for Porous Materials

There are more than 60 research methods for porous structure of solid bodies. In a thorough study [3.80], these methods were classified into eight groups: direct observation, capillary methods, mercury porosimetry, structural-adsorption, hydrostatic weighing, calorimetric, weight-space and standard methods.

The methods of direct observation are especially important because they provide direct information about the size and shape of pores and their connections. In this aspect, they cannot be substituted by other methods. The results of the direct method are an essential compliment to other methods based on some model conceptions. Direct methods are suitable for the study of porosity in a wide range of pore sizes.

Capillary methods imitate the processes occurring in pore filters, permeable membranes and other similar kind of objects.

The method of mercury porosimetry is rather universal and gives information about pore structure for a wide range of pore sizes. The commercial devices for analysis (mercury porosimeters) are available and the treatment of these results is rather simple.

The structural-adsorption method is applicable for the study of adsorbents and catalysts with a strongly developed porous structure.

The methods of hydrostatic weighing were classified into a special group on the basis of their general principle – measurement of the solid body density. These methods are rather simple and provide the wanted precision.

The calorimetric methods of immersion into liquid are based on the thermal effect of wetting of a solid body surface.

The estimates of general porosity through the measurement of volume and weight are simple and fast, though not sufficiently accurate. The method of standard porosimetry is a universal technique. Its essence lies in the comparison of a studied specimen with a control.

The complex use of these methods gives more complete information about the porous structure of a solid body [3.23, 3.42, 3.48, 3.80, 3.107].

3.2.3 Porosity in Coatings

Porosity is inherent in most coatings; its magnitude may vary from 0 to 20% and even higher in some cases [3.4, 3.8, 3.11, 3.75].

Gas-thermal coatings are notable for their diversity in pore shapes. During the coating formation, all types of pores are produced that can be found among a variety of porous dispersed systems. Closed pores, formed by inclusion of hollow particles or gas release during solidification, are different from open pores, passing through material from the surface to the base metal. The formation of open pores is caused by the following reasons: the loose laying of particles-blocks in a growing layer; formation of cavities due to a crack formation, evaporation, splashing, etc. The interconnection of open pores occurs in every possible direction without a preferable orientation in space [3.58].

Depending on the pore shape, open and closed porosity may be distinguished by the void age of open and closed pores in the total volume of the specimen.

The total porosity is defined as the ratio of the sum of open and closed pore volumes to the specimen volume.

Porosity is one of the most important characteristics of a coating. The features of a porous structure and its stability influence the operational characteristics of the coating such as the fatigue [3.10, 3.82], hardness [3.89], electrical and thermal conductivity [3.9], gas permeability [3.66], wear resistance [3.103, 3.114], corrosion resistance [3.64, 3.97], etc.

According to the operational requirements of an item, the coating porosity has to be controlled during spraying. For corrosion-resistant coatings the porosity is a negative feature. An aggressive medium can easily pass through open pores toward the base metal and cause its corrosion. As a rule, a coating operating at high temperatures in an oxygen-rich gas medium has to be of the maximum density.

In corrosion-resistant coatings, the choice of porosity value is caused by several factors. On the one hand, requirements for great hardness, rigidity and wear resistance dictate the need for a low-porosity material. On the other hand, in an operation accompanied by friction between the surfaces, the oil that fills the pores can be supplied steady to the worn out surfaces and make the friction coefficient and linear wear smaller. Therefore, wear-resistant coatings working with lubrication require a special control of porosity during the spraying of layers.

Depending on their size, the pores are subdivided into micropores (radius less than 1.5 nm), macropores (radius more than 100 nm) and mesapores (intermediate size). Classifications by arrangement and shape are the following: through pores, reaching to the base metal, and confined pores. There also exist closed, open, interconnected, blind, and other kinds of pores. A coating is considered porous if the total porosity exceeds the 20% level.

From the numerous research methods for porous structures, only a dozen are used actively to study coatings. In the following sections we will briefly consider some of these methods, and give a detailed description of four of them.

The required accuracy, available devices, number of specimens, peculiarities of coating structure, etc. dictate the choice of a technique used for porosity measurement. The data obtained by these methods gives us a foundation for development of a geometrical model for the porous coating structure.

3.2.4 Method of Hydrostatic Weighing

The method of hydrostatic weighing (according to State Standard 18898) consists of the following operations: weighing in air, closing of the surface pores, weighing in water, impregnation with a working liquid and weighing of impregnated specimens in air and liquid. The first three operations have already been described in the section devoted to the estimation of the apparent density.

When measuring open porosity, the coating must be impregnated with a working liquid by boiling or vacuum treatment. The following liquids may be used for impregnation: industrial oil, xylol, benzyl alcohol, distilled water, or other liquids with a low vapor pressure and a stable density. Distilled water must be previously degassed by means of boiling for 1.5–2 hours.

The boiling impregnation method is the simplest one. The specimens are submerged into boiling liquid and kept there for about two hours. However, this impregnation alone is generally not sufficient. To gain a greater accuracy in testing, a more complete filling of all pores by wetting liquid is recommended. To achieve this goal, the impregnation of a coating must be carried out in a vacuum or under conditions of liquid forced filtration through pores with a further soaking for several hours. The vacuum impregnation setup operates in the following manner. The tested specimens 1 (Fig. 3.4) are placed into the desiccator 2, and a vacuum pump is turned on. After a proper rarefaction is achieved, the pump is turned off and the T-joint cock for liquid supply is opened. The impregnating liquid is forced by atmospheric pressure into the desiccator after which it impregnates the specimens.

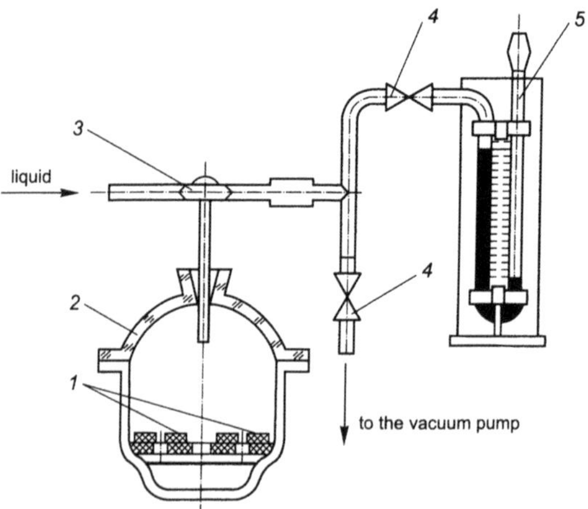

Fig. 3.4. Setup for vacuum treatment. *1* specimens; *2* desiccator; *3* T-joint cock; *4* valves; *5* vacuum gauge

Filling of pores with liquid in a vacuum treatment setup ought to be run after 30–minute exposure in a vacuum at the level of 10^{-2}–10^{-1} Pa. The pore-saturating liquid is fed through the cap gradually, until the specimens are completely covered. For gas-flame and plasma coatings, the impregnation time is more than 20 min, and for detonation-produced coatings – more than one hour.

The impregnation time has a strong influence on the accuracy of results obtained. By Bartenev's information, the porosity of detonation coatings given in most publications was underestimated considerably [3.6]. For example, the total porosity of Al_2O_3 coatings is 6–9%, although many researchers indicate the figures of 0.5–2.0%. Perhaps, this is related to the peculiarities of hydrostatic weighing in estimation of detonation coatings made of aluminum oxide. In this case, the open porosity is rather close to the total porosity; hence correct data require an impregnation time of vacuum-treated detonation coatings up to one hour. If this condition is not satisfied (no preliminary vacuum treatment and the impregnation time is low), then only apparent (conditional) porosity can be calculated, which does not have any physical meaning and remains a function of impregnation time, pore shape, etc.

After the specimens are taken out of the desiccator, the excess liquid is removed by a wrung-out wet cotton fabric.

The total porosity (P_{total}), open (P_{open}) and closed porosity (P_{closed}) are defined (in %) by the following formulas:

$$P_{total} = \left(1 - \frac{\rho_{app}}{\rho_{true}}\right) \cdot 100 , \tag{3.9}$$

$$P_{open} = \frac{(m_2 - m_1)\rho_w}{(m_2 - m_3)\rho_l} \cdot 100 , \tag{3.10}$$

$$P_{closed} = P_{total} - P_{open} \tag{3.11}$$

where the densities are: ρ_{app} – apparent, g/cm^3; ρ_{true} – true, g/cm^3; ρ_w – water density, g/cm^3; ρ_l – density of the impregnant, g/cm^3; m_1 – mass of the specimen weighed in air, g; mass of a soaked specimen weighed in different mediums: m_2 – in air, g; m_3 – in water, g.

The application of standard methods sometimes causes serious errors because of peculiarities in the coating structure and its low thickness. Bartenev put forward a reliable technique which is a variation of the hydrostatic weighing method [3.6].

The author proposed to use a fine string instead of a kapron thread (its weight is not accounted in the method explained above). The maximal diameter of this string is calculated from the formula including the coating weight and the working liquid density. To remove the adsorbed moisture, the specimens must be dried out for 5–8 hours at a temperature of 120–150 °C; after cooling the specimens are weighed on analytical scales with an accuracy of 0.0001 g. During the vacuum impregnation procedure, a special device was used to estimate the substance loss in vacuum treatment and impregnation. To find the true values of m_2, repeated measurements of the mass of an impregnated specimen at a certain interval with subsequent time chart plotting and extrapolation of the curve to zero time are recommended. For a higher accuracy of measurements, it has been proposed that once impregnation is completed, the specimen should be kept in a saturated vapor of the impregnating liquid, and then put into a sealed calibrated box for weighing.

This improved technique allows reduction in the test error to 0.1% for a coating thickness of 0.5 mm, and down to 0.3% for a thickness of 0.1 mm.

The measurement of porosity by hydrostatic weighing is also possible for specimens with coatings non-detached from the base metal. In this case, the technique discussed above must be complemented by a procedure for measuring mass and density of the base metal. During measurement of the impregnated specimen mass (coating + base metal), the surface of the base metal, which is unprotected by the coating, has to be previously dried by a compressed air flow. The open porosity is calculated from a formula with account for the mass and density of the base metal.

Based on an analysis of the experimental accuracy of the discussed characteristics, many authors came to the conclusion that the method of hydrostatic weighing provides reliable determination of open porosity alone. Characterization of coatings and technological optimization using parameters P_{total} and P_{closed} are not always expedient. This arises from the necessity to know the true density of a coating (i.e., the precise chemical and phase composition of the spayed coating). For instance, aluminum oxide has seven modifications different not only in crystalline lattice structure, but also in physical properties including density. Usage of α–Al_2O_3 powder for spraying does not mean that produced plasma or detonation coatings are heterogeneous and have no zones in the γ–Al_2O_3 phase with another density. Calculations reveal that the values of P_{closed} may be sharply different if in place of the true density for γ–Al_2O_3 a similar data for α–Al_2O_3 is used. Spraying of solid alloys is accompanied by decrease in carbon content and cobalt content in comparison with the source powder material, and hence, by a decrease in density [3.6].

Rogozhin et al. presented mathematic relationships showing that for a steady spraying process these desired quantities are statistically reliable. The only sources of errors in finding of the total and closed porosity are the possible inadequacies between the densities of the compact analog and non-porous sprayed material [3.85].

3.2.5 Microscopic Method

The microscopic method of porosity measurement is based on the first stereometric relationship, whereby the share of phase (pores) in the total bulk of alloy (the coating) is equal to a part of the secant passing through this phase (pores) on the microsection surface. In reality, it is requisite that the lengths of segments falling on the pores should be summed up. Then, in order to find the general porosity, the result must be related to the total length of the secants (Rosival's method) (Fig. 3.5). The test accuracy depends on the total length of the secants.

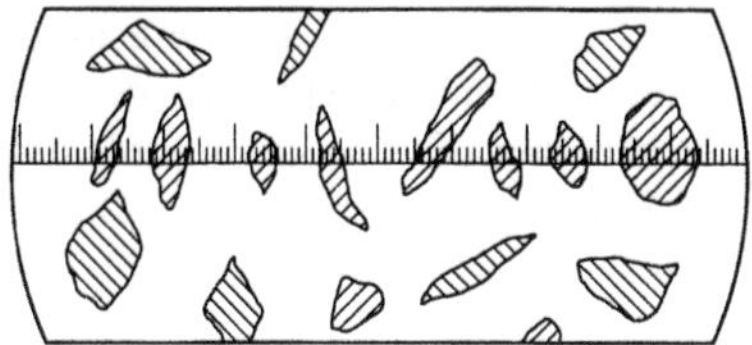

Fig. 3.5. Microscopic method for porosity measurements

The objects of study are the non-etched microsections which must all be analyzed at the same microscopic magnification. The microsection(s) area should be not less than 0.5 cm^2. An eyepiece ruler fitted into the lens can serve as the secant; the axial line of the lens is divided by marks into the hundred arbitrary units. The secants can also be put on the micro-photograph of the structure or on the image displayed on the microscope screen. The following parameters are determined $\sum l, \sum \frac{1}{l}, \sum \frac{1}{l^2}$, where l is a linear size of the pore cross-sections in the secant.

The characteristics of the porous structure obtained by the microscopic method are presented in Table 3.1.

The microscopic method of porosity measurement features at least three serious shortcomings. First, it gives information about porosity only for one random cross-section of a coating and doesn't estimate the coating quality in the bulk. Second, porosity is underestimated since most of the pores with the size less than 1 μm cannot be accounted for in the analysis. Third, the accuracy of porosity estimation is several times lower while the labor consumption is higher than in the method of hydrostatic weighing.

Table 3.1. Characteristics of porous structure

Term	Formula	Dimensionality
Total porosity	$P_{total} = \sum \frac{1}{L_1}$	–
Conditional arithmetic diameter of pores	$\bar{l} = \frac{K \sum l}{m_1}$	mm
Number of pores on the microsection surface	$n_1 = \frac{0.785 \sum l^{-1}}{L_1 K^2}$	item/mm^2
Number of pores in a volume unit	$N_1 = \frac{0.5 \sum l^{-1}}{L_1 K^3}$	item/mm^3
Pore area in a volume unit	$S_1 = \frac{4m_1}{L_1}$	mm^2/mm^3

L_1 is the total length of the secant in arbitrary units of the measuring device; K is the scale of the length unit, mm; m is the number of chords in the microsection plane.

Another serious problem for stereometric study is to prepare a flat undistorted microsection. In processing of brittle, unstable materials among which are ceramic coatings with porosity higher than 10–15% the particles of material crumble out, so the fraction of pores with the size of 1–10 μm increases drastically. On the other hand, small pores can be puttied and disguised by grinding paste and grinding products [3.58].

The technique of microsection preparation was described in [3.14] for specimens with thermal barrier ceramic coatings produced by plasma spraying. The preparation involves the microsection impregnation with a fluorescent epoxy resin, its polishing, sedimentation of a thin interference film and photographing with a metallographic microscope. This technique allows qualitative distinguishing the porosity of the coating itself and the porosity obtained during preparation of a microsection.

3.2.6 Method of Mercury Porosimetry

The method of mercury porosimetry as applied to oxide coatings is comprehensively reviewed in Bartenev's paper [3.6]. Porosimeters provide information about pore size, shape, and size distribution. The operation of the device is based on the determination of the mercury volume penetrating the pores under a stepwise increase of pressure relative to the total volume of a specimen. The mercury surface is to be acted upon since it does not wet most of the materials and fails to fill the pores. In order to fill long and narrow pore channels, a pressure higher than that for globular open pores is required.

A mercury porosimeter (Fig. 3.6) operates in the following manner. A tested specimen is put into the low-pressure chamber 4 with a cell and treated under vacuum, then mercury is introduced into the cell in such a way that it completely covers the specimen; mercury degassing is performed during this procedure. A specified pressure of mercury is automatically set in the cell being kept for a certain length of time to ensure that the mercury fills all the pores with a size larger than critical. For every tabulated value of pressure, the volume of specimen-penetrated mercury is measured by the electric capacitive method.

The volume of mercury filled the coating should be measured in every step of elevating pressure and then related to the total volume of the coating. The pore radius r_{cr}, mm, is calculated from the equation:

$$r_{cr} = \frac{2\sigma \cos\theta}{P_N}, \tag{3.12}$$

where σ is the mercury surface tension, N/mm; θ is the contact angle; and P_N is the pressure, N/mm^2.

For the most part researchers take the contact angle as 140°. This value was obtained for several materials through the use of a new type of a contact protractor that measures the contact angle under near-operational conditions of the mercury porosimeter [3.58].

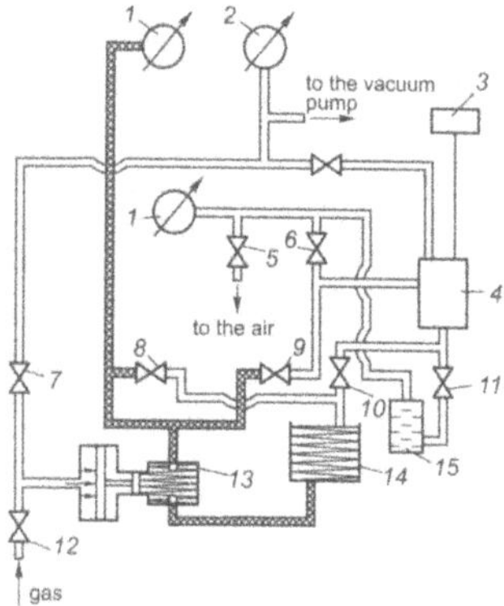

Fig. 3.6. Diagram of a mercury porosimeter. *1* pressure gauges; *2* rarefaction pressure gauge; *3* counter; *4* chamber with a cell; *5* safety valve for low pressure; *6* low pressure valve; *7* pressure reduction valve; *8* safety valve for high pressure; *9* high pressure chamber; *10* oil valve; *11* mercury valve; *12* pressure control; *13* compressor; *14* oil tank; *15* mercury tank

When estimating the applicability of the mercury porosimetry technique to studying porous structures of one or other type of deposited materials, the key factor is the studied structure tolerance for high pressures. In the measurement via the mercury porosimetry technique, the specimens with closed or mercury-inaccessible micro- and submicropores are the most liable to deformation and failure. The sprayed aluminum, copper, and some metal-ceramic compositions belong to this class of materials. Conversely, plasma-deposited ceramic materials that are almost devoid of closed porosity can be investigated in a wide range of pressures.

Mercury porosimetry is one of the simplest and most informative methods for studying porous structures with pore size at an interval from 3.5 nm up to several microns. This study is available due to wide-scale production of automatic and semi-automatic devices from the companies "Carlo-Erba" (Italy) and "Micromeritix" (USA), while the latter manufactures the most perfect device "Autopore 9200".

The Autopore has four independent low-pressure cells and two high-pressure ones. The maximum pressure in a low-pressure cell is 0.19 MPa, so the pores at the interval 0.003–360 μm can be studied. The device is equipped with a microcomputer and a terminal for analysis, data collection and acquisition control. The results are printed out in the form of tables and charts. The latest model of this device is fitted with a PC [3.58].

3.2.7 Exclusion Method

The pore size can be determined by the method of liquid exclusion out of the pores. It is based on the measurement of the pressure required for passing of a gas bubble through a porous specimen. The pores in the specimen are filled with a wetting liquid with a known surface tension coefficient. This method has been certified for powder materials (State Standard 26849).

When applying this method to the coatings with pores having countractions and widenings, only the narrowest sections of these pores are determined. Before testing, a specimen must be impregnated with liquid. This can best be done by impregnation of an evacuated specimen under excessive pressure [3.107].

A specimen saturated with liquid is placed into a special holder; then an additional layer (up to 5 mm) of the same liquid is poured over the specimen surface (Fig. 3.7), or the holder is placed into a vessel containing this liquid. Gas is supplied to the cavity under the specimen, and the gas pressure is gradually increased. Observing the specimen surface, the gas pressure is taken down which results in the opening of at least three surface pores (the same quantity of gas bubble traces can be observed in the liquid layer). This pressure corresponds to the opening of pores with the maximum size. Further increase in under-specimen pressure leads to intense release of the gas bubbles over the whole specimen surface, which is appropriate to the opening of pores of average size (Fig. 3.8).

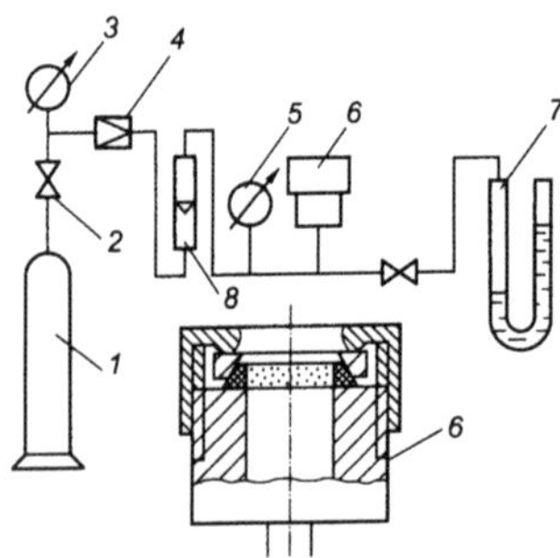

Fig. 3.7. Diagram of a setup for determination of pore sizes using the liquid exclusion method (according to [3.107]). *1* source of excessive pressure; *2* valve; *3* pressure gauge; *4* pressure regulator; *5* pressure gauge; *6* holder with a specimen; *7* U-shaped manometer; *8* flow meter

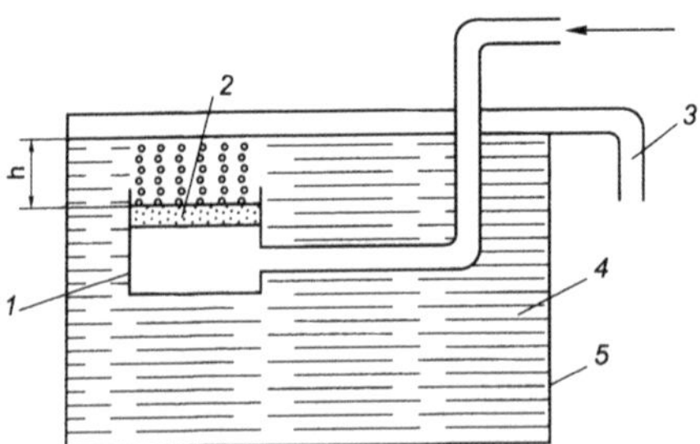

Fig. 3.8. Diagram for the measurement of pore size using liquid exclusion. *1* specimen holder; *2* tested specimen; *3* tube for liquid drain; *4* liquid; *5* vessel for liquid

The maximum and average size of pores d (μm) is calculated from the following formula:

$$d = \frac{4 \cdot 10^3 \delta_t}{P - \rho_t g h}, \tag{3.13}$$

where δ_t is the surface tension, N/m; P is the pressure resulting in the release of air bubbles through the pores of maximum size ($d_{\max}$) or from the entire surface when an average pore size is determined (d_{av}), Pa; ρ_t is the liquid density, g/cm^3; g is the free fall acceleration, m/s^2; and h is the liquid column height, m.

The accuracy of results depends significantly on the rate of pressure growth beneath the specimen. If the pressure grows too fast, the liquid will be pushed out of the pores after the equilibrium pressure is established, which corresponds to the true size of pores. Therefore, that kind of test calls for a very gradual increase in pressure, especially when the opening of pores with maximum size is expected.

The method of liquid exclusion is applied to the measurement of pore sizes in the range 1–250 μm [3.107].

3.2.8 Other Methods of Porosity Measurement

Bulychev and Alekhin are energetically developing a new approach to the measurement of porosity using the parameters of hardness indentation [3.17]. The tests put forward by them can reveal both the total porosity and the character of the pore size distribution. The possibility to find the total porosity is based upon the ability of a porous material to be compressed, which occurs at high compression stresses under indentation. This is accompanied by a variation in the roller height and Young modulus. Using these parameters, the authors obtained empirical relationships for determination of porosity [3.17].

The diagram of forcing an indenter into a coating is influenced not only by the total porosity, but also by the pore size distribution. In this case a pore is considered as a particular case of structural non-uniformity in a two-phase material with zero hardness of the inclusion phase. However, Bulychev and Alekhin noted that non-uniformity of resistance to indentation for tested coatings is determined not only by pores, but also by the structural inhomogeneity of the coating, which consists of several phases varying in hardness.

This technique offers advantages over the other known methods not only in that porosity and mechanical properties are measured simultaneously. The entire complex of material properties can be determined on different parts of a specimen or item, which is essential for study of base metal–coating compositions when the coating is inhomogeneous.

A mathematical model has been proposed for description of the relationship between porosity characteristics and the variation coefficient in microhardness; this model was implemented for two-stage testing [3.17].

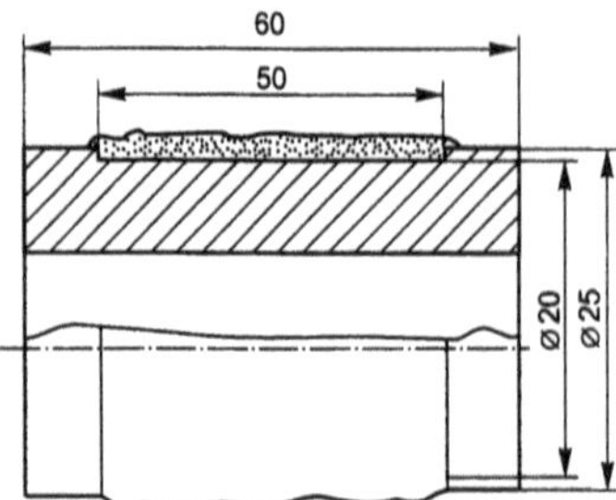

Fig. 3.9. Specimen for porosity measurement by the method of direct weighing

Porosity can also be controlled by the methods of immersion, application of paste and laying down a filter paper. These methods are suitable for galvanic coatings, but have limited usefulness for porosity estimation in case of detonation-produced, plasma and gas-plasma coatings because of their great thickness.

The porosity of plasma coatings sprayed on stainless steel was determined in a 3.5% solution of NaCl. The density of the current taking the path through a specimen when the circuit is closed is proportional to the through-porosity, hence giving the opportunity for quantitative estimates [3.5].

In [3.51], a computer-aided method for coating cross-section image analysis was proposed; this method is remarkable in its simplicity and accuracy in comparison with the method of hydrostatic weighing.

Two rather simple methods are widely used in industrial conditions: comparison with a standard scale and direct weighing. In the first case, the quality of a coating can be estimated by comparing its structure with the standard images of scales. The microsection is studied by a microscope with fixed magnification settings (100, 200, 500) over a length of 20–25 mm. The standard scale is a collection of porous structures estimated by different numbers.

In the direct weighing, porosity is measured on a cylindrical specimen as shown in Fig. 3.9. A cavity is made on the specimen following which the studied material is sprayed on it. After spaying, the coating is ground to make this specimen strictly cylindrical. Given the known sizes of the substrate, the volume of the coating can then be calculated. After grinding, the specimen is weighed and the mass of coating is determined. From the known mass and volume of the coating, the porosity can be calculated provided that the density of the sprayed material is known.

To find the porosity of a thin ceramic coating obtained by physical deposition from vapor, Celis et al. [3.21] proposed a coulometric method.

3.2.9 Elements of Porosity Structure in Gas-Thermal Coatings

The porosity structure was investigated in detail in [3.58]. Results of the optic analysis and scanner electron microscopy for individual sprayed particles and coatings demonstrate that by the mechanism of formation the following types of elements in a porous structure can be distinguished [3.58]:

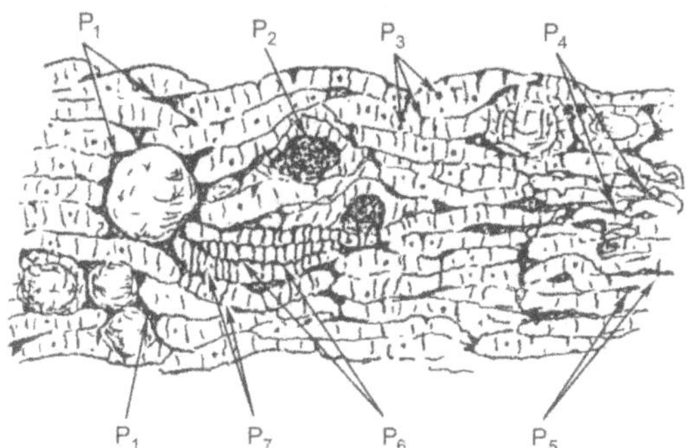

Fig. 3.10. General diagram for elements of the porosity structure in a plasma-sprayed coating material (Reproduced with permission from [3.58])

1. Pores (P_1) (Fig.3.10) produced due to a loose loading of particles in a layer during a single run. These can be either big pores, or micro- and mesopores. The size and fraction of pores of this kind depends on the aggregate state of the material-forming particles, on the overheating degree of melted particles, and on their velocity. The geometry of these pores can be very different.
2. Pores formed due to the interaction of particles with a gas medium. This class of pores is characterized by formation of a wide diversity of elements. The pores of the first type (P_2) are formed during the laying of liquid or repeatedly solidified particles, which contain a gas bubble inside: this bubble remains almost intact after the collision of the granule with the layer material. These cavities are usually linked to bulks of adjacent pores through the cracks and pores.
   The second type of pores of this class is produced by gas release from the liquid material during the laying. In this case along with big pores similar to P_2, micro and mesopores P_3, both isolated and emerging at the particle surface, can be generated. The first type pores have a spherical shape, while the latter are funnel-shaped formations.
3. Pores formed due to particle spluttering after a collision with the base metal and the growing coating. Usually these are a subclass of micropores having a complex shape P_4. The pores formed due to vapor condensation (P_5) are closed to pores originating from spluttering. When a material is highly overheated and intensely evaporates from the surface, the formation of these pores is typical of the inter-layer boundaries and coating-base metal interface, as well as of the spraying spot periphery. These can be both micro- and mesopores [3.58].
4. Pores formed due to peculiarities in the particle material crystallization. For plasma-sprayed particles of oxide ceramics and refractory metals, the dendritic character of crystallization is typical. In this case, two more characteristic types of pores are revealed. The first one is a consequence of dendrite wedging out with an appropriate micro-relief formation. This process results in breaking of the continuousness along the particles interface (P_6). In dendrite crystallization similar violations in continuousness are typical of the particle-substrate inter-

face as well. The dendrite character of crystallization is revealed in the form of mesopores (P_7).

Selected elements of porous structure are localized within different structural elements of the plasma-sprayed materials. Inside the separate particles, the following structures are identified: substructural mesopores (P_7); micro- and mesopores-formed due to the gas release (P_3); macro- and micro-pores resulting from gas-particle interaction and gas release (P_2); and also micro- and submicrocracks in the particle body. The pores located along the particle boundaries include: the micro- and mesopores P_6 formed because of the fragmentation (spluttering, dispersion) of particles after collision with a surface and vapor condensation; macro-, micro-, and mesopores produced by loose packing of particles into a layer; finally, the cracks and microcracks at the points of contact between the particles. The inter-layer interface of material is a low-density zone. In this zone, all types of pores can be found; but the fraction of those formed due to the loose packing of particles and to vapor condensation is higher here.

At the coating-base boundary zone there may be found a specific type of porosity resulting from the presence of loose oxidation products and from reaction of the deposition medium with the substrate. The same is typical of porosity along the inter-layer boundary. Besides, some specific kinds of pores related to the base material erosion, its mixing with the sprayed material or to cracking of the substrate surface layer may be formed here [3.58].

3.3 Coefficient of Gas Permeability

3.3.1 General Overview

Protective coatings should safely preserve base metal when operating in aggressive gases and liquid mediums. Of prime importance is the knowledge of the protective properties of the coatings exposed to aggressive gases at high temperatures.

Gas-permeability is a property of the coating to permit the passage of gas through open pores and cracks. The gas-permeability coefficient is a magnitude, which characterizes the dependence of a given gas volume that passes through the coating in a unit time on the pressure differential in the gas. The gas-permeability coefficient depends on many factors such as porosity, the distribution of pores according to their sizes, the presence of cracks, the temperature and the composition of gaseous atmosphere (dynamic viscosity of the gas) [3.49].

The essence of this method consists in determination of the coating ability to let air pass through in the direction perpendicular to the layers.

This method can be applied to research tests in order to improve the protective coating technologies; it is also used for obtaining data on the permeability of newly developed materials and revealing cracks in the coating.

3.3.2 Specimens

Specially prepared disk-specimens are used for testing. The coating deposited on the end surface of the cylinder is separated either mechanically or by scouring the base metal. In the case of mechanical separation the work piece for a disk-specimen is cut off on a lathe. The specimen consists of the coating and the base metal 0.2–0.3 mm thick. The base metal is removed with emery paper. At a coating thickness equal to 0.05–0.2 mm special specimens with a crater are recommended. The coating is deposited on the disk-specimen opposite the crater side and then the metal is scoured to the coating. There should be not less than three specimens per each coating regime. The diameter of the specimen test surface should be over hundred times and its thickness over ten times larger as compared to the average diameter of a particle of the powder that makes up the specimen.

3.3.3 Equipment

The setup for measuring the gas permeability coefficient of the coatings consists of three main parts: the specimen holder 2, the manometer 8, and the vessel 5 which is 270 mm in diameter and holds 30 liters (Fig. 3.11). The bottom part of the vessel contains a tube with a cock 6; a rubber plug with a glass tube and two cocks 3 and 4 is ground to the neck of the vessel. A special holding device 2 is employed to fix the specimen 1 in the setup (Fig. 3.12).

Test area of the specimen 5 is limited by the small diameter of a rubber annular gasket 2. The specimen is fixed in the holder by means of two elastic rubber gaskets providing a reliable contact. To ensure air tightness of the device, the cracks 3, 4, 6 (see Fig. 3.11) are lubricated by a thin layer of a vacuum lubricant and then ground in.

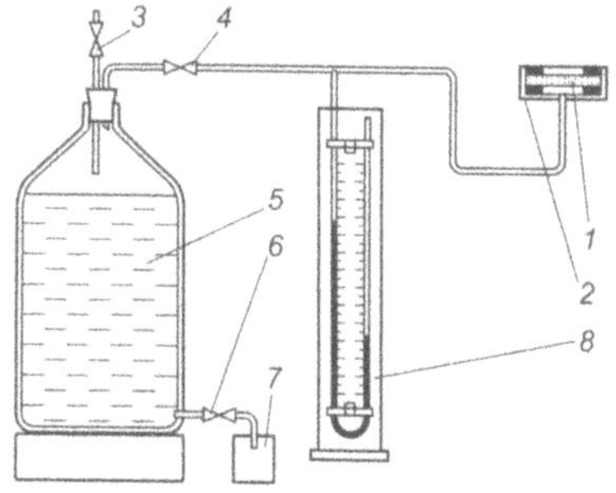

Fig. 3.11. The setup for measuring the gas permeability coefficient. *1* specimen; *2* holder; *3, 4, 6* cocks; *5* vessel; *7* measuring vessel; *8* manometer

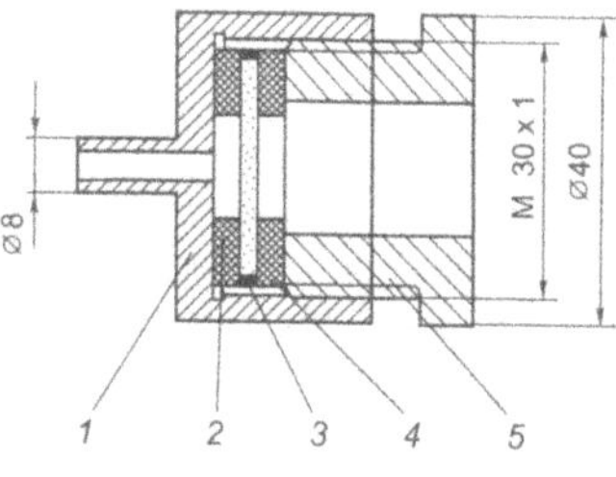

Fig. 3.12. Specimen holder. *1* screw-nut; *2* rubber gasket; *3* seal; *4* specimen; *5* basis

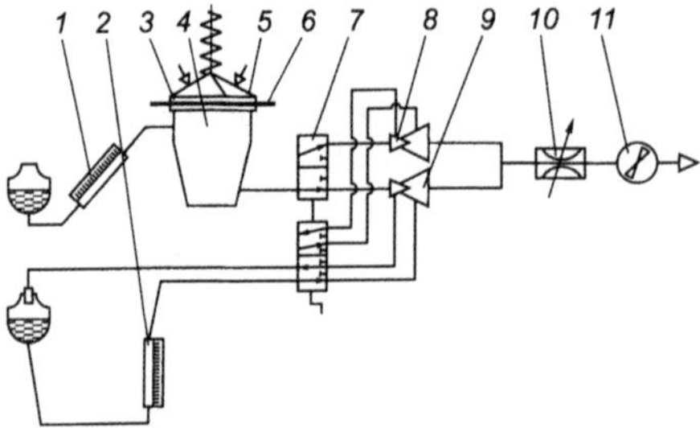

Fig. 3.13. Device for controlling air permeability. *1* rarifaction indicator; *2* differential manometer; *3* clamping ring; *4* rarifaction chamber; *5* accessory table; *6* tested specimen; *7* Venturi tube switch; *8*, *9* air flow meters (Venturi tubes); *10* electric motor with fan

The specimens can also be tested with the help of a multipurpose device which is used for controlling the air permeability of various materials. This device (Fig. 3.13) is provided with a throttle to control the pressure differential across the specimen and an automatic fan with an On/Off system. Specimens may be tested using two of the six sets of accessory tables having orifices of 16.0 and 25.3 mm.

3.3.4 Preparation for Testing

Water is poured into the vessel through the tube with the cock 4 open (see Fig. 3.11). The level of the water in the vessel determines the pressure differential across the specimen.

Before testing, the air tightness of the setup should be checked. For this purpose, a non-porous metal disk has to be inserted into the holder instead of a specimen. If during 5–7 minutes water leakage and a pressure decrease are not observed, the setup can be considered leak-proof.

The thickness of the specimen is measured in three points by a micrometer with an accuracy up to 0.01 mm. To prevent air infiltration, a thin layer of a vacuum lubricant should be applied at the end surface of the rubber gaskets.

Prior to testing, the surface of the specimens should be carefully cleaned, degreased and dried. Before experimentation starts, the specimens should be kept in a desiccator.

Once a specimen is installed in the holder of the device, which was previously tested for air tightness, pressure differentials of 50, 100, 150 and 200 mm of water column are consequently produced. In doing so, the air flow rate is registered. If the ratio between the airflow rate and the pressure differential for all measurements does not exceed 5% of average magnitude, the laminar flow regime may be considered established.

The permeability coefficient is characterized by the slope of a straight line on the "air flow rate–pressure differential across the specimen" coordinates. Such a linear dependence is typical only for laminar flow. At a superposition of a turbulent regime the linear dependence gets broken and the gas permeability coefficient becomes a variable.

3.3.5 Testing

On opening the cock 6 (see Fig. 3.11), water eventually stops flowing to the measuring vessel. Next the cock 3 is opened and, upon achievement of the required stable pressure differential fixed by a manometer, the volume of the water flowing out of the vessel 5 can be determined by means of measuring vessel 7.

The duration of the water drainage is measured by a stopwatch. This duration may vary in the range from several minutes to several hours depending on the coating composition and its structure. The quantity of the drained water is measured with an accuracy up to 1.0 ml.

The airflow rate for the laminar flow regime is calculated as a ratio of the air volume passed through the coating (equal to the volume of drained water) to the drainage time.

The gas permeability coefficient (K) in mm^2 is determined by the formula:

$$K = \frac{Q \cdot \mu \cdot \delta}{S \cdot \Delta P}, \tag{3.15}$$

where Q is air flow rate for the laminar flow regime, mm^3/s; μ is dynamic viscosity , Pa·s; δ is the specimen thickness along the flow, mm; S is the active surface of the specimen cross-section, mm^2; ΔP is the air pressure differential across the specimen for the laminar flow regime, Pa.

The dynamic viscosity of air for the testing temperatures is taken from the table of physical constants.

Assuming that the structure and sizes of the pores[2] in the coating are temperature-independent, the gas permeability coefficient at high temperatures (K_t) can be determined by the formula:

$$K_t = K_{20} \frac{\mu_{20}}{\mu_t}, \tag{3.16}$$

where K_{20} is the gas permeability coefficient at 20 °C; $\frac{\mu_{20}}{\mu_t}$ is the temperature factor (ratio between the air dynamic viscosity taken at the temperature of 20 °C and t).

The gas permeability coefficient of the coating for any gas (K_{gas}) is calculated from the formula which takes into account the ratio between the dynamic viscosity of the gas (μ_{gas}) and the air (μ_{air}).

[2] After determining gas permeability coefficients it makes sense to determine porosity characteristics of the coating.

$$K_{gas} = K_{air} \frac{\mu_{air}}{\mu_{gas}} \tag{3.17}$$

A filtration coefficient, which is independent of dynamic viscosity of the gas, may at times be used to characterize the coating instead of the gas permeability coefficient. More refined techniques for determining the gas permeability coefficient and filtration coefficient in view of possible main errors were developed by Bartenev [3.6].

3.3.6 Results and Discussion

The influence of the shape of pore channels, open porosity, gas pressure and other parameters on filtration coefficient is considered by Bartenev et al. Hereinafter are presented some of results of this study [3.6].

Clear correlation between porosity and permeability values is observed. This circumstance can be used in particular for searching microcracks in the coatings. When analyzing detonation and plasma oxide coatings it was revealed that their gas permeability predominates open porosity by an order or more. Microscopic investigations of coatings revealed the presence of microcracks slightly increasing porosity but essentially raising gas permeability. Permeability of oxide coatings derived by different methods may differ by five orders. But even the most dense detonation coatings cannot safely protect the base metal against corrosion in the specific aggressive gas media.

The filtration coefficient and open porosity are usually correlated by a logarithmic dependence. At the same time for a given value of open porosity of oxide coatings, deposited by various methods (by plasma spraying or detonation), the values of filtration coefficients may differ by 3–10 times. A low filtration coefficient for the detonation coatings is explained by a lower contraction of pore channels as compared with those for plasma coatings. Taking into account that at optimal technological regimes the porosity of plasma coatings is higher than that for detonation coatings by a factor of 2–2.5, the filtration coefficient for the latter coatings will probably be 100 times less.

The permeability is determined to a great extent by the coating technology. When changing the oxygen content in a gas mixture from 55 to 71% the permeability of detonation coatings raises tenfold. In order to obtain oxide coatings with minimal gas permeability it is recommended to utilize gas mixtures with a composition close to the stoichiometric. In this case, a maximal melting of the particles may be achieved. This provides a reduction in porosity since the deposited particles spread better over the surface area between the solid particles previously deposited.

An increase in the coating thickness is accompanied by a diminution of gas permeability. Similarly, gas permeability is dependent on the temperature of preliminary heating of the base metal. It may be supposed that all factors leading to decrease of porosity facilitate a reduction of permeability [3.49].

The authors [3.107] revealed a dependence of the gas permeability coefficients on structural parameters of powder materials for laminar, transient and turbulent flow regimes. When describing theoretically gas flow regularities, the authors considered a model representing the area between the pores as aggregation of capillaries of the various cross sections.

3.4 Permeability of Liquids

3.4.1 General Overview

The liquid permeability coefficient is a magnitude, which characterizes the mass of the liquid flowing through the coating in a time unit depending on the pressure differential across the coating.

Permeability depends on the temperature, nature and viscosity of the liquid; the shape, radius and length of pores; interaction between the coating material and the liquid (i.e. whether the liquid wets the coating or not), etc. The permeability coefficient can be used both for comparing coatings employed as protective ones in a given liquid (melting) [3.7, 3.22, 3.65, 3.74] and when choosing materials for sealing of a certain coating in order to diminish its permeability and porosity.

Two testing methods are considered for protective coatings: quantitative determination of the liquid permeability coefficient, and indirect estimate of permeability.

3.4.2 Quantitative Determination of the Liquid Permeability Coefficient

This method consists in a forced feeding of a liquid in the laminar flow regime through the coating separated from the base metal, determination of the liquid flow rate and pressure differential across a specimen, and calculation of the permeability coefficient.

Specimens in a form of 25 mm diameter disks are separated from the base metal according to the recommendations given in section 3.3.2.

For both methods end surfaces of the specimens should be flat and parallel to each other. Number of specimens should not be less than three.

A working liquid is fed to the coating from a tank by compressed air. The setup applied provides feeding of the liquid to the lower end surface of the specimen at an increasing pressure and observation of the upper end surface state.

The specimens are previously washed in distilled water, wiped by non-fleece tissue, and dried. The thickness of the specimens is measured in three points by means of a micrometer with an accuracy up to 0.01 mm. Immediately before fixing in the holder the specimen should be completely impregnated by the working liquid. For this purpose a vacuum setup may be used (see Fig. 3.4). Then laminar flow of the liquid is controlled.

In both methods the specimens are installed in the holder in such a manner, that the direction of liquid feeding is perpendicular to the layers of the coating.

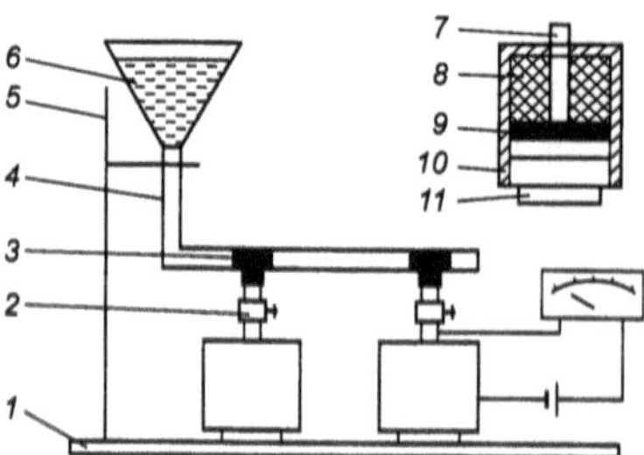

Fig. 3.14. Setup for liquid permeability. *1* basis; *2* cock; *3* T-joint; *4* rubber tube; *5* rack; *6* funnel; *7* brass tube; *8* rubber plug; *9* specimen; *10* cartridge; *11* bush

The liquid permeability coefficient is calculated from the formula (3.15), where Q is the liquid flow rate for the laminar flow regime, ΔP is the liquid pressure differential across the specimen in the laminar flow regime.

3.4.3 Indirect Estimate of Permeability

According to Gotlib, the criterion for permeability when using this testing method is the time it takes for a coating to become conducting after impregnation by electrolyte [3.37].

The disk made of the base metal 25 mm in diameter and 5 mm thick is deposited by a non-conducting coating of the same thickness and with the same subsequent treatment (impregnation, for instance) as was applied to the base specimen.

The setup (Fig. 3.14) allows simultaneous testing of several specimens during sufficiently long time. The setup consists of the basis 1 used for fixing several cartridges. The rack 5 serves for fixing the funnel 6 at a height up to 1 m. 3% water solution of sodium chloride, employed as an electrolyte, is fed through the funnel, rubber tubes 4, glass T-joints 3, cocks 2, and brass tubes 7 inserted into the rubber plugs 8 - to the specimens 9 positioned in the cartridges 10. Thread bushes allow tight pressing the specimens against the plugs.

Countdown of the test duration starts at the initial moment of liquid feeding to the specimens.

The moment of electrolyte penetration through the coating is determined. Indirect estimate of permeability is based on the time it takes for electrolyte to penetrate from the coating surface to the base metal [3.37].

3.5 Electric Strength

3.5.1 General Overview

For non-metallic inorganic coatings used as an electric insulation material, the electric strength should be considered the main property.

The electric strength is an ability of the electric insulation coating to retain high electrical resistance at high electric field intensities. It is characterized by the field intensity, which causes a rapid increase in coating conduction, i.e. the electrical breakdown.

Two types of breakdown can be differed: thermal (due to local heating by electric current) and electric (at a critical concentration of current-carriers).

The thermal breakdown is caused by dielectric and ionic losses at which the heat generation rate is higher than the rate of dissipation. In this case the breakdown is a consequence of thermal instability of material.

At the electric breakdown the discharges are initiated by the local high-voltage fields predominantly in the internal voids of the coating. The discharges cause local destruction of the bridges between pores. Such processes last until the main breakdown channel between electrodes is formed. The electric strength is estimated by the ratio between the breakdown voltage of homogeneous electric field and the specimen thickness.

The breakdown voltage is a minimal voltage which causes the coating breakdown. At a surface breakdown the discharge does not propagate deep into the dielectric but is limited only by the surface layers. Generation of the discharge outside of the specimen is called overlapping. The electric strength of a compact material is always higher than that of similar coatings since the coatings have greater structure defectiveness and an increased content of admixtures.

The electric strength depends on homogeneity of the coating, nature and conditions of the surrounding medium, shape and dimensions of the electrodes. It is also influenced by electric testing parameters such as type of current (alternating or direct), its frequency (industrial or heightened), its character (stepped or smooth), and the rate of current variations. Usually the electric strength diminishes with the temperature increase. Scatter of the electric strength data at room temperature attains 30%. As the testing temperature increases, the scatter of the data grows smaller.

The relative humidity influences the electric strength depending on how moisture absorbed by the testing material or surface moisture influence the dielectric losses and surface conduction.

The electric strength is determined to compare electric insulation materials, spraying technological parameters, indirect estimations of the coating porosity. The tests should be thought of as preliminary, because the breakdown voltage value is dependent on substantial diversity of interacting factors. The electric strength can be finally estimated after on-site tests or based on the operational experience.

The electric insulation coatings are those having the specific electrical resistance equal to 1010 Ohm·m. They are employed for fabrication of insulating gaskets and thermocouple jackets, for protection of inductors in r.f. current plants, in digital-printing and perforating assemblies.

3.5.2 Specimens

The end surfaces of the disc specimens should be flat and parallel to each other. They should not have handling marks, contamination, tool marks, dents, cleavages, and buckling. Mechanical damages appeared in the process of fabrication of the specimens made of thin coatings may have much stronger influence on the test results than the coating structure or electric parameters. The specimens should have round cross-section 25 mm in diameter. In order to prevent overlapping it is desirable to choose the diameter as large as possible.

The number of the specimens should not be less than five per one sort of material or coating regime. If materials used for spraying of coatings fully recover their properties during the tests, than one specimen may be tested many times.

Purity of the contact surfaces processing should be equal to Ra ≤ 2,5–1,25 μm.

When the conditioned specimens are tested under special conditions, it is recommended to fabricate a set of control specimens for testing in the reference medium for comparison.

The electric strength of thick coatings can be determined both across and along the layers. In this case the specimens are fabricated with blind holes to insert special electrodes.

Due to their porosity and specific structure, the coatings display hygroscopicity that reduces their electric insulation properties in the temperature range 20–100 °C at the initial instant of testing. However, heating of the coating during just a few minutes to remove adsorbed moisture results in essential improvement of the electric insulation properties.

3.5.3 Equipment

The breakdown voltage is determined using setups with the schematic shown in Fig. 3.15. The device contains the facility 1 for smooth controlling of voltage, a testing step-up transformer, the chamber 3 that contains the specimen 2 and electrodes. The device should provide pressing of the specimen against the electrodes, continuous voltage recording, and reliable protection of researchers against electrical shock.

The values of the breakdown voltage are read from the scale of meters, which are best placed on the high-voltage side of the electric circuit.

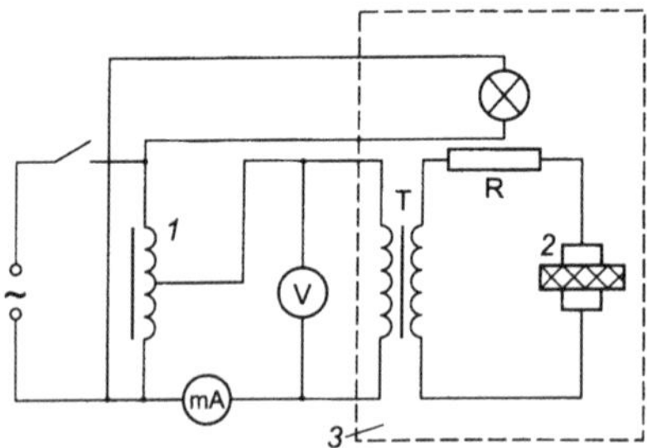

Fig. 3.15. Device for measurement of the breakdown voltage. *1* transformer; *2* specimen; *3* chamber; *T* step-up transformer

There is a variety of industrial plants to measure the electric strength. Apparatuses for insulation testing are designed to determine the breakdown voltage of various materials as well as for cable insulation testing. The highest voltage usually amounts to 50 kV in AC (alternate current) tests, and 70 kV in DC (direct current) tests. The apparatuses are provided with a control console, which is protected by a guard, and a grounding rode to take away an electric charge from the tested specimen and ground the high-voltage terminal.

All commercially fabricated plants have a blocking system to prevent electric current shock. Generation of homogeneous field requires employing electrodes of a special shape whereas usually the electrodes of a simple shape such as disks with rounded edges, spherical, etc. are used. In this case the field is heterogeneous which leads to a breakdown voltage decrease. It is better to employ two cylindrical electrodes of different diameters with rounded edges. To obtain the field close to homogenous, diameter of the lower electrode should exceed that of the upper one by the factor not less than three. The radius of a rounded electrode is chosen depending on the diameter from the following set: (mm): 2.5–50; 1.5–25; 1.0–10. The electrodes are fabricated of copper, brass or noble metals. Roughness of the working surface should be $Ra \leq 0{,}20$ μm. Material and sizes of the electrodes influence the breakdown voltage. Lower breakdown voltages correspond to the electrodes having a larger area. It is typical especially of thin coatings. Probability of the surface breakdown diminishes in accordance with increase of the difference between the electrode and the specimen sizes. The electrodes should have a design that prevents sparking when the voltage is applied.

For special tests the electrodes of other shapes are employed as well. For example, the electric strength was studied in such a system as the needle-plane surface (the needle was positive). The breakdown was realized in 15–30 points for each specimen. This allowed determination of both the average value and dispersion of the electric strength. A knowledge of the dispersion allows judging about the coating homogeneity [3.6].

Since the coatings are promising as an electric insulation material at high temperatures, the variations of electric strength and conduction were investigated for aluminum oxide deposited on nickel plates by the plasma method [3.33]. The electric breakdown of the coating at different temperatures (up to 1600 K) was realized in air between the base metal and a hemisphere electrode pressed to the coating surface. The radius of the electrode hemisphere made of nickel or molybdenum disilicide was chosen in such a manner that the electric field in the breakdown zone was uniform. The voltage applied to the electrodes was increased with the rate of ≈200V/s.

3.5.4 Preparation for Testing

The thickness of specimens is measured in the zone of contact with the electrodes with an accuracy of 0.01 mm before tests.

To prevent the influence of external factors on the results the specimens are normalized and conditioned.

Normalization is a preliminary treatment of specimens in order to eliminate or decrease the influence of the previous coating state primarily in regard to the moisture and temperature. Conditioning is a specimen keeping in specified environment conditions during a certain period of time.

Reference medium is an artificial environment created to compare the results obtained when testing materials in real conditions.

Objectives of normalization and conditioning are as follows: bringing the specimens to the same conditions, reproducibility of the results obtained when testing the same specimens in spite of the previous experiments carried out in various media; determination of influence of special conditions (temperature and moisture) which are encountered when employing coated products. The temperature of normalization varies within the range 50±2 °C, relative moisture of air should not exceed 20%, and the time of keeping is 24 hours. The test specimens are so located that the medium uniformly baths them, while the specimens should not contact either each other or the walls of the heating device. The medium for conditioning and testing is chosen to be similar to the working medium where the tested specimen is going to be used. When conditioning in the liquid both the specimen and the liquid should be preliminary heated up to the same temperature. Impregnation of the coating in the working liquid may be realized in a vacuum plant.

Heat chambers (thermostats) are used for normalizing, conditioning and testing in the temperature range from –80 °C to +400 °C. High temperatures can be reached in the furnaces. In hydrothermal chambers the moisture is checked and controlled along with the temperature.

The working surfaces of the electrodes are carefully inspected, darken areas are smoothed out with abrasive and polished. When mounting the specimens in the device, the electrodes should be pressed to the specimen by central and coaxial compressive force (deviation of the electrodes against coaxiality should not exceed 0.5 mm).

Pressure of the electrodes should be enough to provide reliable contact in the "electrode #1–specimen–electrode #2" system.

3.5.5 Testing

The electric strength is found at alternating, constant and impulse voltages. When determining the electric strength at the alternating voltage applied, the character of the voltage increase up to the breakdown point is to be specified. The electric strength is measured at smooth or stepped voltage rise. When the voltage rises smoothly, it should be increased in such a manner that the breakdown would come in 10–20 s since the initial moment of the voltage rise.

Generally in the beginning the value of the breakdown voltage of a material is unknown. Therefore, several preliminary tests are required to determine the rate of voltage rise.

In the case of stepped loading, 40% of the voltage, preliminary determined in the process of a smooth voltage increase, is applied first. Thereupon the voltage is increased step-by-step with the time delay of 20 s at each stage, and the voltage growth at each subsequent stage should not exceed 10% as compared with the previous step. The transition time from one step to another should be equal to 1–2 s. The breakdown may be evidenced either by a real physical failure of the coating, or by rapid current increase in the circuit. If during the tests surface breakdown or overlapping takes place, it is recommended to pour into the chamber either transformer oil (at a temperature not higher than 90 °C), or silicon-organic liquids at high temperatures.

The number of breakdowns should not be less than five. If the particular results differ from each other more than by 15%, than the number of breakdowns should be increased up to 10, and the electric strength may be determined as an average of all measured values.

Tests of electric strength of the specimens made of aluminum oxide were conducted at the voltage equal to 600±30 V applied during 3 s [3.33]. The electrode was pressed to the specimen from the coated layer side with a stress equal to 300±10 N. Voltage growth rate did not exceed 100 V/s. The measurements were carried out at least in five points spread over the entire surface of the coated layer. The electric strength (E_{st}) in kV/mm is determined by the formula:

$$E_{st} = \frac{U_{br}}{\delta}, \tag{3.18}$$

where U_{br} is the breakdown voltage, kV; δ is the coating thickness in the testing zone, mm.

The correction factor $\alpha > 1$ is introduced when calculating the electric strength at the breakdown in a non- homogenous field, i.e.

$$E_{st} = \alpha \frac{U_{br}}{\delta}. \tag{3.19}$$

The value of α varies depending on the electrodes' shape and dimensions as well as the distance between them.

The results of the tests conducted using the electrodes of different shape and dimensions at unequal temperatures with the specimens of various thickness and other conditions being not equal cannot be compared.

3.5.6 Results and Discussion

The aluminum, magnum, beryllium oxides, as well as aluminum, boron, silicon nitrides possess of good dielectric properties. The breakdown voltage of the electric insulation coatings, with all other factors being equal, is a maximum at the minimal porosity. The electric strength is dependent also on the character of pores distribution according to their sizes, the method and spraying technology, purity of original powder, temperature, etc. [3.6, 3.49]. Besides, the coatings possess of increased structural defectiveness and admixture content as compared to the compact material, which also tells on their electric strength level. It is assumed that the breakdown voltage value and the thickness of the ceramic electric insulation coating are correlated as follows [3.49]:

$$U_{br} = a \cdot \delta^n , \tag{3.20}$$

where a, n are constants depending on the deposition method and thermal treatment of the coating.

Considering aluminum oxide as an example, the authors [3.6] showed that the electric strength of coatings just slightly depends on their thickness, but is determined by the method of deposition, and, consequently, by the coating structure (2 kV/mm – for gas-flame powder coating; 30 kV/mm – for detonation coating at an automatic deposition). The reduction in porosity due to sintering and impregnation by aluminum raises the electric strength by the factor of 1.5–2.

The research results of electric insulation characteristics for aluminum oxide plasma coatings have shown [3.33] that the electric strength within the range of practically used thicknesses (0.1–1.0 mm) amounts to 6–12 kV/mm at the normal temperatures, and with increasing temperature it decreases smoothly down to 1–2 kV/mm at 1250 °C.

The breakdown of heterogeneous ceramic coatings occurs due to conduction of the gas filled in one of the pores. In a dielectric the strength of the field near the pore in which the breakdown occurred rapidly increases. As this takes place, either the dielectric or one of the neighboring pores breaks down. In the case when two neighboring pores located near each other in the dielectric layer break down, the electric field strength between them exceeds the average value by the factor of 6–100. Therewith the effective thickness of the coating decreases thus causing development of an essentially heterogeneous field, which leads to the coating breakdown.

The electric strength of the pores rapidly increases with reduction in their size. For example, when reducing the size of the pores from 500 to 10 μm E_{st} increases from 13.6 to 40 kV/mm [3.6]. During the experimental studies of the electric strength many various pores fall under the electrode. An electric charge that causes the breakdown of the coating develops in the largest pore. In other words, the electric strength of the coating is determined by the size of the largest pores. Therefore, according to Bartenev's data, the electric strength of the detonation coatings is considerably higher than that of the gas-flame or plasma ones.

It is shown in [3.32] that at temperatures above 1000 K the electric breakdown of Al_2O_3 plasma coatings is of a thermal nature, while at lower temperatures it is electric by its character. The conditions of transition to the thermal breakdown are mainly defined by two parameters: the electric conduction and thermal conduction.

In [3.108] it is found that the dielectric strength of the aluminum oxide coating, deposited on the copper substrate, does not change in argon in the temperature range 20–427 °C.

3.6 Bulk and Surface Resistivity. Electric Conduction

3.6.1 General Overview

The electrical resistance is the quantity which characterizes counteraction of the coating to electric current flow. The electrical conduction is the reciprocal of the resistance.

Current flowing through the coating can be divided into two types: surface current (flowing through the thin surface layer and conditioned by the presence of the dissolved salts), and bulk current (flowing directly through the bulk of the coating perpendicularly to its surface). The surface resistance is the ratio between the voltage and the current flowing over the coating surface between two electrodes located on the same side. The bulk resistivity and surface resistivity are the ratios between the voltages and corresponding current densities. The coatings having bulk resistivity less than 10^7 Ohm mm are considered electrically conducting.

The electrically conducting coatings are deposited on pinpoint contacts, which are electrical inlets in vacuum devices in fabrication of electrical circuits according to stencils. They are employed also for the resistive layer in electric heaters, i.e. as heat-liberating elements.

The magnitude of the resistivities depends on porosity, quantity of admixtures and oxides, anisotropy, and the presence of cracks as well as possible phase transitions in the coating caused by temperature changes. The majority of electric characteristics including resistance are influenced to a considerable extent by the external factors such as humidity, temperature, and medium.

It should be particularly noted that during the recent years the intent interest of researchers from many countries is focused on a study of electrical characteristics of thin film coatings deposited by various methods [3.13, 3.29, 3.31, 3.40, 3.45, 3.93, 3.99, 3.110, 3.112, 3.113]. Several principally different techniques using various types of specimens are employed for measurements of the electrical resistance.

3.6.2 Specimens

Specimens with a diameter not less than 40 mm are made of gas thermal coatings in the same manner as for the permeability tests (see 3.3.2). Both of the end (contact) surfaces of the specimens are ground. Sometimes special specimens having the shape different from the disc are used. Thus, Kudinov [3.56] deposited the coating of 0.5–1.0 mm in thickness on the surface of a tube specimen with the substrate of NiCr and, subsequently, sprayed current-collecting and guard rings of nichrome 0.05 mm thick at the top (Fig. 3.16).

Kopylov [3.52] suggested an interesting technique for determination of the transient (contact) electrical resistance on the boundary between the deposited coating and the base metal. The base metal with the deposited coating is fixed in the cell (Fig. 3.17) between copper electrodes and holders of a helical clamp; a compressive force (constant during the measurements) of the clamp is transmitted by means of spherical surfaces.

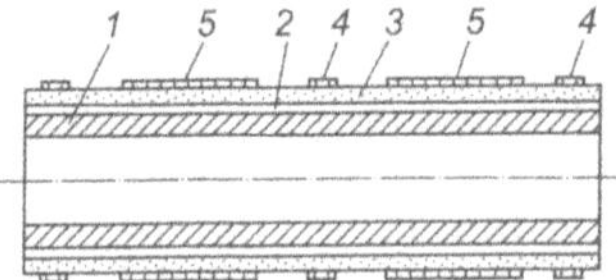

Fig. 3.16. Specimen for testing of electric conduction of coatings: (according to [3.56]). *1* tube-substrate; *2* sublayer of NiCr; *3* tested coating; *4*, *5* guard and measuring electrodes respectively

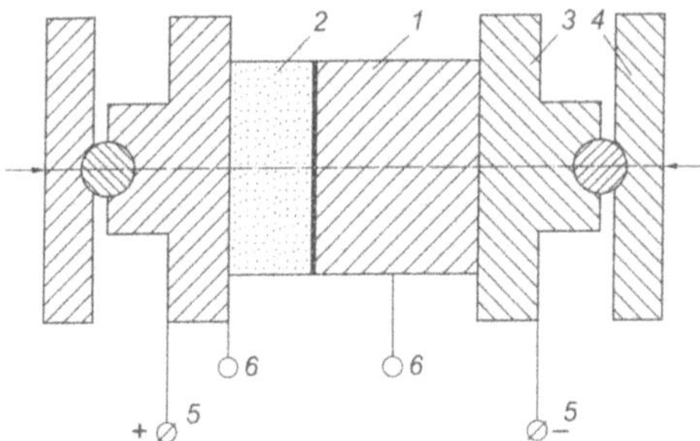

Fig. 3.17. Arrangement of a cell for measurements of the contact electrical resistance (according to [3.52]). *1* base metal; *2* coating; *3* electrode; *4* holder; *5* current outlets; *6* potential outlets

The current outlets are fixed by means of terminal joints; both potential outlets are welded to the specimen and the electrode respectively. The transient electrical resistance is measured through the use of a resistance double bridge. In addition, the purity of the electrode end surfaces, the coating and the base metal are examined; the contact load and the influence of a stray pick-up are to be checked as well.

The internal resistance of flat and tube specimens made of thick multilayer coatings may be determined by the ratio between the voltages and the current flowing between two cylindrical or conical electrodes, located in the blind holes of the coating and having the axes parallel to each other and perpendicular to the coating layers.

When producing the specimens of galvanic coatings, a polished stainless steel or other material providing week cohesion with the substrate is used as the base metal.

Required configuration of the specimens is attained with the help of photographic templates or simply etched off. In order to prevent errors in determining the resistivity, the specimens, when fabricated, should have a uniform thickness along the whole length. The simplest way of getting depositions satisfying to the criteria required is to deposit the layer on a rotating cylindrical cathode made of titanium or stainless steel with further cutting along the element of the cylinder.

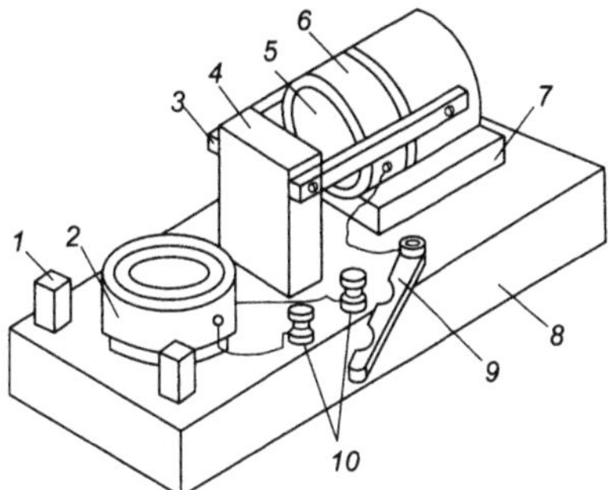

Fig. 3.18. Setup for determination of resistivity. *1* auxiliary supporting poles; *2* measuring and guard electrodes; *3* runners; *4* main pole; *5* potential electrode; *6* load; *7* supporting pole; *8* frame; *9* switch; *10* terminals

At least three specimens should be prepared for one deposition regime. If the specimens undergo conditioning and are tested under special conditions, than for comparison it is advisable to fabricate a set of control specimens to investigate them in the reference medium.

3.6.3 Equipment

A special setup is recommended for testing disc specimens (Fig. 3.18).

The mechanical portion of the setup consists of three main parts: the frame 8, poles 1, 4, 7 and runners 3 with the load 6 providing pressure (9.8 kPa) of the potential electrode 5 on a specimen. The electrode device (Fig. 3.19) consists of potential and measuring electrodes which are fixed on the runners 3 and the frame 8 of the setup respectively (see Fig. 3.19). The electrodes are made of annealed aluminum, tin or lead foil 0.005–0.02 mm thick.

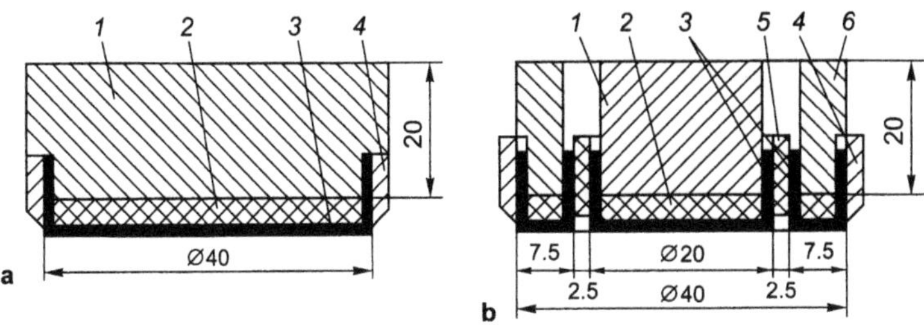

Fig. 3.19. Electrode device (titles of the electrodes are given for the case of bulk resistivity measurements): **a** potential electrode; **b** measuring and guard electrodes. *1* metal holder; *2* gasket made of soft rubber; *3* aluminum foil; *4* metal ring; *5* teflon ring; *6* ebonite support of the guard ring electrode

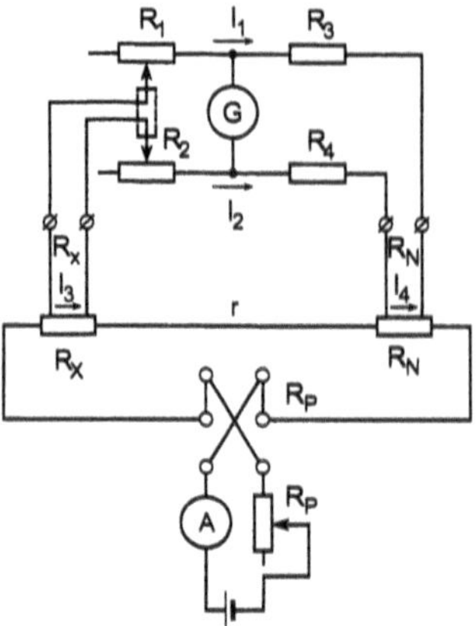

Fig. 3.20. Schematic of the double bridge

The resistance of the coating may be estimated by direct or indirect measurements. Megohmmeters and teraohmmeters are employed predominantly for the direct measurements. The magnitude of the resistance in that case is read directly from the scale of a meter.

Among indirect methods the most widespread is the method based on measuring of a current flowing through the specimen at a fixed voltage. For these measurements ammeters are employed since they allow determination of the direct current with a small error.

The measurements by the double bridge method require detachment of the coating from the base metal. This is usually attained by means of sensitive DC bridges or milliohmmeters employed in the four-contact circuit which eliminates the influence of the resistance of the contact points. The schematic of the double bridge is shown in Fig. 3.20. In this schematic the measured resistance R_x is joined with the reference resistance R_N by means of an ammeter, a power supply, and the regulation resistor R_P. This branch of the circuit is parallel to the second one consisting of the resistors R_1 and R_3, and the third one which includes the additional resistors R_2 and R_4. The resistance of R_1, R_2, R_3, and R_4 is much higher than that of connecting wires and contact points. R_N should be selected to have the value close to R_x. These resistances should be much smaller than R_1, R_2, R_3 and R_4.

When choosing the conditions at which $R_1/R_2 = R_3/R_4$, it may be assumed that

$$R_x = R_N \frac{R_1}{R_3}. \tag{3.21}$$

Determination of the resistance reduces to balancing the bridge by means of R_1–R_2 variations. In so doing the values of R_3 and R_4 remain constant during the whole time of measurement.

The double bridge method is quite well known and universal. However, it is applicable only to the coatings having a thickness within the range 5–50 μm. At smaller thicknesses the obtained values of the electrical resistance are essentially overstated due to high porosity. For thick coatings the electrical resistance may be overstated mainly because of high roughness of the surface.

3.6.4 Preparation for Testing

In order to improve the contact between the electrode and the specimen it is recommended to lap them following the preliminary application of a thin layer of petrolatum, transformer oil or organic-silicon liquid over the surface. The choice of the lubricant is conditioned by the testing temperature: petrolatum is used at the temperatures from –40 °C to +180 °C, organic-silicon liquids and oils – from -60 °C to +250 °C.

The thickness of the coating is measured by a micrometer with the error not more than 0.01 mm at four points on the testing surface of the specimen: in the central point and at equal distances from it. The measured results are averaged.

3.6.5 Testing

When determining the bulk resistivity, the following consequence should be kept in testing. A prepared specimen is installed on the measuring and guard electrode 2 (see Fig. 3.18) being pressed by the potential electrode 5 with the load 6. The latter is transferred to the specimen from the supporting pole 7 by means of rotation round the main pole 4. The electrodes, the specimen and the load are installed in such a manner that their longitudinal axes coincide with each other, and the end surfaces are horizontal.

The terminals of the electrode device are short-circuited by the switch a minute prior to the tests in order to level the potential on the electrodes. Than portable contact points of a meter are set into the terminals. The switch opens the terminals of the electrode device, and then a measuring voltage is applied to the electrodes. Reading of the measured data is taken 3 seconds after the voltage supply. In order to prevent random error it is advisable to test one specimen twice by rotating it along the axis at a certain angle. The schematic of the testing setup is shown in Fig. 3.21.

The control specimens are tested in the reference medium with the following parameters: relative humidity of air is 65%, temperature is 20 °C, atmospheric pressure is 760 mm of mercury column.

The specimens obtained through the use of each deposition technique are tested employing the same device at the same geometrical relative position of the specimens and the electrodes, and under identical external conditions.

The bulk resistivity (ρ_v) in Ohm·mm and the surface resistivity (ρ_S) in Ohm are calculated from the following formulas:

$$\rho_V = P\frac{D_0^2}{4\delta}R_B, \tag{3.22}$$

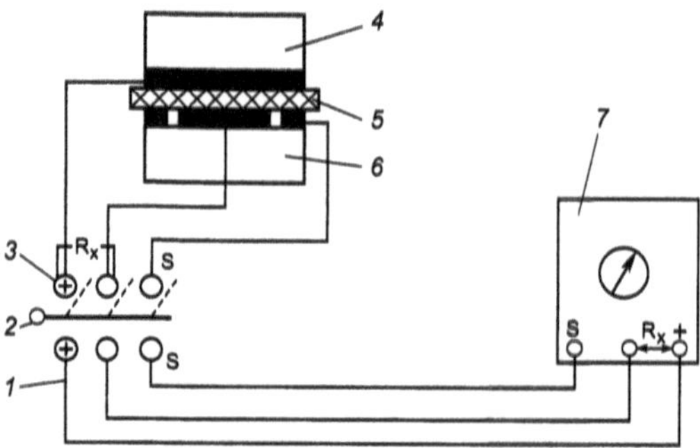

Fig. 3.21. Device for determination of electric resistivity (titles of the electrodes are given for the case of bulk resistivity measurements). *1* portable contact points of the meter; *2* switch; *3* contacts of the meter; *4* potential electrode; *5* specimen; *6* measuring and guard electrodes; *7* meter

$$\rho_S = P\frac{D_0}{g}R_S \,, \tag{3.23}$$

where δ is the coating thickness, mm; R_B is the bulk resistance, Ohm; R_S is the surface resistance, Ohm; g is the width of the gap between the measuring and potential electrodes, mm; D_0 is the average diameter of the measuring or potential electrodes calculated respectively from the formulas:

$$D_0 = \frac{d_1 + d_2}{2} \,, \tag{3.24}$$

$$D_0 = \frac{d_1 + d_3}{2} \,, \tag{3.25}$$

where d_1 is the diameter of the measuring electrode, mm; d_2 is the inner diameter of the guard electrode, mm; d_3 is the inner diameter of the potential electrode, mm.

3.6.6 Results and Discussion

The electric conduction of a dielectric can be represented as the following sum [3.57]:

$$\sigma(T) = \sigma_A + \sigma_C + \sigma_P \,, \tag{3.26}$$

where σ_A is the conduction of admixtures; σ_C is the intrinsic ionic component of conduction; σ_p is the conduction of the ionized material (dielectric and gas in pores) under the action of high voltage or ionizing radiation. Therefore, to decrease the electric conduction of ceramic coatings in air reduction both in material admixtures and porosity of the coating is to be sought for. Therewith it is shown by Kudinov and Ivanov that in vacuum and argon medium the porosity of plasma ceramic coatings does not effect the electric resistivity [3.57].

Coatings of Al_2O_3 having a thickness of 5 mm were obtained by means of high frequency dispersion [3.108]. 99,90% pure powder was deposited on a copper disc 45 mm in diameter. For better adhesion of aluminum oxide with the copper the intermediate layers of nickel and titanium were deposited. The coating had an amorphous structure with the electric conduction up to 10^{-15} Ohm/cm at the room temperature and about 10^{-11} Ohm/cm at 550 K.

The electric resistance and other electric properties of oxide (Al_2O_3, ZrO_2, Y_2O_3, Cr_2O_3) gas thermal coatings have been studied in the context of deposition parameters, and chemical and granulometric compositions of the deposited powders as well [3.77].

The influence of ionizing radiation on the conduction of coatings was studied using a special setup in the temperature range of 20–800 °C; in case of gamma-neutron radiation the study was carried out in the experimental channel of the reactor [3.57].

The experimental results reveal that the electric conduction of all studied plasma deposited coatings increases due to reactor radiation. The authors suppose that the observed increase in electric conduction, besides the direct impact of radiation, is caused by ionization of helium penetrating into pores of the coating. The rate of this effect is not yet estimated [3.57].

In recent studies many authors examined the possibility of obtaining superconductive films on a various substrates, and have measured their electric characteristics [3.68, 3.71, 3.88, 3.109 et al.].

The measurements of electric resistivity of several coatings display anisotropy of the electric conduction. Numerical values of the electric resistance measured in two mutually perpendicular directions, as noted by Zverev et al., differ in 2–3 times. The existence of boundaries between the layers and deformed particles in the direction perpendicular to the axis of deposition leads to the significant decrease in conduction. At the same time, the conduction in the direction parallel to the axis of deposition approaches to that of a sintered solid alloy since in this case conduction of separate layers plays a determinative role. [3.115].

It should be noted that the conduction of metal coatings is 2–10 times worse than that for cast metal. The main reason of this fact is the presence of oxides in the bulk of a coated layer [3.39].

3.7 Thermal Conductivity

3.7.1 General Overview

The thermal conductivity of a material is characterized by the amount of heat transferred in unit time through a unit area of a unit of length thick plate with the temperature difference of 1 degree between its surfaces.

From the physical point of view, the phenomenon of thermal conductivity is a transfer of kinetic energy. In metallic crystals the heat energy is in general case transferred by two types of carriers, conductivity electrons and oscillations of the crystalline lattice (phonons). Respectively, electronic (λ_e) and lattice (λ_{ph}) components of thermal conductivity are distinguished. In metals and alloys the heat transfer by conduction electrons is the dominating mechanism of thermal conductivity, as the lattice thermal conductivity of pure metals is usually low (about 30 times smaller) in comparison with the electronic component. This explains why high thermal, as well as high electric conductivity, is a characteristic feature of the metallic state. The thermal conductivity of non-metallic solids (ion and covalent crystals) caused by oscillations of the crystalline lattice is by 1–2 orders of magnitude lower than that of metals [3.63].

Generally the thermal conductivity of metals is a sum of lattice and electronic components:

$$\lambda = \lambda_{ph} + \lambda_e \,. \tag{3.27}$$

Heat transfer in coatings is more complicated than in compact materials; it is determined by their laminated structure, the presence of a large number of pores, cracks and boundaries between the particles and proceeds through the following mechanisms [3.57]:

- by the electrons in the body of the metallic particles composing the coating, and also in the adhesion zones where strong metallic bonds between the particles were formed during deposition (λ_e);
- for non-metallic coatings (e.g., oxides) of more importance is the phonon conductivity in the particles and in the adhesion zones between the particles (λ_{ph});
- by the thermal conductivity of the gas trapped in the coating pores (λ_g);
- by the radiant heat transfer in pores if the coating is heated to a high temperature (λ_l).

Since the inter-particle boundaries are not completely covered by zones and sources of seizure, and the rates of heat transfer by λ_g and λ_e mechanisms are low, the overall thermal conductivity of the coating calculated as

$$\lambda = \lambda_e + \lambda_{ph} + \lambda_q + \lambda_e \,, \tag{3.28}$$

is much lower than the conductivity of compact materials. As the temperature increases, the contribution of all factors into the thermal conductivity of the coating varies, which explains the complex temperature dependence, differing from the similar dependence for compact materials. Furthermore, due to anisotropy in properties of sprayed materials, the thermal conductivity along the coating layer and normal to it are different.

Thermal conductivity is an operating characteristic of heat-insulation coatings. Besides heat-insulation coatings blocking the heat flow, heat-shielding coatings are often used to protect the elements of construction against thermal effects mainly caused by the heat absorption. The typical heat-insulation materials are AL_2O_3 and ZrO_2, with the densities of 3.99 and 5.60 g/cm^3 respectively.

When choosing among heat-insulation coatings, certain contradictions are to be resolved. Generally together with the thermal protection the coating must ensure protection against gas corrosion, i.e. it should have a sufficient heat resistance. While in the first case it is desirable to apply a porous refractory coating, in the second one the densest coating is preferable. Increase in the coating thickness causing improvement of the heat insulation and thermal stability has a negative influence on the coating-base metal bond strength. The search for optimal ways of increasing heat insulation without sacrificing its thermal stability and bond strength is one of the important tasks when choosing and developing a deposition technology for protective coatings.

The thermal conductivity of coatings is determined by several factors such as the chemical composition of the deposited material, the structure of the coating, the structure of the "coating–base metal" boundary (a loose boundary impedes the heat transfer), the coating thickness.

The thermal conductivity of the coatings should be considered a structure-sensitive characteristic. It depends on the absolute values of porosity, the size of microcracks, the presence of oxides, the size of particles, the area of the contact spot between the particles, etc.

A correct study of the coatings requires analysis of the factors influencing the thermal conductivity [3.38, 3.43], which is the reason for intense studies of thermal conductivity, mainly of ceramic thermal insulation coatings [3.76].

Steady-state methods became the most popular in the field. Their essence is that the temperature at certain points of the tested rod does not change during the experiment. In steady-state methods the ends of the tested rod across which the heat flows are maintained at different, but unchangeable during the process temperatures; thus, the temperature of any point depends only on its coordinates, but not on time. This results in a settled heat flux. The steady-state methods are divided into absolute and relative.

The following absolute method can be used for metals and metal coatings at low and room temperatures. One of the ends of a tested specimen is electrically heated with a certain power (P), whereas the temperature of the other end remains constant. With a sufficiently good thermal insulation of the side surface of the specimens all the power P may be considered completely transferred through any cross-section S of the rod. The settled temperature difference $T_1 - T_2$ between two cross-sections being the distance l apart can be measured by means of two regular thermocouples. The thermal conductivity is calculated from the following equation:

$$\lambda = \frac{Pl}{S(T_1 - T_2)} \tag{3.29}$$

and refers to the average temperature $(T_1+T_2)/2$.

To determine thermal conductivity at medium temperatures (up to 500 °C) the tested specimen is fixed between the heater and water calorimeter which serves as a cooler. The heating of the water in the cooler provides means to estimate the amount of heat Q, transferred through the specimen over a certain time interval.

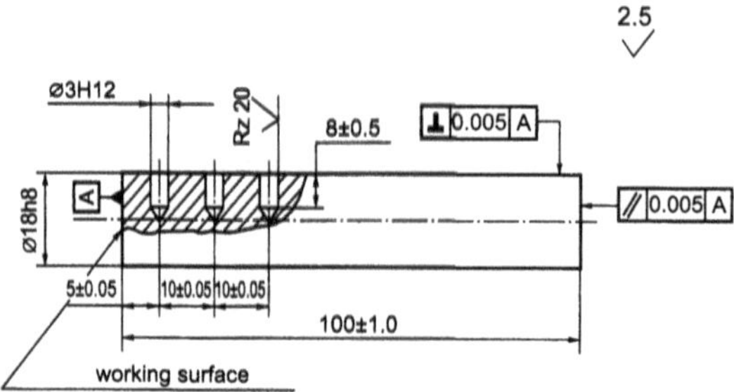

Fig. 3.22. Test specimen

By measuring the settled temperature difference between the points being at the known distance from each other using thermocouples, and knowing the rod cross-section, it is possible to calculate the thermal conductivity of the specimen.

Thermal conductivity λ can also be determined by relative methods in which the absolute value of thermal conductivity is not determined directly, but rather is compared with the known thermal conductivity of a previously studied specimen and is subsequently calculated from a proper equation.

The following sections contain the description of several methods and a detailed consideration of the steady-state absolute method for estimation of the thermal conductivity of coatings.

3.7.2 Specimens

At least three copper or bronze specimens with coatings of equal thickness deposited on the end surface in the same technological regime are used (Fig. 3.22).

Half a diameter deep openings to accommodate the spherical heads of thermal converters are drilled in the specimen. For more accurate estimate of the temperature distribution in the "specimen–counter-specimen" system a junction of thermal converters should be also embedded into the "base metal–coating" boundary and sprayed down directly into the coating (if the thickness of the latter permits) [3.49].

The end surface of the specimen with a coating must be flat and parallel to the coating-base metal boundary. The end surface must be free of swellings, cracks and deep scratches, and the side surface should be free of chips and burrs. To improve the contact it is recommended to lap the coating and counter-specimen surfaces.

3.7.3 Equipment

A setup for estimate of the coating thermal conductivity, whose scheme is shown in Fig. 3.23, consists of the following constructive parts. To generate a heat flux in the "base metal–coating–counter-specimen" system the electric furnace is used. The heaters (spirals) in this furnace are winded to provide specimen heating only in the upper part of the furnace. In other words, the furnace consists of two functionally different parts, with spiral heaters located in the upper part, and asbestos thermal insulation and thermocouples for measuring the temperature across the length of the specimen in the lower part.

The task unit and the temperature regulator ensure the temperature stability in the setup. A potentiometer allowing simultaneous connection of all thermal converters is used for automatic measurements of thermoelectromotive force. The cooler required for counter-specimen cooling and determination of a heat flux through the coating consists of two thermally insulated vessels (thermos). Water is fed into the internal vessel. The water temperature at the inlet and outlet of the thermos is measured by copper-constantan thermal converters. The water flow rate can be estimated using measuring cylinders with a holding capacity from 1.0 to 1.5 l by measurements of the filling time with a stopwatch.

To provide the required contact between the working end surfaces of the counter-specimen and the specimen, the force *P* not less than 500 N should be applied to the latter. To improve the test accuracy the force *P* can be increased. This setup allows determination of transverse thermal conductivity of the coating.

The longitudinal thermal conductivity of aluminum and zircon oxide plasma coatings detached from the base metal was measured using the following installation (Fig. 3.24) [3.57].

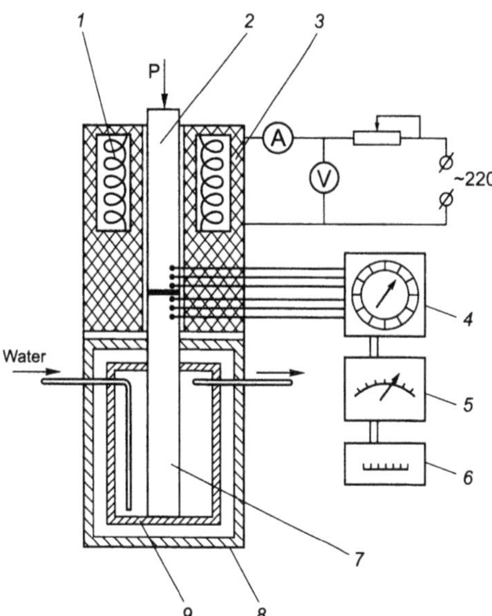

Fig. 3.23. Setup for determination of the thermal conductivity of coatings. *1* heater; *2* specimen; *3* electric furnace; *4* potentiometer; *5* task unit; *7* counter-specimen; *8* thermos; *9* internal vessel of the thermos

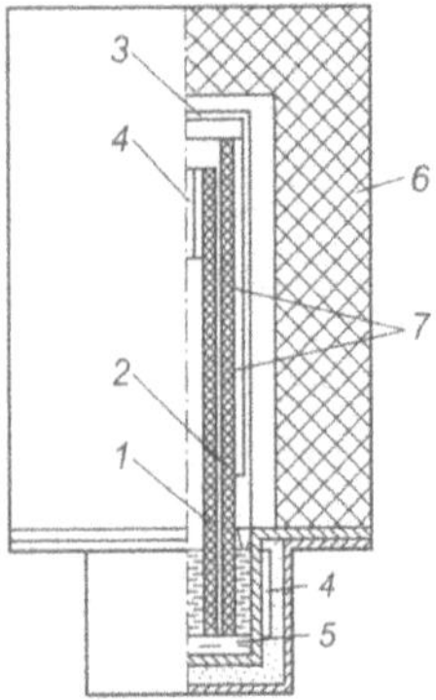

Fig. 3.24. Arrangement for measurement of thermal conductivity by the steady-state absolute technique (from [3.57]). *1* tested specimen; *2* safety device; *3* shields; *4* heaters; *5* eutectic melt; *6* silica fiber insulation; *7* thermocouples

A specimen of cylindrical shape with the length of 100 mm and the wall thickness of 1mm was installed in such a manner that its top was heated by one of the electric heaters, while another end was in a eutectic melt. Application of the safety device, shields, and silica-fiber insulation, as well as measurements of the heat flux along a rather considerable length provided better compliance with the steady-state conditions.

Relative methods of measurement are less common than absolute steady-state methods. They ensure simplification achieved through comparison of the temperature fields in the tested coating and a standard material, previously studied. The thermal conductivity of coatings is estimated through the measurement of the total heat flux against sintered quartz used as a standard specimen. Its thermal conductivity was determined many times, and it possesses of high stability and reliability in the temperature range from 100 to 1700 K.

In the experimental setup (Fig. 3.25) a disk specimen with a thickness of 3–4 mm and a diameter of 23–25 mm was placed between standards of sintered quartz [3.18]. The specimen was manufactured from the base metal and polished on either side. The thermal conductivity was measured under condition of radiation heating from silit rods.

To reduce the radial heat loss, the system consisted of the specimen and quartz disks was surrounded by three concentric protective rings made of asbestos cement and the quartz-sand filling. The temperature difference in the steady-state regime was detected by four platinum – platinum-rhodium thermocouples. The system of the specimen and thermocouples was placed on a copper cooler and pressed by a weight to reduce the transition contact resistance between the specimen, standards and thermocouples. The thermal insulation provides the difference between the heat fluxes through the first and the second standards not higher than 4%.

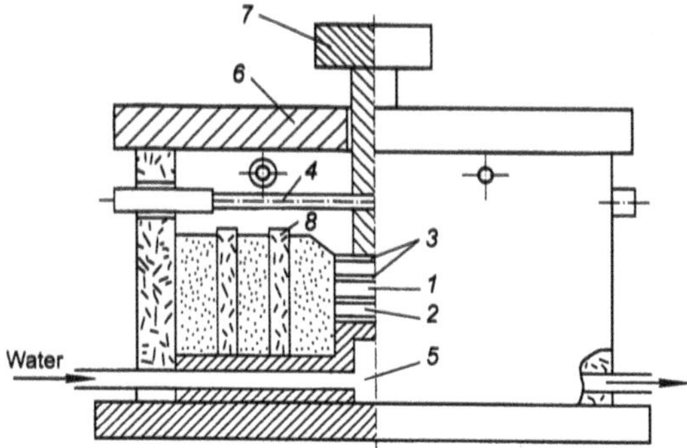

Fig. 3.25. Setup for measurement of thermal conductivity by the steady-state reference method (according to [3.115]). *1* specimen; *2* standard (sintered quartz); *3* thermal transformer; *4* silit rods; *5* coolers; *6* cap; *7* weight; *8* rings

Among the non-steady-state methods the technique of the regular thermal regime, though rarely used because of high errors, may be pointed out. However, this is one of the simplest techniques for estimation of thermal conductivity of multilayer coatings on specimens of different shapes.

The essence of the method is that a coated specimen is placed into a molten metal (aluminum), following which the thermocouples measure the temperature in the melt and in the center of the specimen. Considering the specimen as a calorimeter, the sum of the effective thermal resistance of the coating and the surface resistance of the specimen is determined. With a preliminary measured coating thickness, the thermal conductivity can be calculated by the formulae from [3.57].

Compo Therm Surface Technology Ltd company (Germany) proposed the Xenonpulse – System XP20 for determination of such material properties as the temperature conductivity and thermal conductivity. The control of measurements can be carried out using specimens with the minimal diameter of 9 mm. An energy impulse from a lamp flash is applied to the top surface of the specimen, whereupon a growth of temperature is detected on the opposite side using a high-velocity pyrometer. This system might be used for quality control of metallic, ceramic, amorphous and other plasma-spray coatings [3.3].

3.7.4 Preparation for Testing

All the specimens have to be normalized before testing on the setup (Fig. 3.23). The distance between the centers of holes for thermocouples must be measured with slide calipers for each specimen. The coating thickness is determined and the total area of the coating is calculated at three or four points uniformly distributed over the surface. Then the results are averaged with an accuracy of 0.1 cm^2.

Chromel–alumel (chromel–copel) or other kinds of thermal converters are rigidly fixed be wedges in the drilled holes of the specimen.

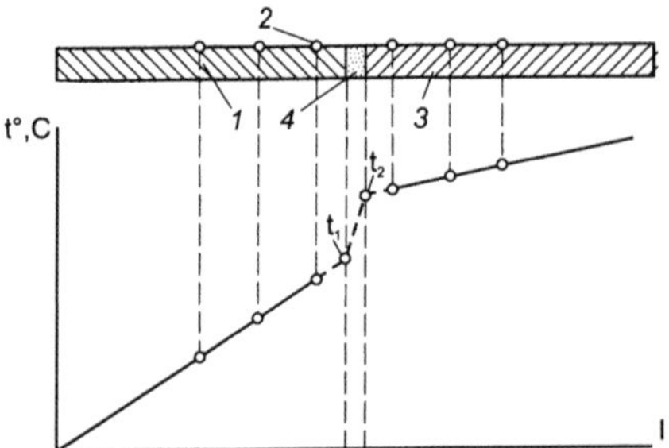

Fig. 3.26. The longitudinal temperature distribution in the "base metal–coating–counter-specimen" system. *1* counter-specimen; *2* places for thermo-transformers; *3* base metal; *4* coating

The specimen with a sprayed coating and the counter-specimen are mounted coaxially in the working space so that their faceplates are tightly pressed to each other, and their common axis coincides with the direction of load application and is normal to the coating surface. To provide a better thermal contact a thin layer of the fine powder (the same material) may be put between the specimen and counter-specimen.

In order to avoid a short-circuit between heaters, the side surface of the specimen may be wrapped with asbestos fiber.

3.7.5 Testing

A required heating temperature of the electric furnace is fixed using a voltage regulator (auto-transformer). The temperature of the external surface of the coating must be ultimately close to the operational temperature of the coated article. The electric furnace is heated up gradually, with the rate not more than 500 °C/hour, to avoid thermal stresses and disruption of continuousness in the "base metal–coating" composition. The heating temperature should not cause structural and phase transformations of the base metal and coating unless this has been stated a research goal.

In parallel with the heating process the water flow rate through the cooler is measured. The water flow rate must be kept constant during the testing.

Then the steady-state conditions for the heat flow are checked up. The flow may be considered steady state if during 60 minutes the temperature variations in the specimen and counter-specimen do not exceed ±5 °C.

If the proper heating of the specimen and steady-state conditions are provided, the reading of thermo-transformers can be taken. Three temperature readings are taken at every point of the specimen every 20 minutes. Simultaneously, the water temperature is recorded at the inlet and outlet.

Based on the data received, the temperature distribution in the "base metal–coating–counter-specimen" system can be plotted. The temperatures on the inner and outer surfaces of the coating are determined from the graph by extrapolation (Fig. 3.26). The thermal conductivity (λ) in W/(m·K) units is calculated from the formula

$$\lambda = \frac{Q \cdot \delta}{S(t_2 - t_1)}, \tag{3.30}$$

where Q is the heat flux through the coating, W, $Q = c \cdot V \cdot \Delta t_W$; C is the specific thermal capacity of water equal to $4.19 \cdot 10^3$ J/(kg·K); V is the mass flow rate of the water passing through the thermos, kg/s; Δt_W is the temperature elevation of the water in the cooler, °C, defined as $\Delta t_W = t_{OUT} - t_{IN}$, where t_{IN}, t_{OUT} are the water temperatures at the inlet and outlet of the cooler, °C; δ is the coating thickness, m; S is the coating area, m^2; t_1, t_2 are the temperatures on the inner and outer surfaces of the coating respectively, °C.

The calculated value of thermal conductivity is related to the average temperature as $t_{av} = (t_1 + t_2)/2$.

The errors in determination of thermal conductivity are caused by non-compliance with the conditions of a steady-state heat flow, a wrong estimation of the temperature distribution in the "base metal–coating–counter-specimen" system, and the influence of the contact thermal resistance between the specimen and the thermal converter.

3.7.6 Results and Discussion

In the general view, variations of thermal conductivity with increasing temperature can be expressed by the formula

$$\lambda_t = \lambda_0 (1 + \alpha T), \tag{3.31}$$

where α is the temperature coefficient of thermal conductivity, negative for compact materials since heating causes thermal conductivity to decrease.

Another pattern is observed in the case of gas-thermal coatings. With growth in temperature, a certain rise in the thermal conductivity takes place. This unusual behavior can be explained by activation of λ_g and λ_e mechanisms.

Analysis of numerous literature data on the thermal conductivity of gas-thermal coatings revealed that the results presented by different authors are inconsistent. It is difficult to state something definite about the relationship between structural characteristics and thermal conductivity of coatings, with the exception of the very general statement to the effect that an increase in porosity decreases the thermal conductivity. In [3.19] the authors tried to link the structural parameters of Al_2O_3 plasma-spray coating with its thermal conductivity. The temperature curve of thermal conductivity was plotted for the interval of 200–900 °C. Then the influence of microcracks, spots of contact between the particles, particle sizes and other structural parameters on the thermal conductivity was analyzed using a PC. This research demonstrated that it is possible to control the thermal conductivity through microstructural variations of the plasma coatings.

Most of the authors studied the influence of deposition regimes and the chemical composition of materials on their thermal conductivity.

It has been shown that for the plasma-spray coatings of Al_2O_3–Al and ZrO_2–Al type the thermal conductivity is 1.17–1.84 and 0.76–1.50 W/m·K, respectively, and it depends on the aluminum content [3.81]. The minimal value of λ was obtained for a coating deposited with the powder of ZrO_2–(15–20) mass % Al. An increase in thermal conductivity was observed with growing thickness of the coating; this can be explained by the enhanced contact interaction of materials in the layer or at the coating–base metal boundary due to high temperature of the "base metal–coating" composition in the course of deposition.

An attempt was undertaken to make theoretical calculations of the thermal conductivity for plasma coatings of Al_2O_3 and ZrO_2 as a function of their porosity through application of the structural model of a disperse substance [3.57]. This model is based upon an assumption that the dispersed material consists of flakes (bricks) laid down in a honeycomb manner. Based on the calculations of thermal resistance of several layer areas and summation of their thermal conductivities, a formula for the total thermal conductivity of the system was derived (for porosity of the system less than 50%). It was established that zirconium dioxide has the calculated thermal conductivity very close to the experimental data, and for aluminum oxide the discrepancy between the theory and experiment is only 19%.

In the book by Khasui, some practical problems of deposition of thermo-insulation coatings with wire-mesh reinforcement were considered [3.49]. The best choice for deposition is the materials with a high melting temperature and a low thermal conductivity (to reduce thermal transfer to the substrate), and ceramics meet these requirements.

Usually the thickness of a coating from aluminum oxide is about 0.025–1.25 mm. In the case of drastic heat supply to the surface of this coating a high temperature gradient developing over the coating cross-section leads to a possible failure of the coating. Besides, thick ceramic coatings are sensible to impact mechanical loads. To prevent the failure of this type coatings they have to be reinforced. A wire-mesh of corrosion-resistant steel is covered by a thick layer of Al_2O_3 and ZrO_2. The coating was not subjected to any additional treatment after deposition. These reinforced coatings were used in rocket nozzles [3.49].

When measuring thermal conductivity with steady-state technique it is also possible to determine the temperature drop on the substrate-coating interface and estimate directly the effect of the contact thermal resistance.

In most cases the study of thermal conductivity by the steady-state methods was performed for oxides. This is explained by difficulties in measurement of the temperature differential on the surface of the metallic coatings.

Along with λ, the value of temperature conductivity is also important in engineering; it is defined as follows:

$$a = \frac{\lambda}{\rho c}, \tag{3.32}$$

where λ is the thermal conductivity; ρ is the density; and c is the thermal capacity.

The coefficient of temperature conductivity in thermal processes is a characteristic of the temperature change rate. The higher the value of a, the less the temperature difference between two areas in the body under the same heating/cooling conditions. Unfortunately, the study of temperature conductivity has not received sufficient attention so far [3.86].

3.8 Temperature Coefficient of Linear Expansion

3.8.1 General Overview

The temperature coefficient of linear expansion (TCLE) characterizes a relative increment in the length of a specimen caused by the increase in its temperature from the lower to the upper limit of the temperature interval as related to the length of this interval. Depending on the value of a selected interval the temperature coefficients of linear expansion can be determined within either minimal or prescribed intervals.

The average TCLE in the prescribed temperature interval will be either greater or smaller than the coefficient determined in the minimal or any other temperature interval. Therefore the substitution of one coefficient for the other when calculating or selecting the coating and base metal materials is not always justified. The TCLE depends not only on the temperature interval of testing, but also on the heating rate, the coating porosity, the size of sprayed powder particles, and the presence of phase conversions in the tested interval. Furthermore, the TCLE is influenced by the coating anisotropy; therefore, the tests should be arranged in view of the main anisotropy axes so that the latter coincide with the axes of the tested specimens. A knowledge of the TCLE is necessary: for optimum choice of materials in the "base metal–intermediate layer–coating" system [3.87]; for calculation of the level of residual stresses [3.44]; for studies of phase conversions in the coating material [3.91]; for prediction of the base metal–coating bond strength level at elevated temperatures [3.73]; for elucidation of the feasibility of the thermal treatment of sprayed coatings [3.94]; for a substantiated choice of materials when formulating multi-component powders [3.78]; for analysis of the thermal fatigue failure [3.27, 3.92]; for studying wear processes [3.62].

In the design of a coating the chemical composition and the layer thickness should be chosen in such a manner as to "equalize" the thermal expansion of the base metal and the non-metallic inorganic coating. This is achieved due to a "stepwise" expansion in the intermediate layers that impedes the peeling of a coating under conditions of the heat impact, for example.

Tentatively, a reliable operation requires the TCLE difference of the coating and the base metal less than $2 \cdot 10^{-6}$ K^{-1}. It is especially important to comply with this condition in the case of brittle non-metallic inorganic coatings. The magnitudes of TCLE of the coating and the base metal are correlated with each other for critical component parts operating at high temperatures, such as gas turbine blades, for instance.

The TCLE estimation method is based on heating of a specimen made of the coating detached from the base metal, and further measuring its length, which varies depending on a temperature that lies in the range 20–900 °C.

3.8.2 Specimens

The shape and dimensions of the specimens are selected depending on the design of a measuring device. The specimens of square cross-section with the length of (50±5) mm having the quadrate side of (4.00±0.5) mm are preferable (Fig. 3.27). Surfaces of the specimens are ground. The presence of cracks, separation and cavities is impermissible.

3.8.3 Equipment

The equipment should furnish variations in temperature in stead-state or non-steady-state regimes, as well as thermostatting within the selected temperature intervals.

The system that transmits expansion of a specimen for further indication should be provided with the compensation of inherent temperature expansion. It is desirable to fabricate the transmission system and the facility for specimen fixing in the device from materials with the minimal TCLE, like sintered quartz, for example. An automatic curve-recording unit would be appropriate for measuring the length change.

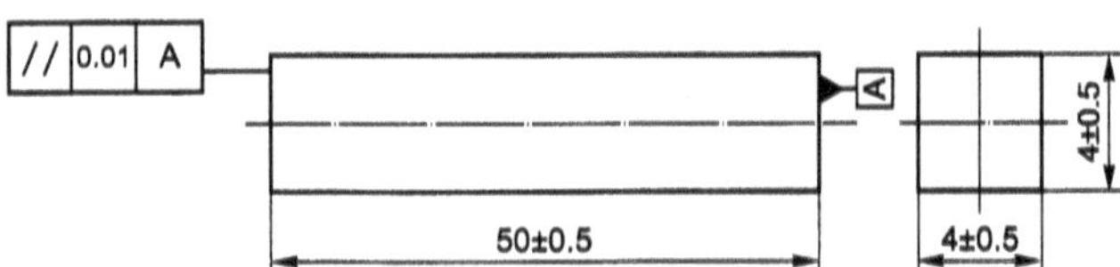

Fig. 3.27. Specimen for measuring the temperature coefficient of linear expansion of the coatings

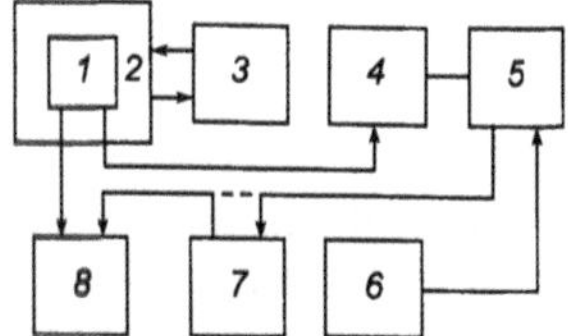

Fig. 3.28. Schematic of the automatic dilatometer. *1* specimen; *2* thermostat; *3* thermal regime control unit; *4* displacement meter with an electric converter; *5* adding circuit; *6* equalization unit; *7* functional voltage converter; *8* two-coordinate electron potentiometer

It is requisite that the TCLE is measured by means of the instruments allowing maintenance of the temperature at a given constant level with an error 0.5 °C or less in the temperature range of 20–900 °C. The temperature metering error should not exceed 3 °C. The head of a thermocouple is located at the midpoint in the immediate vicinity (not more than 0.5 mm) of its surface.

Dilatometers, special instruments of mechanical, optical or electrical type, are employed for recording the length variations. In the dilatometers of the first type the linear displacement is fixed by an indicator or a pen moving along a diagram paper attached to a rotating drum. In the second type instruments the measurement is carried out either via direct usage of various cathetometers or microscopes, or by means of an optical arm, when translational motion caused by the expansion of a specimen is converted into rotary motion, which is fixed by a light flash moving along the scale.

There are several designs of dilatometers, where the linear displacement is converted into an electric signal with the aid of photoelectric or electron tube devices, for example, or by means of various sensors of tensometric, inductive or capacitive type. Based on these converters, the automatic programmable dilatometers as well as the dilatometers for fixing galloping processes at a rapid heating or cooling have been created. Fig. 3.28 shows the functional layout of an automatic dilatometer.

Dilatometers are equipped with specimen temperature control facilities. Sometimes they are also provided with devices assigning a certain heating (cooling) or isothermal self-control program. Temperature measurements in dilatometers are carried out using thermocouples. Besides, in many models dilatometric pyrometers, i.e. reference specimens, characterized by smooth changing of the expansion coefficient when heating or cooling, are also employed. The temperature of the reference specimens may be exactly determined based on their expansion (constriction).

The advantages of such dilatometer pyrometers are their inertia-free character and more precise estimation of the temperature of a specimen heated under the same conditions as the reference one; this is achieved thanks to the registration of the temperature averaged along the whole specimen length rather than at a single point. The dilatometric pyrometers (or etalons) are the integral parts of the so-called differential dilatometers, in which expansion of the reference and tested specimens are registered simultaneously. Differential dilatometers are used more often. Dilatometers that register the expansion of a specimen alone are referred to as simple, being used more rarely.

The TCLE of the coatings may be determined employing widespread dilatometers of capacitive and interference types. Working principle of the capacitive dilatometer is based on a change in capacitance, when one of the condenser plates is bound with the tested specimen and shifts in proportional to its extension. The interference dilatometer is based on a mirror displacement in Linnik instrument and measurement of the interference pattern caused by specimen lengthening on heating.

A dilatometer of "DKM–1" type, which is intended for investigations into the temperature deformation of non-metallic materials and provides continuous recording of a dilatometer curve, is useful in testing non-metallic inorganic coatings at the temperatures below 1200 °C.

The setup is comprised of a working chamber, a specimen heating system, a deformation-measuring instrument, and automatic heating control and thermal deformation recording systems. The specimens can be studied in a medium with controlled composition. The heating system consists of a heating device, a transformer and a voltage buncher. The specimen is heated according to the program assigned by a diagram tape of the controlling device. The deformation measuring system consists of an alundum tube with recesses to fix the specimen for more uniform heating, a measuring piston made of alundum ceramics, and a clock-type indicator. Resistance transducers glued on a cramp are used for continuous recording of a signal, which is amplified by a tensometric amplifier and registered by an electron potentiometer.

A "DKM–1" type dilatometer may serve according to both a simple scheme, when the specimen expansion curve is registered in time, and differential scheme, when the difference between the standard and specimen expansions is registered as a temperature function.

The most exact results are obtained in studies by means of differential dilatometers, which exclude errors caused by lengthening of holding parts of the device.

In the Chevenar differential dilatometer the lengthening of a specimen 1 (Fig. 3.29) and the standard 4, located in the quartz tubes 2 and 3, is transmitted through the rods 5, 6, 11, 12, and 13 to the points 9 and 11 of the invar plate 10, which has a shape of a right triangle and is fixed stationary at the point 7. This point serves as a center of rotation when moving the points 9 and 11. A long-focal concave mirror 8 reflects the light beam from a lighting source to a screen. To heat the specimen and the standard, the electric tube furnace is moved up to quartz tubes 2 and 3. Before starting the tests, a chamber is moved up to the head of the dilatometer from the side of the mirror 8. The chamber contains a cassette for photographic recording and a bulb, which are installed on the opposite side.

A light beam from this bulb passes through the lens and falls on a photographic plate. When moving the mirror 8, the light spot moves accordingly indicating the curve on the photographic plate.

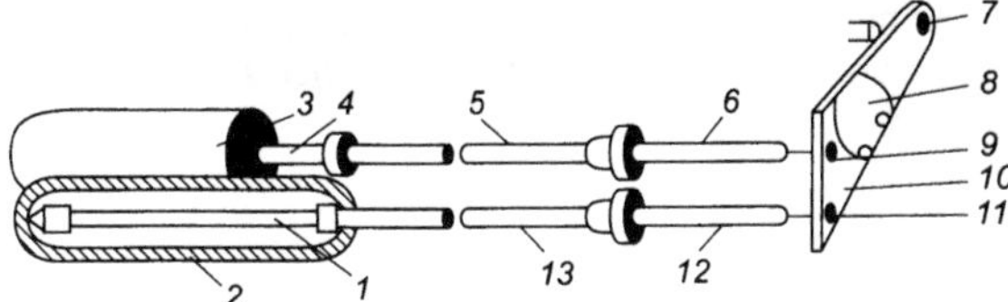

Fig. 3.29. Schematic of a measuring cell of the Chevenar differential dilatometer. *1* specimen; *2*, *3* quartz tubes; *4* standard specimen made of pyros; *5*, *13* quartz rods; *6*, *12* invar rods; *7*, *9*, *11* points of fastening; *8* mirror; *10* invar plate of right triangle shape

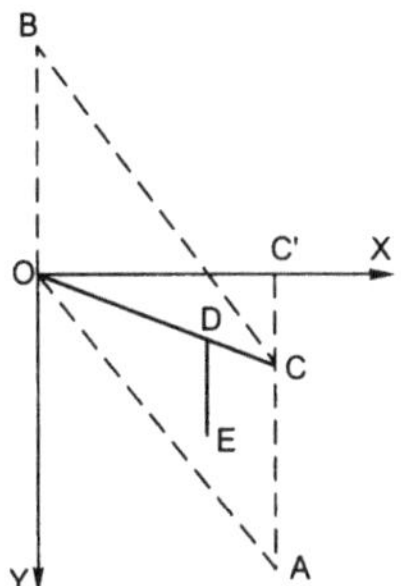

Fig. 3.30. Curve recording scheme for the Chevenar dilatometer

Fig. 3.30 shows the dilatometer curve-recording scheme. If it is assumed that during heating the standard specimen is expanding, then the plate 10 with the mirror 8 (see Fig. 3.29) would rotate around the axis 7–11, and the (x, y) light spot would move on a screen from the 0 point along the dotted line AO (see Fig. 3.30).

If only the specimen expands, then the plate 10 with the mirror 8 will rotate around the axis 9–11, and the light spot will move on the screen upward along the dotted line OB (see Fig. 3.30). In case of simultaneous expansion both of the specimen and the standard, the light spot shifts along the resultant of the mentioned trajectories (solid line OC, see Fig. 3.30). This line is slopped to the down right side because the TCLE of a pyros standard[3] (the alloy with the known coefficient of linear expansion, which is constant within the tested temperature interval) is higher than that of the majority of materials. If at some temperature contraction of the specimen takes place, then the light spot on the screen will move downward (DE segment, see Fig. 3.30). The shift of the light spot along the horizontal direction OX is proportional to the pyros temperature, therefore, given its TCLE, the horizontal axis can be graduated in the terms of temperature.

[3] A material with the known TCLE that is constant in the tested temperature range may be used as a standard instead of pyros.

The range of the tested temperatures is limited by the thermal stability of quartz, and these temperatures are a maximum for the instruments where poles, rods and tubes are made of quartz.

3.8.4 Preparation for Testing

Measuring errors of a dilatometer in a proper temperature range are determined prior to the tests. The TCLE of a reference dilatometer measure (made of monocrystal corundum or other material) certified in a proper manner is determined first. If corrections exceed the admissible values, then the use of a given device is impermissible. In that case additional adjustment and inspection are required. The errors of an instrument should be checked after every 25 cycles of measurement, as well as after dilatometer repairing.

3.8.5 Testing

The TCLE of the coatings may be determined in steady-state and non-steady-state heating regimes. In the steady-state heating regime the specimen's exposure time at a selected temperature should not be less then 30 min in the temperature range 20–100 °C; 20 min – in the range 100–300 °C, 10 min – for the temperatures above 300 °C. Heating rate in the non-steady-state regime should not exceed 0.5 °C·min^{-1} in the range of 20–80 °C, and 2 °C·min^{-1} – at the temperatures above 80 °C[4].

Each specimen is tested just once. The minimal testing interval amounts to 100 °C.

The TCLE ($\overline{\alpha}_{t_2-t_1}$) in K^{-1} is calculated by the formula:

$$\overline{\alpha}_{t_2-t_1} = \frac{\Delta l}{l_0 \Delta t} + \overline{\alpha}_{K(t_2-t_1)}, \tag{3.33}$$

where Δl is a length variation in the given temperature range, mm; l_0 is the initial length of the specimen, mm; Δt is the difference between final (t_2) and initial (t_1) temperatures, K (°C); $\overline{\alpha}_{K(t_2-t_1)}$ is the average meaning of the correction coefficient of a device for the given temperature interval (t_2–t_1).

3.8.6 Results and Discussion

The results of study of the α coefficient for plasma coated aluminum oxide and zirconium dioxide in the temperature range of 20–800 °C have been presented by

[4] Accelerated heating is accompanied by distortion of the obtained results owing to a non-uniform heating along the specimen's length

Virnik et al. [3.106]. The measurements were carried out along the direction of the coating formation (i.e. perpendicularly to the layer) and along the layer. In this particular case the anisotropy of the properties was not disclosed. A slight difference in the magnitudes of the α coefficient lies within the measurement accuracy.

Rangaswamy et al. revealed the dependency of the TCLE on porosity studying coatings of Al_2O_3 and Al_2O_3+13% TiO_2 [3.83]. Powders of a various granular composition were deposited. When testing the specimens with particles of 15–33 μm, the total porosity amounted to 5–6%, whereas the particles less than 15 μm conditioned the total porosity below 2%.

It is evident for both coatings, that the greater is porosity, the higher is the TCLE. This effect can be explained based on the following considerations: first, heating activates cracking caused by the developed porous structure; second, the energy of residual stresses disengages. Probably, these residual stresses lead to further microcracking and thus – to TCLE extension [3.100].

Introduction of metallic admixtures to the ceramic coatings allows controlled changing of the α coefficient of a coating to a considerable extent.

3.9 Thickness

3.9.1 General Overview

The thickness of a coating is a normal distance between the surfaces of the base metal and the external layer of the coating, i.e., the layer, whose surface makes contact with the environment. The local (i.e. at a given point), minimal, maximal, and average thicknesses are distinguished.

Since time consumption and prime cost of the coating production are proportional to their thickness, it is necessary to determine its optimal size, which meets the demands made on a coating at the minimal production cost. [3.16, 3.25].

For practical purposes, it is recommended to determine the average thickness and uniformity of the coating thickness. The same characteristics are appropriate for the technical specifications. The minimal thickness of protective coatings should be regulated with consideration of the medium corrosiveness and required time of protection. Significant variations in the coating thickness are observed at their deposition on the profiled parts. In this case, the minimal coating thickness should be also controlled at some areas. If the reference documentation does not contain demands on the tolerance range limitation, the maximal coating thickness can be increased without a change in the minimal thickness at a constant level of the base metal-coating bond strength and compatibility of parts.

Udler analyzed in detail the problems pertaining to a change in size of car parts after coating deposition. The author cited the thickness of cadmium, nickel, and chromium coatings generally used in the automotive industry and demonstrated possibilities of varying these parameters [3.102].

Being a technological characteristic, the thickness affects such an important service property as the bond strength between the coating and the base metal

[3.41]. An increase in thickness above the optimal leads not only to deterioration (a decrease in the bond strength), but also to economic expenses.

The coating thickness influences the lifetime of nickel alloys at high-temperature extension [3.96], erosion rate [3.15], mechanical characteristics of thin films [3.67], wearing processes [3.28], residual stress values [3.60], fracture toughness [3.75], etc.

Boving et al. consider the thickness one of fundamental characteristics of coatings [3.12]. Due to uniformity of the coating thickness distribution over the item surface, stability of the deposition regime can be estimated. In particular, knowledge of the coating thickness is required to dose the powder consumption.

The average coating thickness in microns is determined as the simple average of the local thickness values. Thickness uniformity (S) in microns is calculated by the formula:

$$S = \delta_{max} - \delta_{min}, \tag{3.34}$$

where δ_{max}, δ_{min} are the local maximal and minimal thicknesses (in microns), respectively.

The most efficient solution to the problem of the choice of the coating thickness on parts can be obtained due to the analytical-calculation method.

The maximal and minimal thicknesses of the treated coatings are caluclated and used for inspection of the chosen technological process conformance to technical conditions of part manufacturing, its purpose and operation conditions, which provide the specified service life. These values are compared with the permissible ones, and the acceptibility and quality of the technological process can be estimated by the results of comparison [3.47].

Due to the numerous combinations of coatings and substrates, the coating thickness cannot be measured by one method only and, hence, using one device. For various "coating-part" combinations, the most appropriate methods and equipment should be chosen depending on the type and shape of the coating and part, the required accuracy and measurement duration, and economic expediency. Permissibility or inadmissibility of failure of the coating or the whole part may become the decisive condition.

When controlling the coating thickness, it should be considered that the coating thickness on the items, especially the profiled ones, is not the same at various points of the surface.

Numerous methods were developed for coating thickness measurements. The comparison allows selection of the most appropriate non-destructive methods to control the finished articles under the working conditions. These methods are as follows: magnetic, eddy-current, and radioisotope ones. The magnetic method is recommended for control of various non-magnetic coatings deposited on the ferromagnetic substrate; the eddy-current method is recommended for control of non-magnetic non-metallic coatings on the weak-magnetic substrate; and the radioisotope method can be used for any coatings.

Other methods are not currently competitive with those mentioned above. These methods are mainly laboratory, and the metallographic method is the chief one.

3.9.2 Metallographic Method

This method is based on thickness measurements using optic microscopes and metallographic lateral microsections. The relative measurement error is ±10%. Disadvantages of this method are relatively high laboriousness and the possibility of measurement at a random cross-section only.

Specimens are cut out directly from the part; otherwise control specimens made according to the same reference documentation are used. Specimens assigned for different tests, e.g., for the metallographic structural analysis, can also be used.

Specimens should be made in accordance with recommendations for metallographic microsections. Special attention should be paid to prevent coating peeling and chipping during grinding. If there is no clear boundary between the coating and the base metal, etching can be used for the higher contrast.

Microsections with broken edges must be rejected. The plane of a microsection should be normal to the coating surface. The angle lap technique widely used in metallography is recommended for measurements of the thin coating thickness with increased accuracy (see Chap. 1). The polishing surface should be changed in such a manner as to make up a small angle with the specimen surface (Fig. 3.31). As a result, when the polished surface intersects with the specimen surface, "extension" of the coating cross-section is observed. This "extension" is proportional to cosecant of the surface slope.

The angle of 5°44′ gives the tenfold "increase" in the coating thickness. This angle can be obtained directly at grinding of a specimen, but it is more reasonable to incline the specimen when mounted into a holder by approximately 6°. This is performed by means of a metal insertion of a given thickness located parallel to the "oblique" surface. Therefore, it is easy to calculate the "extension" due to microscopic measurement of a visible insertion thickness.

The optic method based on measurements of the ledge length between the coating edge and the base metal by means of a metallographic microscope, is a variation of the metallographic method. This method is applicable to thickness measurement for the coatings of 1–40 μm with the reflection coefficient not less than 0.3.

The ledge is obtained in coating formation due to isolation of a small area of the base metal. The coating thickness is measured after removal of an insulating material. The ledge can also be obtained via dissolution of a small coating area with preliminary isolation of the rest part of the surface. The method error is ±10%.

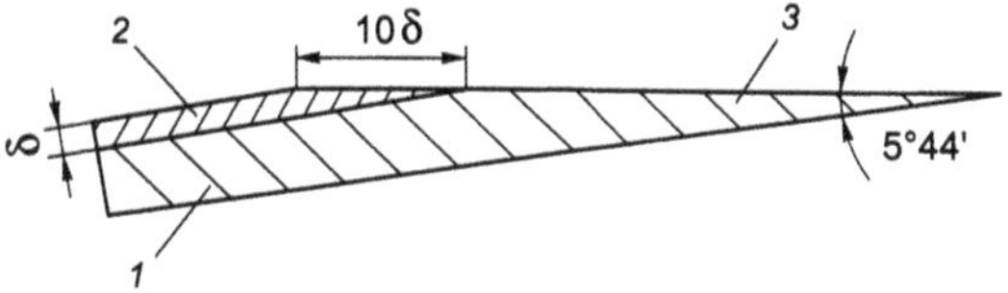

Fig. 3.31. Scheme of angle lap fabrication. *1* base metal; *2* coating; *3* plane of a microsection; δ coating thickness

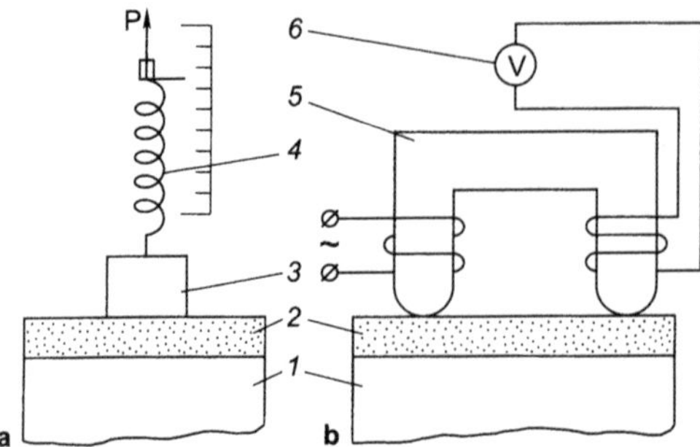

Fig. 3.32. Arrangement of coating thickness measurement by the detachment (**a**) and induction (**b**) magnetic methods. *1* base metal; *2* coating; *3* magnet; *4* spring; *5* bipolar piece; *6* recording device; *P* force of the magnet pullout from the coating surface

Any metallographic microscopes may be used for the thickness measurement. At least 10 measurements are carried out along a microsection. Depending on the coating thickness, the following magnifications are recommended: 500–1000 times up to 20 μm and 200 times above 20 μm.

3.9.3 Magnetic Method

Devices, which use the magnetic method for measurement of the coating thickness, are classified into the following groups:

- devices with constant magnets, whose pullout force is measured by the spring dynamometers (the detachment method);
- devices with electric magnets, whose pullout force is measured by a change in magnetizing current (the induction method).

The detachment magnetic method (Fig. 3.32a) is based on measurement of the force, which should be applied to the magnet to pull it away from the coating surface 2 on the base metal 1, using the spring 4. The pullout force of the magnet correlates with the coating thickness. This method perfectly proved itself under the working conditions in mass production of articles [3.104]. In order to determine the coating thickness the graduated curves for the standard specimens with a known coating thickness are preliminary plotted. The effect of the coating cleanness and structure as well as of thermal treatment and chemical composition of the base metal on measurement results should be attributed to disadvantages of the detachment magnetic method. This method is used to estimate the thickness of non-magnetic coatings deposited on the ferromagnetic substrate, it can also be used for the cases, when magnetic properties of materials differ dramatically. Some devices based on this method are produced serially and characterized by an easy design and portability. The measurement limits for these thickness gauges are 0–2000 μm. The highest measurement error is ±10%, measurement duration is 5–6 s.

In some models the electric magnet replaces the constant magnet, and the force is measured by variations in magnetizing current intensities rather than by the spring dynamometers.

The induction magnetic method (Fig. 3.32b) consists in determination of a magnetic flux through the bipolar piece 5. The voltage measured by the voltmeter 6 logarithmically depends on the coating thickness. This method is used only in the case, when the magnetic permeability of the coating is significantly less than that of the base metal. A single polar piece can also be used as the working probe, but, if this is the case, the measurement error will be greater. Most devices based on the induction magnetic method have transportable sensors – probes, which can measure the coating thickness on hard-to-reach areas of complex parts and in the holes. The measurement range of these devices is from 0 to 10000 μm, a measurement error is 5–10%, and surface roughness of the coating should not exceed Rz20 μm. Devices with point and digital indications are available.

Magnetic thickness gauges can be used, when magnetic permeability of coating and substrate materials differ drastically (in 50–300 times). At the same time, the electric conductivity may be the most diversified.

To increase the accuracy by means of magnetic thickness gauges, it is reasonable to choose the devices with two contact points. Depending on the test conditions (the coating thickness and roughness, the probe curvature radius, etc.), the value of a local thickness can be by 30% less when the single-point devices are in use.

The local coating thickness is defined as the simple average of three measurements, during which the contact points must be within a 10×10 mm square. If magnetic thickness gauges with two contact points are used, one of the contact points must be within the above square, and measurements should be carried out according to the scheme (Fig. 3.33). Values of the local thickness are determined not less than at ten points.

For coarse roughness, a distance between the magnet and ferromagnetic substrate increases, while the efficient interaction surface decreases thus leading to a great error in thickness measurements.

In controlling the coatings with intermediate layers or layers of variable composition, the error of the magnetic thickness gauges becomes 2–3 times greater. Besides, the error of the magnetic thickness gauges increases with a rise in thickness of the measured layer. Therefore, the measurement of thick coatings requires preliminary calibration of devices.

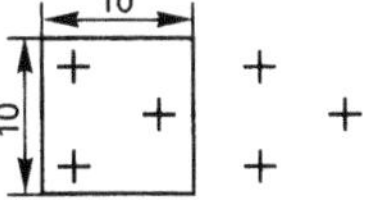

Fig. 3.33. Scheme of measuring by the magnetic two-point thickness gauge. "+" – the contact point of the thickness gauge

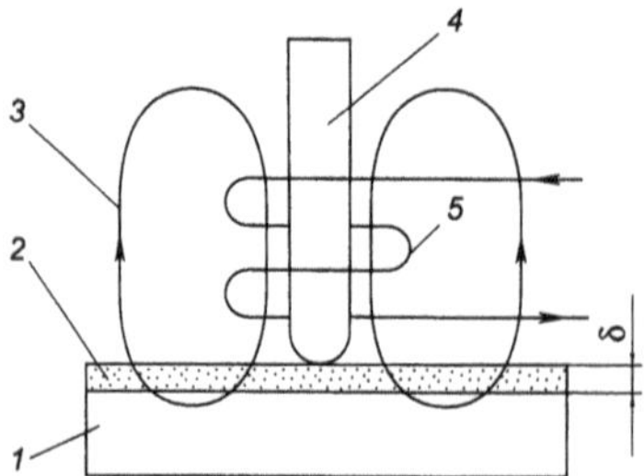

Fig. 3.34. Schematic of an eddy-current thickness gauge: *1* base metal; *2* coating; *3* magnetic field lines; *4* contact rod; *5* probe coil

3.9.4 Eddy-Current Method

The eddy-current thickness gauges are used for non-magnetic coatings and base metals at considerable difference in their electric conductivity.

The principle of different interaction of the coating and substrate materials with the high-frequency current is used in the eddy-current thickness gauges (Fig. 3.34). High-frequency current passes through the coil 5, which is rigidly connected to the contact rod (probe) 4, contacting with the coating 2; as a result, the coating thickness δ determines a distance between the coil and conducting material of the substrate 1. Eddy currents induced by the high-frequency field 3 in the substrate material exert a reverse effect on the electromagnetic field of the coil. The value of this effect depends on the coating thickness and leads to a change in the full resistance of the coil. The resistance change is measured by an electronic circuit and subsequently converted into the readings of the coating thickness.

The eddy-current thickness gauges are used to control the thickness of coatings with electric conductivity, significantly different from that of the material of parts (e.g., copper coatings on titanium), non-metallic films on metal (oxide films on aluminum), and metal coatings on non-metals. The measurement range and error of these devices are almost the same as those of the magnetic thickness gauges. The probes of devices based on the impulse eddy-current method may be of the contact and non-contact type. Despite the complexity of adjustment, the impulse eddy-current thickness gauges allow the control of the coating thickness with an accuracy of ±5%.

A disadvantage of the devices based on the eddy-current method is the impossibility to have a single scale for various combinations of coating and base metal materials. Each combination of the coating and article materials requires an appropriate device calibration. Another shortcoming is the necessity of having a non-magnetic substrate.

3.9.5 Radiometric Method

There are two methods for radiometric control of the coating thickness.

The thickness gauges based on registration of the changes in the flux intensity of reverse scattering of beta-radiation depending on the measured coating thickness are available.

In beta-decay, some portion of radiation falling on a substance is absorbed by the latter, while another portion scatters after reflection. At first, the flux of reflected beta-radiation increases with a rise in thickness of the reflecting substance layer, following which it remains unchanged. The largest substance layer, at which the flux of reflected beta-radiation reaches its maximal value, is called the thickness of saturation 1. The saturation thickness depends on an emitter and reflector's nature:

$$l = \frac{100E^{3/2}}{\rho}, \tag{3.35}$$

where ρ is the density of material; E is the source energy.

With the coating, the flux of reflected beta-radiation would depend on the thickness of substrate and coating metal, and their ordinal numbers in the periodic table of elements. The schematic of registration method for the flux of reverse scattering of beta-radiation is shown in Fig. 3.35. To measure the coating thickness the following conditions should be kept:

- the substrate thickness should be greater than the saturation thickness;
- the coating thickness on the substrate should be less than the saturation thickness;
- the difference in ordinal numbers of the coating and substrate materials should be less than 2–4 units, and the greater is the difference, the more accurate are the measurement results obtained by the given method.

Radioactive isotopes with different penetrability are used as the source of beta-radiation.

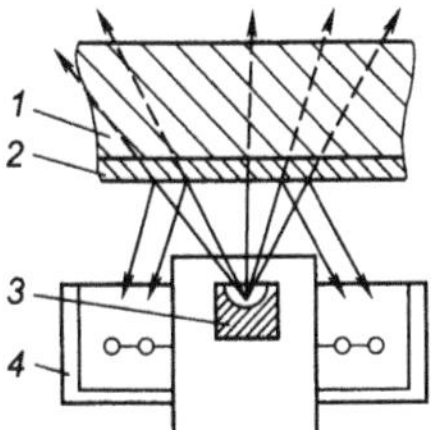

Fig. 3.35. The registration schematic for the flux of reverse beta-radiation scattering. *1* substrate; *2* coating; *3* emitter; *4* radiation indicator

Reverse scattering is registered by various digital-registration devices: ionizing chambers, Geiger and scintillation counters.

Advantages of this method are as follows [3.104]:

- no necessity of achievement of the direct contact with the controlled part;
- long service life of radiation sources;
- convenience for developments involving automated control processes.

Together with these advantages, the radiometric methods and devices for coating thickness measurements have several disadvantages, which limit the fields of their application. They are as follows [3.104]:

- thorough and relatively complex calibration required for each pair of coating and substrate materials; therefore, graduated diagrams and fabricated preliminary specimens are necessary to measure various type parts;
- special precautions for the work with radioactive substances;
- hindered measurements in hard-to-reach places or with very small parts.

Another radiometric method, the roentgen fluorescent one, is based on a change in intensity of roentgen radiation excited by an isotopic source. It is known that intensity depends on the number of radiating atoms, i.e., on the mass (thickness) of a coating. On excitation of coating atoms, the only measure of its thickness is the intensity of fluorescent radiation. On excitation of atoms of a steel substrate, the thickness of a coating layer is estimated by the absorption of fluorescent radiation.

The radiation thickness gauges are used for non-contact control of the coating thickness (for instance, at sheet or tube zinc plating).

The measurement error for radiometric devices is ±10% from the top scale reading.

3.9.6 General Remarks on Thickness Measurements by Non-destructive Methods

The coating thickness can be measured on a part either during assembling or at a finished product. Zones for local thickness measurements should be chosen within accessible areas without knurl or surface defects, and located at a distance not less than 5 mm from edges, fins, corners, holes and contacts between the parts.

A particular calibration curve must be plotted for every "base metal–coating" composition. For this purpose the standard measures with a known coating thickness should be used.

Depending on the purpose of a calibrated thickness gauge, the coating and substrate materials are chosen for the standard measures, and these materials must be similar to those used in the controlled article. Technological regimes of spraying are settled to be similar to those for the controlled parts; the coating should be uniform and perfectly bound with the base metal.

The edge effect, the surface curvature radius, an increased roughness, and variations in the physical-chemical properties and structure of the base metal and coating influence the device readings.

The edge effect of coating thickness measurement is a distortion of the informative parameter of a gauge input signal at the edges of a coated area or at the areas, where the surface shape changes.

The essential feature required of the devices applied for coating thickness measurements is to exclude the effect of various geometrical and physical parameters of the controlled parts on device readings. Among geometrical parameters, the shape of a controlled surface characterized by the curvature radius is of the prime importance. Surface curvature increases the efficient distance between the probe and a controlled article within this area, which breaks the relationship between a device reading and the thickness [3.104].

A similar effect takes place for measurements at the article edges or within transition areas from one surface shape to another. To reduce this effect, the contact surface of the probe is made spherical or cylindrical with a small curvature radius.

In some cases, reduction in the surface curvature effect is achieved by a special magnetic circuit, which reduces or limits the scattering flux. But it should be noted that a decrease in the contact area reduces device sensitivity.

The effect of the surface roughness on device readings should also be assigned to geometrical parameters. The influence of the roughness error on the measurement error is more considerable, especially at thin coating thicknesses (up to 5 μm). When it is small, the substrate thickness also affects the readings. If the substrate thickness is greater than 1–2 mm, this factor does not influence the device readings. Physical properties of the coating and substrate are also important for measurement of the magnetic coating thickness on the magnetic substrate. The effect of chemical composition on magnetic properties of the substrate, the influence of all types of thermal treatment and hardening in a surface layer after some types of mechanical treatment should also be considered.

Usually, reduction or complete elimination of the effect of geometrical and physical parameters on device readings is provided by the following methods:

- corrections by means of tables, diagrams, graphical charts;
- proper design of the probe;
- device calibration for each series of parts of the same type.

Easiness of operation, possible automation of a technological process and operative control may be attributed to advantages of the non-destructive methods. The modern thickness gauges allow the usage of computers both for control of the measurement process and calculation of the average thickness, which should be kept up depending on the certain parameters of the series of articles. Connection of a microprocessor to the thickness gauges allows estimation of the root-mean-square deviation, the variation coefficient and other statistical parameters.

3.9.8 Other Methods of Coating Thickness Measurements

The simplest method of direct measurement is based on determination of dimensions of a component before and after the coating deposition. Slide calipers and micrometers are used for that purpose. The helical micrometers with friction clutch, providing a certain pressure on a measured component, give more precise results. The measuring error amounts to ±0.01 mm. The method is applied to estimation of thick coatings.

When using optimeter, the accuracy increases significantly. In this case a point or a series of points are chosen for performing measurements on determination of geometrical sizes before and after the coating deposition.

Gravimetric method provides for direct weighing and measuring of the coated surface of a specimen or any article. The thickness of the coating is determined according to a difference between the masses of a specimen before and after the electric deposition. The average coating thickness corresponds to the ratio between the deposited coating volume and the surface of the coated product similarly as in the scouring method.

Gravimetric method is usually employed to determine the average thickness of a galvanic coating on small parts with simple geometry, which results from difficulty of measuring surface area of coated items. The relative error of this method is ±10%.

The weighing sensor for controlling thickness and deposition rate of the coatings, obtained by vacuum deposition in electron-beam setups, has been proposed in [3.70]. The mechanism of equal arm balance has been used in this sensor, which employs also a differential transformer and an elecromagnetic coil, placed in a vacuum chamber, separated from a working space by a water-cooled screen. Using the compensation measuring method in the sensor, up to 50 experiments without depressurization of the working chamber may be carried out.

The essence of the thermal electric method is in the measurement of a thermal emf generated when heating a coated specimen. This emf is caused by the difference in thermal electric properties and conductivity of the coating and base metal.

The method is applied to metal coatings with the thickness up to 50 mm and at a difference between the specific thermal emf of the coating and that of the base metal equal to at least 20 mV/deg.

Fig. 3.36 shows the schematic of a setup for measuring thickness of the coatings using the thermal emf method.

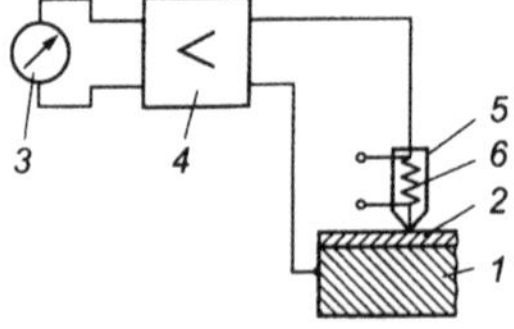

Fig. 3.36. Schematic of the thermal electric method for coating thickness measurements. *1* base metal (substrate); *2* coating; *3* electric meter; *4* direct current amplifier; *5* probe; *6* heating element of the probe

The following conditions should be met when carrying out the experiments:

- individual and careful calibration of the metering device should be fulfilled using proper measuring specimens for each combination of coating and substrate material;
- factors, having influence on the measuring accuracy of the coating thickness (such as temperature of the heater, its contact area with a part, etc.) should be constant;
- a portion of the probe-heater, having a direct contact with the measured part, should be made of the same material as the coating. Otherwise, the additional source of thermal emf may appear in the contact zone between the probe and the coating.

If the above conditions are obeyed, the relative error of the method amounts to ±15%.

The experimental methods for non-destructive measuring of the thickness of coatings deposited on a metal substrate using diffractometers are given in [3.34]. The procedures are based on measurements of attenuation of integral intensity of diffraction lines in the substrate. Preliminary tests were carried out by measuring aluminum and TiN coatings deposited on medium-carbon steel. The results obtained are in a good agreement with the direct methods data. The described methods allow measuring the thickness of the coatings in the range from decimal fractions to tens of microns.

The essence of the capacitive method is implies that the base metal, the dielectric coating, and the electrode applying to the part make up an electric condenser. The capacitance of this condenser depends on a distance between the capacitor plates, i.e. on the coating thickness.

Based on a complex solution to the conductivity and thermal elasticity problems for cooling of turbine blades with a ceramic coatings, Tretyachenko et al. [3.98] suggested a theoretical method for selection of a local thickness. This allows, in particular, reduction in temperature levels of thermal stresses in the blades of a gas turbine engine both for stationary and non-stationary regimes.

Among the recent developments devoted to coating thickness measurements, the "Positector 6000" model designed by Graham and White Instrument Company should be noted here. This facility is a probe instrument for coating thickness measurements [3.3]. Together with compactness, this instrument features simultaneous self-calibration according to the standards of the National Standard Institute (NIST), completely replaceable probing heads, availability of measurements of the coating thickness on the substrates of magnetic and non-magnetic alloys with automatic switching, availability of a special probing head for a steady-state measurements of the coating on a convex and concave details. The instrument allows measurement of the coating thickness up to 4400 μm with an accuracy of ±1% [3.3].

Elecktromatik Equipment Company (USA) manufactures Check Line TSF and TSN devices designed for checking accuracy of instruments used to measure the thickness of coatings made of magnetic and non-magnetic alloys. These devices fully meet the standards developed by the National Standard Institute (USA). Each of the sets consists of three coated plates having a precisely specified thickness, and one uncoated "origin" plate [3.1].

3.10 Roughness measurement

Surface *roughness* is defined as an aggregate of surface irregularities with relatively small widths distinguished within the base length. Post-machining surface roughness is primarily a geometric cutter mark distorted by plastic and elastic deformation and by the action of process system vibrations associated with cutting. Roughness is a principal surface geometric characteristic, which affects the visual appearance, thermal conductivity [3.79], wearing-out processes [3.20, 3.24, 3.35], contact endurance [3.69], base metal – coating composition fatigue [3.101], and corrosion resistance of surfaces [3.54].

Six roughness parameters are established:

R_z – height of irregularities by ten points;
R_a – arithmetic mean profile departure;
R_{max} – overall height of profile irregularities;
S_m – average width of profile irregularities;
S – average width of local profile peaks;
t_p – relative reference profile length.

Surface roughness is standardized and estimated based on one or more of these parameters.

The standards set forth the minimum requirements on surface roughness irrespective of their production or treatment methods. This allows wide application of standard requirements and methods for evaluation of roughness parameters both to machined surfaces and to those measured immediately after spraying [3.30].

Surface roughness can be controlled either qualitatively by checking against reference specimens, or quantitatively through the use of special instruments for determination of roughness parameters. Instrumental methods have enjoyed the widest application due to their higher accuracy and objectivity.

References

3.1 Hand-held coating thickness gauge (1996) PED: Prod and Ind Equip Dig, Febr:20
3.2 NIST-traceable coating thickness standards (1998) Mod Mach Shop 70:368,270
3.3 Thermal diffusivity as the basis of a production process control system (1991) Powder Met Int 23:240
3.4 Agamal Vishal, Marder Arnold R (1996) Simulation of chromized coating processing and microstructure. Mater Charact 36:35–42
3.5 Arata Yashiaki, Ohmori Akira, Li Chang-Siu (1988) Electrochemical method to evaluate the connected porosity in ceramic coatings. Thin Solid Films 156:315–325

3.6 Bartenev SS, Fed'ko YuP, Grigorov AI (1982) Detonation coatings in engineering industry (in Russian). Mashinostroenie, Leningrad
3.7 Bernacchi E, Ferrero A, Gariboldi E, Korovkin A, Pontini G (1996) Coatings in aluminum die casting parts and steel forming tools. Metal Sci and Technol 14:3–11
3.8 Berman VA, Mayorova LA, Sukhova EE (1984) Porosity of chrome carbide detonation coatings (in Russian). Nauchnaya Mysl, Kiev
3.9 Bjomeklett A, Haukeland L, Wigren J, Kristiansen H (1994) Effective medium theory and the thermal conductivity of plasma-sprayed ceramic coatings. J Mater Sci 29:4043–4050
3.10 Bomas H, Mayr P, Kurth B (1997) Fatigue properties of steel coated with TiN by a PVD process. Mater and Manuf Processes 12:17–27
3.11 Borisov YuS, Kharlamov YuA, Sidorenko SL, Ardatovskaya EN (1987) Gas-thermal Powder Coatings (in Russian). Nauchnaya Mysl, Kiev
3.12 Boving HJ, Hintermann HE, Julia C (1992) Characterization of CVD and PVD coatings. Adv Mater Technol Int:81–84
3.13 Boxman RL, Goldsmith S (1993) Mass and surface conductivity gain on polymer surfaces metallized using vacuum arc deposition. Thin Solid Films 236:341–346
3.14 Brindley WJ, Leonhardt TA (1990) Metallographic techniques for evaluation of HSBF thermal barrier coatings. Mater Charact 24:93–101
3.15 Bromark Michael, Larsson Mats, Hedengvist Per, Olsson Mikael, Hogmark Sture (1992) Influence of substrate surface topography on the erosion resistance of TiN coated tool steels. Tribology 11:153–160
3.16 Bull SJ, Rickerby DC, Gent JT (1991) Quality assurance assessment of thin films. Surface Eng 7:145–153
3.17 Bulychev SI, Alekhin VP (1990) Testing of materials by continuous indentation (in Russian). Mashinostroenie, Moscow
3.18 Buzovkina TB, Men' AA (1972) On the use of sintered quartz as a standard for comparative measurement methods of thermal conductivity (in Russian). Eng Phys J 23:669–672
3.19 Buzovkina TB, Sokolova TV, Obukhov AP (1972) The effect of structural parameters and temperature on the efficient thermal conductivity of plasma-deposited aluminum oxide (in Russian). Thermal Physics of High Temperatures 10:395–399
3.20 Cartier M, McDonnell L, Cashell EM (1991) Friction of tungsten carbide–cobalt coatings obtained by means of plasma spraying. Surface and Coating Technology 48:241–248
3.21 Celis JP, Dress D, Maesen E, Roos JR (1993) Quantitative determination of through-coating porosity in thin ceramic physically vapor-deposited coatings. Thin Solid Films 224:58–62
3.22 Chan HC, Liu ZY, Chuang YC (1993) Degradation of plasma-sprayed alumina and zirconia coatings on stainless steel during thermal cycling and hot corrosion. Thin Solid Films 233:56–64
3.23 Cheremskoi PG (1985) Research Techniques for Porous Solid Bodies (in Russian). Energoatomizdat, Moscow
3.24 Coil BF, Sathrum P, Fontana R, Peyre JP, Duchateau D, Benmaiek M (1992) Mechanical properties of (Ti,Al)N films prepared by arc vaporation for cutting tool application. Vide, Couches Minces 48:112–114
3.25 Crostack HA, Jahnel W, Meyer EH, Selvadurai V (1996) Developments in nondestructive testing of surface coatings. Weld World 37:114–120

3.26 Curtis CL, Gawne DT, Priestnall M (1994) The processing and electrical properties of plasma-sprayed yttria-zirconia. J Mater Sci 29:3102–3106

3.27 Cushnie K, Bell JA, Smith GD (1991) Thermal barrier coatings increase applicability of Ni-Co-Cr-iron based alloy. Ind Heat 58:30,32–35

3.28 De Bruyn K, Celts JP, Roos JR, Stals LM, Van Stappen M (1993) Coating thickness and surface roughness of TiN-coated high speed steel in relation to coating. Wear 166:127–129

3.29 Demichelis F, Pirri CF, Tresso E (1993) Degree of crystallinity and electrical transport properties of microcrystalline silicon-carbon alloys. Phil Mag B 67:331–346

3.30 Deuis RL, Yeliup JM, Subramanian C (1997) Aluminum composite coatings produced by plasma transferred arc surfacing technique. Mater Sci and Technol 13:511–512

3.31 Dietrich S, Schneegans M, Moske M, Samwer K (1996) Investigation on metallurgical properties for VLSI applications. Thin Solid Films 275:159–163

3.32 Dudko DY, Primak AV, Falkovsky NI (1982) Electric insulation characteristics of plasma-sprayed coatings of aluminum oxide at high temperatures (in Russian). Poroshkovaya Metallurgia 2:82–86

3.33 Dudko DY, Primak AV, Ivanov VM, Kuzovitkin VF (1992) Electrical properties of plasma-sprayed coatings made of shpinel materials. In: Borisenko AI(ed) Corrosion-Resistant Coatings (in Russia), Nauka, St. Peterburg, pp 81–84

3.34 Edneral NV, Ivanov AN, Kosyak GN, Fomicheva EI (1993) Radiographic measurement of the coating thickness (in Russian). Defectoscopy 9:41–44

3.35 Erdemir A, Switala M, Wei R, Wilbur P (1991) A tribological investigation of the graphite-to-diamond-like behavior of amorphous carbon films ion beam deposited on ceramic substrates. Surface and Coating Technology 50:17–23

3.36 Goswami J, Raghunathan L, Devi A, Shivashankar SA, Chandrasekaran S (1996) Chemical vapor deposition of thin copper films using a new metalorganic precursor. J Mater Sci Lett 15: 573–575

3.37 Gotlib LI (1990) Plasma spraying (in Russian). Scientific and Engineering Information Institute of Chemical Machine Building Press, Moscow

3.38 Graebner JE, HarlneU TM, Miller RP (1994) Improved thermal conductivity in isotopically enriched chemical vapor deposited diamond. Appl Phys Lett 64:2549–2551

3.39 Griffen LA (1988) Peculiarities and prospects for development of electric heaters with surface-distributed heat release. In: Griffen LA (ed) Electric heaters made of composite resistive materials (in Russian). Institute of Materials Technology Problems AN SSSR Press, Kiev pp 3–11

3.40 Hammer P, Helmbold A, Rohwer KC, Meissner D (1991) Electrical characterization of plasma-deposited hydrogenated amorphous carbon films. Mater Sci and Eng A 139:334–338

3.41 Hasugama Hiroki, Shima Yukari, Baba Koumei, Wolf Gerhard K, Martin Herbert, Stippich Frank (1997) Adhesive and corrosion-resistant zirconium oxide coatings on stainless steel prepared by ion beam assisted deposition. Nucl Instrum and Meth Phys Res B 127–128:827–831

3.42 Havsky J, Alien AJ, Long GG, Herman H, Berndt CC (1997) Characterization of the closed porosity in plasma-sprayed alumina. J Mater Sci 32:3407–3410

3.43 Hollander John (1991) Use of ceramic coatings to enhance performance of metal furnace components. Ceram Eng and Sci Proc 12:152–161

3.44 Huntz AM, Lebrun SL, Boumara A (1990) Relation between the residual stresses and the high-temperature oxidation resistance of superalloys protected by plasma-sprayed coatings. Oxide Metals 33:321–355

3.45 Jouan PY, Lemperiere G (1994) Influence of low energy ion bombardment on the properties of TiN films deposited by magnetron sputtering. Thin Solid Films 237:200–207

3.46 Kerkush IR, Melnik PI (1997) Properties improvement of plasma-sprayed iron coatings by chromium plating. Poroshkovaya Metallurgia 40:230–231

3.47 Kharlamov YuA (1982) The effect of geometrical errors of a rough work piece on thickness non-uniformity of gas-thermal coatings after dimensional processing (in Russian). Welding Engineering 7:32–33

3.48 Kharlamov YuA (1990) Prediction of porosity of powder coatings (in Russian). Poroshkovaya Metallurgia 12:36–41

3.49 Khasui A (1975) Deposition techniques (in Russian). Mashinostroenie, Moscow

3.50 Kim Soo-Hyun, Chung Deuk-Seok, Park Ki-Chul, Kim Ki-Bum, Miu Seok-Hong (1999) A comparative study of film properties of chemical vapor deposited TiN film as diffusion barriers for Cu metallization. J Electrochem Soc 146:1455–1460

3.51 Kobayashi Akira, Bessho Nobuyori, Kurihara Setsu, Arata Yoshiaki (1989) Porosity measurement of ceramic coating by image processing method. Trans JWRI 18:25–30

3.52 Kopylov VI (1973) Method for estimation of cohesion between the substrate and coating (in Russian). Fizika I Khimia Obrabotki Materialov 5:72–81

3.53 Kostikov VN, Shesterin YuA (1978) Plasma Coatings (in Russian). Metallurgia, Moscow

3.54 Kovensky IM, Povetkin VV (1999) Physical metallurgy of coatings (in Russian). Joint Venture "Internet Engineering", Moscow

3.55 Kreye H (1998) A comparison of HVOF systems-behavior of materials and coating properties. Galvanotekhnika 89:3006–3013

3.56 Kudinov VV (1977) Plasma coatings (in Russian). Nauka, Moscow

3.57 Kudinov VV, Ivanov VM (1981) Plasma depositing of high-melting coatings (in Russian). Mashinostroenie, Moscow

3.58 Kudinov VV, Pekshev PYu, Belaschenko VE, Solonenko OP, Safiulin AP (1990) Plasma-sprayed coatings (in Russian). Nauka, Moscow

3.59 Kudryavtsev YuP, Dependence of the plasma coating density on parameters of plasma atomizer (in Russian). Izv. VUZov, Mashinostroenie 11:175–178

3.60 Kuroda S (1992) Fundamental phenomena in spray deposition of surface coatings. Res Activ:9–10

3.61 Li CC (1980) Characterization of thermally sprayed coatings for high temperature wear protection applications. Thin Solid Films 73:59–77

3.62 Lin Jen Fin, Li Tzuen Ren (1993) Analysis of the friction and wear mechanisms of multilayered plasma-sprayed ceramic coatings. Wear 160:201–212

3.63 Livshitz BG, Kraposhin VS, Linetsky YaL (1980) Physical properties of metals and alloys (in Russian). Metallurgia, Moscow

3.64 Lizandier MF, Lanza E, Sebaoun A, Giround A, Guiraldeng P (1992) Characterization of porosity of ceramic coatings assigned to applications in sea-water with the evolution of their electrochemical behavior under friction Wear 153:387–397

3.65 Longa Y, Takemoto M (1994) Laser processing high-chromium nickel–chromium coatings deposited by various thermal spraying methods. Corrosion (USA) 50:827–37

3.66 Lyasnikov VN (1994) Properties of plasma-sprayed powder coatings. J Adv Mater 1:381–387

3.67 Ma Dejun, Xu Kewei, He Jiawen (1997) Numerical simulation for measuring yield strength of thin metal film by nanoindentation method. Trans Nonferrous Metals Soc China 7:66–68

3.68 Mahajan S, Wen JG, Ito W, Yoshida Y, Kubota N, Liu CJ, Morishita T (1994) Growth and superconductivity of C-axis in plane aligned $YBa_2Cu_3O_7$ films fabricated by the self-template method. Appl Phys Lett 65:3129–3131

3.69 Middleton RM, Huang PJ, Wells MGH, Kant RA (1991) Effect of coatings on rolling contact fatigue behavior of M50 bearing steel. Surface Eng 7:319–328

3.70 Mischenko BP, Osechkov PP, Novichenko LF (1985) The weight meter of coating thickness and growth rate (in Russian). Problemy Spetsialnoy Elektrometallurgii 4:51–55

3.71 Miyazawa Hajime, Hotta Katsuyoshi, Hirose Haruo, Mureikawa Masao (1992) Obtaining of superconductive films by gas-thermal coating (in Japanese). J Jap Soc Powder and Powder Met 39:748–751

3.72 Monaghan DP, Teer DG, Laing KC, Logan PA (1994) The state-of-the-art in thin protective coatings. Surface Technol Int June:2–5

3.73 Mueller Andrew, Wang Ge, Rapp Robert A, Courtright Edward L, Kircher Thomas A (1992) Oxidation behavior of tungsten and germanium-alloyed molybdenum disilicide coatings. Mater Sci and Eng A 155:199–207

3.74 Ohmori Akira, Zhou Zhan, Inoue Katsunori (1992) Penetration behavior into connected porosities of plasma-sprayed Al_2O_3 coatings with liquid Mn_+. Trans JWRT 21:302–303

3.75 Papyrin AN, Tushinsky LI, Plokhov AV, Bolotina NP, Bol AA, Alkhimov AP (1992) New Materials and Technologies. Theory and Practice of Reinforcement in Extreme Processes (in Russian). Nauka, Novosibirsk

3.76 Pawlowski L, Fauchais P (1992) Thermal transport properties of thermally sprayed coatings. Int Mater Rev 37:271–289

3.77 Pawlowski Lech (1991) Applications and properties of thermally sprayed oxide ceramics. Powder Met Int 23:357–362

3.78 Pawlowski Lech (1991) The properties of plasma-sprayed aluminum–aluminum oxide cermets. Surface and Coating Technology 48:219–224

3.79 Peterson GP, Fletcher LS (1990) Measurement of the thermal contact conductance and thermal conductivity of anodized aluminum coatings. Trans ASME J Heat Transfer 112:579–585

3.80 Plachenov TG, Korosentsev SD (1988) Porosimetry (in Russian). Khimia, Leningrad

3.81 Pustotina SR, Novikov IN, Soloviev BN, Glukhova AK, Grechishkina AI (1983) Development and study of properties of multilayered thermal-insulating coatings made from thermal-reacting ceramic-metal composites deposited by plasma spraying In: Shultz MM (ed) Anticorrosion coatings (in Russian). Nauka, Leningrad, pp 122–127

3.82 Rakitsky AA, De los Rios ER, Miller KJ (1994) Fatigue resistance of medium carbon steel with a wear resistant thermal sprayed coating. Fatigue and Fract Eng Mater and Struct 17:563–570

3.83 Rangaswamy S, Herman H, Safai S (1980) Thermal expansion study of plasma-sprayed oxide coating. Thin Solid Films 73:43–52

3.84 Rogozhin VM, Akimova LV, Kutakova EE (1985) Measurement technique for porosity and density of coating by the hydrostatic weighing (in Russian). Poroshkovaya Metallurgia 6:63–65

3.85 Rogozhin VM, Bobrov GV, Amelchenko NA (1985) On the applicability of different methods of the hydrostatic weighing for estimation of coating properties (in Russian). Poroshkovaya Metallurgia 4:65–67

3.86 Rudajevova A (1993) Thermal diffusivity of plasma-sprayed coatings of ZrO_2 with 8wt% Y_2O_3 and ZrO_2 with 25wt% CeO_2. Thin Solid Films 223:248–252

3.87 Saint-Jacques RG, Bordeaux F, Stansficid B, Veilleux G, Zurak WW, Lakhsasi A, Boucher C, Moreau C (1992) Enhanced resistance of plasma-sprayed TiC coatings to thermal shocks. J Nucl Mater 191–194:465–468

3.88 Salvato M, Attanasio C, Coccorese C, Maribato L, Prischepa SL (1994) Superconducting and structural properties of BSCCO thin films by molecular beam epitaxy. Cryogenics 34:859–862

3.89 Sampath S, Gansert R, Herman H (1995) Plasma-spray forming ceramics and layered composites. JOM: J Miner, Metals and Mater Soc 47:30–33

3.90 Sampath S, Tiwari R, Credmundsson B, Herman H (1991) Microstructure and properties of plasma-sprayed consolidated two-phase nickel aluminides. Scr Met and Mater 25:1425–1430

3.91 Schutz HG, Globmann T, Stover D, Buchkremer HP, Jager D (1991) Manufacture and properties of plasma-sprayed Cr_2O_3. Mater and Manuf Processes 6:649–669

3.92 Shaw Leon L, Barber Brent, Jordan Eric H, Gell Maurice (1998) Measurements of the interfacial fracture energy of thermal barrier coatings. Sci Mater 39:1427–1434

3.93 Silvestre C, Hauser JR (1996) Time-dependent dielectric breakdown measurements on RPECVD and thermal oxides. Thin Solid Films 277:101–114

3.94 Smagorinski M, Tsantrizos P, Grenier S, Entezarian M, Ajersch F (1996) The thermal plasma near-netshape spray forming of Al composites JOM: J Miner, Metals and Mater Soc 48:56–59

3.95 Taylor TA, Appleby DL, Weatherill AE, Griffiths J (1990) Plasma-sprayed yttria-stabilized zirconia coatings: structure–property relationship. Surface and Coating Technology 43–44:470–480

3.96 Tchizhik AA, Getsov LB, Rybnikov AI, Malashenko IS (1995) Creep studies of the EB PVD coatings with a ceramic layer. Thin Solid Films 270:243–246

3.97 Tjong SC (1996) Performance of laser-consolidated plasma-sprayed coatings on Fe–28Mn–7A1–1C alloy. Thin Solid Films 274:95–100

3.98 Tretiyachenko GN, Gribkov YuA, Karpinos BS,Samuleev VV (1994) The method for reduction of the thermostressed state level of turbine-gas engine blades through the choice of optimal thickness distribution of the ceramic thermal-insulating coating (in Russian). Problemy Prochnosti 1:62–67

3.99 Trottier C Michael, Gregory Otto J, Burbank Kenneth A (1992) Dielectric stability of oxides formed on NiCrAlY-coated substrates. Thin Solid Films 214:253–259

3.100 Tushinsky LI, Plokhov AV (1986) The Study of Structure and Physical-Mechanical Properties of Coatings (in Russian). Nauka, Novosibirsk

3.101 Tushinsky LI, Plokhov AV, Sindeev VI (1996) Protection properties of surface layers after high-energy impacts (in Russian). Novosibirsk State Technical University Press, Novosibirsk

3.102 Udler David M (1997) Achieve required CpK values in plated machined parts. Mod Mach Shop 69:80–85

3.103 Usmani Saifi, Sampath Sanjay (1996) Erosion studies on duplex and graded ceramic overlay coatings. JOM: J Miner Metals and Mater Soc 48:51–54

3.104 Valitov AM, Shilov GI (1970) Devices and methods for control of coating thickness. Handbook (in Russian). Mashinostroenie, Leningrad

3.105 Vinokurov GG, Bolotina NP, Larionov VP (1993) Density calculation of gas-thermal coatings (in Russian). Fizika I Khimia Obrabotki Materialov 1:96–100

3.106 Virnik AM, Morozov IA, Podzey AV (1970) Concerning the estimate of residual stresses in coatings deposited by plasma spraying (in Russian). Fizika I Khimia Obrabotki Materialov 4:53–58

3.107 Vityaz PA, Kaptsevich VM, Sheleg VK (1987) Porous powder materials and their products (in Russian), Vysshaya Shkola, Minsk

3.108 Vuoristo PJM, Telama AK, Matyla TA, Kettunen PO (1985) Electrical insulating properties and thermal stability of R.F.-sputtered alumina coatings. Thin Solid Films 126:43–49

3.109 Wang I Yen, Niarchos D, Tsakalakos T (1993) Bi–Sr–Ca–Cu–O superconducting films by plasma sputtering of nanocrystalline compound targets. Nanostruct Mater 2:81–90

3.110 Wang M, Schmidt K, Reichelt K, Jiang X, Hübsch H, Dimigen H (1992) The properties of W–C films deposited by reactive pf sputtering. J Mater Res 7:1465–1472

3.111 Wang Yinglong (1993) Friction and wear performances of detonation-, gun- and plasma-sprayed ceramic and cermet hard coatings under dry friction. Wear 161:69–78

3.112 Wirz Ch, Blatter A, Hauert R (1992) Properties of films prepared by thermal coevaporation of Cr and Ti in nitrogen. Thin Solid Films 214:63–67

3.113 Yamamichi Shintaro, Yabuta Hisato, Sakuma Toshiyuki, Miyasaka Yoichi (1994) (Ba+Sr)/Ti ratio dependence of the dielectric properties for $(Ba_{0,5}Sr_{0,5})TiO_3$ thin films prepared by ion beam sputtering. Appl Phys Lett 64:1644–1646

3.114 Yi Maozhong, Zhang Xianlong, Ji Gengshun, Zheng Jihong, He Jiawen (1997) Erosion wear of AlSi-graphite and Ni /graphite abradable seal coatings. Trans Nonferrous Metals Soc China 7:99–102

3.115 Zverev AI, Sharivker SYu, Astakhov EA (1979) Detonation sputtering of coatings (in Russian). Sudostroenie, Leningrad

4 Investigation of Protective Properties

4.1 Heat Resistance

4.1.1 Terms and Definitions

Heat resistance is a property of material to withstand chemical surface degradation (oxidation) when exposed in idle state to hot air or other gas. Heat-resistant steels, alloys and coatings are required to perform reliably at temperatures above 550 °C. Oxidation processes are generally divided into two classes: solid oxidation (uniform or non-uniform) occurring over the entire external surface, and local oxidation which takes place in separate areas. Local corrosion areas are sometimes called pits.

The State Standard GOST 21910 sets up terms and definitions for heat-resistance characteristics of metals. For uniform corrosion (oxidation) the following basic notions are to be used. *Specific mass loss (growth)* is a reduction (gain) in mass within the considered interval of time referred to unit surface area. *Uniform corrosion depth* is an average size of the surface layer removed as a result of corrosion within the considered interval of time. *Average mass loss (growth) rate* is a ratio of the difference between specific mass losses taken in the beginning and in the end of the considered interval of time to the length of the interval. *Average rate of corrosion penetration* is a ratio of the difference between uniform corrosion depths taken in the beginning and in the end of the considered interval of time to the length of the interval.

Heat resistance against local corrosion is characterized by the following basic parameters. *Maximum depth of surface corrosion pits* is the largest value of the pit depth over the entire set of measurements. *Average depth of surface corrosion pits* is the arithmetic average of all measured depths. *Pitting level* is a ratio of the number of corrosion pits to the value of the uniform corrosion depth. *Density of surface corrosion pits* is a ratio of the number of surface corrosion pits to the surface area. *Average growth rate of the maximum corrosion pits* is a ratio of the difference between the deepest corrosion pits taken in the beginning and in the end of the considered interval of time to the length of the interval.

4.1.2 High-Temperature Gas Corrosion of Metals

High-temperature corrosion is caused by thermodynamic instability of materials under the given pressure and temperature. Any chemical reaction between metal and oxygen may be regarded as a typical case in point. The resulting oxide film can be either porous and brittle or nonporous and solid; in both cases the processes of gas corrosion are heterogeneous and proceed at the gas metal interface.

Formation of a porous film proceeds via the following stages [4.71, 4.78]:

- oxygen transport to the surface;
- adsorption of oxygen molecules at the metal surface;
- metal-oxygen chemical interaction with formation of oxides;
- removal of the corrosion products.

The rate of a porous film growth rate remains constant, and the slowest stage of the process is the chemical interaction.

The process of solid protective film formation can be divided into the following separate stages [4.71, 4.78]:

- transition of metal in the form of ions and electrons from the metal phase into the oxide: $Me \leftrightarrow Me^{n+} + ne^{-}$;
- migration of metal ions and electrons in the oxide layer;
- oxygen transport to the oxide film-oxygen interface;
- oxygen adsorption at the surface of the oxide film;
- ionization of the absorbed oxygen: $O - 2e^{-} \rightarrow O^{2-}$;
- migration of oxygen ions in the oxide layer;
- chemical interaction forming oxide.

The rate of the process is determined by the rate of one of these stages depending on ambient conditions, thickness and quality of the forming film. As the temperature goes up, the corrosion rate begins to obey the laws of diffusion. The thicker the film and the higher its quality, the sharper diffusion nature of the process reveals itself. Therefore, the better the protective properties of the film, the slower it grows. Sometimes film growth proceeds with strong logarithmic decay.

In order to have protective properties, an oxide film must meet the following requirements [4.71]:

- to be solid and flawless;
- to have good adhesion with metal;
- to have temperature coefficient of linear expansion (TCLE) close to the value typical for metal;
- to be chemically inert against the specified aggressive medium;
- to have high hardness and good resistance to wear.

If the formed oxide film is porous, brittle and has poor adhesion with metal then, even if the film is inert to the given aggressive medium, it will not be protective.

At the later stages of oxidation the rate of reaction depends on whether the thick oxide film remains solid or, as the film growth proceeds, it develops cracks and pores reducing the protective qualities of the film. Films of reaction products are often brittle and low-plastic, and the mode of crack formation depends on whether the film experiences stretching as the growth proceeds, or the crack is formed during compression [4.71].

Iron forms a number of chemical compounds with oxygen such as FeO, Fe_3O_4 and Fe_2O_3. At temperatures below 570 °C the oxide layer consists of two oxide zones: Fe_3O_4 and Fe_2O_3 having complex crystal structure. The rate of diffusion in this comparatively thin layer is rather low. At temperatures above 570 °C the oxidized layer consists of three oxides: FeO, Fe_3O_4 and Fe_2O_3, with FeO being the main one. Under these conditions an abrupt rise of the oxidation rate as a result of accelerated atomic diffusion through the elementary crystal lattice is observed.

The problem of producing heat-resistant alloys consists in preventing FeO formation and creating conditions beneficial for Fe_2O_3 formation. In this case the rate of oxidation becomes much lower due to slower diffusion through the complicated lattice of Fe_2O_3.

Since chromium, aluminum and silicon have large chemical affinity to oxygen, upon introduction of one of these components into steel the oxidation process results in formation of dense oxides, such as Cr_2O_3, Al_2O_3 or SiO_2, which are hardly permeable to diffusion. The thin film of these oxides protects the metal from further oxidation. Thus, a way to produce heat-resistant alloys is to alloy steel with chromium, aluminum or silicon. The higher the working temperature of a piece, the larger the content of these elements forming a stable oxide film is required.

At the same time many even alloyed steels intended for operation under high temperatures do not provide the necessary heat-resistant properties. Such refractory metals as molybdenum and tungsten are readily oxidized due to formation of volatile oxides. The scale formed on tantalum and niobium surfaces cracks and comes off the surface. Alloys based on nickel and cobalt, being utilized as refractory materials, do not always provide the necessary resistance to oxidation and are thus not heat-resistant enough. All these problems can be overcome using coatings which, although have poorer mechanical properties than the base metal, provide adequate protection against oxidation [4.10, 4.78].

An optimum selection of source materials and deposition techniques must provide maximum scaling resistance, sufficient coupling strength, minimum open porosity and gas permeability, similarity of the temperature expansion coefficients of the coating and the base metal.

The main task in investigation of heat-resistant coatings is evaluation of the role of coatings in the protection of the base metal. The criteria for heat-resistance can be used for comparison of protective properties of the coatings under study, prediction of the life time of the "base metal–coating" composition in actual operating environment under high temperatures and oxidizing conditions, establishing the mechanisms of gas corrosion [4.1, 4.10, 4.59, 4.70]. Heat-resistance characteristics of protective coatings in many cases dictate durability and reliability of critical products and constructions at high temperatures. The most practical methods for determination of heat-resistance of metals are standardized in the State Standard 6130.

4.1.3 Determination of Heat-Resistance of Coatings by Mass Change and With Direct Measurements

To test the heat-resistance a specimen is exposed to the given air or gas medium at a constant temperature for a period of time sufficient for establishing the mechanism of gas corrosion. The heat-resistance characteristics here can be determined by three methods:

- specimen mass growth (method A);
- specimen mass reduction (method B);
- direct measurement of the corrosion depth (method C).

In the methods A and B a coating is exposed to corrosion, and its heat-resistance characteristics are determined from mass growth as a result of oxidation or from the difference between the masses of the specimen before the oxidation and after the removal of the corrosion products from the specimen surface, respectively. In the method C heat resistance is determined from the change of linear dimensions of the specimen.

The methods rely on the following assumptions: the coating material rather than the base metal is mainly subjected to corrosion; the forming oxides do not evaporate; no intense coating-base metal inter diffusion takes place during testing.

The choice of the specific heat-resistance measurement method is governed by testing conditions and by the base and coating materials. Method A is expedient to use for any coatings during large-scale production tests and in revealing oxidation kinetics. It should also be used if the forming oxides are difficult to remove. Method B is best suited for testing metal coatings with easily removable oxidation products. Method C is recommended for coatings formed at high temperatures in non-uniform oxidation conditions or when the first two methods cannot be used.

For more reliable evaluation of the protective properties of a coating parallel application of one of the weighing methods and method C is recommended.

The following principle can be used as a guide when choosing specimen dimensions and geometry: for improved accuracy of results large specimens with flat, maximum area surface should be taken. In practice the specimens are usually prepared from 3±0.2 mm thick, 30–60 mm long and 20–30 mm wide rectangular plates cut out of the base metal. Prior to deposition the existing burrs are mechanically removed by grinding and edges are rounded with radius about 1.5 mm. The thickness of the base metal is measured in 3–5 points using a micrometer. The obtained results are averaged. In the methods A and B the coating is deposited uniformly on each of the six faces of the plate. In method C the coating can be deposited on one side only.

In order to improve testing accuracy it is recommended to finish the surface of specimens to equal roughness.

At least three specimens should be used for each variant of deposition. Base metal specimens without coating are used as controls.

The installations for testing heat-resistance should provide automatic temperature control and stabilization in separate points of operating volume. External heating of specimens should be used, since electric current passing through a specimen causes significant temperature gradient not only along its length but also over its cross-section as a result of non-uniform current density distribution due to different conductivities of the base metal and the coating. Electron, plasma and high-frequency current heating are also not recommended.

A number of specialized furnaces and other heaters for testing compact specimens are commercially available.

For high-temperature specimen testing in aggressive gas media a heater must

- have enhanced heat-resistance,
- provide the required frequency range and sterility of heating conditions,
- perform without sublimating and corroding,
- and, obviously, provide stable heating.

All these requirements are met in heaters using focused radiant energy. Several such installations have been created in Ukraine. A furnace with specimen heating by radiation focussed by elliptic-cylindrical closed-type surfaces was designed in the Ukrainian Institute of Strength Problems Research [4.21]. The furnace provides heating rates at least 100 °C per second and is capable of producing high temperatures (up to 1800 °C) on the surface of the specimen. A graphite emitter fitted at one of the focuses of the elliptic-cylindrical heater serves as a source of radiant energy. Inside the operating chamber of the optical furnace the zones surrounding the emitter and the specimen are separated, which makes possible specimen heating in an active gas medium.

Several solar stations conventionally called SGU (SGU–1, SGU–2M, SGU–3, etc.) and the "Kaskad" station designed in the Ukrainian Institute of Materials Technology Problems allow studying the influence of various factors including chemical composition and working medium pressure on the efficiency of protective coatings [4.21, 4.51]. The main advantages of solar stations are: easily-controlled high temperatures (up to 3500 °C) obtained in chemically pure (free of heater burning fragments) media; easy access to and possibility of the visual inspection of the specimen in the course of experiment; ease of maintenance; harmlessness and safety for personnel. In addition, the heat in solar stations is applied from one side only, which in most cases conforms to real working conditions of products with protective heat-resistant coatings.

Fig. 4.1 shows a schematic diagram of one of the stations. Its optical system consists of two floodlight-type parabolic mirrors 2 m in diameter and two flat heliostats, one of which tracks the Sun movement and reflects radiation flow onto horizontally-axial concentrator, while the second heliostat, installed at an angle of 45° with the horizontal axis, irradiates the vertically-axial mirror. Despite certain energy losses on heliostats, such a design is versatile and should provide the most beneficial conditions for precision experiments on heating of stationary objects placed in the horizontal or vertical plane.

All solar concentrating stations can be divided into high-temperature (2500–3000 K), middle-range (1500–2000 K) and low-temperature stations (500–800 K) [4.51].

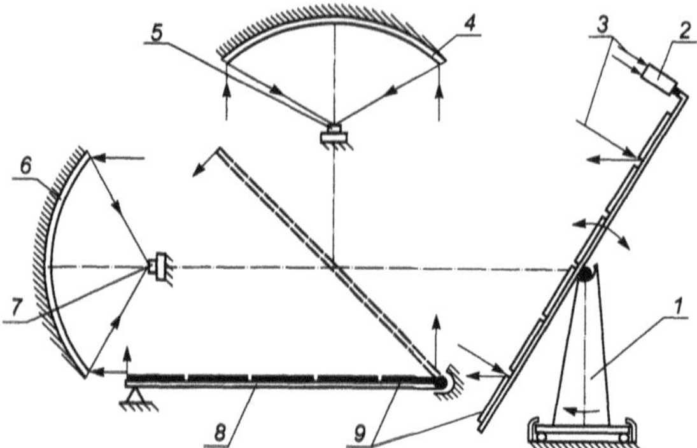

Fig. 4.1. Schematic diagram of double-position solar station (adopted from [4.51]). *1* heliostat for Sun movement tracking in azimuth & zenith; *2* optical sensor of the tracking system; *3* sun beams; *4* vertically-axial concentrator; *5*, *7* specimen with traversing probe; *6* horizontally-axial concentrator; *8* auxilary heliostat; *9* (0.5x0.5) m flat bevels

In testing by methods A and B the specimens should be weighed either periodically or continuously. For continuous weighing a specimen or a crucible with specimens placed inside a shaft furnace is suspended by a wire made of heat-resistant steel from the balance beam replacing one of the balance pans (see Fig. 4.2). The balance with the crucible, wire and specimens is set to zero. The beam of the balance 2 is rigidly connected with the core of a differential transformer. Once the weight of the crucible 6 placed inside the furnace 5 has changed, the balance beam is inclined forcing the core to move inside Kt1 and Kt2 coils which are connected with R1 and R2 resistors in a bridge circuit. The potential difference induced in the bridge by the moving core is amplified by amplifier 4 which then passes the signal to the recorder 3. A disadvantage of the scheme is creation of magnetic forces which have negative influence upon the weighing accuracy.

A photoelectric recorder does not suffer from this drawback. A light beam coming from the illuminator is incident on a mirror fitted on the balance beam. After reflecting from the mirror, the beam illuminates two photoresistors through an optical system. Once the balance beam is positioned horizontally, both photoresistors are equally illuminated so the output current is not generated. When the balance beam inclines because of load change the distribution of illumination for each photoresistor changes. In the extreme only one of the photoresistors is illuminated. The difference in illumination conditions generates output current which after amplification is passed to a recorder.

Other design schemes for continuous specimen weighing have also been suggested. One of the instruments is fitted with an automatic recorder for measuring the mass of the specimen with the accuracy of $2 \cdot 10^{-4}$ g at temperatures below 1300 °C. In order to remove side effects the following precautions are taken: the heating furnace slides along vertical guides using electric drive, the specimen suspension system is made of platinum wire, while the balance is fitted with a water-cooled protective shield.

The continuous weighing method has the advantage of avoiding coating damages (cracks, cleavages) caused by repeated cooling and heating of specimens and is free from errors due to gradual degradation of crucibles.

All specimens are thoroughly examined before the test. The coating should be homogeneous and firmly coupled with the base metal without cleavages, swells, cracks and other defects.

For methods A and B the thickness of the coating is determined by measuring the difference of the thicknesses of the specimen with micrometer before and after deposition. Edge length of each specimen is measured with the accuracy of at least ±0.1 mm. The total area of the coating surface is determined from averaged results of the measurements. For method C the average thickness of the coating is determined.

The specimens are washed in benzine or acetone and in alcohol, dried in a drying chamber and cooled in desiccator down to room temperature. The specimens are taken out of desiccator and weighed immediately before testing. Weighing accuracy for methods A and B must be ±0.1 mg or better.

In methods A and B the specimens are weighed together with crucible. The latter must be annealed to constant mass and stored in a sealed package or in the desiccator until the specimens are loaded.

The specimens in the crucible are arranged so as to provide free access for air, which means they should not touch either each other or the walls of the crucible. The testing accuracy is higher if specimens are put with their narrow side downward so that equal spacing be left between them and the walls of the crucible.

In order to estimate heat-resistance in specific conditions the heating temperature and gas medium parameters are chosen close to the actual operating values. If the purpose of the study is prediction of the durability of critical constructions, all operating conditions should be reproduced with great care.

The environment (ambient temperature and air composition) should remain constant during the entire testing period. The set temperature around the specimen should be maintained using forced circulation, for example. However, the movement of the gas medium during the test should favor oxidation processes and prevent erosion.

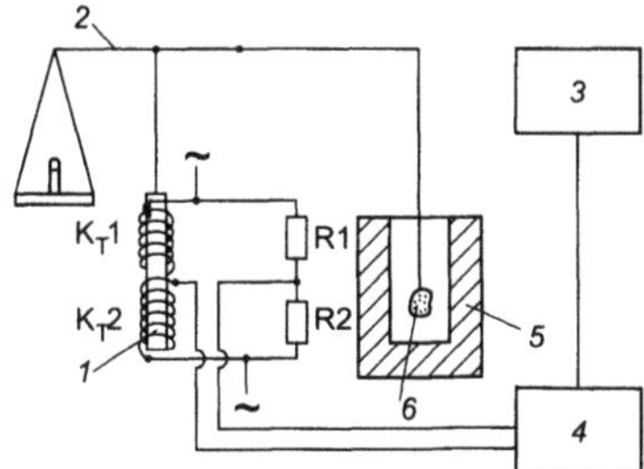

Fig. 4.2. Heat-resistance testing set-up for continuous specimen weighing. *1* core of the differential transformer; *2* beam of the analytical balance; *3* recorder; *4* amplifier; *5* laboratory furnace; *6* crucible with specimens

If the research only seeks to preliminary estimate heat-resistance, then, taking into account long duration and high labor expenditures of the testing process, testing conditions can be made more stringent as compared to those of actual operation. As one of the options, the temperature can be increased provided the oxidation mechanism remains the same.

In order to estimate corrosion penetration depth by the A method, the proportionality coefficient between the mass growth and mass reduction of the specimen is sometimes calculated. The coefficient is determined only once for each coating, temperature and specific gas medium conditions. The product of the coefficient and the specimen mass growth gives the conventional specimen mass reduction value which can be used to determine the thickness of the coating layer subjected to corrosion.

The crucible with specimens is placed into the working volume of the cold furnace. The duration of the test should be measured from the moment when the temperature of the specimens reaches the preset value. The dependence of the test duration on the expected service life of coatings is given in Table 4.1.

The rate of heating should be chosen experimentally for each "base metal–coating" composition so as to prevent formation of cracks on the surface of the coating. The specimens having surface cracks developed as a result of heating to working temperature should be rejected and changed. In method A crumbling oxides must be preserved for further weighing together with the crucible and specimens.

If periodical rather than continuous weighing is preferred in methods A or B, the specimens are sampled after 5, 10, 20, 50, 100, 200, 500, 1000, 2000, 5000, and 10000 hours of testing. For high testing temperatures the specimens can be weighed according to another time series, in smaller intervals, the only requirement being that the numbers in the series form a geometric progression, while the total test duration must be sufficient for revealing the mechanisms of oxidation. Periodicity of specimen sampling and the number of points depend on tasks and purposes of a given study, as well as on the coating thickness and corrosion depth.

In case of periodical weighing specimens are cooled together with the furnace or in still air down to room temperature. The rate of cooling is chosen taking into account thermal stability of the coating. Similar to heating, if after cooling any coating cracks are found on a specimen upon visual inspection, the rate of cooling should be reduced.

Table 4.1. Dependence of the test duration on the expected service life of material (according to State Standard 6130)

Service life of material [hours]	Minimum test duration [hours]
Above 100000	10000
50000 to 100000	5000
25000 to 50000	3000
10000 to 25000	2000
Under 10000	20% of the service life

In testing by method B the products of oxidation should be completely removed from the coating surface. This can be done in three ways: mechanically (using brush, rubber or by shaking in vibrating metal sieves); chemically (by dissolving in solution of a proper composition); electrochemically (applying direct current in a suitable electrolyte). The mechanical method is used as an ancillary technique for removal of upper layers of oxidation products or for their loosening without damaging the coating. After the mechanical treatment the remaining oxidation products are removed by chemical or electrochemical methods.

After the oxidation products have been removed, the coating is thoroughly examined using an x10 magnifying glass. The oxides brought up by testing should be distinguished from those formed as a result of heating during spraying. The presence of the latter on the coating surface is not an indication of spoilage.

In method C the place for cutting the cross-sections is selected after oxidation products have been removed from the surface of the specimen. The zones with maximum corrosion depth are located with the help of a magnifying glass. The cross-sections on these zones are cut out so that their planes pass through the maximum number of larger pits. The depth of corrosion pits in method C is estimated in three sections using a metallographic microscope. At least ten measurements along the length of each cross-section should be taken. The obtained values are averaged. Depending on the average depth of corrosion pits the following magnification factors are recommended: x500 to x1000 magnification with the average depth below 20 μm; x200 magnification with the average depth above 20 μm. When visual or metallographic examination is used, two corrosion zone are differentiated: full corrosion zone which should be removed and partial corrosion zone which is taken into consideration in testing.

Corrosion depth is determined from the thickness reduction by corrosion with due account of the partial oxidation zone. If there is no full oxidation zone, partial oxidation zone is measured from the edge of the cross-section to the base (non-oxidized) coating structure.

State Standard 21910 defines the following heat-resistance characteristics for uniform corrosion, that are determined in methods A and B: specific mass loss (q) or growth (q_*) measured in kg/m^2 (g/cm^2, g/m^2, mg/cm^2); uniform corrosion depth (h) measured in m (mm, μm); average rates of mass loss (V_q) and growth (V_{q*}) measured in kg/(m^2·day) [g/(m^2·day), g/(m^2·hour), g/(cm^2·day), mg/(cm^2·day)]; average rate of corrosion penetration (V_h) measured in m/day (mm/year/, μm/hour).

$$q = \frac{m_o - m}{S}, \tag{4.1}$$

$$q_* = \frac{m_* - m}{S}, \tag{4.2}$$

where m_0 is the mass of the specimen before corrosion test; m is the mass of the specimen without oxidation (corrosion) products after testing for a specified period of time; m_* is the mass of the specimen together with oxidation (corrosion) products after testing for a specified period of time; s is the area of the oxidized corroding surface of the specimen (strictly speaking, the area of the projection of the actual rough specimen surface. To account for roughness, a coefficient of surface geometry can be introduced).

$$h = \frac{q}{\rho} = \frac{m_o - m}{\rho \cdot s}, \tag{4.3}$$

where ρ is the density of the coating;

$$\overline{V_q} = \frac{q_2 - q_1}{t_2 - t_1} = \frac{m_2 - m_1}{s(t_2 - t_1)}, \tag{4.4}$$

$$\overline{V_*} = \frac{q_{*2} - q_{*1}}{t_2 - t_1} = \frac{m_{*2} - m_{*1}}{s(t_2 - t_1)}, \tag{4.5}$$

where $t_2 - t_1$ is the duration of the corrosion test;

$$\overline{V_h} = \frac{h_2 - h_1}{t_2 - t_1} = \frac{q_2 - q_1}{\rho(t_2 - t_1)}. \tag{4.6}$$

In method C the uniform corrosion depth is defined as the difference between the coating thickness before (δ_0) and after testing (δ_1) (see Fig. 4.3).

Heat-resistance for pitting corrosion is characterized by the following parameters: maximum depth of surface corrosion pits (h_{pmax}) measured in m (mm, μm); pitting corrosion level K_p; density of surface corrosion pits (ρ_p) measured in m^{-2}, cm^{-2}, mm^{-2}; average growth rate of the maximum corrosion surface pits (V_{max}) measured in m/day (mm/year/, μm/hour).

$$\overline{h}_p = \frac{1}{N}\sum_{i=1}^{N} h_{pi}, \tag{4.7}$$

where N is the number of measurements; h_{pi} is the depth of a separate corrosion surface pit.

$$K_p = \frac{\overline{h}_p}{h} = \frac{\rho \cdot S \sum_{i=1}^{N} h_{pi}}{N(m_o - m)}, \tag{4.8}$$

$$\rho_p = \frac{N_p}{S}, \tag{4.9}$$

where N_p is the number of corrosion pits on the surface of a material.

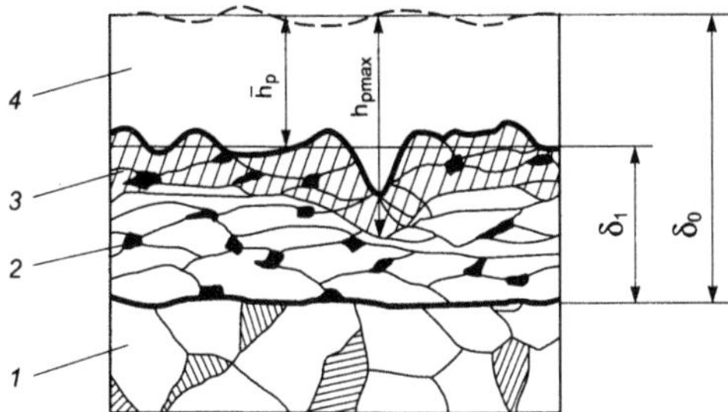

Fig. 4.3. Measurement scheme for determination of local corrosion parameters. *1* base metal; *2* coating; oxide zones: *3* local; *4* full (to be removed prior to testing)

$$\overline{V}_{p\max} = \frac{h_{p\max 2} - h_{p\max 1}}{t_2 - t_1}. \quad (4.10)$$

Volume or mass of absorbed or regenerated gas, changes in physical properties of coating, etc. can be used as indirect indexes to characterize heat resistance. In this case a relation between such indexes and the characteristics of gas corrosion and heat resistance should be established beforehand.

4.1.4 Estimation of Durability of Coatings With Electron-Probe X-ray Microanalysis

An interesting technique to determine the time of the decrease of alloying element concentration in the coating to its level in the base metal was employed in the work of Kolomytsev [4.32]. The method is primarily applicable to diffusion coatings and has been standardized. The corrosion durability of a protective coating (according to State Standard 9.312) in the method is evaluated from changes of the contents of the main components determining heat-resistance of the coating.

The content of elements in the "base metal–coating" composition is determined using electron-probe X-ray microanalysis. For this purpose the chemical composition of the coating in the zones located at varying distances from the surface is determined before the test and after testing for time interval τ. The content of elements is determined using step-by-step scanning of the specimen placed under the electron probe in the direction perpendicular to the specimen side surface in microsection. The step size is 1 μm. The number of scanning steps (analyzing points) is determined by the overall thickness of the coating and the adjacent base metal layer (at least 50 μm).

The content of the main components in the coating C_c determined from the electron-probe microanalyzer measurements is then plotted as a function of τ on the $\lg C_c - \lg\tau$ coordinates.

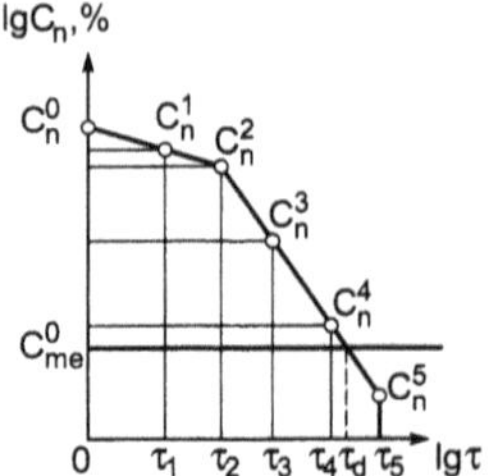

Fig. 4.4. Estimation of durability of protective coatings (τ_d) according to State Standard 9.312. C_c content of main components in the coating; C^0_{me} content of main components in the base metal

The coating is considered to have lost its protective properties if the content of the main heat-resistant element has reached the level of its content in the base metal (C^0_{me}) (see Fig. 4.4).

4.1.5 Other Methods for Determination of Heat-Resistance of Coatings

Acknowledging certain advantages of the weighing method, especially when combined with coating thickness measurement and quantitative evaluation of structural changes, Nikitin [4.49], at the same time, points to the following limitations of the technique: incorrect results can be obtained when comparing coatings from metals with strongly differing densities; the method cannot be applied to coatings with partial sublimation of the forming oxide; the results become unreliable in the conditions of intense base metal-coating interdiffusion.

Regarding high-temperature corrosion, Nikitin has developed a method for predicting durability of protective coatings, which, in his opinion, is free from shortcomings inherent in the standard weighing method. The testing is carried out either at temperatures higher than operational or in a more hostile medium. Under these conditions the coating degrades much faster. The criterion for heat-resistance is the coating durability τ_n, which is the most universal and direct characteristic of the protective properties. The exposure interval is progressively increased by a certain time, e.g. 200 hours. Before each weighing corrosion products are removed from the coating surface, and specific mass loss of the specimen is estimated. Another option is to estimate specific mass growth of the specimen after each exposure without removing corrosion products. The results of the measurements, specific mass loss g_1 (in the first case) or mass growth g_2, are then plotted versus τ in the logarithmic coordinates. Both plots usually yield a straight line (see Fig. 4.5) until the moment of coating failure τ_n. The failure of the coating is accompanied by sharp increase in oxidation rate (inflection point on the curve). The lowest exposure period corresponding to the beginning of intense oxidation is taken as the coating durability.

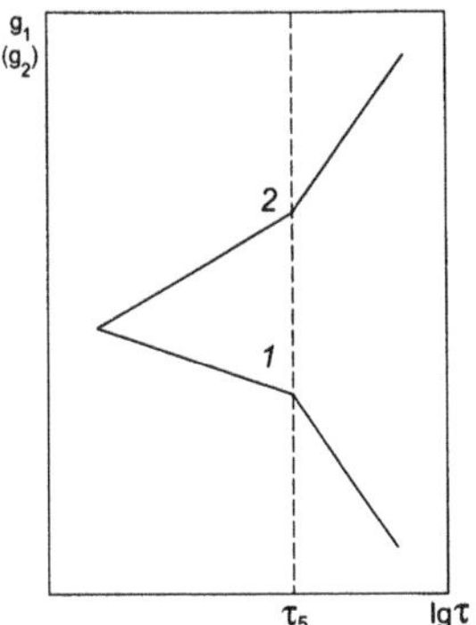

Fig. 4.5. Kinetic dependencies of the heat resistance of protective coatings. *1* with oxidation products removal; *2* without oxidation products removal

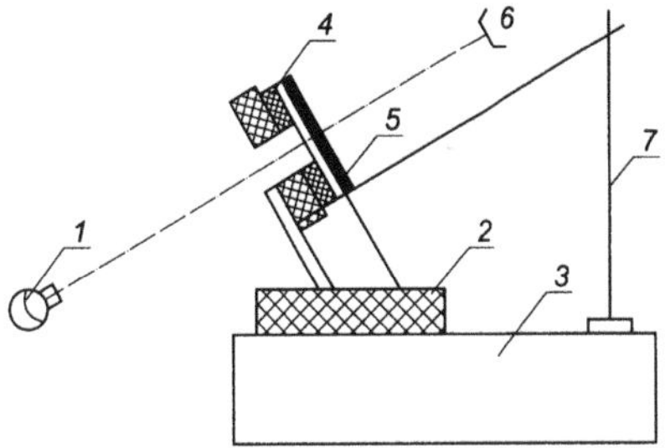

Fig. 4.6. Setup for heat-resistance testing using burning-through technique (adopted from [4.31]). *1* lens; *2* refractory material; *3* wooden support; *4* asbestos; *5* specimen; *6* burner; *7* post

One of the simplest and most effective methods for estimating heat-resistance is the burning-through technique described by Khasui [4.31]. A specimen (a plate 1mm thick with coating on one of the wider faces) is placed before the window of a refractory support (see Fig. 4.6). The flame of an acetylene burner fixed on a post is directed at the middle part of the specimen with coating projecting through the window. The time needed to burn a hole through the base metal-coating composition is taken as a measure of heat-resistance.

In [4.15] is suggested a new technique for estimation of heat-resistance of protective coatings for gas turbine parts developed by General Electric. The authors present examples of microstructures of coatings deposited by various methods (Al_2O_3, MeCrAlY, etc.) after laboratory testing and after operating in field conditions.

When choosing best-formulated "base metal–coating" compositions, it is advisable, along with studying solid coatings, to estimate heat-resistance functions of the coating with mechanical defects, such as peeling, cracking, etc. [4.12]. This problem has not yet been solved.

To gain more information about the behavior of heat-resistant coatings to be used in various constructions it is expedient to perform complex tests including: heat-resistance tests, microscopic studies [4.70], X-ray analysis [4.2, 4.4, 4.45], calculation of closed, open and general porosity [4.31], estimation of permeability [4.43], calculation of temperature coefficients of linear expansion (TCLE) [4.1, 4.28, 4.46] and others.

4.1.6 Protection of Base Metal With Heat-Resistant Coatings

Quite often the attempts to enhance heat resistance of materials using purely metallurgic techniques do not bring any positive results. The latest achievements in the field of base metal protection clearly demonstrate, that protecting critical constructions operating at temperatures above 550 °C with coatings is the most real and perspective way of structural strength improving. Protective coatings can be prepared from various heat-resistant materials, such as metals and alloys, metal-ceramic compositions, high-melting oxides, nitrides, carbides, etc. [4.13, 4.29, 4.50, 4.58, 4.64].

The mechanisms of protective action of oxide films formed on the surfaces of metal coatings and heat-resistant alloys are similar in many respects, and well-developed principles of steel alloying can be used to advantage when choosing the composition of a particular heat-resistant coating.

However, the mechanisms of oxidation of the base metal–coating composition are slightly different from oxidation mechanisms of compact materials due to characteristic properties of the coating structure: the presence of pores, cracks [4.9, 4.72]; oxide inclusions formed in the course of deposition [4.44]; branched loose boundaries between layers, particles, etc. In order to improve the heat-resistant properties of the coatings, sometimes the composition of the oxidation media is also changed along with optimizing the formulation of the base metal and the coating. Hermetic sealing, introduction of oxygen absorbers into operating space, special inhibiting and passivating agents are commonly used for this purpose.

The rate of oxidation usually follows linear, logarithmic or parabolic law. In some cases other dependencies are possible for complex systems.

Oxide films formed during oxidation of metal coatings slow down gas corrosion only if they are sufficiently solid and firmly bonded to the base. For porous coatings surface oxidation processes may be accompanied by base metal corrosion caused by permeating of the oxidizing medium through open pores. This may results in spontaneous peeling of the coating, since formation of loose oxides at the base metal-coating interface reduces the bond strength.

The condition of the specimen surface (its roughness) also influences the heat-resistance characteristics, mainly because rough mechanical processing creates a considerable difference between the geometrically measured and the actual reaction surface of the specimen.

It is known from practice, that heat-resistant coatings can be deposited by various methods, but it is diffusion coatings that have the best protective properties.

Among these, the best "diffusion coating–base metal" compositions are silicide coatings on refractory metals, molybdenum and tungsten, and aluminum coatings on heat-resistant nickel- and cobalt-based alloys [4.73]. In the former case the surface is covered with silicon that diffuses inside forming molybdenum and tungsten silicides ($MoSi_2$ and WSi_2). When oxidized, these compounds form oxides rich in silicon which provide high oxidation resistance. In the latter case aluminum diffuses into the upper layers of heat-resistant alloys forming NiAl and CoAl. These compounds then form protective Al_2O_3 films upon oxidation [4.10].

The most important gas thermal coatings are coatings of the MeCrSiAlY type (Me stands for one or more elements of the Fe–Ni–Co group). Yttrium is introduced in small quantities to improve scale adhesion. At present these coatings are the best developed protective systems for heat-resistant alloys [4.10]. The behavior of these coatings under oxidation is described by the same general laws as for alloys, i.e., they provide base metal protection by selective oxidation forming aluminum, chromium and silicon oxides. However, oxidation of coatings is a more complicated process because of their small thickness. Coating-base interdiffusion leads to depletion of chromium, silicon and aluminum and ingress of undesirable elements from the base metal into the coating.

In order to prevent access of the ambient atmosphere to base metal, heat-resistant coatings operating at comparatively low temperatures are treated by special sealing materials. Currently there are no materials applicable for sealing high temperature coatings, and their heat resistance is improved by special technological procedures such as spraying underlayers, using multilayer coatings [4.38, 4.47], thermal treatment, using refractory compounds.

Spraying of an underlayer, in particular, of molybdenum can be effectively used to prevent detachment of the coating from the substrate under heating and to improve heat resistance properties of ceramic coatings. Combined deposition of oxides and metal yields solid coatings having enhanced oxidation resistance. Thus, products with a multiplayer coating produced by flame spraying have been reported. The multiplayer coating consisted of 6 molybdenum layers (with average layer thickness 0.05 mm) alternating with 5 aluminum oxide layers (with average layer thickness 0.18 mm). Molybdenum layers provided high erosion resistance and good adhesion, while Al_2O_3 layers ensured the required heat-resistance [4.31]. Thermal treatment assists in formation of a diffusion layer on the coating-base interface which improves density and heat-resistance of the coating.

An important factor having large effect on coating heat-resistance is the melting point of the deposited material. Thus, zirconium dioxide coating (melting point 2700 °C) has proved to have higher heat-resistance as compared to aluminum oxide + titanium oxide coating (melting point about 2000 °C).

Interaction of coated material with high-temperature chemically active gas flow may cause either development or healing of coating defects depending on the relation between gas and solid products formed as a result of chemical reactions. Typical interaction examples are given in Fig. 4.7 [4.67].

If oxidation leads to formation of gas products only (see Fig. 4.7a), then empty spaces appearing near the bases of through pores and cracks under coating with time may merge to form through channels or zones completely scaled off the coating surface. From this moment the supply of the oxidizer to under-coating material is determined by convective under-coating oxidizer flows generated due to pressure gradient over the surface rather than by gas diffusion through pores and solid material. This is a nonstandard corrosion mechanism that should be considered a failure in construction operation. Continuation of work under these conditions usually causes mechanical cracking and carrying away of a piece of the coating.

If excessive volumes of solid and hard oxides are produced as a result of interaction with the oxidizer gas flow (see Fig. 4.7b), a mechanical coating failure (breaking open) followed by carrying away of the coating by the flow takes place. If the relation between gas and solid oxidation products is such that the amount of solid products is sufficient for filling empty places left after carrying away of gas products or is slightly greater, the material has the tendency to self-healing of pores and defects (see Fig. 4.7c). Ejection of oxides out of pores results in formation of crusts on the coating surface. It is desirable to have oxide temperature slightly higher than surface temperature, which would lead to melting of oxide, its spreading over the surface and healing of microdefects. For this purpose the melted oxide of the protected material should properly wet the coating [4.67].

The influence of the residual stresses in plasma-sprayed MeCrAlYTa-type coatings on high-temperature oxidation resistance has been investigated in [4.28]. Prior to testing the heat-resistance, stretching stresses were present in the coating mainly because of the difference between temperature coefficients of linear expansion of the coating and of the base metal (heat-resistant nickel-based alloy). After the oxidation tensile stresses in the coating are reduced due to structure changes occurring as a result of coating-base element interdiffusion. The best heat-resistance has been achieved for alloys with CoNiCrAlYTa coating.

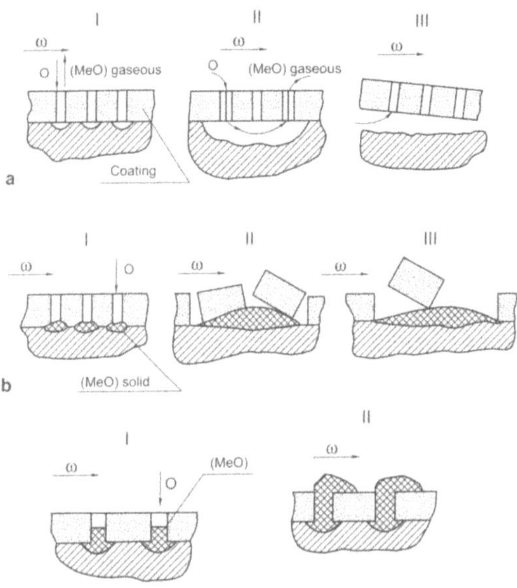

Fig. 4.7. Typical oxidation mechanisms of coated materials (adopted from [4.67]). Oxidation products: **a** gaseous; **b** solid; **c** in different aggregate states. *I–III* different stages of oxidation

Liu Yung and Nafesan developed models for simultaneous oxidation-erosion of coatings of alloys and metals [4.39]. The models are based on the assumption that oxidation is controlled by diffusion of metal ions through the growing oxide film, while erosion is a result of uniform removal of the material with a constant rate. Model computations were performed for several double alloys used as coating materials and forming Cr_2O_3 or Al_2O_3 films during oxidation. All computations were made using experimentally determined parabolic constants for the oxidation rate of the alloys. The models enabled to plot the graphs allowing easy determination of the conditions for prevailing oxidation or erosion.

Ckuse with co-authors [4.14] put forward an algorithm to compute coating durability based on the analysis of the accumulation of damages caused by oxidation and cyclic loads. A coating consisting of ZrO_2–7%Y_2O_3 layer and a NiCrCoAlY underlayer plasma-sprayed on nickel-based alloy was studied. The quality of the life prediction model was estimated by trial runs of coatings in full-scale testing. The relative difference between the actual and the predicted periods of coating lifetime before peeling did not exceed 30%.

4.2 Thermal Stability

4.2.1 General Overview

Thermal stability is a property of coatings to preserve adhesion–cohesion strength under sharp thermal impacts. A single cycle of such impacts characterized by high temperature gradient inside a specimen is called a thermal shock or, in case of lower gradient, a thermal change. Thermal cycling is a process of sequential effects by periodically repeated thermal changes (heating–cooling) within the chosen temperature range at a certain heating–cooling rate.

Thermal stability of the base metal–coating composition is determined by the following factors: thermal effect conditions (temperature gradient, heating and cooling rates), relationship between average values of TCLE within the pre-set base-coating temperature range [4.6, 4.16, 4.60], mechanical coating properties (strength, fracture toughness) [4.8, 4.11, 4.24, 4.42, 4.65], coating–base metal bond strength [4.35] and coating porosity [4.70].

A difference between the average values of coating and base metal TCLE in thermal cycling may give rise to considerable stretching stresses which lead to adhesion-cohesion failure of the coating.

Porosity has the following degrading effect: pores as stress concentration points reduce coating resistance to cracks. Besides, air oxygen finding its way to the base metal through open pores oxidizes its surface which causes coating peeling over the loose layer.

4.2.2 Specimens

Flat metal specimens (20.0±0.5)x(10.0±0.2)x(5±0.1) mm in size are used for testing. Coating of the given thickness is deposited on the wider face of the specimen with due allowance for mechanical processing. Occasionally specimens of different shapes are also applied. In [4.33] a specimen having the shape of a wedge prism which enables to model real loading conditions of edge areas of a blade was chosen for investigation. After mechanical processing external surface should have no defects (cracks, scratches, cleavages, scaling) visible with bare eye. At least five specimens should be used for each variant of deposition.

4.2.3 Equipment

Muffle furnaces and electric furnaces providing heating temperature of at least 1000 °C are used for heating specimens. The temperature in the furnace working space must be automatically adjusted with the error of ±5% or less. In order to provide identical conditions both for heating and cooling of specimens, a special stainless steel cassette should be used.

If rapid cooling is planned, a cold-water tank or a liquid-nitrogen vessel fitted with removable metal net basket for fast immersion of specimens is required. Temperature deviations inside the tank should not exceed ±5 °C from the given value within the whole testing period. If forced cooling by air flow is scheduled by the investigation program, it is recommended to use a special device providing stable position of specimens during air cooling as well as identical cooling conditions, i.e. frontal or side air inflow onto the specimen surface.

Thermal stability of coatings is also determined in the conditions of single-sided heating by quartz lamps, graphite heaters or plasma burners. Single-sided heating increases testing temperature up to 1500 °C and higher, providing high heating rates and, to a large extent, an approximation to actual operating conditions of the coating.

A special installation for investigation of the efficiency of heat-resistant coatings under rapid thermal changes has been described in [4.41]. A specimen and a graphite emitter are placed at the focuses of the optical furnace of an elliptic-cylindrical radiant energy concentrator. The optical furnace has copper hollow walls and covers, which allows efficient water cooling. Single and cyclic thermal changes in heating device of the installation are provided by reciprocally moving shutters providing radiation flow diaphragming for the specimen. Using automatic shutter control system, it is possible to adjust thermal flux according to the desired pre-set program.

A serious disadvantage of the above methods is the difficulty of determination of the onset of coating failure, since monitoring can be performed only by visual inspection.

An analysis of currently existing nondestructive control methods prompted Dekhtyar and his co-authors to turn to the acoustic emission method, which is suitable for solving thermal stability problems. The investigations have shown that at temperatures below 1500 °C it is preferable to use a stationary furnace providing stable temperature field, and periodically insert the specimen into the furnace working zone [see Fig. 4.8].

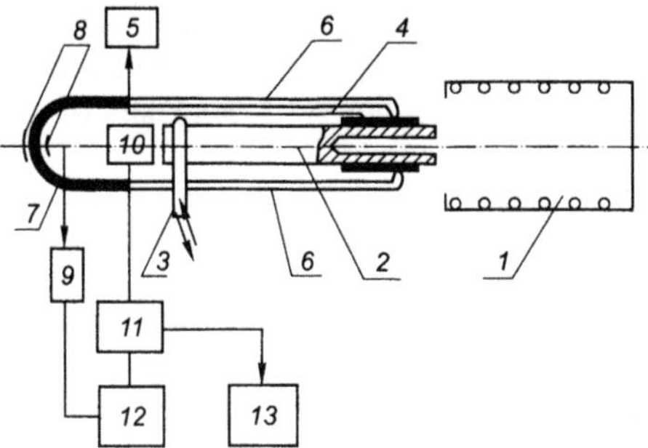

Fig. 4.8. Installation for thermal cycling (adopted from [4.17]). *1* furnace; *2* specimen; *3* holder/cooler; *4* thermocouple; *5* recorder; *6* quartz prods; *7* flat spring; *8* resistor strain gauge; *9* strain gauge station; *10* acoustic emission sensor; *11* acoustic emission equipment set; *12* recorder; *13* cycle counter

Such a design of the installation allows producing any desired temperature modes by varying furnace temperature, specimen insertion rate and specimen exposure time inside the heat zone and outside the furnace. Permanent furnace operation mode allows to considerably simplify temperature controlling equipment. The specimen is fixed in a reciprocally moving holder serving at the same time as a cooler for the trailing end of the specimen, where the acoustic emission sensor is pressed. The specimens are pieces of cylindrical acoustic waveguides manufactured together with a tubular or solid coated working part. Synchronous recording of all acoustic emission parameters, deformation of the specimen and its temperature in cycles allows dynamical studies of coating failure during thermal cycling [4.17].

Comparison of the recorded results of acoustic emission, deformation and visual inspection demonstrated that acoustic emission method allows determination of the initial moment of cracking and intense development of cracks long before coating failure becomes noticeable. This makes the acoustic emission control technique extremely promising for prediction of thermal stability of coated parts immediately in the field conditions.

4.2.4 Testing

The temperature range of thermal cycling and the rates of specimen heating and cooling are chosen close to actual temperature conditions of operation of the coated products.

During thermal shock testing the specimens inserted in a cassette or put down-edge on a heat-resistant support are placed inside the electric furnace. After heating and ten-minute exposure at the given temperature the specimens are thrown off down-face into a running-water tank into the metal net basket that is vigorously shaken to prevent formation of the steam jacket around the heated specimens. After cooling the specimens are wiped and visually inspected. Damaged specimens are not used for further testing, while the good ones are thoroughly dried up before each subsequent heating. Thermal cycling is repeated until coating failure is achieved on each specimen.

For nonmetallic inorganic coatings instead of water cooling it is expedient to use softer thermal cycling modes such as cooling together with the furnace, cooling in still air or in air flow and so on. Barvinok estimated thermal stability of plasma sprayed coatings by the number of thermal cycles required for the coatings to give indications of failure during heating inside a furnace followed by cooling with compressed-air [4.7]. Testing was terminated if neither cracks nor cleavages developed in the coating after 100 thermal cycles. In another work [4.33] specimen testing mode simulated startup-shut down of an engine with 60 s intervals of heating and cooling. Yar-Mukhamedov conducted the most stringent testing of composite coatings in two variants: in cast iron melt (1450±50 °C) and aluminum melt (680±50 °C). Thermal cycling consisted of immersing the specimen into a relevant melt for 30 s followed by water-cooling [4.76].

A microscope with x100–x250 magnification can provide additional information on coating failure kinetics. It can be used to reveal microstructural crack formation pattern at cross-sections of the specimens prepared after a specified number of thermal cycles [4.48].

After each thermal cycle the number of specimens with the damaged coating is determined. The break of adhesion-cohesion bond in the "base metal–coating" composition shows itself as formation of cracks perpendicular to the interface, coating scaling off the base metal, breaking of bonds between separate layers or particles, and microcracking inside the particles.

The criterion for thermal stability is the number of cycles-thermal changes required to develop coating damages visible with bare eye or detected at a lateral cross-section during metallographic studies. As described above, nondestructive control facilities using acoustic emission can be used to register the onset of damage during heating–cooling. The arithmetic average of all results obtained by concurrent determination of thermal stability for the entire batch of specimens is taken as testing result. The maximum temperature range, within which no damages were detected at the given cooling rate on working coating surfaces or at cross-sections of the batch of specimens in study, can be taken as another criterion for thermal stability.

4.2.5 Results and Discussion

The review [4.27] summarizes the data on structure and properties of ceramic thermal barrier plasma sprayed coatings. Generally a thermal barrier coating is made of two layers: the external layer of ZrO_2-based ceramics and the internal layer of MeCrAl-based alloy. Partially stabilized ZrO_2 oxide with cubic-lattice phase as the main phase component has the best thermal stability. Stabilization of high-temperature cubic modification is achieved by adding Y_2O_3, MgO or CaO. In field conditions the most frequent coating failure is peeling of ceramics off the metal underlayer. It is caused by the difference between TCLE of ceramics and of MeCrAlY-based alloy, and by oxidation of the underlayer forming new Al_2O_3 and Cr_2O_3 oxides on the layer interface which significantly reduces the compound bond strength.

Verstak [4.74] studied the effect of the initial condition of powders and parameters of plasma spraying on the structure of coatings and their thermal cycling resistance in the temperature range 1100±20 °C to 900±20 °C. The studies were carried out for ZrO_2 powders partially stabilized by one or more magnum, yittrium, or zerium oxides with underlayers from coatings made of NiCoCrAlY-based alloy. Structure and thermal stability of the coating are mainly influenced by the chemical composition and manufacturing technique of the powder. The main factors of the spraying technology are the degree of melting of the powder particles in plasma jet, and the surface temperature of the substrate during spraying. The content of the ZrO_2 tetragonal phase capable of conversion during spraying or thermal cycling rather than phase composition of ceramics is of prime importance for attaining high thermal stability proper.

In [4.77] an aluminum oxide coating was deposited on austenite and ferrite-pearlite steel using the gas-flame technique. Ni–10%Al was used as the underlayer. After thermal cycling three types of failure were observed: cracking, swelling and peeling. Ultrasonic control was employed to reveal underlayer quality. It was found that the number of thermal cycles the coating can stand before failure is mainly determined by the presence of the underlayer.

Kravchuk and his co-authors [4.33] came to the conclusion that among the many investigated compositions the CoCrAlY-based coating with external zirconium dioxide layer 30–50 μm thick is the most effective. The presence of coating reduces the rate of temperature changes in the heating–cooling half-cycles and, as a consequence, the material is less thermally loaded as compared to unprotected specimens. Deformation conditions of base material are also favorably influenced by the field of residual stresses caused by the difference of mechanical characteristics of base and coating materials, which is especially large in the case of zirconium dioxide, and by the deposition technology.

Tawancy et al. [4.66] investigated failure mechanism of a thermal barrier coating on a heat-resistant superalloy with percent composition: Cr – 9.0; Al – 5.5; Co – 10;W – 10; Ta – 2.5; Hf – 1.25; Ni – the rest. First a 50 μm thick aluminide underlayer, and then a partially stabilized $ZrO_2 + Y_2O_3$ 200 μm thick underlayer were deposited on rods 8 mm in diameter. Specimens having length 10 mm were exposed to air at 1010 and 1150 °C and cooled to 23 °C periodically in 24 hours intervals. Peeling of the protective coating was caused by the failure of Al_2O_3 layer formed during oxidation of aluminide and containing inclusions enriched in chromium and hafnium. Internal oxidation of hafnium diffusing from the substrate caused occurrence of stresses destroying the oxide layer.

4.3 Corrosion and Corrosion Resistance

4.3.1 Classification, Terms and Definitions of Corrosion Processes

Corrosion is degradation of the "base metal–coating" composition as a consequence of chemical or electrochemical interaction with corroding medium.

Corrosion processes are differentiated by:

- reaction mechanisms of metal-medium interaction;
- type of corroding medium;
- type (geometrical pattern) of corrosion failures on the surface or in the bulk of material;
- character of additional effects the material is subjected simultaneously with attack of the corroding medium.

By reaction mechanism of interaction two main types of metal corrosion can be distinguished: chemical and electrochemical. A special type of corrosion is microbiological and radiation corrosion.

By the type of aggressive medium participating in material degradation processes gaseous, atmospheric, underground corrosion, corrosion in solutions of electrolyte and non-electrolyte liquids, and some others are distinguished.

By behavior of material surface or by modification of its physicochemical properties corrosion is divided into general, local, and selective, irrespectively of the actual material-medium interaction that took place [4.78].

Corrosion spreading over the whole surface of material is called general or continuous corrosion. Continuous corrosion is divided into uniform and non-uniform (see Fig. 4.9a,b) depending on the depth of the corrosive surface failure. Continuous corrosion is the least dangerous type since the material of which a device or one of its parts is made only slightly looses its strength properties.

Under local corrosion only separate portions of material surfaces fail, and thus local corrosion has unequal failure degree. The most typical types of local corrosion are spot, pitting, pointed, subsurface, and intercrystalline corrosions.

Spot corrosion is a comparatively shallow failure of separate portions of the material surface (see Fig. 4.9c).

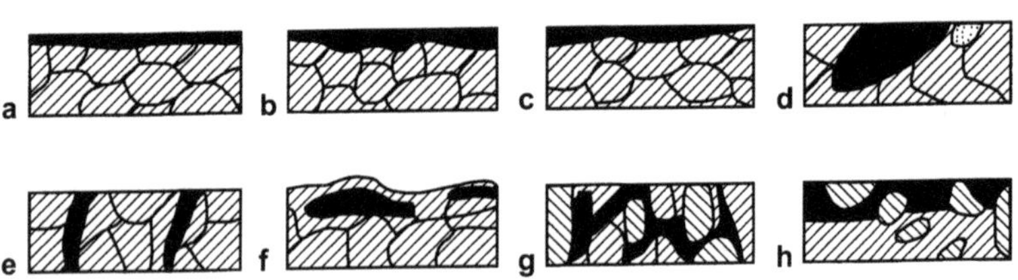

Fig. 4.9. Types of corrosion: **a** uniform; **b** non-uniform; **c** spot; **d** pitting; **e** pointed; **f** subsurface; **g** intercrystalline; **h** structure-selective

Pitting differs from spot corrosion by greater depth of penetration into the material layer (see Fig. 4.9d).

Pointed corrosion is a failure of material in the form of point defects developing into through defects (see Fig. 4.9e).

Subsurface corrosion generally occurs when protective coating is broken in separate zones. Metal failure then takes place predominantly under surface, and corrosion products are mainly concentrated inside the metal (see Fig. 4.9f).

One of the most dangerous types of local corrosion is intercrystalline corrosion, which propagates along the less stable boundaries of metal grains without causing their failure (see Fig. 4.9g).

Local corrosion is more dangerous than continuous type as it may considerably deteriorate the strength properties of a product [4.78].

Selective corrosion is observed for alloys containing several structural components. This type of corrosion is called structure-selective (see Fig. 4.9h).

Among the types of corrosion which differ by the nature of additional effects the following are to be highlighted:

- fretting corrosion which is a failure of material under simultaneous action of corroding medium and frictional forces;
- corrosion cracking occurring under simultaneous action of corroding medium and either external or internal mechanical stresses with formation of transcrystalline or intercrystalline cracks;
- corrosion under stress developing under simultaneous action of corroding medium and either constant or variable mechanical stresses;
- corrosion fatigue which means a decrease of the fatigue limit of material due to simultaneous action of cyclic tensile stresses and corroding medium.

Corrosion fatigue limit is defined as the maximum mechanical stress which still does not cause metal failure after simultaneous action of a pre-determined number of variable load cycles and the given corrosion conditions. Corrosion fatigue limit depends both on the corrosion rates and on the relation between the corrosion rates and accumulation of fatigue damages.

Contact corrosion is an accelerated corrosion of one metal upon contact with another metal in the presence of aggressive medium. Materials with metallic coatings are a special case of heterogeneous bimetallic compositions, and during their operation a problem of local corrosive failure in coating-base metal contact zone caused by occurrence of micro- and macro electrochemical corrosion systems is often encountered. This type of material degradation is a special case of contact corrosion that is especially dangerous for product operating in aggressive media.

The term "corrosive process" should be used for the corrosion process itself while its result is better described by the term "corrosion failure". Corrosion products are chemical substances formed as a result of interaction between the coating and corrosion medium. Corrosion losses are the amount of metal converted into corrosion products within a certain time interval. Corrosion resistance is the ability of a coating to resist the effect of corroding medium. Corrosion losses from unit surface area per unit time are characterized by corrosion rate. Corrosion penetration rate is the depth of corrosion failure advance per unit time.

4.3.2 Corrosion of Protective Coatings

Depending on the protective properties of the coating either corrosion behavior of the coating itself or of the base metal-coating composition as a whole is investigated. The latter is caused by penetration of the corroding medium to base metal with ensuing electrochemical reaction which results in failure of both materials. Reliable protection of base metal can be provided only if the coating is solid (without pores and cracks) and blocks the access of aggressive medium to the substrate. In composite and multilayer coatings the development of electrochemical corrosion is complicated by numerous contacts between chemically different materials with resulting formation of a multielectrode corrosion system.

At the heart of contact corrosion lies electrochemical interaction between two metals. The metal having more negative corrosion potential acts as anode and thus fails at a higher rate. Cathode metal having more positive corrosion potential may either avoid failure or fail at a much lower rate as compared to anode. Different metals are not equally effective as anodes or cathodes which is determined by formation and decomposition of oxides, the effect of oxygen, hydrogen and other active gases on the course of electrochemical processes, and a lot of other conjugated processes. Failing in aggressive media, materials with anode-type coatings provide protection for base metal. For such coatings defects and piecewise deposition are acceptable owing to their much higher corrosion rate as compared to base metal. Materials with cathode-type coatings have very high corrosion resistance indexes in aggressive mediums in themselves. For these materials defects and discontinuity flaws of deposition cannot be tolerated since, having more electrically positive corrosion potential, they promote accelerated failure of base metal [4.61].

Lower corrosion resistance of coatings as compared to their compact material counterparts may be attributed, in the first instance, to the defects of the coating structure, the presence of pore channels connected with each other and with environment; nitride and oxide inclusions forming a developed grid along particle boundaries. Furthermore, residual internal stresses also have negative influence. It is highly desirable to know not only the sign and value of surface stresses, but also the pattern of stress distribution in the coating and base metal. Drawing on the analogy between corrosion and more well-studied galvanic failure processes which occur under cyclic loading in aggressive media, it may be expected that gas-thermal coatings will probably also reduce corrosion fatigue resistance of steel.

In testing of porous coatings non-uniform corrosion is generally observed. The presence of large-size pores promotes formation of local corrosion sources. The coating zones having defects and inclusions are the first to dissolve. As a result, the porosity increases and corrosion processes are further stimulated. Through open pores aggressive medium reaches the base metal surface [4.40].

The formation of corrosion products on the base metal-coating interface increases tensile stresses, since the volume of the formed products frequently exceeds the volume of the reacted metal. Under these conditions cracking and peeling of coatings become quite possible.

Internal (composition, structure, residual stresses, surface condition, etc.) and external (temperature, pressure, medium composition, speed of specimen movements through the medium, etc.) factors all influence the rate, type and propagation of corrosion [4.3, 4.22, 4.23, 4.36, 4.75].

Anticorrosive coatings must be homogeneous, have minimum porosity, uniformly cover the base metal surface, and reliably couple with base metal [4.18, 4.25, 4.26, 4.34, 4.52].

4.3.3 Purposes, Methods and Special Equipment for Corrosion Testing

It should be noted that the majority of corrosion tests are regulated by international or state standards. The quality of testing can be guaranteed only by strict observance of the normative documents. In case these regulations are not observed in their entirety including minute details the distorted results can be obtained.

The purposes of testing are: obtaining of corrosion resistance and coating protective ability characteristics; prediction of the base metal–coating composition service life in actual field conditions; studies of the kinetics of corrosion process.

The following corrosion testing methods are commonly used: the method of witness specimens; electric resistance measurements; electrochemical measurements; non-destructive testing; analytical testing; special tests.

The frequently used method of witness specimens not only provides comparatively easy determination of mass loss (growth) but also makes possible investigation of corrosion products, estimation of the number and location of pittings as well as visual interpretation of corrosion failure patterns. The method of electric resistance measurement is based on recording the change in electric resistance due to reduction of the cross-section and loss in electric conduction of the specimen. The method is applicable to conducting materials only. Electrochemical methods allow judgment not only about the average corrosion rate but also about the "instantaneous" rate in conducting medium. However, they are not suitable for aggressive non-electrolyte mediums. Application of ultrasound and eddy currents for non-destructive corrosion testing yields too approximate, though operative, results. Analytical methods are based on chemical analysis of corrosion medium samples.

Among special methods the tests for corrosion cracking, corrosion under stress and corrosion fatigue, i.e. the test for corrosion under additional effects, should be mentioned [4.53, 4.63].

Corrosion cracking and corrosion under stress can be studied using a special installation [4.30] (see Fig. 4.10) allowing to vary the temperature of aggressive medium and mechanical stress.

The specimen 7 fixed in the special holder 3 is loaded by the calibrated spring 6 protected by the housing 5. Aggressive medium is supplied to the vessel 1 made of corrosion-resistant steel. The temperature and type of medium effect (liquid, vapor or air-vapor phases) are adjusted by turning on the electric heater 8 and the cooler 2. The level of solution (below the specimen, or, given waterline, over the specimen) and testing temperature are monitored by the level gauge 9 and the thermal converter 3, respectively.

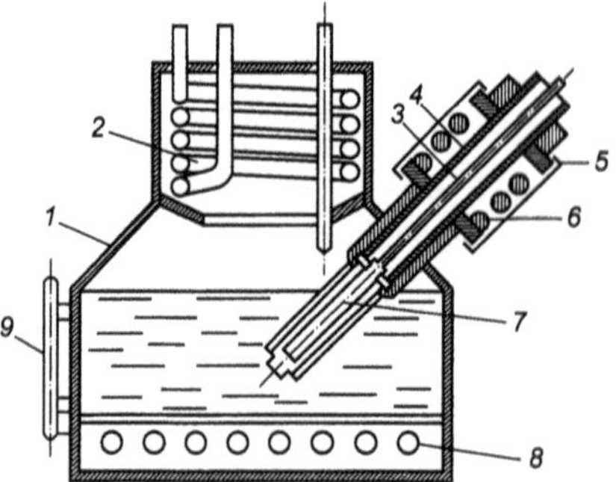

Fig. 4.10. Installation for corrosion cracking and stress corrosion testing (adopted from [4.30]). *1* vessel; *2* cooler; *3* thermal converter; *4* specimen holder; *5* protective housing; *6* calibrated spring; *7* specimen; *8* electric heater; *9* level gauge

Another installation [4.57] allows estimating corrosion fatigue limit for coated specimens in liquid aggressive media under symmetric bending at repetition frequency 2800 cycles per minute.

Many commercial machines are equipped with corrosion chambers for corrosion fatigue testing.

Crack depths (lengths) under corrosion cracking and corrosion fatigue are measured using optical microscopes or non-destructive inspection devices.

4.3.4 Atmospheric Corrosion Testing

Special attention in the deposition of protective coatings is paid to the problem of providing normal operation of products in various natural conditions. Therefore the products should pass the relevant climatic tests.

Atmospheric corrosion testing of coatings is carried out at special climatic testing stations. Depending on the atmospheric environment climatic testing stations can be divided into three types:

- ground-based (for any climatic regions on the land);
- waterside (near the shorelines of oceans, seas and water reservoirs);
- floating and stationary above-water (surface) stations (in atmospheric environment and sea water of water areas and fresh water of water reservoirs).

Ground-based and waterside stations should preferably be located within representative and/or extreme places of climatic regions. Surface stations should be located onboard the specially adapted boats sailing in the regions specified by the specimen testing program. Stationary surface stations are built in coastal areas of oceans, seas and other water pools on hydraulic-engineering structures (trestles, stationary platforms, etc.) or pontoons which are kept in place using anchors. The stations are fitted with equipment for performing a complex of meteorological observations including monitoring air temperature, intensity and duration of precipitation, intensity and duration of solar irradiation, atmospheric pressure, wind velocity and direction, concentration of ozone, sulfur dioxide, chlorides and dust, sea- and fresh water analysis.

The values of atmospheric corrosion aggression parameters in many places of Russia located in different climatic zones (cold, moderate, warm, wet and hot) are defined by State Standard 9.039. The influence of two factors, surface humidity and air pollution with corrosion-active agents such as sulfur dioxide, chlorides, ammonia, nitrogen oxides, is estimated. For each locality the Standard presents information on duration of surface humidification by phase and adsorptive moisture films given in hours per year. It also describes the techniques for determination of sulfur dioxide, chloride and ammonia content in the air. Atmospheric corrosion aggression parameters can be used to determine corrosion losses, establish field condition groups and develop accelerated testing techniques.

Atmospheric field conditions of coated products are divided into four groups: easy, moderate, hard, the hardest. This classification is based on severity of exposure to the following factors determining atmospheric corrosion aggression: air pollution by corrosion-active agents, microclimatic region where the given product is used, conditions of the product location in the open air, under a shed, or in closed premises. The coated specimens should be continuously tested for at least two years. The specimens are fixed on racks at a given angle or vertically, in the chosen direction and with due account of the wind rose. The specimens are examined daily within first ten days, weekly over the first month, and then in 2, 3, 6, 9, 12, 24 months and so on depending on the testing program.

4.3.5 Specimens for Atmospheric Corrosion Testing

Four types of specimen are used for atmospheric corrosion testing: flat specimens, parts and assembly units, product or construction scale models, and products. The appearance, shape, and size of a specimen are chosen depending on the purpose of testing. The specimens should emulate the principal product properties such as substrate and coating materials; their coupling; surface condition; coating thickness and deposition technique, etc.

Flat specimens in the shape of metal plates with dimensions 150x100x2.5–5.0; 100x50x2.5–5.0; 50x50x2.5–5.0 mm coated on all faces are the most convenient. In order to improve testing accuracy the coatings may be finished to equal roughness. Once the coating has been deposited, the edges of the plates are rounded.

The total surface area of the specimens under test should be at least 50 cm^2. If narrow faces of the plates cannot be coated, they are protected by a layer of chemically stable lacquer, primer, or enamel. At least three specimens are prepared for each variant of deposition. Reference specimens are stored under conditions excluding corrosion, e.g. in desiccators or in polyethylene bags filled with silica gel. Along with plates made of base metal, the specimens of the coating material can also be used as controls.

4.3.6 Equipment for Accelerated Atmospheric Corrosion Testing

Equipment for accelerated atmospheric corrosion testing should automatically maintain the preset test parameters such as temperature, relative air humidity, sulfur dioxide and saline fog concentration, etc.

Thermal chambers, thermal moisture chambers, sea (saline) fog chambers, thermal vacuum chambers and other kinds of special equipment are used for this purpose.

The temperature is usually controlled by periodic switching electric heaters or coolers on. The specified air humidity is maintained by one of the three techniques: evaporation from open free water surface; air circulation through humidifier; steam feed. The required water content (the number of moisture drops per unit volume) and dispersity of sea (saline) fog are sustained by periodic spraying of salt solutions using a sprayer or a centrifuge of an aerosol machine. A commonly used solution for fog spraying in chambers has the following composition, g/l: 27 – sodium chloride NaCl, 6 – magnum chloride $MgCl_2$, 1 – calcium chloride $CaCl_2$, 1 – potassium chloride KCl. At least 90% of particles getting into the chamber should have the size 1–5 μm, with 75% of these particles falling in the range 2–4 μm. The number of particles in the chamber should be in the range $(2–4)\cdot10^5$. Fog dispersity is estimated using an optical microscope by inspecting an object-plate with fog drops precipitated on its surface.

The specified concentration of sulfur dioxide is sustained by periodically supplying the gas from a cylinder into a special chamber. For moisture precipitation on specimen surfaces the latter are cooled during the second part of the test cycle. The condensing moisture must precipitate uniformly on all surfaces under investigation [4.20].

The test for periodic immersion is performed using "corrosion wheel" or "rod beam" devices consisting of an electrolyte bath and a mechanism which provides alternating immersion and lifting of specimens fitted inside the bath (see Fig. 4.11).

Corrosion source depths can be measured by one of the following three techniques: at longitudinal cross-section using a microscope fitted with a calibrating fine adjustment screw (by depth of focus); with an indicator depth meter; at lateral cross-section using a microscope.

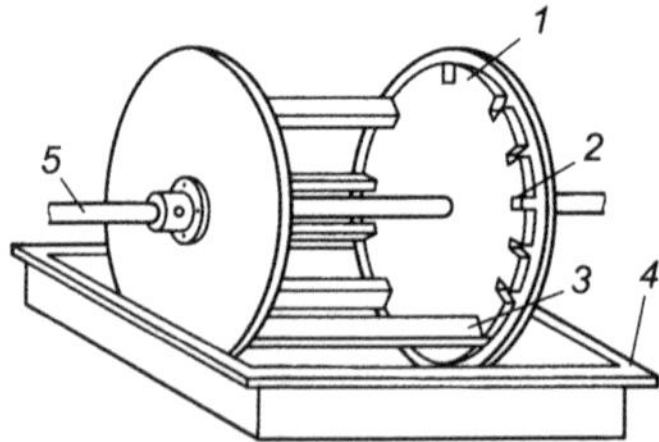

Fig. 4.11. Device for accelerated testing by periodic immersion in electrolyte. *1* rotating disk; *2* slots for fixing the specimens; *3* specimen; *4* electrolyte bath; *5* rotating shaft

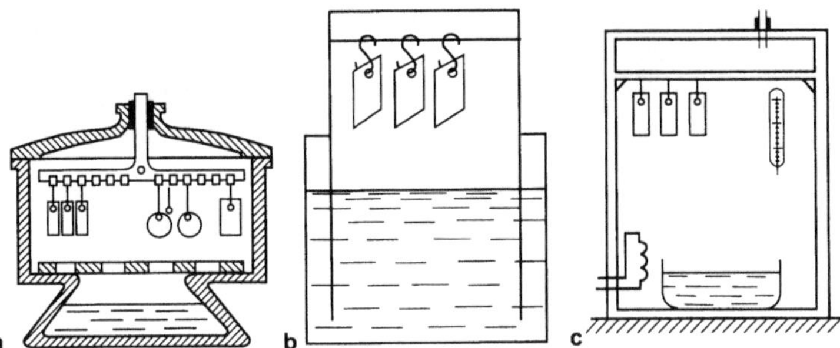

Fig. 4.12. Basic setups for damp atmospheric testing. **a** desiccator; **b** hydrostat; **c** simple moist chamber

The principal equipment for studying continuous, pointed, intercrystalline and spot corrosions includes desiccators, hydrostats and moist chambers (see Fig. 4.12).

As described above, laboratory tests are conducted in artificially created and thoroughly monitored conditions. In some cases multi-channel equipment using continuous recording of the measured values is used [4.20].

4.3.7 Preparation and Conduction of Accelerated Atmospheric Corrosion Tests. General Requirements

Acceleration of laboratory tests is achieved by deliberately creating harder corrosion conditions including temperature elevation, aggressive medium concentration increase, moisture precipitation, etc. It is necessary to provide circulation of the aggressive medium and renewal of its composition. If the latter is impractical, periodic replacement of the medium should be provided. The concentration of the solutions of aggressive medium prepared beforehand should be checked by titration or by density analysis prior to feeding to the equipment. In order to provide stable testing conditions, simultaneous studying of specimens with strongly chemically different coatings at the same setup is not allowed. The specimens are visually examined prior to testing. The coating should be uniform, without any cracks, cleavages, burrs or other defects. The specimens are fixed in the slots of the support or hung on a rack so as to exclude their interaction and provide identical effect of the aggressive medium on every surface.

In the case of forced interruption of testing the specimens are rinsed in distilled water and dried by compressed air. To improve testing accuracy and reduce interaction between the specimens it is recommended to periodically interchange their positions.

Creation of harder testing conditions by elevating temperature and increasing the corrosion medium activity should not cause changes in the corrosion mechanism. The following extrapolation of the obtained results to the area of operating temperature and aggressive medium concentrations will allow to judge about the corrosion resistance of the coating.

Below are discussed the standardized methods of accelerated atmospheric corrosion testing (as defined by State Standard 9.038) and their foreign analogues.

4.3.8 Method of Testing Under the Effect of Neutral Saline Fog

In this method the process of corrosion is accelerated by elevating the ambient temperature and introducing sodium chloride solution into the atmosphere. The specimens are placed inside a chamber so that drops of solution do not trickle down on the specimens positioned below. The construction of the chamber should provide uniform internal conditions and free circulation of the fog around every specimen. The chamber is heated up to the temperature of 35 °C and is exposed to saline fog produced in an aerosol machine. The recommended duration of test is 2, 6, 24, 96, 240, 480, 720 hours and so on.

This method is the softest of all techniques using saline fogs so it is mostly used to check the corrosion resistance of organic coatings. It is also frequently used for cathode protective coating testing.

In the USA tests are performed in a similar fashion except that a solution containing 50 g/l of NaCl and acidified to pH = 3.2–3.5 by acetic acid is used. The tests are conducted at T = 308 K, cycle duration is 24 hours with the total of 3 cycles.

The ASTM standard (USA) specifies testing in saline fog chamber using 200 g/l NaCl solution at T = 288–298 K as the working medium.

In British manuals a similar test method is designated as ASS (Acetic Acid Salt Spray); it is used to test moderately resistant coatings and anodized aluminum.

4.3.9 Method of Testing Under the Effect of Acidic Saline Fog and Copper Chloride (CASS-Method)

This is a harder variant of the ASS-method. Adding copper dichloride (26 g/l of $CuCl_2x2H_2O$) to acetic salt solution accelerates corrosion process. The initials CASS stand for "Copper Accelerated Acetic Acid–Spray" Test.

The tests are conducted in a chamber at the temperature of 323 K for 16 hours; the solution is sprayed inside the chamber and deposits on horizontally-positioned specimens; solution consumption is 1.5–2.0 ml/h. The air inside the chamber is preheated and humidified by passing through distilled water with temperature (328±3) K. After each cycle the specimens are rinsed with water at 320 K, dried up and investigated.

According to reports from laboratories in the USA, one CASS testing cycle (keeping inside the chamber for 16 hours plus cooling for 8 hours) is equivalent to field operation in heavily polluted industrial atmosphere for a year [4.62].

The method is used for determination of protective properties of copper–nickel–chromium, nickel–chromium and anodic oxide coatings on aluminum and its alloys [4.62].

4.3.10 "Corrodcote" Testing Method

This method is extensively used in American automotive industry. The products are not immersed but covered with paste consisting of 0.99 g of ferrous chloride $FeCl_3{\cdot}6H_2O$, 0.21 g of copper nitrate $[Cu(NO_3)]_2{\cdot}6H_2O$, 6.0 g of ammonium chloride NH_4Cl, 180 g of kaolin and 300 ml of distilled water. The paste is spread as uniformly as possible over the surface under test to form an even layer 0.1–0.2 mm thick, then the specimens are dried up and placed in a condensation chamber with internal temperature 38 °C and relative air humidity 80–90%. If corrosion pattern is not clear enough, the specimens are additionally tested in saline fog after removing the paste [4.62].

The duration of the testing cycle is 16 hours unless another period of time is specified by the testing program. Once a cycle has been completed, specimens are taken out of the chamber and the paste is completely washed away with soft brush or sponge in cold running water.

The method is used for determination of protective properties of copper-nickel-chromium and nickel-chromium coatings on steel and zinc alloys.

One testing cycle is equivalent to field operation for 8 months in sea atmosphere or one CASS testing cycle.

4.3.11 Method of Testing at Elevated Relative Humidity and Temperature Without Moisture Condensation

The specified relative air humidity inside the chamber is created by blowing air humidified with distilled water. Air humidification with salt or acid solutions is not allowed.

Fans should provide air circulation in the chamber at a speed of 1 m/s or less.

After heating the specimens up to a specified temperature, the relative humidity of (93±3)% is created. The temperature and humidity are maintained constant for the whole testing period.

According to DIN50017 the following variants of the test are possible:

- temperature 40 °C and relative air humidity 100% (the onset of corrosion failure is registered);
- 8 hours at temperature 40 °C and relative air humidity 100% and 16 hours at temperature 20 °C and relative air humidity 100% (the degree of corrosion damage is estimated after completing the testing cycles);
- 8 hours at temperature 40 °C and relative air humidity 100% and 16 hours at temperature 20 °C and relative air humidity 75% (the same parameters are estimated).

4.3.12 Method of Testing at Elevated Relative Humidity and Temperature With Periodic Moisture Condensation

The test consists of continuously repeated cycles each having 24 hours duration. Each cycle consists of the following stages. Initial temperature of (25±2) °C and relative humidity of at least 95% are created inside the chamber. The temperature is elevated up to (40±2) °C during (3±0,5) hours. The relative humidity during this period should be at least 95% excluding the last 15 minutes when it should be at least 90%. At this time moisture is condensed on the surfaces of the specimens.

The temperature of (40±2) °C is sustained inside the chamber for (12±0,5) hours from the beginning of a cycle at the relative humidity of (93±3)% excluding the first and the last 15 minutes when it should be 90 to 100%. The moisture should not condense on the surfaces of the specimens during the last 15 minutes.

The temperature is lowered to (25±2) °C during 3–6 hours at the relative humidity at least 95% excluding the first 15 minutes when it should be at least 90%.

According to DIN 50016 24-hour cycles can be conducted at T = 29 °C and T = 40 °C and relative humidity 83% and 92%, respectively.

4.3.13 Method of Testing Under Continuous Effect of Sulfur Dioxide Without Moisture Condensation

The specimens are placed inside a chamber made of sulfur dioxide-resistant material. The chamber is fitted with an external gas feed device which is positioned at the height at least 50 mm from the chamber bottom providing uniform feed of gas and preventing gas flow from direct hit at the specimens.

The testing modes are set as follows: temperature (25±2) °C; relative humidity (75±5)%; sulfur dioxide concentration (75±15) mg/m^3. Sulfur dioxide is fed into the chamber as soon as the specified temperature and relative humidity have been set. Sulfur gas concentration is sustained constant and continuously monitored using gas analyzer.

The flow of the gas-air mixture in the working space of the chamber should be sufficient to provide 3–5 renewals of the mixture per hour.

The specified relative air humidity inside the chamber is created by blowing air humidified with distilled water.

4.3.14 Method of Testing at Elevated Relative Humidity and Temperature Under Continuous Effect of Sulfur Dioxide and Periodic Moisture Condensation

Corrosion in humid atmospheres is enhanced by introduction of aggressive gases. When performing tests by Kesternich (according to DIN 50018), which basically simulate behavior of materials in various industrial atmospheres, from 0.2 to 2 l of sulfur dioxide is added to the working volume of 300 l. In this test, besides comparatively resistant metals, all metal coatings are liable to corrosion to a different extent, so with time the substrate is also affected [4.62].

According to State Standard 9.308 the testing is cyclic with continuously repeated tests by one of the two modes. Cycle duration is 24 hours from the beginning of closed chamber heating.

By the beginning of each cycle the amount of water specified by the test program is poured in the water bath at the chamber bottom and the chamber is closed snugly. Then (2±0.2) g/m^3 of sulphur dioxide is introduced into the chamber.

The first mode: After introducing sulfur dioxide the chamber is heated up to (40±2) °C during 90 minutes, then this temperature is sustained constant during the test. After 24 hours the heating is turned off, the chamber is opened, and the water bath is drained.

The second mode: After introducing sulfur dioxide the chamber is heated up to (40±2) °C during 90 minutes, then this temperature is sustained constant during 8 hours. After this the chamber is opened, the water bath is drained, and the drawn specimens are hold for 16 hours at room temperature and relative humidity not higher than 75%.

4.3.15 Method of Testing at Periodic Immersion in Electrolyte

The test is performed using a "corrosion wheel" or "rod beam" device consisting of an electrolyte bath and a mechanism which provides alternating immersion and lifting of specimens fitted inside the bath. The device should provide complete immersion of specimens into electrolyte.

The volume of electrolyte in the bath is set depending on the surface area of specimens at the level 30–50 cm^3 of electrolyte per 1 cm^2 of the surface area.

The specimens are periodically immersed in electrolyte for 10 minutes and taken out of electrolyte for 50 minutes, respectively. If testing has to be interrupted, the specimens should stay in the air.

The solution is changed every 15 days of testing.

The method of periodic immersion is free from disadvantages of testing in saline fog chamber. In particular, it emulates the periodic processes of drying and wetting to which specimen surfaces are usually exposed during testing in natural conditions.

When testing automobile decorating parts, NaCl and $CaCl_2$ solutions with added sodium sulfate and thiosulfate are generally used; corrosion resistance is judged after 288 testing cycles.

4.3.16 Evaluation of Corrosion Damage

Evaluation of corrosion behavior can be quantitative, semi-quantitative (in points) and qualitative [4.19].

Table 4.2. Corrosion index (K_m) of base metal–coating composition

Corrosion damage	Affected surface [%]	Point (K_m)
Change of coating color	≤ 50	1
	> 50	2
Coating loosening	≤ 50	3
	> 50	4
Base metal corrosion	≤ 1	4
	1-3	5
	3-5	6
	5-10	7
	10-30	8
	30-50	9
	>50	10

The main quantitative index is the time of reaching the specified (permissible) degree of corrosion damage of the coating. If calculation of quantitative indexes is for some reason impossible or impractical, semi-quantitative or qualitative parameters can be used.

Visual inspection reveals qualitative changes such as surface tarnishing without any visible corrosion products; presence and character of the layer of coating corrosion products (uniform, non-uniform, solid, loose, peeling), its color.

Semi-quantitative changes in the appearance of the base metal–nonmetallic coating composition can be characterized by conventional scale points (corrosion index K_m) (see Table 4.2.).

According to State Standard 9.908 the corrosion behavior of the base metal–coating composition depending on corrosion type is quantified by the following indexes (see Table 4.3).

Table 4.3 also defines rate (differential) corrosion indexes corresponding to the corrosion effects (integral corrosion indexes). If the corrosion effect is a linear function of time, the corresponding rate index is determined by a ratio of the corrosion effect change within a definite time interval to the length of this interval.

If corrosion effect is a non-linear function of time, the corresponding corrosion index is determined by graphic or analytical methods. Corrosion resistance indexes marked with asterisk (*) can be determined either graphically, using time dependence of the corresponding corrosion index as illustrated in Fig. 4.13, or analytically from its empirical time dependence $y = f(\tau)$ by finding τ_{perm} corresponding to the permissible (specified) value Y_{perm}.

Indexes of corrosion resistance effected by mechanical factors including residual stresses are marked in the table with double asterisk (**) and can be determined immediately during testing.

If during testing different types of corrosion are observed, corrosion and corrosion resistance indexes corresponding to each type should be determined separately.

Table 4.3. Corrosion behavior indexes of materials

Corrosion type	Main quantitative indexes of corrosion and corrosion resistance		
	Corrosion effect (integral corrosion index)	Rate (differential) corrosion index	Corrosion resistance index
Continuous	Corrosion penetration depth Mass loss per unit surface	Linear rate of corrosion Mass loss rate	Time of corrosion penetration to permissible (specified) depth* Time of mass reduction by permissible (specified) value*
Pointed	Surface damage degree		Time of reaching the permissible (specified) degree of damage *
Pitting	Maximum pitting depth Maximum pitting diameter in the mouth Degree of surface damage by pitting	Maximum rate of pitting penetration	Minimum time of pitting penetration to permissible (specified) depth* Minimum time of reaching the permissible (specified) pitting diameter in the mouth* Time of reaching the permissible (specified) degree of damage *
Intercrystallyne	Corrosion penetration depth Degradation of mechanical properties (relative elongation, reduction of area, impact elasticity, point of maximum load)	Corrosion penetration rate	Time of penetration to permissible (specified) depth* Time of degradation of mechanical properties to permissible (specified) level*
Corrosion cracking	Depth (length) of cracks Degradation of mechanical properties (relative elongation, reduction of area)	Cracks growth rate	Time to appearance of the first crack* Time to specimen failure** Safety-stress level** (conventional limit of long-duration corrosion stress**) Threshold stress intensity coefficient for corrosion cracking**
Corrosion fatigue	Depth (length) of cracks	Cracks growth rate	Number of cycles to specimen failure* Conventional corrosion fatigue limit** Threshold stress intensity coefficient for corrosion fatigue**
Layer corrosion	Degree of surface affection by peeling Total length of cracked faces Corrosion penetration depth	Corrosion penetration rate	

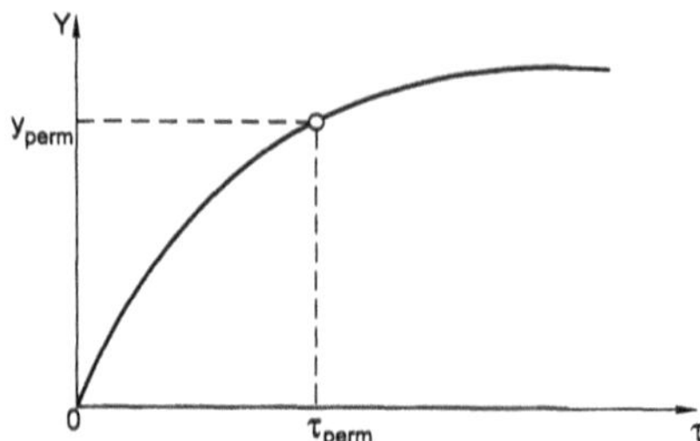

Fig. 4.13. Corrosion effect (integral index) as a function of time τ

Integral point corrosion index can be determined by two techniques: planimetring and using a square grid. In the first technique the surface of each spot is measured with a planimeter. In the second method the surface of a flat specimen is crossed by lines 1 to 5 mm apart from each other depending on dimensions of the prevailing corrosion spots. The degree of corrosion damage G in percent is then calculated from the following equations:

$$G = \frac{\sum_{i=1}^{n} S_i}{S} \cdot 100, \tag{4.11}$$

$$G = \frac{n_*}{N_*} \cdot 100, \tag{4.12}$$

where S_i is the area of the i-th spot, mm^2; n is the number of spots; S is the surface area of the specimen, mm^2; n_* is the total length of line segment falling in the corrosion spots, mm; N_* is the total length of all lines on the surface of the specimen.

The depth of intercrystalline corrosion penetration into base metal can be determined by standard procedures for austenite steels. Maximum pitting corrosion penetration depth is calculated as the arithmetic average of maximum pittings. The extent of averaging depends on the number (n) of pittings on the surface: for $n < 10$ 1–2 pittings are measured, for $n < 20$ – 3–4 pittings, and for $n > 20$ 5 pittings are measured. The degree of coating surface pitting damage is expressed in the form of the fraction of surface occupied by pittings in percent.

If a coating scales off the base metal, corrosion index K_x can be calculated from the following equation:

$$K_x = \frac{10}{\lg(100 - S_{sc})}, \tag{4.13}$$

where S_{sc} is the ratio of the surface where the coating scaled off the substrate to the total estimated surface area, %.

If corrosion proceeds with a constant rate, the two corrosion resistance indexes, time to mass reduction per unit surface by permissible (specified) value (τ_m), and time of penetration to permissible (specified) depth (τ_l), both measured in years, are calculated from the next equations:

$$\tau_m = \frac{\Delta m}{V_m}, \tag{4.14}$$

$$\tau_l = \frac{\Delta l}{V_l}, \tag{4.15}$$

where Δm is mass loss per unit surface, kg/m^2; Δl is corrosion penetration depth, m; V_m is mass loss rate, kg/m^2·year; V_l is linear rate of corrosion, m/year.

Weighing method is rather simple but not reliable enough if solid corrosion is accompanied by pitting corrosion.

If this is the case, it is possible to obtain objective information on corrosion behavior of a material from concurrent data both on mass loss automatically averaged over the whole surface, and maximum pitting depths which can play the main role in this case.

Only results which have been obtained for the same corrosion type under similar testing conditions should be considered comparable. If during testing of different coatings several types of corrosion are observed, then the obtained values of quantitative indexes are not statistically homogeneous and cannot be compared.

Corrosion processes are random in nature and generally yield large scatter in measured values. This is especially true for low rate indexes of corrosion and short testing duration [4.37].

As has already been mentioned, many methods for testing the base metal–coating composition are regulated by the appropriate national and Russian standards included in the Unified System of protection against corrosion, aging and biological damages. The Unified System provides the specified level of quality of products and materials by means and methods of protection against corrosion, aging and biological damages with due account of country defense requirements and competitiveness of products on the world market. Some of more than 120 standards included in the Unified System have immediate relation to "base metal–coating" composition testing and are therefore referred to in the present chapter.

4.3.17 Electrochemical Methods for Evaluating Corrosion

Along with techniques aimed at obtaining the dependence of a corrosion index on time, there are other methods of evaluating corrosion [4.5, 4.68, 4.69]. The most common among them are the galvanostatic and potentiostatic electrochemical methods.

In the galvanostatic method polarization curves are taken in the course of corrosion of the metal under test. Potential measurement is performed using any convenient technique, e.g., with a cathode voltmeter or a potentiostate. The character of polarization curves is indicative of the polarizability of the metal under test as a cathode or anode as well as of the probable role of cathode and anode reactions in oxidation. Analysis of the changes in metal potential with acidity and composition of the solution, nature of additives, temperature and other factors provides important information regarding the influence of these factors on the rate of corrosion, allows to determine the stages limiting corrosion rate and develop the most effective corrosion protection methods.

The method of galvanostatic curves is not the best technique for studying metals liable to passivation, since it does not represent the actual alternating pattern of anodic reactions. For these cases it is more convenient to use potentiostatic polarization curves. A typical potentiostatic curve plotted from the data of the dependence of current vs. time for several values of potential is shown in Fig. 4.14. As is seen from the figure, the rate of metal dissolution increases when the potential is biased positively within the (E_a–E_p) range. Once passivation onset potential E_p has been reached, the process abruptly slows down, anode reaction rate falls to i_p and the complete metal passivation occurs.

Once overpassivation potential E_{op} has been reached, the metal starts dissolving with linearly growing rate, although in this case electrode reactions differ from the reaction in the region of anode activity (E_a–E_p). In the state of overpassivation the metal, as a rule, dissolves forming ions of higher valency. Sometimes, this leads to the appearance of the second passive region or to onset of the anode oxygen release. The results of the potentiostatic studies reveal regions of active or passive metal state depending on testing conditions and also substantiate expediency and techniques of corrosion protection.

4.3.18 Protective Properties of Galvanic Coatings

The kinetics of cathode and anode reactions responsible for the process of corrosion is strongly influenced by the structure of the coating.

The structure of the coatings has especially large effect on hydrogen overvoltage and on the rate of coating dissolution in acidic aggressive mediums. For alloys the protective properties of the coating also depend on chemical and phase composition of resulting coatings.

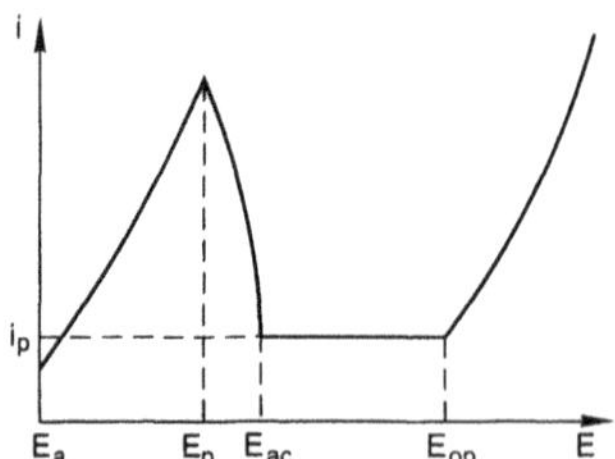

Fig. 4.14. Typical anodic polarization curve: E_a active anode potential; E_p passivation onset potential; E_{ac} activation potential; E_{op} overpassivation potential; i_p passivation current density

Thus, electrodeposited nickel–cobalt alloys containing up to 32% of cobalt corrode in diluted sulfuric acid (0.5n solution) at a rate close to that of pure nickel (see below).

Cobalt content in alloy [%]	–	14	22	32	47	50
Corrosion rate [g/(m^2·h)]	0.41	0.42	0.43	0.45	0.58	0.63

This can be explained by the fact that nickel which is more liable to passivation as compared to cobalt prevents development of corrosion within the given range of content. However, as the content of cobalt is further increased, corrosion rate sharply increases due to structural changes unfavorable for corrosion resistance.

In such coatings the texture is preferably formed along the <210> axis. A considerable increase of the extent of twin boundaries and grain boundaries having enhanced adsorption ability for water, anions, and hydrogen and providing sufficient numbers of corrosion active centers is also observed. Nickel electrodeposited coatings having similar structure and <210> texture are usually characterized by minimum corrosion resistance in acid media [4.54].

Raichevsky [4.56] investigated the influence of chemical and phase composition and texture of nickel-cobalt alloys on the corrosion electrochemical behavior in acid media (1M solutions of sulfuric and hydrochloric acids). It was demonstrated that in these conditions Ni–Co alloys dissolve by both chemical and electrochemical mechanisms simultaneously. The rate of chemical dissolution is 2–3 times higher than the rate of electrochemical dissolution and does not depend on electrode polarization within a wide range of potentials. A 20 to 70 % increase of cobalt content in coatings increases the rate of chemical dissolution. Nevertheless, corrosion resistance of coatings is basically determined by phase, rather than chemical, composition. Coatings with 20% of cobalt are single-phase solid nickel solutions with face-centered cubic lattice which have <100> texture and the lowest rate of dissolution.

Coatings with 50% cobalt crystallizing with <110> texture are also characterized by good corrosion resistance. Dissolution rate of the two-phase coatings containing 70% of Co is almost an order of magnitude higher (β-phase with face-centered cubic lattice and α-phase with face-centered close-packed lattice and <110> and <1120> textures, respectively).

Single-phase coatings containing more than 80% of Co and crystallizing with face-centered close-packed lattice and <1120> texture dissolve at the maximum rate.

These coatings have approximately the same rates of corrosion and chemical dissolution as pure electrodeposited cobalt with face-centered close-packed lattice and <1120> texture.

The basic mechanisms of the influence of phase composition on corrosion resistance of binary electrodeposited coatings were also verified for Ni–Fe alloys. It was demonstrated, that electrochemical alloying of nickel coatings with iron to the concentration up to 35%Fe basically does not reduce corrosion resistance. Therefore bright decorative protective nickel electrodeposited coatings can be fore bright decorative protective nickel electrodeposited coatings can be replaced by cheaper Fe–Ni ones.

As iron content in the alloy goes up to 50%, the corrosion rate of the coating gradually increases. Further growth of iron content in the alloy leads to formation of two-phase coatings, where, besides the phase of nickel solid solution with face-centered cubic lattice, the phase of iron solid solution with body-centered cubic lattice also occurs.

This causes sharp reduction of coating corrosion resistance: corrosion potential shows negative bias of 0.13 V, corrosion rate increases by an order of magnitude, while the rate of anode dissolution in the active region grows by a factor of 10^2 [4.56].

Noticeable reduction of corrosion resistance of coatings containing more than 50% Fe can be attributed not only to increase of dispersity and defectiveness of structure but also to formation of a phase with "loose" crystalline lattice in the alloy.

It has been calculated that in dense face-centered cubic lattice packing of Ni 74% of the total volume is taken up by the combined volume of metal atoms and 26% is taken by the interstitial voids between them, while for body-centered cubic lattice of Fe the corresponding figures are 68% and 32%, respectively. Therefore binding energy between metal atom is higher in face-centered cubic lattice than in body-centered cubic lattice. This circumstance leads to different adsorption ability of the lattices for water molecules and ions and affects corrosion-electrochemical behavior of the given coating. As Fe content goes up, corrosion resistance of coatings from Co–Fe alloys also goes down, corrosion rate falling exponentially as pH of the medium is increased. This indicates that Fe–Co alloys dissolve mostly via the chemical mechanism.

Electrodeposited alloys of metals belonging to Fe subgroup selectively dissolve in acid media. This follows from the analysis of partial dissolution rates of separate alloy components. Thus, during dissolution of ferric–cobalt alloys with iron content of 22, 34 and 50%, cobalt goes into solution faster: its relative content in solution is 3-6 times higher than in the alloy. Selective dissolution of cobalt proceeds at corrosion potential and, in the case of chemical dissolution, at cathode polarization.

It should be mentioned here, that in the process of ferric–cobalt coatings dissolution the coefficient of selectivity Z reduces. For example, at the beginning of dissolution of a coating with 34% Fe $Z = 3$, while after 30-minute dissolution $Z = 1$. Such specific behavior of coatings during corrosion is apparently connected with the fact that, as a result of their anomalous dissolution, their surface is enriched with iron, and with time the partial dissolution rates of alloy components become equal.

Judging by the recent publications, researchers have shown great interest in electrodeposited coatings from zinc-based alloys. It was demonstrated that alloying zinc electrodeposited coatings with tin, chromium and especially with nickel and cobalt considerably enhances their corrosion resistance. The table below shows experimental corrosion resistance data obtained by testing such coatings in saline fog chamber for 24 hours [4.55].

Alloying element		Fe	Co	Ni	Sn	Cr	Mo
Alloying component content, mass [%]	–	3.12	2.49	4.76	97.5	0.62	0.59
Surface area of the corroded coating [%]	95	90	30	50	50	70	80

Among the electrodeposited alloys mentioned above zinc-nickel alloys are most commonly used as protective coatings. They replace both toxic and expensive cadmium and cadmium-titanium coatings. Besides that, depositing zinc-nickel coatings substantially reduces hydrogen pickup and embrittlement of high-strength steels.

It should be noted, that zinc-nickel coatings consisting of γ-phase with 10–16% of nickel have the best protective ability independently of the nature and composition of aggressive medium. In such coatings there are practically no internal stresses, which, probably, complicates the development of corrosion processes [4.54].

References

4.1 High-temperature testing for intermetallic coatings (1995) Metallurgia 62:380

4.2 Ansart F, Ganda H, Saporte R, Thaverse JP (1995) Study of the oxidation of aluminum nitride coatings at high temperature. Thin Solid Films 260:38–46

4.3 Ashary AA, Tucker RC (1990) Electrochemical and long-term corrosion studies of several alloys in bare condition and plasma-sprayed with Cr_2O_3. Surface and Coating Technology 43–44:567–576

4.4 Atic M, Zarzycki J, R'Kha C (1994) Protection of ferritic stainless steel against oxidation by zirconia coatings. J Mater Sci 13:266–269

4.5 Baldwin KR, Bates RJ, Arnell RD, Smith CJE (1996) Aluminum magnesium alloys as corrosion resistant coatings for steel. Corros Sci 38:155–170

4.6 Bartlett Andrew H, Maschio Roberto Dal (1995) Failure mechanisms of a zirconia–8 wt% yttria thermal barrier coating. J Amer Ceram Soc 78:1018–1024

4.7 Barvinok VA (1990) Stressed state control and properties of plasma-sprayed coatings (in Russian) Mashinostroenie, Moscow

4.8 Bernacchi E, Ferrero A, Gariboldi E, Korovkin A, Pontini G (1996) Coatings in aluminum die casting parts and steel forming tools. Metal Sci and Technol 14:3–11

4.9 Bianco R, Rapp RA, Smialek JL (1993) Chromium and reactive element modified aluminide diffusion coatings on superalloys: Environmental testing. J Electrochem Soc 140:1191–1203

4.10 Birks N, Meier GH (1987) Introduction to high-temperature oxidation of metals (in Russian). Metallurgia, Moscow

4.11 Chan HC, Liu ZY, Chuang YC (1993) Degradation of plasma-sprayed alumina and zirconia coatings on stainless steel during thermal cycling and hot corrosion. Thin Solid Films 233:56–64

4.12 Cheruvu Narayana S, Carr Tomas J, Dworak John, Coyle James (1996) The in-service degradation of corrosion-resistant coatings. J Miner, Metals and Mater Soc 48:34–38

4.13 Chevelier S, Bonnet G, Colson JC, Larpin JP (1998) Influence of a reactive element oxide coating on the high temperature oxidation of chromia-formed alloys. J Chim Phys et Phys-Chim Biol 95:2083–2101

4.14 Ckuse T A, Stewart S E, Oktiz M (1988) Thermal barrier coating life prediction model development. Trans ASME J Eng Gas Turbines and Power 110:610–616

4.15 Connor Jeffrey A, Connor William B (1994) Ranking protective coatings. J Miner, Metals and Mater Soc 46: 35–38

4.16 Cushnie K, Bell JA, Smith GD (1991) Thermal barrier coatings increase applicability of Ni–Co–Cr–iron based alloy. Ind Heat 58:30,32–35

4.17 Dekhtjar LI, Veinberg VE, Berman SH, Loskutov VS (1983) Evaluation of heat resistance of gas thermal coatings by the method of acoustic emission. In: Kudinov VV (ed) Theory and practice of gas thermal spraying of coatings (in Russian). Intersectoral Technologies Department Press, Dmitrov, pp 71–74

4.18 Du HL, Datta PK, Lewis DB, Burnell-Gray JS (1996) Enhancement of oxidation/sulphidation resistance of Ti and Ti–6Al–4V alloy by HfV coating. Mater Sci and Eng A 205:199–208

4.19 El-Sawy S M (1985) Guidelines for corrosion testing of coated and uncoated metals. Anti-Corros Meth and Mater 32:16–18

4.20 Fokin MN, Zhigalova KA (1986) Methods for corrosion tests of metals (in Russian). Metallurgia, Moscow

4.21 Frantsevich IN, Dvernjakov VS, Pasichny VV (1970) System of high temperature solar plants IAM of AS USSR intended for investigation of serviceability of heat-resistant coatings (in Russan). Zaschitnye Pokrytia na Metallah 3:43–52

4.22 Garcia MC, Escudero ML, Lopes U, Macias A (1996) The corrosion behavior of laser treated Ni–P alloy coatings on mild steel. Corros Sci 38:515–530

4.23 Govindaraju Madhav Rao, Molian PA (1994) Enhancement of wear and corrosion resistance of metal-matrix composites by laser coatings. J Mater Sci 29:3274–3280

4.24 Harmsworth PD, Stevens R (1992) Phase composition and properties of plasma-sprayed zirconia thermal barrier coatings. J Mater Sci 27:611–615

4.25 Hashimoto K, Habazaki H, Akiyama E, Yoshioka H, Kim J–H, Park P–Y, Kawashima A, Asami K (1996) Recent progress in corrosion-resistant new alloys prepared by sputter deposition. Sci Repts Res Inst Tohoku Vniv A 42:99–105

4.26 Hasugama Hiroki, Shima Yukari, Baba Koumei, Wolf Gerhard K, Martin Herbert, Stippich Frank (1997) Adhesive and corrosion-resistant zirconium oxide coatings on stainless steel prepared by ion beam assisted deposition. Nucl Instrum and Meth Phys Res B 127–128:827–831

4.27 Herman H, Shankar NR (1987) Survivability of thermal barrier coatings. Mater Sci and Eng 88:69–74

4.28 Huntz AM, Lebrun JL, Boumara A (1990) Relation between the residual stresses and the high-temperature oxidation resistance of superalloys protected by plasma-sprayed coatings. Oxide Metals 33:321–355

4.29 John W, Hecht Ralph J (1987) The durability and performance of coatings in gas turbine and diesel engines. Mater Sci and Eng 88:321–330

4.30 Kharitonov NP, Sandler NG, Troshin NA (1983) Application of organic silica material to protection of pipelines made of austenite steel against corrosion cracking. In: Schulz NM (ed) Anticorrosion coatings (in Russian). Nauka, Leningrad, pp 257–261

4.31 Khasui A (1975) Deposition techniques (in Russian). Mashinostroenie, Moscow
4.32 Kolomytsev PT (1979) Heat-resistant diffusion coatings (in Russian). Metallurgia, Moscow
4.33 Kravchuk LV, Semjonov GR, Ovcharenko OYu, Yakovchuk KYu, Rabinovich AA (1992) Gas dynamic tests of multicomponent heat-resistant coatings with the outer layer in thermal cycling. In: Borisenko AI (ed) Corrosion resistant coatings (in Russian). Nauka, St.-Petersburg, pp 226–229
4.34 Kreye H (1998) A comparison of HVOF systems-behavior of materials and coating properties. Galvanotekhnika 89:3006–3013
4.35 Kurushima T, Ishizaki K (1993) The thermal shock endurance of a plasma-sprayed ZrO_2 coating on steel. Mater and Manuf Processes 8:467–474
4.36 Lamesle P, Steinmetz P (1995) Growth mechanisms and hot corrosion resistance of paladium modified aluminide coatings on superalloys. Mater and Manuf Processes 10:1053–1075
4.37 Levy Alan V (1988) The erosion-corrosion behavior of protective coatings. Surface and Coating Technology 36: 387–406
4.38 Lih W, Chang E, Wu BC, Chao CH (1991) Effect of bond coat preoxidation on the properties of ZrO_2–Y_2O_3–Ni–Cr–Al thermal-barrier coatings. Oxide Metals 36:221–238
4.39 Liu Yung Y, Nafesan (1988) Oxidation-erosion of metals and alloys: models, verification and prediction. Surface and Coating Technology 36:407–417
4.40 Lizandier MF, Lanza E, Sebaoun A, Giroud A, Guiraldeng P (1992) Characterization of porosity of ceramic coatings assigned to applications in sea-water with the evolution of their electrochemical behavior under friction Wear 153:387–397
4.41 Lyashenko BA, Rishin VV, Tovt VN (1977) Techniques for studying carrying capacity of materials with protective coatings (in Russian). Problemy Prochnosti 9:123–125
4.42 Logan HL (1966) The stress corrosion of metals. A Wiley–Interscience Publication, New York
4.43 Longa Y, Takemoto M (1994) Laser processing high-chromium nickel-chromium coatings deposited by various thermal spraying methods. Corrosion (USA) 50:827–837
4.44 Longa Yrene, Shinya Masanoby, Takemoto Mikio (1994) Coatings of aluminide intermetallic compounds on steel utilizing a hybrid technique of spraying and IR-Laser fusion. Mater and Manuf Processes 9:495–505
4.45 Lou Hanyi, Tang Youjung, Sun Xiaofen, Guan Hengrong (1996) Oxidation behavior of sputtered microcrystalline coating of superalloy K17F at high temperature. Mater Sci and Eng A 207:121–128
4.46 Mueller Andrew, Wang Ge, Rapp Robert A, Courtright Edward L, Kircher Thomas A (1992) Oxidation behavior of tungsten and germanium-alloyed molybdenum disilicide coatings. Mater Sci and Eng A 155:199–207
4.47 Nesbitt J A (1990) Thermal response of various thermal barrier coatings in a high heat flux rocket engine. Surface and Coating Technology 43–44:458–469
4.48 Newar GM, Nusier SQ, Chaudhury ZA (1998) Damage accumulation mechanisms in thermal barrier coatings. Trans ASME J Eng Mater and Techol 120:149–153
4.49 Nikitin VI (1982) Method for prediction of service life of protective coatings (in Russian). Fizika I Khimia Obrabotki Materialov 3:95–99
4.50 Parker D (1991) Thermal barrier coatings – taking the heat. Turbomach Int 32:42–44

4.51 Pasichny VV (1995) Investigation into physical-technical properties of materials and coatings using concentrated solar power (in Russian). Poroshkovaya Metallurgia 7/8:160–172

4.52 Peteres John A (1996) New approaches to surface engineering. Metallurgia 63:408–409

4.53 Phys-Jones T N, Cunningham T P (1990) The influence of surface coatings on the fatigue behavior of aero eng materials. Surface and Coating Technology 42:13–19

4.54 Povetkin VV, Kovensky IM, UstinovshchikovYuI (1992) Structure and properties of electrodeposited alloys (in Russian). Nauka, Moscow

4.55 Proskurkin EV, Popovich VA, Moroz AT (1988) Sherardizing. Reference book (in Russian). Metallurgia, Moscow

4.56 Raichevsky GM (1989) Corrosion electrochemical behavior of galvanic coatings. Outcomes of science and engineering. Corrosion and protection against corrosion (in Russian). Nauka, Moscow

4.57 Ryabchenkov AV, Ovsyankin VV, Vidanov AP (1973). Method for determination of crack development kinetics in specimens with protective coatings during corrosion fatigue failure (in Russian). Zavodskaya Laboratoria 10:1250–1253

4.58 Rogers PM, Hutchings IM, Little JA (1992) TiN coatings for protection against combined wear and oxidation. Surface Eng 8:48–54

4.59 Sacre S, Wienstroth V, Feller HG, Thomas LK (1991) Influence of the oxide morphology on the oxidation resistance of an Ni–Co–Cr–Al–Y coating material. Mater Sci and Eng A 145:123–126

4.60 Saint-Jacques RG, Bordeaux F, Stansfield B, Veilleux G, Zurak WW, Lakhsasi A, Boucher C, Moreau C (1992) Enhanced resistance of plasma-sprayed TiC coatings to thermal shocks. J Nucl Mater 191–194:465–468

4.61 Semyonov AP, Kovsh IB, Petrova IM, Arkhipov VE, Birger EM (1992) Methods and means for hardening of surfaces of machine parts by concentrated energy fluxes (in Russian). Nauka, Moscow

4.62 Simon G, Toma M (1985) Applied techniques for surface treatment of metal materials. Reference book (in German). Carl Hanser Verlag, Munchen

4.63 Sonobe Masaro, Shizawa Kazuaki, Motobayashi Kou (1997) Corrosion resistance and corrosion fatigue strength of carbon steel coated with chromium nitride by multistage. JSME Int J A 40:436–444

4.64 Srinivasan V (1994) High-temperature corrosion and erosion in gas turbine engines – where do we stand? J Miner, Metals and Mater Soc 46:34–35

4.65 Taniguchi Shigeji, Shibata Toshio, Yamada Takayuki, Liu Xianghuai, Zou Shichang (1993) High-temperature oxidation resistance of TiAl improved by IBED Si_3N_4 coating. ISIJ International 33:869–876

4.66 Tawancy HM, Sridhar N, Abbas NM, Rickerby D (1998) Failure mechanism of a thermal barrier coating system on a nickel-base superalloy. J Mater Sci 33:681–686

4.67 Terentyeva VS, Lebedev PD, Sychev VA, Korotkov VP, Golovin VD, Nizovitina LF, Fedyaev MYu, Eryomina AI (1987) Multifunctional coatings in heat-loaded structural members. In: Borisenko AI (ed). Production and application of protective coatings (in Russian). Nauka, Leningrad, pp 89–94

4.68 Tjong SC (1996) Performance of laser-consolidated plasma-sprayed coatings on Fe–28Mn–7A1–1C alloy. Thin Solid Films 274:95–100

4.69 Tjong SC, Ku JS, Wu CS (1994) Corrosion behavior of laser consolidation chromium and molybdenum plasma-sprayed coatings on Fe–28Mn–7Al–1C alloy. Sci Met et Mater 31:836–839

4.70 Tushinsky LI, Plokhov AV, Sindeev VI (1996). Protective properties of surface layers after high-energy impacts (in Russian). Novosibirsk State Technical University Press, Novosibirsk

4.71 Uhlig HH, Revie RW (1985) Corrosion and corrosion control. An introduction to corrosion science and engineering. A Wiley–Interscience Publication, New York

4.72 Vaidya RU, Zurek KN, Castro R, Wolfenden A, Hosman SL, Subramanian KN (1994) Alumina coated Ti–25Al–10Nb–3V–1Mo for improved oxidation resistance. J Adv Mater 26:16–22

4.73 Veksler YuG, Lesnikov VP, Kuznetsov VP, Paleeva SYa, Rozhko AL (1985). Influence of heat-resistant coatings on heat resistance of nickel alloys in aggressive gas flows. In: Borisenko AI (ed). Termally stable coatings (in Russian). Nauka, Leningrad, pp 15–20

4.74 Verstak AA (1990) On quality control of plasma-sprayed coatings made of zirconium dioxide ceramics. In: Irinin IA (ed) Development and usage of technological equipment and materials for gas thermal spraying of protective coatings (in Russian). Belorussian Institute of Scientific and Engineering Information Press, Minsk, pp 23–24

4.75 Wirz Ch, Blatter A, Hauert R (1992) Properties of films prepared by thermal coevaporation of Cr and Ti in nitrogen. Thin Solid Films 214:63–67

4.76 Yar-Mukhamedov ShH (1989) Thermal cyclic resistance of some types of composition coatings in melts of cast iron and aluminum (in Russian). Zaschitnye Pokrytia na Metallah 23:78–81

4.77 Yokofa Osamu, Dohi Hiroshi, Watanabe Takao, Sato Morihide, Sugawara Yasunori (1992) Failure of gas plasma aluminum oxide coatings in thermal cycling (in Japanese). J High Temp Soc 18:321–327

4.78 Zhukov AP, Malakhov AI (1991) Fundamentals of physical metallurgy and corrosion theory (in Russian). Vysshaya Shkola, Moscow

5 Determination of Bond Strength Between a Coating and Base Metal

5.1 Mechanism of Coating Joint With the Base Metal

At present there is no unified term specifying the bond strength between the base metal and coating referred to a unit of their common surface. The following terms are in most common use: adhesion, adhesive bond strength, coating-substrate bond strength, and adhesion strength of coating joint with a substrate. Obviously, such terminological uncertainty of one of the most important coating characteristics causes confusion both in special literature and technical documentation in studies of coating properties under industrial conditions.

The term adhesion (adherence, sticking together, seizure) and its derivatives do not always correctly characterize the numerous reasons that cause a base metal to join with a coating. In our opinion, the more accurate term estimating the specific strength of this joint is the bond strength between the coating and base metal, or the coating bond strength.

In deciding on a particular coating material and method, generally the level of bond strength is estimated first. This characteristic is one of the main criteria for determination of both the application field and operational characteristics of the coating, especially when a coated product serves as a load-carrying structure. It is clear that prior to depositing protective coatings, significant attention has to be given to their special properties. Despite the fact that in most cases bond strength is the main characteristic that proves the coating efficiency, development of theoretical methods for its calculation has just started.

The calculation methods for serviceability of coated parts with consideration of the random character of alternating loads are suggested in the work of Kalmutsky [5.19]. The "adhesion safety coefficient" is introduced there. This coefficient is the ratio of the minimal probabilistic bond strength at a given risk level of coating peeling to the maximal possible value of stresses in the part determined by statistical methods. At the same time, even though the realization of the developed principles allows enhancement of the bond strength more than by a factor of three, the regularities discovered were tested only for "steel substrate–electrolyte coating" composition. It seems likely that probabilistically statistical analyses and technical forecasting of the bond strength for plasma and detonation coatings will be hampered by a more complex structure of these latter in comparison with the electrolytic ones.

It was assumed earlier that for the majority of the spraying methods the base metal–coating joint occurs due to mechanical bonds, and preliminary surface preparation (sand blasting, in particular), increasing roughness contributes to enhancement of mechanical bonds through jamming of deformed deposited particles in the base metal relief. Now it is felt that along with mechanical interaction chemical bonds and Van der Waals forces established in spraying determine the bond strength.

However, importance of the latter is insignificant for the bond strength enhancement. As for chemical interaction, it may be the governing one. High bond strength is observed in spraying high-melting coatings on metals with a lower melting temperature. As this takes place, two materials with different chemical composition and properties mix which leads to the high bond strength between the coating and the metal. Preliminary surface preparation before coating is needed not only for the required relief formation on the metal surface, but also for an increase in the contact area and activity intensification [5.8, 5.16, 5.42].

The reasons determining the level of bond strength will be discovered on the basis of systematic and thorough investigations into the base metal–coating boundary using modern methods of structural research [5.13, 5.40].

The bond strength between the coating and the metal depends not only on the bond character; the level of internal stresses that is dependent of the coating thickness is also of great importance. An increase in the coating thickness is usually accompanied by the growth of residual stresses, which reduce the bond strength of the coating. Kudinov et al. [5.23] suggested a scheme, which correlates the thickness of a plasma sprayed coating with the bond strength (see Fig. 5.1). Being provided with some assumptions and refinements, this scheme may cover other coating types as well.

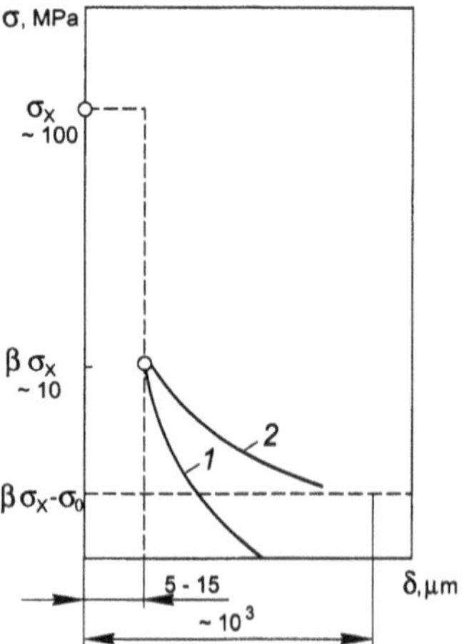

Fig. 5.1. Scheme of variations in the bond strength between the coating and base metal depending on the coating (by [5.23]). *1* coatings from brittle materials, *2* coatings from plastic materials

Point σ_x corresponds to the coating bond strength in chemical interaction (sticking of coating particles to the base metal), i.e. to the maximal bond strength possible in spraying. Point $\beta\sigma_x$ characterizes the bond strength of one particle layer; β is the coefficient equal to the ratio of the area of zones where chemical interaction (sticking) occurred to the total area of the base metal surface. The value of β depends on the spraying method and sprayed material nature. Thus, in plasma spraying of metals and with high-melting composites $\beta \approx 0.1$; in detonation spraying of plastic metals when the modes are optimal $\beta \rightarrow 1$. The thickness of a single layer is 5–15 μm. Spraying of the next layer initiates residual stresses σ_x increasing with a rise in the number of layers and the coating thickness σ. Therefore, in actual practice a decreased bond strength $\beta cx - \sigma_0$ is observed almost without exception. An increase in the coating thickness resulting from the next layers spraying leads to a growth of σ_0. If brittle materials are deposited, at $\sigma_0 > \beta cx$ the coating will spontaneously detach from the base metal (the curve 1 will cross the horizontal line $\beta cx - \sigma_0$). In the case of a plastic coating, residual stresses partially relax remaining less in value than $\beta cx - \sigma_0$, which leads to stabilization of the coating bond strength [5.23].

New, at times unexpected, approaches to the problem of the bond strength increase are developed now. Thus, Lyashenko et al. [5.28] reason that a considerable bond strength between the coating and base metal should not be considered a guarantee because a crack in the non-peeled coating is a stress concentrator which can significantly decrease the structural strength of the "base metal–coating" composition. The optimal strength between adhesion and cohesion strengths was determined. This ratio provides coating peeling before the critical stressed state is generated in the substrate material.

Along with residual stresses the following factors can have a negative influence upon the bond strength: strong bonds between the coating and base metal materials formed at separate contact areas only; the bond strength in the welding zone lower than that of a compact material either due to macro- and micro defectiveness of a bond formed or because of slight mutual penetration of the coating and base metal materials. Besides, when spraying coatings by certain methods there may be some other typical reasons decreasing or increasing bond strength [5.12, 5.17, 5.20, 5.29].

At present determination of the bond strength of coatings with the base metal is performed using special equipment and original techniques designed to provide qualitative and quantitative estimation of this important parameter under various conditions [5.6, 5.7, 5.14, 5.18, 5.25, 5.31].

The main demands that should be satisfied when fabricating specimens for being tested by all methods are presented below.

Before testing the working surfaces of specimens are prepared by means of degreasing, sand blasting, chemical etching or some other technique provided by technology.

Mechanical or magnetic devices, which allow simultaneous mounting of several test pieces for coating spraying, are recommended to fabricate specimens. This, on the one hand, provides the same conditions for every variation of spraying and, on the other hand, facilitates fabrication of large specimen batches.

Coatings are deposited in stationary modes of equipment operation avoiding overheating, which causes change in the structure and properties of the base metal. Specimens with cracks and peelings as well as with considerable difference in thickness should be rejected. The results obtained by different investigation methods are incomparable.

Valli formulated the demands on the ideal estimation method for the bond strength between the coating and substrate [5.41]: complete exclusion of the influence of other coating and substrate features on the sought-for quantity; simplicity of measurements and result interpretation; test mobility; (100%-control of the output production); PC-based date recording and processing; non-destructivity of the method applied and usage of a real product as a specimen.

A lot of methods were developed for the coating bond strength estimate. However, there is no one satisfying to all the above requirements. The majority of the methods for bond strength determination cause deformation of the coating and substrate thereby limiting the possibility of their application to precise measurements. Nowadays, the mechanical methods based on measurement of the load required for coating detachment from the substrate have received wide acceptance in engineering.

5.2 Pin Method

5.2.1 General Overview

One of the main techniques for determination of the bond strength between the coating and base metal is the pin method. A specimen is a washer with a hole, where a pin is installed to make its end surface flush-mounted relative to the plane of the washer bottom. A coating is deposited on the common surface of the pin end and the washer after an appropriate preparation. The specimen is tested by means of pin removal from the washer and strength recording. When the pin is detached from the coating, the ratio of the maximal load to the area of the pin end is determined. This ratio is a quantitative characteristic of the coating bond strength. A cylindrical pin is rare in application, and now it is used only for the estimate of galvanic coatings.

Usually, technological difficulties in fabrication of the "cylindrical pin–washer" pair do not provide the necessary mating accuracy of these parts. Therefore, in the clearance between the parts there is a coating area subjected to the effect of maximal stresses leading to its failure at lower forces than those required for its detachment from the pin. If the mating between the pin and the washer is tight, the friction and Van der Waals forces considerably effect the measurement accuracy.

Nowadays, the coatings are often tested using the modernized specimens with the pin and the washer hole in the shape of a cone (Fig. 5.2). In the absence of friction forces such a shape of the pin decreases the mating clearance and increases measurement accuracy.

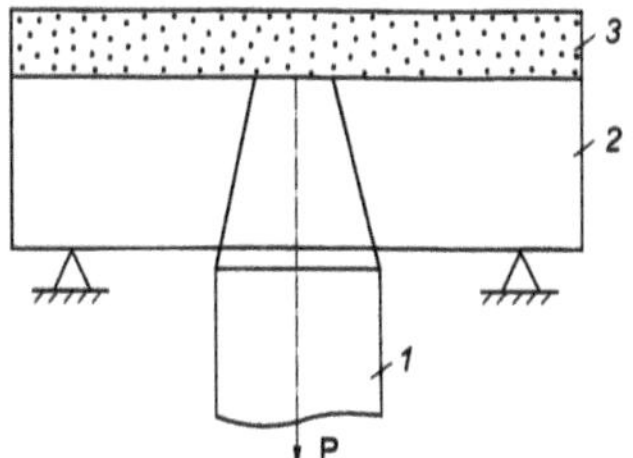

Fig. 5.2. Specimen for tests of the bond strength between the coating and base metal by means of the cone pin method. *1* pin, *2* washer, *3* coating

The "pin shift–applied force" relationship (Fig. 5.3) is considered by Bartenev [5.2] and Khasui [5.21].

Before loading the backlashes are taken up. Then, a fast growth of loading at insignificant plastic deformation of the pin is observed. Point A corresponds to the beginning of pin detachment from the coating, and this is accompanied by a decrease in load. When testing a relatively plastic coating, at first, the latter starts exfoliating from the pin edges. With a decrease in area of the coating contact with the pin end, the load decreases until the moment of complete detachment. For the brittle, weakly deformable coating, the pin detachment may occur simultaneously over the whole surface of the pin end. This leads to a more drastic load drop, and inclination of area 4 increases. It is hard to ground theoretically the pin method due to complex stressed state of the coating at loading. The relationship between simultaneous shear and bend stresses and the coating bond strength determine the character of the coating failure.

Rogozhin et al. [5.35] suggested a mathematical model for the process of coating loading up to its failure in detachment testing by the pin method. The values of adhesion and cohesion strength for the given type of failure were calculated using the suggested formulas. This method slightly increases their absolute values, though doubles the accuracy.

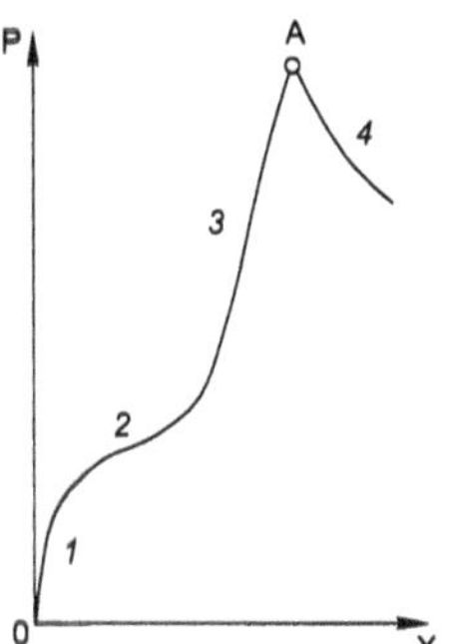

Fig. 5.3. Character of load change (P) at pin removal (by [5.2]). Portions of: *1*, *2* backlashes taking up; *3* elastic strain; *4* load drop; x pin shift

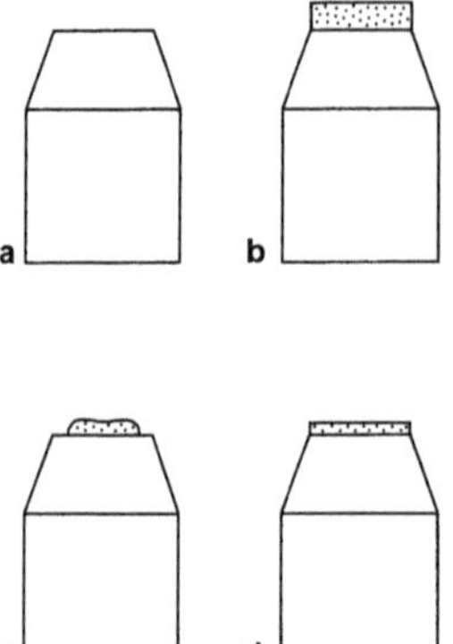

Fig. 5.4. Types of coating failure on tests by the pin method

5.2.2 Types of Coating Failure at Pin Removal

There are for types of failure (see Fig. 5.4). The pin end detaches from the coating precisely along the interface (see Fig. 5.4a). For such "clean" detachment the coating bond strength is determined only by the normal stress. An opposite case of coating lateral failure (break through) along the pin perimeter is observed at high shear stresses (see Fig. 5.4b). For this case, the breaking load, obviously, cannot be used for determination of the bond strength.

For the third case, detachment along the interface occurs only at some areas, i.e., a mixed failure is observed, which takes place partially on the coating and partially along the interface (Fig. 5.4c).

The case of internal failure is predominantly possible in stratified coatings, when a uniform part of the coating stays over the whole surface of the pin end (Fig. 5.4d), while the part of the coating above the pin remains intact.

Many researchers determined dependence of the bond strength on the specimen sizes. Thus, Lyashenko et al. [5.28] determined experimentally dependencies of the bond strength of detonation coatings on the pin and washer diameters (Fig. 5.5).

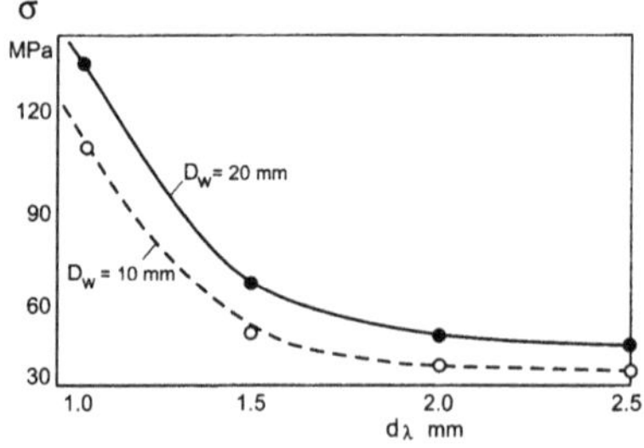

Fig. 5.5. Dependence of the bond strength between the coating and base metal on the pin (d_λ) and washer (D_w) diameters (by [5.28])

As the pin diameter d increases from 1 to 2.5 mm, the bond strength decreases. At $d > 2$ mm, the bond strength is kept almost constant. A decrease in the washer diameter also leads to some decrease in the bond strength. It is clear from Fig. 5.5, that the dashed curve is always lower than the solid curve, however, the difference between them is insignificant and does not exceed the measurement error. The authors made a conclusion that the pin and washer with respective diameters of 2 and 20 mm should be used [5.28].

We should note that recently the failure mechanism typical of the given test method attracts more and more attention. This can be explained by the fact that until now, despite its disadvantages, the pin method is the most commonly employed technique, being the only one used so far for determination of the bond strength in case of detonation coatings.

Earlier the data obtained on the pin method basis was considered a first approximation, but now we can speak about an increase in reproducibility and result stability. This becomes possible due to understanding of the coating failure processes and improvement of the test methods.

The detachment character of failure can be obtained through variations in the coating thickness or pin end diameter. Puzryakov et al. [5.33] suggested to use the ratio of the pin radius r to the coating thickness δ as the correctness criterion. Calculations demonstrated that the pin method of bond strength determination may be used only for low r/δ ratios (≤ 2.0).

At different values of r/δ this method is applicable only for the coatings, whose adhesion strength is considerably higher than the bond strength with the base metal. Interesting diagrams obtained by the authors allow determination of the real value of stresses generated in the coating at loading depending on the above ratio.

5.2.3 Conic Pin Method

Results of works dealt with the study of coating failure mechanisms by the pin method demonstrated increased accuracy and reliability of this method. However, together with these problems, development of standard specimens and research setups for the pin method is very important. As a rule, specimens shown in Fig. 5.6 are used. Drafts of the pin and washer are presented in Fig. 5.7.

Rules of specimen fabrication for practical tests are the following:

1. The pin is made from the same material and at the same structural state as a part, whose bond strength with the coating should be determined.
2. For pin h7, tolerance of the cone diameter in any cross-section should correspond to washer H7.
3. The washer is made from steel of any quality. A surface for coating deposition, consisting of the pin and washer faceplates, is ground after assembling and mounting of the pin up to roughness Ra < 1.25 μm. It should be taken into account that both the diameter of the pin end and inner diameter of the washer increases in grinding, and therefore an allowance for grinding should be made in their fabrication. The value of this allowance depends on the thickness of a grinded-off layer. After grinding, the pin is pull out, burrs are smoothed out, and the specimen is assembled again.

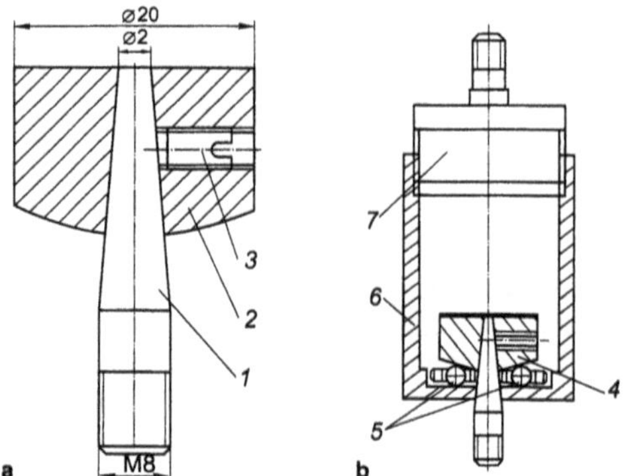

Fig. 5.6. Scheme of a specimen (**a**) and a device (**b**) for testing of the bond strength between the coating and base metal using the conic pin method. *1* pin; *2* holder; *3* lock screw; *4* assembled specimen with a coating; *5* ball bearing; *6* vessel; *7* vessel lid

4. Heating of the specimen in fabrication should not cause structural changes in the material. Treatment allowances, parameters, regimes and treatment sequence should minimize hardening, exclude local heating of specimens, especially at the working surface of the pin.
5. More often, the surface is prepared by the shot-sand blasting. It is necessary to check the possibility of pin free removal from the washer when unscrewing the stopper. The pin is washed in alcohol or benzene. Then the diameter of the pin end is measured with an error of 0.01 mm, the specimen is assembled again, and the stopper fixes the pin. The coating should be uniformly deposited on the working surface without cracks, differences in thickness and other defects, if it is not foreseen by research tasks.

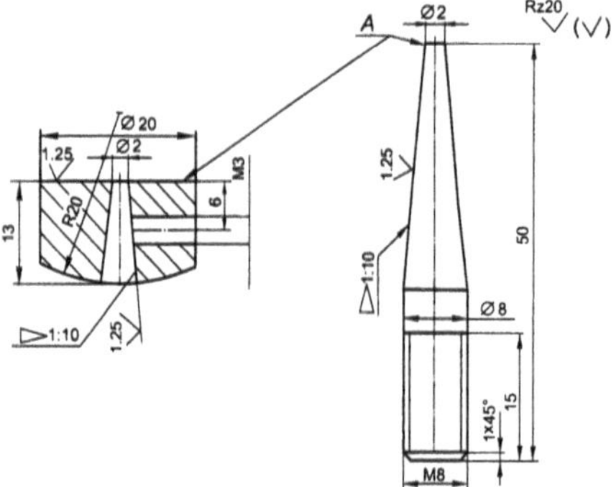

Fig. 5.7. Pin specimen for tests of the bond strength between the coating and base metal: *A* coated surface

Scientific developments of the last years gave rise to some achievements in estimates of the bond strength between the coating and base metal by the pin method. However, many problems are unsolved yet.

1. There are no accurate methods for calculation of the complex stressed state in a local area. This is caused by the fact that during testing in the coating and along the coating–base metal boundary the stresses of three types appear simultaneously: normal, shear and bend stresses. Thus, coating failure begins at lower loads, and this leads to conservative results.
2. Usually, the detachment character of failure is possible only for coatings thicker than 0.3–0.4 mm. For testing of thinner coatings, a decrease in the pin diameter (less than 2 mm) is not always possible due to technological difficulties in fabrication of specimens.
3. A change of the pin shape from a cylindrical to a conic one reduces, but not eliminates completely, a clearance between the pin end and upper hole of the washer. The clearance size varies from units to tens of microns. A very small clearance is also undesirable because this leads to generation of considerable Van der Waals forces, and pin and washer welding or riveting by high-speed particles. At a large clearance, the overlapping coating brings higher uncertainty to the complex-stress state and test results.
4. There is no correlation between the coating porosity and the scatter of bond strength values. It is shown that the higher the coating porosity and the less the pin diameter, the greater scatter of experimental values of bond strength.
5. When it is necessary to determine the effect of after-coating treatment on the bond strength, additional difficulties arise. At high temperatures of thermal treatment the pin and the washer may weld over the contact surface. When the coating is saturated with polymers, solder, etc., the saturating material penetrates the clearance between the pin and washer. Welding and sticking of the pin and washer introduce errors into the testing results.
6. The use of a conic pin requires measurement of its end surface area, and the bond strength is determined by this parameter at each test. After each test the pin and washer end surfaces are ground again to remove remains of the coating, which results in a change in geometrical sizes. When a coating is made of difficult-to-machine materials, the pin and washer can hardly be used repeatedly because it is very difficult to remove the coating remains.
7. Diversity and interaction of factors, which determine the bond strength of the coating and base metal, and the absence of the standard methods impede and sometimes make impossible comparison and use of data obtained for specimens with different sizes and shapes. The problems of specimen standardization are not solved yet, especially in the field of geometrical sizes and shapes of the pin.

5.2.4 Improvement of the Pin Method

Regardless the obvious advantages of the pin method, which motivated its wide usage, this method has some shortcomings. Therefore, the means are searched that could increase the accuracy, reproducibility, and also expand the sphere of the method application. Among various methods, we could single out the four main lines. Firstly, the variations of technique improvement were proposed that use a smaller clearance between the pin and washer. Secondly, some authors recommend to estimate the bond strength between the coating and base metal from energy estimation, i.e., to find the specific work on disruption of bonding between the coating and metal. Thirdly, the improvement of the pin method due to reasonable choice of the pin size and especially its shape is actively developed. Fourthly, usage of high-temperature coatings demands their testing at high temperatures. Some typical variations of determination of coating bond toward the mentioned lines of investigation are considered below.

Even with the most thorough fabrication of a specimen, a clearance between the pin and the washer still exists. For a small clearance in a conic space, the mated surfaces approach within a distance when significant Van der Waals forces start developing (Fig. 5.8a). A considerable clearance creates the possibility of either penetration of high-speed particles causing jamming of the pin in the washer (Fig. 5.8b), or flowing of melted sprayed material inside that often results in welding of the specimen's parts (Fig. 5.8c). All these factors cause additional forces fixing the pin, which takes additional effort to overcome them. This must give overrated resulting values of the bond strength between the coating and base metal.

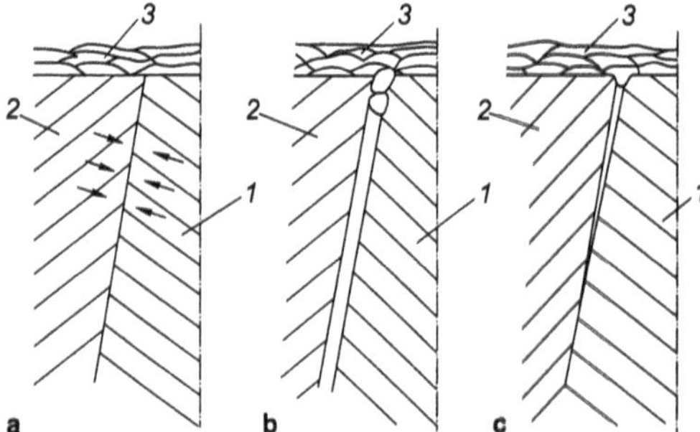

Fig. 5.8. Scheme of possible obstacles to pulling out of a pin: **a** Van der Waals forces; **b** penetration of solid particles into the clearance; **c** fusion of the pin and holder by a molten coating. *1* pin; *2* washer; *3* coating

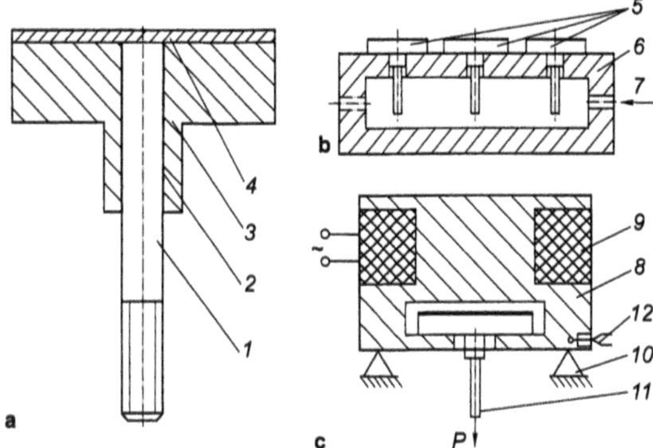

Fig. 5.9. Schemes of the specimen (**a**), fitting for spraying (**b**), and the unit (**c**) for testing of the bond strength between the coating and the substrate by the method of a pin heat fixing: *1* pin; *2* solder in clearance; *3* holder; *4* coating; *5* specimen; *6* cooling device; *7* cooling agent; *8* body of testing unit; *9* heater; *10* machine's traverses; *11* pin; *12* thermocouple; *P* load

To eliminate the shortcomings of the present technique there was proposed a method consisting in replacement of the conic pin by a special cylindrical pin. The pin 1 is fixed by solder 2 poured into the clearance between the pin and the holder 3 (Fig. 5.9a). The clearance is minimal. The solder can be different depending on the required testing temperature.

Before coating deposition, a specimen has to be treated in a special way. The pin and the hole are tinned, and then the pin is soldered into the holder. The working surface of the specimen 5 is ground and exposed to sand-jet treatment. After that specimen 5 is put into the fitting for metal spraying (Fig. 5.9b).

Compressed air is fed through the pipe connection to provide cooling of specimens during coating deposition.

The specimen with a sprayed coating is taken out of the unit and placed into the testing block (Fig. 5.9c), consisting of the body 8 with the heater coil 9. After that the block is heated up to the testing temperature controlled by the thermocouple 12. After a 15-minute delay, sufficient for complete melting of the solder, the specimen is loaded up to fracture.

The resulting specimen has several advantages.

1. Solder prevents coming of the sprayed material into the clearance and provides optimal conditions for coating formation above the annular pin-washer joint.
2. During testing the molten solder in the clearance works as lubricant and exclude friction between the pin and the washer.
3. Application of a cylindrical pin allows repeated tests of the same specimen without changes in the calculated tearing-off area.

A shortcoming of this method is that testing cannot be run at low temperatures.

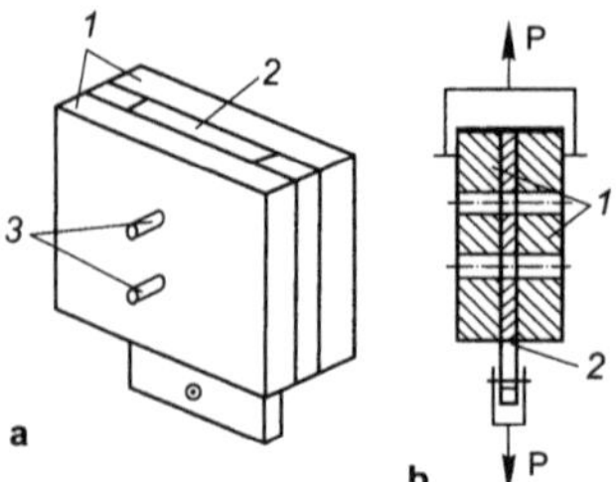

Fig. 5.10. General view (**a**) and testing scheme (**b**) of a specimen with a rectangular pin. *1* holder; *2* pin; *3* fixing rods; *P* load [courtesy of Bataev V.A.]

To make the design simpler and cheaper, a specimen with a rectangular pin was proposed (Fig. 5.10a). The side-frames make up a rectangular channel, and the pin may slide in it. During spraying the pin is fixed by two cylindrical rods 3. Then the specimen surface is treated with a shot-blasting machine, after which the surface is coated. Once the rods are removed, the specimen is put into the tearing machine (Fig. 5.10b).

Several installations are known that have been designed to test the bond strength of the coating with base metal at high temperatures. In spite of some differences, these installations have a common operation principle: a specimen is heated in vacuum up to the required temperature with subsequent loading.

5.2.5 Method of Annular Pin

Rogozhin et al. [5.36] came to a conclusion that the probability of fracture along the base metal–coating interface becomes higher with increasing ratio of the perimeter of the coated top cross-section of the pin to its area. From analysis of cross-section with different shapes (Fig. 5.11) and their influence on the character of fracture, we see that annular shape is the best (Fig. 5.11c). In this case the condition for guaranteed failure is the following:

$$D_1 - d_1 \le 4\sigma \frac{\tau_{max}}{\sigma_{max}} . \qquad (5.1)$$

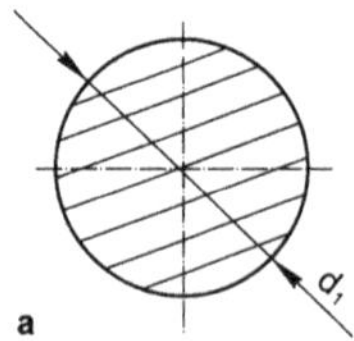

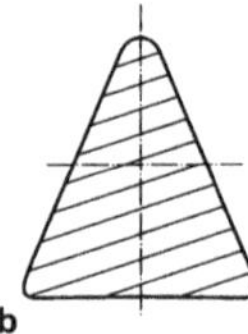

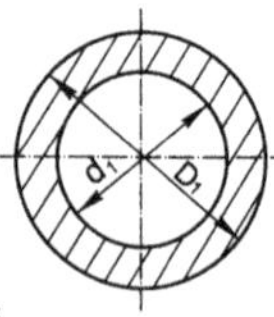

Fig. 5.11. Shapes of pin cross-section: **a** round; **b** triangle; **c** annular. *D1*, *d1* inner and outer diameter of the pin

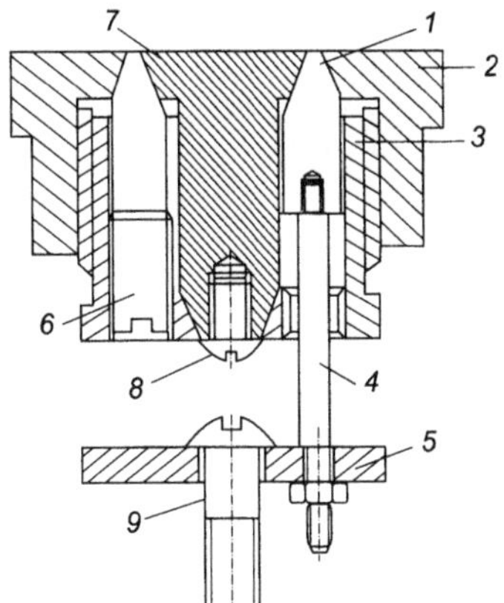

Fig. 5.12. Specimen with the annular pin (by [5.36]). *1* pin; *2* cartridge; *3* nut; *4* stud; *5* plate; *6*, *8* screws; *7* rod; *9* bolt

Therefore, in this case the failure pattern does not depend on the absolute value of the pin diameter, as it was for the circular shape (Fig. 5.11a), i.e. the test allows determination of the averaged bond strength of the coating which is more important in practice.

A specimen with the annular shape of the pin (Fig. 5.12) consists of a cartridge with a nut screwed in. Three screws provide the contact between the top conic surface of the pin and the cartridge. The rod is fixed into the central hole of the nut. The main part of the specimen is an annular pin (Fig. 5.13).

During grinding of the pin, the working area becomes larger. Therefore, after grinding and before spraying of the coating, the inner and outer diameters have to be measured again for correct calculation of the ring area.

The coating is sprayed simultaneously upon at least three specimens installed into sockets of the clamping device; the coating thickness must be higher than 0.3 mm. The thicker the coating, the higher the probability of "pure" tear-off. After the coating has been sprayed on, the screws are unscrewed (see Fig. 5.12) being replaced by studs, which are attached to the plate by three nuts. While testing in a tearing machine, the top of the specimen is put into the gripping mechanism. The velocity of the gripping mechanism motion at gradually increasing loading is 10–20 mm/min.

The advantages of the annular shape of the top cross-section in comparison with a regular round shape are the following: a smaller data scattering because of structural heterogeneity smoothed over a sufficiently large area; estimation of a more essential averaged strength instead of the local one; more uniform loading distribution during testing; higher probability of the pin detachment by the "pure tear-off" scheme.

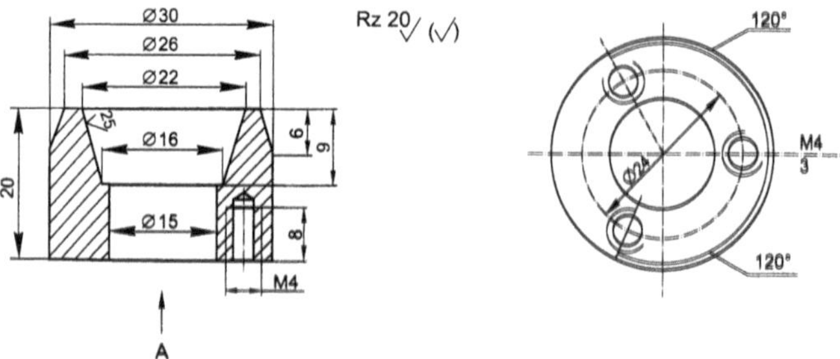

Fig. 5.13. Draft of the annular pin. Limit deviations are by *h7* for shafts and by *H7* for holes

In addition to installations considered above, there are many various setups for determination of the bond strength between the coating and base metal by the pin method [5.33–5.36]. The most urgent problems in this field are as follows:

1. further development of theoretical works on calculation of the complex-stressed state in the coating in pin detachment, and the choice of optimal sizes and shapes for the pin on the basis of the above calculations;
2. development standards for the tests using equipment and specimens, which provide high accuracy and reproducibility of results;
3. application of this method to the area of negative temperatures with test parameters reflecting the real field conditions for parts operating in Siberia and North regions.

5.3 Adhesive Method

5.3.1 General Overview

All modifications of the adhesive method can be combined in two schemes. Fig. 5.14a shows a test scheme of cylindrical specimens, expressly fabricated. After proper preliminary preparation one top of the specimen is deposited by a coating. Counter-specimen of a similar diameter is bonded to the coating by the end. A ratio between the force P, at which the coating separates from the base metal along their boundary during extension, and the surface area of the end of the specimen is a criterion for the bond strength.

A scheme shown in Fig. 5.14b is destined for testing coatings deposited on real products. A cylindrical counter-specimen is bonded to the coating deposited on a flat surface of an item. Tensile stress is applied to the counter-specimen until the coating separates along the boundary with the base metal (destruction test).

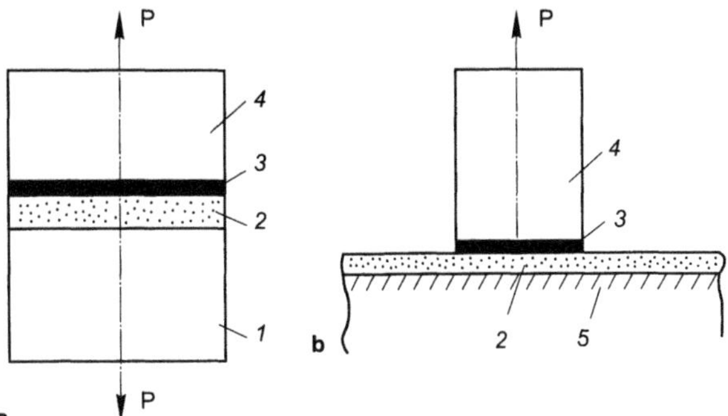

Fig. 5.14. Determination of the bond strength with the base metal by the adhesion method: **a** for specimen; **b** for real part. *1* specimen; *2* coating; *3* adhesive; *4* counter-specimen; *5* base metal; *P* load

Let us consider some peculiarities of the above tests. Here breaking force depends on a scale factor in the same way as for the pin method. Khasui [5.21] has shown, that when coatings are plasma sprayed on specimens of various diameters placed under the equal conditions, the obtained values of bond strength differ radically from one another. Reduction in specimen diameter from 55 down to 25 mm causes double drop in breaking force. This is probably due to complexity of the coating detachment process. Usually detachment of a coating does not proceeds over the whole contact area simultaneously, even if it takes place along the coating-base metal boundary. First of all, detachment starts developing from an edge of the specimen, and subsequently propagates to the central part. Therefore, the bond strength of a given coating, determined for a specimen with a larger diameter, will be greater than that for a thin specimen (edge effect). Edge effect diminishes with increasing diameter of a specimen.

When testing specimens, the same kinds of destruction as for the pin method may take place (see Fig. 5.4). In particular, the character of failure depends on adhesive penetration depth. Bartenev with co-workers [5.2] used the adhesive method for estimation of the bond strength of Al_2O_3 detonation coating with niobium as the base metal. A glass bath was used as an adhesive. Glass penetration depth into the coating was controlled by micro-X-ray spectral analysis. Increase in impregnation depth was accompanied both by enhancement of breaking stress (Fig. 5.15), and change in breaking character. At insignificant penetration depth of the glass into the coating detachment was characterized by exceptionally cohesive character, corresponding to a small value of stress. Increasing impregnation time, and hence, penetration depth of the glass into the coating was accompanied by increase in breaking stress and change of detachment character, first to mixed and than to adhesion one. Breaking stress was maximal in the process, representing actual value of the bond strength between the coating and base metal.

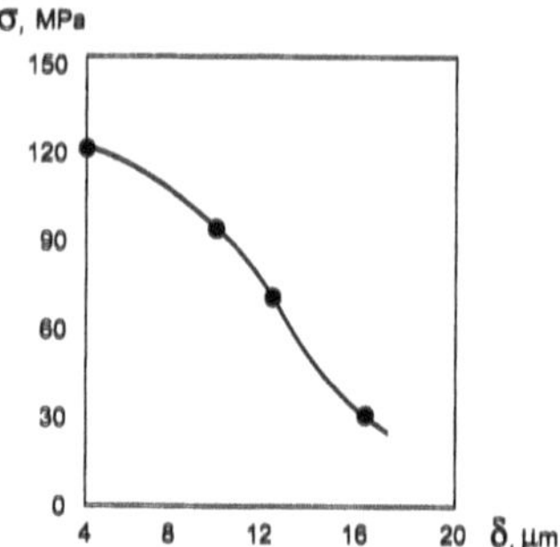

Fig. 5.15. Dependence of breaking stress of Al_2O_3 coating (σ) on the thickness of a layer (δ), not impregnated by glass (according to [5.2])

Therefore, when employing adhesive method, selection of the optimal diameter of a specimen (in order to diminish edge effect), and also control of the depth of adhesive impregnation are required.

5.3.2 Specimens and Testing

Depending on construction of tearing machine holders, specimens can have either threads or special turned heads for mounting. Usually, the effect of shear and bend stresses appears in the cases of non-coaxial adhesion of specimens, non-uniform coating thickness or incorrect loading.

There are some installations, whose operation principle is based on the scheme shown in Fig. 5.14b. A cylindrical specimen is glued to a surface of a coated article. After adhesive is consolidated around the specimen end, a turning is cut in the coating down to the base metal. The scheme of an experimental setup is shown in Fig. 5.16. For non-destructive tests under industrial conditions, a specimen is loaded by a preset force. If the coating does not detach, the spraying may be considered satisfactory.

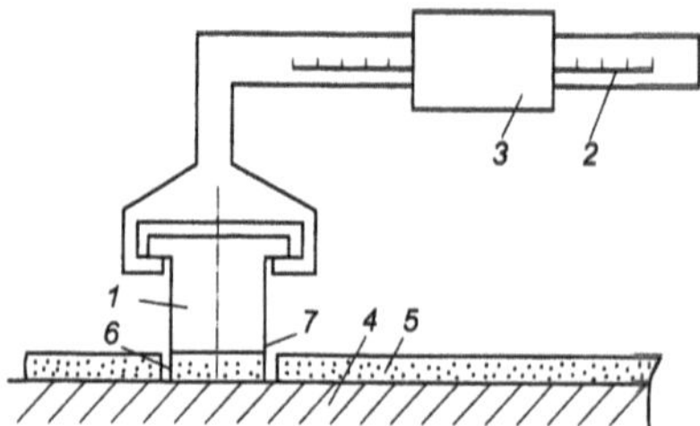

Fig. 5.16. Setup for determination of the bond strength of the coating and base metal. *1* specimen; *2* arm ruler; *3* load; *4* part; *5* coating; *6* turning; *7* glue

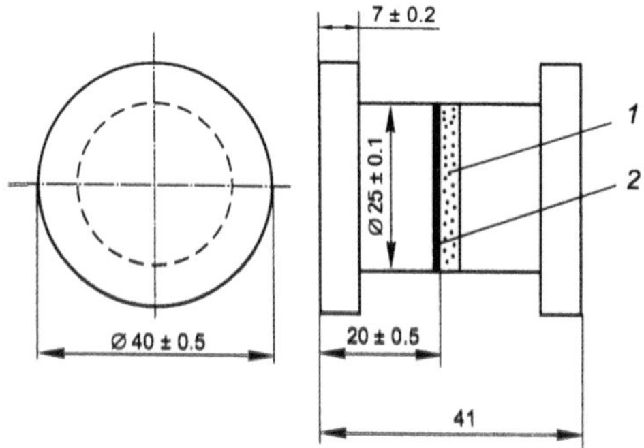

Fig. 5.17. Draft of a glued specimen. *1* glue line; *2* coating

To implement the scheme shown in Fig. 5.14a specimens (Fig. 5.17) consisting of two similar parts (the specimen itself and the counter-specimen) should be used. The working end of one part is uniformly coated. Its surface should be smooth, normal to the longitudinal axis of the specimen and parallel to the base surfaces of heads.

To glue the specimen parts, epoxy adhesives (10–16 MPa), polyurethane adhesives (22–35 MPa), and phenolpolyvinylacetal adhesives (26–60 MPa) can be used. In parenthesis the tearing strength is presented.

The following circumstances should be considered when obtaining an adhesive joint.

Essentially non-porous joint can be obtained using epoxy adhesives because they do not contain solvents. Durable polyurethane adhesive, which perfectly fills clearances, is toxic. Adhesives may solidify at room temperature, but maximal strength characteristics (even for a shorter period) can be reached at the high-temperature mode of gluing.

To increase the strength of adhesive bond, the coating and base metal surfaces should be treated by sand blasting and degreased by benzene or acetone. The surfaces prepared in such a manner should be covered by an adhesive not later than in 30 min. Mechanical presses, vacuum bags, and autoclaves are used for pressing during solidification.

It is necessary to consider that penetration of adhesive into the base metal through open pores of the coating overrates experimental results. If the depth of impregnation is insufficient, the possibility of cohesion fracture or detachment "along the glue" increases.

It is likely that numerous attempts to improve the adhesive method in combination with theoretical developments will soon allow assignment of the adhesive method to the most objective ways for estimation of the bond strength between the coating and base metal [5.3, 5.27].

To prevent misalignment and provide reliable centering, it is recommended to equip the test setup with a device, which works as a universal joint (see Fig. 5.18). This will assist in elimination of possible shear and bend stresses only due to generation of the tension stress.

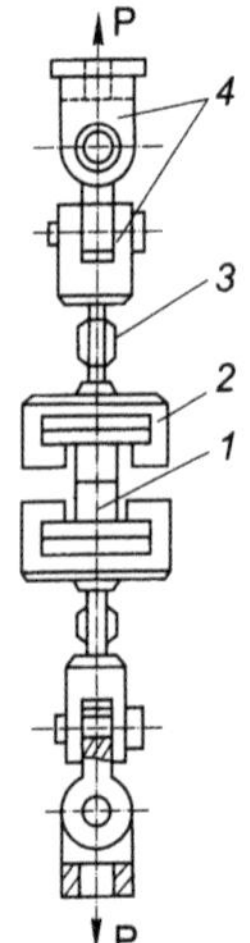

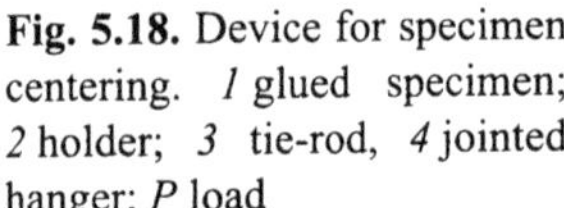
Fig. 5.18. Device for specimen centering. *1* glued specimen; *2* holder; *3* tie-rod, *4* jointed hanger; *P* load

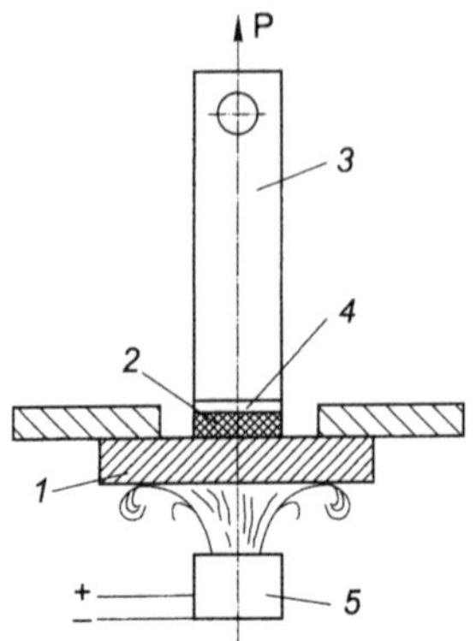

Fig. 5.19. Determination of temperature dependence for the bond strength of the coating and base metal (by [5.24]). *1* base metal; *2* coating; *3* cylindrical specimen; *4* adhesive; *5* plasmatron; *P* load

Disadvantages of the adhesive method are as follows.

1. The coating bond strength depends on specimen sizes. The higher bond strength corresponds to a larger specimen. Standard experimental specimens are developed only for some metal gas-thermal coatingsThe correct test is only possible, when the strength of adhesive bond in detachment exceeds the bond strength of the coating and base metal. At the same time, the strength of most adhesives does not exceed 60 MPa. Therefore, at strong bonds between the coating material and metal, detachment occurs along the glue line. Bartenev et al. demonstrated that when testing detonation coatings from aluminum oxide, separation takes place along the glue line in 90–95% of cases [5.2].
2. Adhesive penetration through the open pores to the base metal overrates experimental results. On the other hand, insignificant depth of impregnation provides for cohesion fracture of the coating. To overcome this contradiction, there are no theoretical and technical solutions yet. For every certain case, the impregnation depth is chosen by the cut-and-try method.

To determine the temperature dependence for the coating bond strength, Kudinov et al. [5.23, 5.24] suggest heating the base metal by a plasma jet (see Fig. 5.19) from the side opposite to the coating. A high temperature gradient is formed over the coating thickness and the temperature of the "coating–adhesive" transition zone remains equal to the room temperature (for some certain period of time). Simultaneously the coating is torn away from the base metal.

5.4 Shear Method

5.4.1 General Overview

In contrast to the pin method in this case detachment of the coating from the base metal occurs due to shear stress. When testing cylindrical specimens (Fig. 5.20a), a coating is deposited on the central part of their side surface. The base metal punch and matrixes are made in a way that provides their movement relative to each other with minimal friction. Being loaded, the coating detaches over the whole side surface of the base metal. The coating of flat specimens (see Fig. 5.20b) is deposited on certain areas by means of a mask. Under the action of the force *P* a knife separates a coating section over the whole area of its contact with the base metal in one cut.

In both cases a ratio of the maximal force *P* to either coated cylindrical surface area or contact area between the coating material and a flat section of the base metal is the criterion for the bond strength.

5.4.2 Equipment and Specimens

A special precision technique has been developed for studying contact processes between plasma-sprayed particles and metal. Koudinov with co-authors determined the bond strength by cutting separate particles using a special device (see Fig. 5.21). A cylindrical specimen 10–15 mm in diameter and 15–20 mm in length with a polished end sprayed through a mask with separate particles is fixed in the chuck. The micro-cutting knife is brought under a particle by moving the lever and rotating the chuck. The diameter of a tested particle varies within 100–600 μm. The cup is loaded with a powder by releasing the stopper until the particle is detached from the base metal.

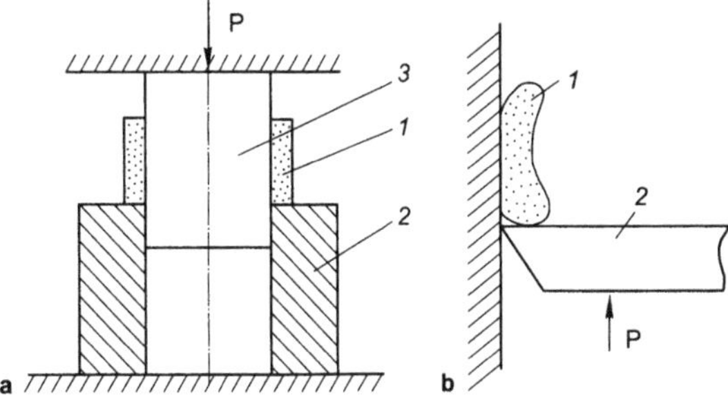

Fig. 5.20. Determination of the bond strength between the coating and the base metal by the shear method: **a** on a cylindrical specimen; **b** on a flat specimen. *1* coating; *2* matrix; *3* base metal; *P* load

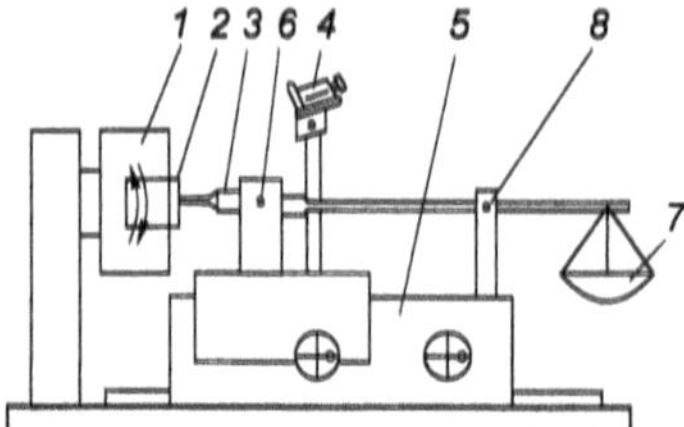

Fig. 5.21. Installation for determination of the bond strength of separate particles (according to [5.23]). *1* revolving chuck for fixing a specimen; *2* specimen with deposited particles; *3* lever with micro-cutting knife; *4* binocular magnifier; *5* anvil; *6* lever axis; *7* load cup; *8* lever stopper

The force P is calculated taking into account the relation between the lever arms. The bond strength is determined by the ratio between the maximal force and the area of chemical interaction, diameter of which is measured using an optical microscope. The installation described was employed in many works for laboratory studies of interaction kinetics between the coating particles material and the base metal. It was also applied to studying influence of various factors upon bond strength.

To estimate the bond strength of an integral coating it is advisable to use a method based on a cylindrical specimen testing scheme (see Fig. 5.20a). A specimen suitable for this method is shown in Fig. 5.22. In fabricating specimens the following conditions are to be met.

1. A punch should be made of the same material and in the same structural state as the product, which is to be coated and tested for bond strength.
2. The matrix should have the hardness not less than HRC 60.
3. When a coating is being sprayed on the working surface of the punch, all other surfaces should be protected.
4. The coating thickness is recommended to be 0.1 mm; the end surfaces of the coating are to be in parallel with the base of the punch.

The tests can be carried out employing any machine allowing capture of the maximal load under which the coating shearing takes place. When installing specimens, it is required to maintain alignment between the punch and a matrix and also provide coincidence between their axes and the axis of the load applied.

Grutzner and Weiss [5.11] described a new method for shear testing employing a press. Plate of a hard material used as a knife was attached to the upper press tool. The distance between the "base metal–coating" boundary and the loading point was adjusted in a way allowing testing different coating layers. The following advantages of the new method are highlighted: the loading can be selected in accordance with the service conditions of the coating; the tests are simple and inexpensive; statistical processing is feasible. Shear testing results for Al_2O_3 and Y_2O_3 coatings are also presented in [5.11].

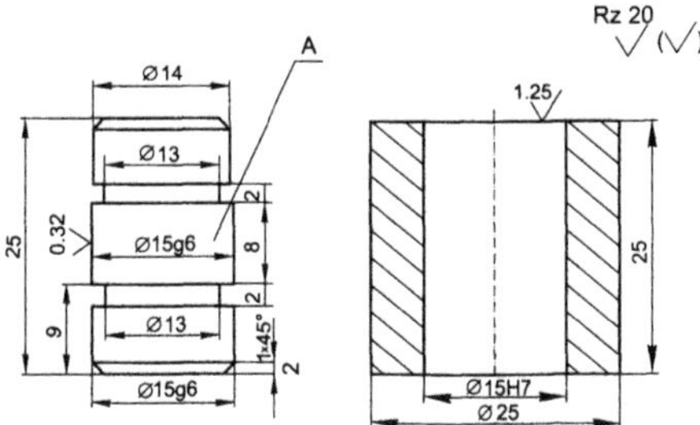

Fig. 5.22. Layout of a specimen for determination of the bond strength between the coating and the base metal by the shear method. *A* surface used for coating; maximum tolerance of dimensions (not indicated) H12, h12±IT12/2

A specimen is loaded with gradual loading increase up to its failure. After shearing of the coating the punch is inspected to determine the nature of the failure. The test is considered to be valid if detachment of the coating took place along the boundary with the base metal.

It should be noted, that all schemes for determination of the coating-base metal bond strength have a number of shortcomings that may introduce errors into the actual results. When using the scheme *a* (see Fig. 5.20), the major problem lies in securing alignment between the coating and the punch during the whole loading time. If this condition is violated, the matrix edge is deflected from its movement along the coating-base metal boundary thus cutting down the coating material from one side and the punch material from another. Alignment may be violated due to change in shape and dimensions of the punch when its surface is being prepared for deposition or under affection of high temperature particles changing the relief of its cylindrical part.

Correct determination of bond strength in implementation of the scheme *b* is hindered by the following reasons. First, along with tangential stresses bending stresses will be applied to the detachment zone, which makes stressed state pattern even more complicated.

Secondly, it is rather difficult to estimate the actual area of chemical interaction spot between materials of the coating and base metal. As was established, the spot structure is very non-uniform. The contact surface between the coating and base metal is filled with seizure centers (sticking). The dislocation density of these centers may differ.

Thirdly, application of the mask influences the coating formation character. This causes a difference between structure and properties of the coating major part as compared to its periphery and distorts actual strength values.

The above-noted disadvantages do not mean abandonment of the further development of the method for determination of shear bond strength between the coating and base metal. At the same time it should be noted, that the method is of limited utility as compared with the pin and adhesive methods.

5.5 Fracture Toughness on the "Base Metal–Coating" Boundary

5.5.1 General Overview

This method consists in determination of fracture toughness characteristics of "base metal–coating" compositions when detaching the coating along the common boundary during eccentric tension of special rectangular specimens with the edge cut. Critical stress intensity factor – a major force characteristic – is a measure of the bond strength. It is determined during the extension process by recording load-displacement diagrams (for more details concerning fracture toughness characteristics see Chap. 8).

With a knowledge of fracture toughness characteristics on the "base metal–coating" boundary and having measured the crack length by one of the nondestructive control methods, it is possible to calculate the breaking stress value. It is assumed that the plastic strain on the boundary between the base metal and coating is constrained to the limit. The method is proposed by Berndt and McPherson [5.4, 5.5, 5.30].

5.5.2 Specimens and Tests

Glued specimens with the edge cut are used for tests (Fig. 5.23). To initiate adhesive fracture it is expedient to cut furrows 1.0–1.5 mm in depth along the whole length of the boundary between the base metal and coating (Fig 5.24).

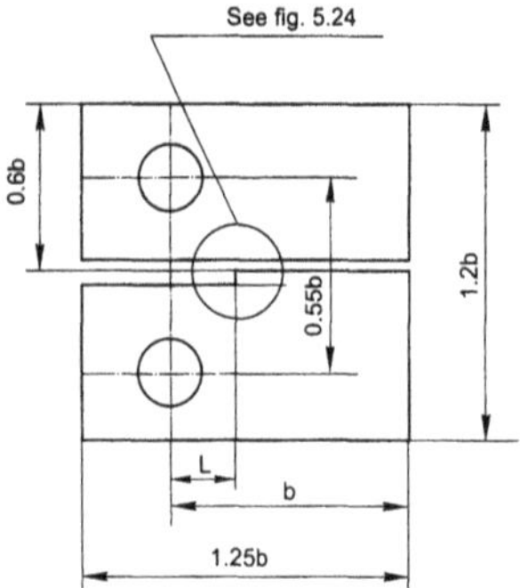

Fig. 5.23. Layout of a rectangular specimen with an edge cut for eccentric extension tests

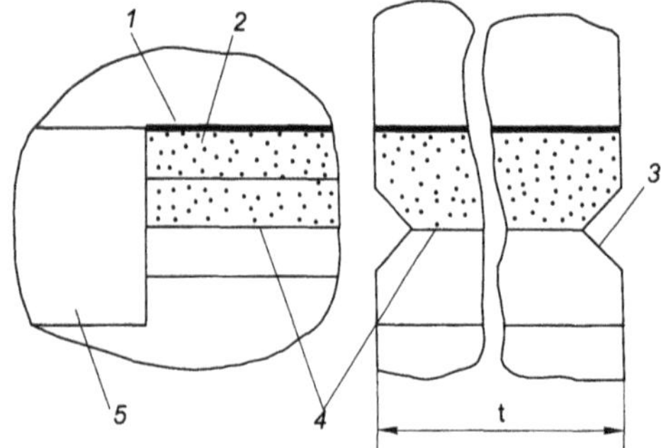

Fig. 5.24. Enlarged picture of a part of a specimen with the edge cut. *1* glue line; *2* coating; *3* furrows to initiate fracture along the boundary between the base metal and coating; *4* "base metal–coating" boarder; *5* edge cut

In contrast to the tests applicable for metal specimens with the edge crack at the eccentric extension, when estimating the bond strength between the coating and base metal, it is not necessary to apply (to initiate) fatigue cracks. The tip radius of a crack extending along the "base metal–coating" boundary will be determined not by deposition (initiation) conditions, as it would have been for metal specimens, but will depend on geometry of pores and micro-cracks already existing on the boundary.

The length of a specimen, its thickness, the distance between the load application axis and the end of the specimen, and the coating thickness should be measured with accuracy not exceeding 0.1 mm. Adhesive remainders in the joint area should be smoothed out.

Displacement of the edge cuts are measured by sensors made of spring steels and provided with resistive-strain tensors glued on the surface. Before testing, the sensor springing elements are fixed at the cut on a specimen by means of attached prisms.

The specimens are tested until they are broken at the rate of loading equal to 1.0–2.0 mm/min, providing recording of "load P–edge cut displacement V" diagram. The design load on a specimen is determined according to the state diagram.

After splitting the specimen in two parts, the type of fracture can be classified. The same situations as for the pin method are possible.

Critical stress intensity factor K_{Ic}, MPa·m$^{1/2}$ is calculated from the formula:

$$K_{Ic} = \frac{P_Q}{t\sqrt{b}} \cdot Y , \tag{5.2}$$

where P_Q is the design load on a specimen, MN; t – the specimen thickness, m; b – the distance between the load axis and the end of a specimen, m; Y – correction function for the III-type specimen, see Eq. (8.4).

Power characteristic of the toughness J_{Ic} may be determined if necessary.

Heintze and McPherson studied ZrO_2-based plasma coatings completely or partly stabilized by Y_2O_3 or CeO_2 additions [5.15]. The specimens were loaded until appearance of a crack either inside a specimen or on the "base metal–coating" boundary. Three kinds of failure were observed: adhesion, cohesion, and a mixed one. In the last case areas of the adhesively failing coating remained on the substrate. The results show that the cohesion fracture toughness is greater than that for adhesion. The highest rate of fracture toughness was achieved for ZrO_2–CeO_2 coatings, which contained essential part of a transformed tetragonal phase.

A modification technique allowing separation of interface adhesion and fracture toughness of the proper coating has been proposed in [5.37]. Two half-parts of a specimen were fabricated first. Then a coating was sprayed along the prospective cutoff plane, after which both parts were glued (soldered) to each other. Next, notches were made either along the interface or in the middle of the coating thickness. The notches predetermine whether adhesion or fracture toughness of the coating is to be tested. Afterwards, both sets of specimens were tested at static extension with the subsequent estimation of fracture toughness K_{Ic}.

5.6 Scratch Method

5.6.1 General Overview

The essence of the method is that a steel or diamond needle is passed over the coating surface at the increasing vertical load until the coating is detached from the substrate (Fig. 5.25). This moment is registered either visually, or automatically using an electric or optic converter. This method is more precise for a relatively hard substrate meeting the following condition: $\delta < r$ (δ – coating thickness; r – needle radius). The substrate surface together with a deposited coating is plastically strained under the action of the needle.

Shearing force F that arise in the process is maximal at the recess edge where the coating breaking-off takes place. Relationship between the load P applied to the needle and the shearing force F looks like following [5.22]:

$$F = \frac{P}{\pi a \sqrt{r^2 - a^2}}, \qquad (5.3)$$

where a is a half of the scratch width.

The method is used as comparative and can be recommended for development of technological modes applied to preparing surfaces for spraying and for process optimization in case of single-type substrates. Simplicity and quickness of the method makes it advisable for studying deposition parameters and selecting coatings with the maximal bond strength. Valli [5.41] assumes that today the scratch method is the only one suitable for practical estimations of adhesion of thin and hard tribological coatings deposited in vacuum.

In Russia the scratch method developed earlier by Khruschev to determine hardness is widely used for estimating properties of thin and hard coatings. Adhesion estimation results obtained by the scratch method deviate within 10–20%, which should be taken into account when comparing adhesion of various coatings. Besides adhesion, the following factors effect on the critical normal force value: the hardness of a substrate, the coating thickness, the surface quality, the coating hardness, the rate of loading, the indenter shape, and the friction between the indenter and a coating [5.1, 5.26, 5.32].

The described method extends performance capabilities of the automatic measuring process by means of modern measuring computation engineering.

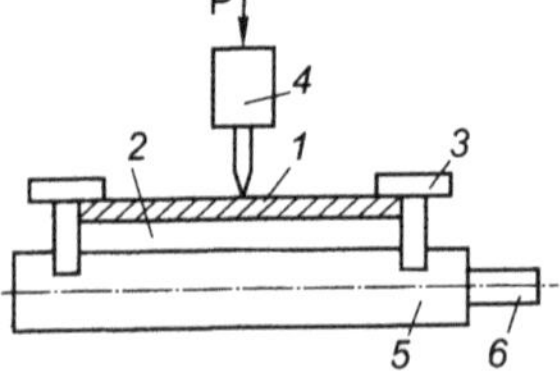

Fig. 5.25. Scheme for determination of bond strength by the scratch method. *1* coating; *2* substrate; *3* fixing screws; *4* needle support; *5* table for fixing specimens; *6* lead screw

5.6.2 Specimens

When scratching a specimen with a thin coating, the value of the critical normal load rises with increasing hardness of the substrate. This relationship may have the nonlinear character. Therefore, when a comparative estimation of adhesion is required, coatings should be deposited on substrates with the same hardness. Increasing coating thickness causes rise in the critical normal load. When roughness of the substrate surface increases, the obtained results become unstable. As a consequence, in order to obtain reproducible values of the critical normal load, coatings should be sprayed on the specimen's surface with roughness $Ra < 0.25$ μm.

The value of the critical normal load varies depending on changing coating hardness at other parameters being constant. It was revealed, in particular, that in case of molybdenum nitride coatings the highest value of the critical normal load was obtained for the coatings with the hardness equal approximately to 15.0 GPa (the coatings with hardness from 5.0 to 30.0 GPa were examined). Further increase in hardness leads to reduction in critical normal load values. Apparently, this is caused by enhancement of the coating brittleness rather than by reduction in adhesive bond strength with the substrate [5.39].

5.6.3 Equipment

When testing different hard coatings, application of a diamond indenter allows maintenance of the friction effect at a constant level. Three-edged (Khruschev-Berkovich), or four-edged (Vickers) diamond pyramids, as well as diamond cones are often used as indenters instead of a needle. There is a well-known indenter, which differs from a pointed needle by its body-of-revolution shape with the toroidal working surface. The indenter is fixed on a holder enabling rotation round its axis. In case the working surface is worn, the indenter is rotated round the axis through a certain angle and fixed then in a new position. Thus, the whole working surface of the tore is used for scratching, which prolongs durability of the tool.

There is a familiar method, which consists in use of high-voltage electric discharge as a tool for cutting flutes. A coated specimen is positioned between two electrodes being energized by a power source. While scanning by a high-voltage electric spark discharge over the coating surface, a reference grid is created. Then, upon measuring the broken sections, the bond strength with the base metal is determined as the ratio of the total area of destroyed areas to the unbroken portion of the coating. It is shown in [5.10, 5.38, 5.41] that the bending radius of the indenter's point exerts an influence on the critical normal load values. When testing hard coatings, a diamond indenter with the bending radius of the point equal to 0.2 mm is generally applied. As the indenter wears out, the radius changes. This causes variation of the stressed state both in the substrate and coating. As a result, the values of the critical normal load, determined by means of the worn indenter may differ significantly from those obtained using a new indenter. In case of good cohesion between the coating and substrate, loads on indenter are high enough to cause its quick wear. Therefore, periodical control of indenter status is required, especially when a scratch is formed at high loads.

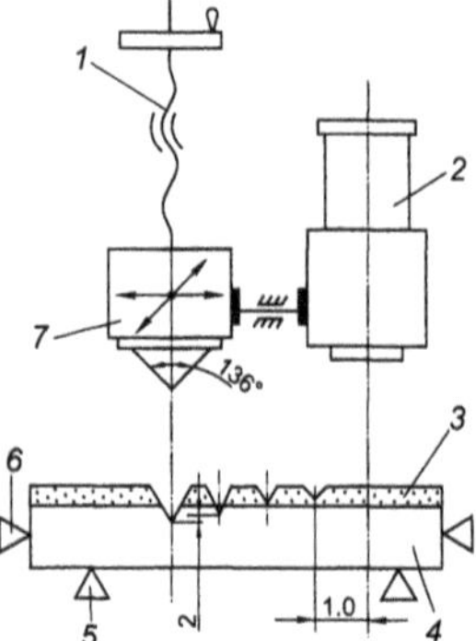

Fig. 5.26. Scheme of the setup for determination of the bond strength between the coating and base metal by the scratch method. *1* indenter feeding mechanism; *2* microscope; *3* coating; *4* specimen; *5* sample stage; *6* clamping screw; *7* mandrel

In a number of cases scratching under the constant load on inclined metallographic section surface (with the inclination angle of 1–3°) is more informative. In these tests the influence of the coating thickness on the critical normal load can be easily estimated. In case of multi-layer coatings the estimation of the role of layers with different mechanical properties is also possible [5.38].

The experimental setup may be designed on the basis of a metallographic microscope. A diamond pyramid with the apical angle equal to 136° was used as a scratching indenter (Fig. 5.26). The indenter is fixed in a special mandrel 7, which can move vertically by macro- and micro-transmission mechanisms 1. Constant of a micro-transmission vernier amounts to 2 μm. A specimen (or a small part) 4 with a coating 3 is fixed by fastening screws 6 on the sample stage 5 of the microscope 2, which has a mechanism of reciprocally perpendicular displacement in a horizontal plane with vernier constant equal to 0.1 mm. The equipment should both provide smooth increase in load when pressing the indenter into the coating, and maintain constant level of the load applied during the scratch process.

5.6.4 Preparation and Testing

Prior to scratching the coating surface should be checked for being flat and free of soiling. Surface roughness should not be lower than $Ra = 0.25$ μm. A coated specimen should be installed on a sample stage of the instrument in order to exclude flexure and shifting during testing. The surface should be perpendicular to the axis of the scratching indenter. At good quality of the deposited plasma-ionic coating and a high bond strength with the substrate, plastic strain of the coating and substrate during scratching can be observed (Fig. 5.27).

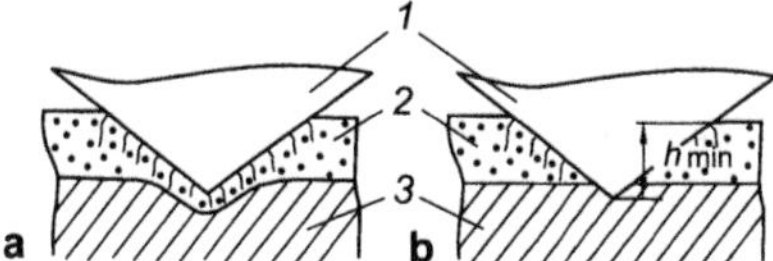

Fig. 5.27. Destruction scheme of plasma-ionic TiN coating with a satisfactory (**a**) and unsatisfactory (**b**) bond strength between the coating and substrate. *1* indenter; *2* coating; *3* base metal; h_{min} mark depth

Coating detachment or injury may be revealed by means of optical and electron microscopy, acoustic emission, measurements of friction. It is necessary to note, that this method can be used to estimate not only the adhesion bond strength between the coating and substrate, but the whole set of working characteristics of the coating, including brittleness, magnitude and sign of residual stresses, correlation between hardness and elastic moduli of the coating and substrate, and many others. As was already mentioned above, the scratching method, in spite of some disadvantages, is the major method for assessment of vacuum coatings. Some of other less popular methods are briefly considered below.

5.7 Testing Methods for Vacuum Coatings

A small thickness and high adhesion bond strength between the coating and substrate produced by vacuum methods hampers determination of quantitative adhesion characteristics. Since coating deposition modes, when properly selected, provide the adhesion bond strength comparable with the substrate strength or even exceeding it, the most of the well-known methods for quantitative estimation of adhesion are unacceptable for such coatings. At best, these methods ensure only revealing defective coatings with low adhesion caused by violation of coating deposition modes or by the influence of some other factors.

Brief review of the well-known estimation methods of adhesion bond strength for such coatings is given later [5.37, 5.38, 5.41].

A coating breaking-off method based on the effect of electromagnetic waves is described below. A current passed through an electrically conducting coating on a specimen, made of dielectric and placed into external electromagnetic field, causes generation of a force, normal to surface, that tends to detach the coating from the substrate. Unfortunately, this method can be applied only to electrically conducting coatings on dielectric substrates.

Coating adhesion can be estimated employing methods based on inertial forces applied to the coating. These involve centrifugal and ultrasonic methods.

In the centrifugal method a small rotor sprayed with a coating is suspended in vacuum by magnetic field being rotated with gradually increased frequency by means of electromagnets. The force applied to the coating is calculated by rotational velocity at the instant it was broken-off. This method cannot be used for parts or instruments of a complex geometry as well as for parts made of nonmagnetic materials.

Ultrasonic methods are based on the effect of inertial forces applied to a coating. They can be employed only for coatings possessing of low adhesion with the substrate. In a number of cases in order to reveal defective coatings (with low adhesion) parts are placed into an ultrasonic bath used for cleaning surfaces.

A number of laser methods for estimation of coating adhesion have been developed recently. In one of them a high-energy impulse laser beam effects on a specimen from the side of the substrate. Generated shock wave passes through the interface material between the coating and substrate thus causing exfoliation of the coating at a sufficiently high-powered impulse. By increasing smoothly the impulse power, the critical value characterizing the bond strength between the coating and substrate can be determined. The main constraint of the method is that the coating must consist of separate sections with diameter less than that of the laser beam.

The laser nondestructive control method consists in the impulse impact of a laser beam on the coating surface with registration of acoustic waves on the rear side of the substrate. The method enables to detect defective adhesion areas with the width of 2.0 mm.

Fig. 5.28 shows the scheme of a device which design is based on the principle described.

The method employing an impulse laser beam generating a thermal wave is outlined in literature as well. Micro thermocouples or an infrared radiation detector are used to register the thermal wave. This method is also nondestructive.

Laser methods for adhesion estimation have a number of advantages as compared with the other ones. These methods are nondestructive for coatings having high adhesion with a substrate. Moreover, employment of these methods provides a temperature gradient in the surface layer similar to that appearing when actual products such as cutters, for instance, are used in the field conditions. For qualitative estimation of adhesion an adhesive tape can be used, which is glued to the coating being detached from the substrate together with it. Prior to gluing the adhesive tape, a grid of longitudinal and lateral scratches cutting the coating surface into a number of equal quadrates can be plotted.

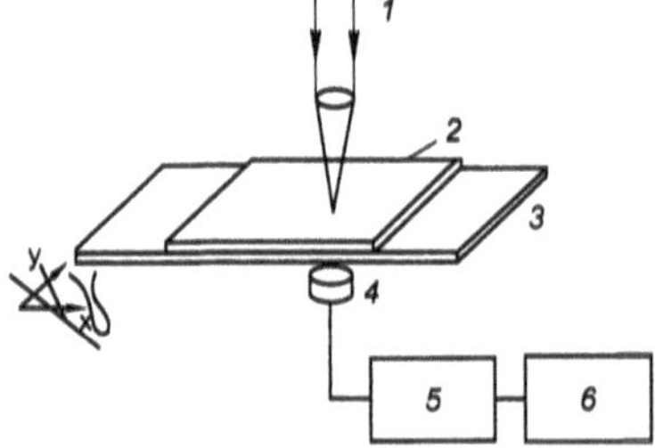

Fig. 5.28. Scheme of the nondestructive laser method for control of the bond strength between the coating and substrate (by [5.41]). *1* laser impulse; *2* coating; *3* base metal; *4* transducer; *5* computer; *6* display

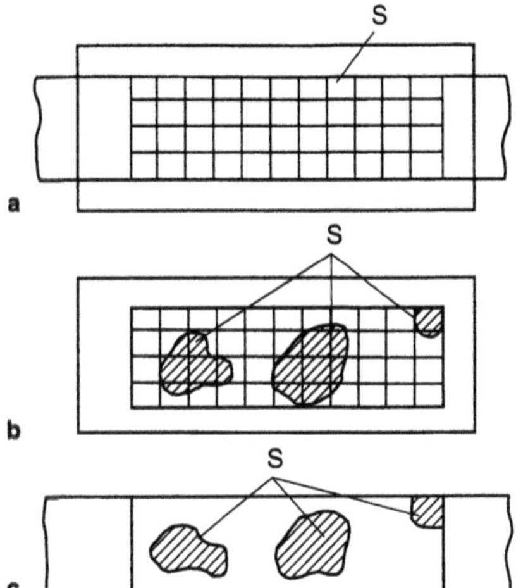

Fig. 5.29. Scheme for estimation of adhesion using an adhesive tape: (according to [5.22]): **a** substrate with a coating and a glued adhesion tape; **b** substrate from the coating side after detachment of an adhesive tape; **c** adhesive tape with a partly detached coating

Once the tape together with the coating is detached, adhesion K_C (%) can be estimated by the ratio between the area S_{det} of the detached coating and the total area S_{tot} of the section with the tape glued:

$$K_C \approx \frac{S_{det}}{S_{tot}} \cdot 100\,. \tag{5.4}$$

This method is simple and does not require high skills. It is mainly applied to qualitative estimation of slightly pronounced adhesion [5.22]. In some cases a coating detaches even when drawing the scratch grid (the grid incision method). Adhesion is estimated qualitatively by the detached coating surface area (Fig. 5.29).

A great number of other methods for estimating adhesion either cause the failure of parts, or cannot be applied to real objects. Thus, for example, when extending coated specimens, the adhesion bond strength can be estimated quantitatively but in doing so the specimen fails. Such methods give a pictorial comparative view of the adhesion bond strength, but cannot be applied to controlling real parts. Rapid cooling from a high temperature (thermal shock) causes appearance of high stresses between the coating and the substrate, but it is hard to make any theoretical analysis based on the results obtained. Further development of paths for new methods of estimating thin coatings adhesion, including nondestructive ones, is outlined in [5.1, 5.22, 5.41].

5.8 Indentation Method

There are two modifications of the indentation method. For the first modification an indenter is pressed into a wide surface of a platelet specimen. For the second modification the indenter penetrates the boundary between the base metal and coating on a cross microsection. The first variant enables estimation of coating damage stages depending on plastic deformation degree of a base metal.

The Brinnel press is employed to force a hardened steel ball-shaped indenter 10 mm in diameter into the flat specimen surface opposite to the coated surface under a certain load (Fig. 5.30). This results in convexity leading to damages on the coating. The specimen is strained at an increasing load, which is selected depending on a substrate material. The strain and overall extension of damages (cracks) on the coating are measured as well as the diameter of a print (strain value), which is metered with the help of the Brinell magnifier. The plates with coatings sprayed on a wide side are employed as specimens (Fig. 5.31). A specimen is installed in a socket of the holding table and pressed to its surface by means of a special mandrel, which prevents the specimen against bending and being drawn into the orifice. Based on experimental data, the overall extension of defects is plotted as a function of the print diameter.

Sglavo and Dal Maschio [5.39] developed a method based on measuring dimension of a print appearing when pressing Vickers pyramid on the boundary between the coating and base metal.

After measuring micro-cracks formed on a boundary, calculations in the framework of basic concepts of fracture mechanics are performed. The suggested method was used to determine adhesion characteristics of ZrO_2–CeO_2–Y_2O_3 coating, deposited by plasma spraying on a steel substrate with the Ni–CoCrAlY sublayer. Thickness of the coating and the sublayer amounted to 1500 μm. Good agreement between the results obtained by the present method and other well-known older methods used in [5.39] has been revealed.

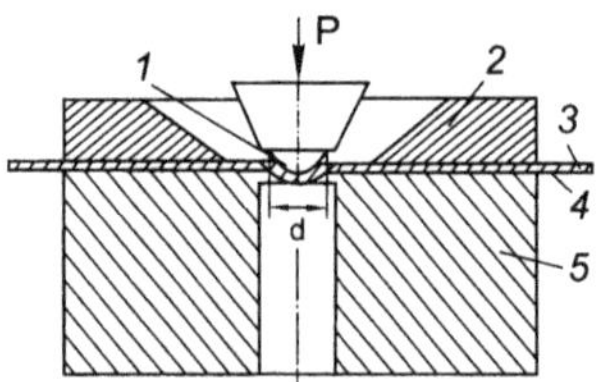

Fig. 5.30. Scheme for determination of the bond strength between the coating and base metal by the indentation method. *1* ball-shaped indenter; *2* mandrel; *3* specimen; *4* coating; *5* matrix; *P* load

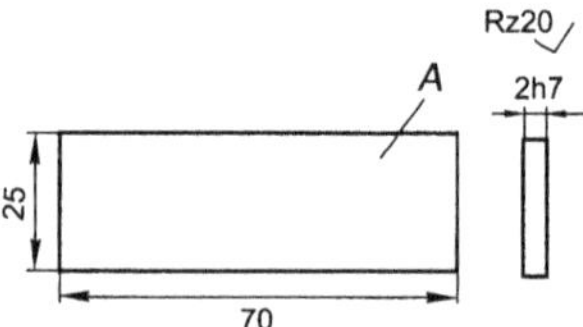

Fig. 5.31. Draft of a specimen used for determination of the bond strength between the coating and base metal by the indentation method. *A* coated surface

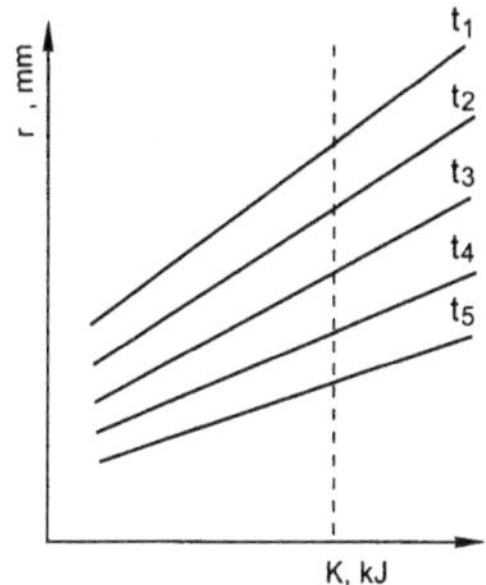

Fig. 5.32. Dependence of the total length of coating defects (l_d) on strain energy for different test temperatures ($t_5<t_4<t_3<t_2<t_1$)

Adhesion strength of Cr_3C_2–25% NiCr coating deposited on various steels by supersonic gas-thermal spraying was studied in [5.9]. Authors determined the critical load when indenting pyramid, which led to initiation of crack on the boundary between the coating and substrate. It has been shown, that the critical load value does not depend on the coating thickness if internal stresses are previously withdrawn by annealing.

References

5.1 Agethen R, Lugscheider E (1988) Adhesion strength of layers obtained by thermal deposition (in German). VDI Ber 702:197–213

5.2 Bartenev SS, Fed'ko YuP, Grigorov AI (1982) Detonation coatings in engineering industry (in Russian). Mashinostroenie, Leningrad

5.3 Baton HE, Novak RC (1986) Coating bond strength of plasma-sprayed stainless steel. Surface and Coating Technology 27:197–202

5.4 Berndt CC, McPherson R (1979) The adhesion of flame and plasma-sprayed coatings. A literature review. Australian Welding Research 1:75–85

5.5 Berndt CC, McPherson R (1981) A fracture mechanics approach to the adhesion of flame and plasma-sprayed coatings. Mechanical Engineering Transactions M E 6:53–58

5.6 Chalker PR, Bull SJ, Rickerby DS (1991) A review of the methods for the evaluation of coating-substrate adhesion. Mater Sci and Eng A 140:583–592

5.7 Coil BF, Sathrum P, Fontana R, Peyre JP, Duchateau D (1992) Mechanical properties of (Ti,A1)N films prepared for cutting tool application. Vide, Couches Minces 48:112–113

5.8 Cookson C, Cookson CJP (1977) The influence of surface preparation on the bond strength of powder metal sprayed coatings. Welding and Metal Fabrication 45:501–503

5.9 Demarecaux P, Lesage I, Chicof D, Mesmacgue G (1994) An examination of the validity of the interface indentation test: application to thermal sprayed coatings. Vide, Couches Minces 50:524–527

5.10 Drory MO, Thouless MD, Evans AG (1988) On the decohesion of residually stressed thin films. Acta Met 36:2019–2028

5.11 Grutzner Heinrich, Weiss Horst (1991) A novel shear test for plasma-sprayed coatings. Surface and Coating Technology 45:317–323

5.12 Guy D (1997) Thermal spray coatings helps aluminum adhesive bonding. Adv Mater and Process 152:12–13

5.13 Hasugama Hiroki, Shima Yukari, Baba Koumei, Wolf Gerhard K, Martin Herbert, Stippich Frank (1997) Adhesive and corrosion- resistant zirconium oxide coatings on stainless steel prepared by ion beam assisted deposition. Nucl Instrum and Meth Phys Res B 127–128:827–831

5.14 Heinke W, Leyland A, Matthews A, Berg G, Friedrich C, Broszeit E (1995) Evaluation of PDV nitride coatings using impact, scratch and Rocwell–C adhesion tests. Thin Solid Films 270:431–438

5.15 Heintze GN, McPherson R (1988) A further study of the fracture toughness of plasma-sprayed zirconia coatings. Surface and Coating Technology 36:125–132

5.16 Howes CP (1994) Thermal spraying: processes, preparation, coatings and applications. Weld J 73:47–51

5.17 Hu MS, Evans AG (1989) The cracking and decohesion of thin films on ductile substrates. Acta Met 37:917–925

5.18 Ighal M, Ducarrar M, Leiogeais M, Garden J (1992) An experimental approach to the analysis of the adhesion properties of coatings deposited on steel substrates. Thin Solid Films 22:271–276

5.19 Kalmutsky VS (1980) Strength and reliability of parts with metallic coatings (in Russian). Problemy Prochnosti 9:96–101

5.20 Kerkush IR (1997) Property improvement of plasma sprayed iron coatings bychromium plating. Poroshkovaya Metallurgia 40:230–231

5.21 Khasui A (1975) Deposition techniques (in Russian). Mashinostroenie, Moscow

5.22 Kostrzhitsky AI, Karpov VF, Kabanchenko VG, Solovyova ON (1991) Operator handbook on vacuum coating deposition set-ups (in Russian). Mashinostroenie, Moscow

5.23 Kudinov VV, Pekshev PYu, Belaschenko VE, Solonenko OM, Safiulin AP (1990) Plasma-sprayed coatings (in Russian). Nauka, Moscow

5.24 Kudinov VV, Statsura VV, Nirkov IM (1967) Determination of the bond strength of coatings at elevated temperatures (in Russian). Zavodskaya Laboratoria 2:232–234

5.25 Kuper A, Clissold R, Martin PJ, Swain MVQ997) A comparative assessment of three approaches for ranking the adhesion of TiN coatings onto two steel. Thin Solid Films 308–309:329–333

5.26 Laugier MT (1987) Adhesion and toughness of protective coatings. J Vac Sci and Technol A 5:67–69

5.27 Li CC (1980) Characterization of thermally sprayed coatings for high temperature wear protection application. Thin Solid Films 73:59–77

5.28 Lyashenko VA, Tsigulev OV, Kuznetsov PB (1987) Whether it is always necessary to enhance adhesion strength of protective coatings (in Russia). Problemy Prochnosti 5:70–74

5.29 Lyasnikov VN (1994) Properties of plasma-sprayed powder coatings. J Adv Mater 1:381–387

5.30 McPherson R (1981) The relationship between the mechanism of formation, microstructure and properties of plasma-sprayed coatings. Thin Solid Films 83:297–310

5.31 Müller D, Cho JR, Fromm E (1993) Measurement of the adhesion of TiN and Al coatings by fracture mechanics tests. Thin Solid Films 236:253–256
5.32 Perry AJ, Vally J, Steinmann PA (1988) Adhesion scratch testing: a round-robin experiment. Surface and Coating Technology 36:559–575
5.33 Puzryakov AF, Eremichev AI, Garanov VA (1984) Failure mechanism of deposited coatings in determination of adhesion strength by the pin method (in Russian). Poroshkovaya Metallurgia 4:94–98
5.34 Rickerby DS (1988) A review of the methods for the measurement of coating-substrate adhesion. Surface and Coating Technology 36:541–557
5.35 Rogozhin VM, Shustov AV, Zhemkova EB, Podobryazhnikh AN (1990) Adhesion strength control of gas-thermal coatings. Zavodskaya Laboratoria 10:62–64
5.36 Rogozhin VM, Smirnov YuV, Petrov VYa (1982) Determination of adhesion strength of gas-thermal coatings. Poroshkovaya Metallurgia 7:87–91
5.37 Schweitzer KK, Ziehl MM, Schwamiuger C (1991) Improved methods for testing bond and intrinsic strength and fatigue of thermally sprayed metallic and ceramic coatings. Surface and Coating Technology 48:103–111
5.38 Semyonov AP, Kovsh IB, Petrova IM, Arkhipov VE, Birger EM (1992) Methods and means for hardening of surfaces of machine parts by concentrated energy fluxes (in Russian). Nauka, Moscow
5.39 Sglavo VM, Dal Maschio R (1990) Adhesion testing of plasma-sprayed ceramic coatings on metals by an indentation technique. Eur Appl-Res Repts Nucl Sci and Technol Sec 7:1487–1494
5.40 Usmani Saifi, Sampath Sanjay (1996) Erosion studies on duplex and graded ceramic overlay coatings. JOM: J Miner, Metals and Mater Soc 48:51–54
5.41 Vally JA (1986) A review of adhesion test methods for thin hard coating. J Vac Sci and Technol A 4:3007–3014
5.42 Zhang Zhiming, He Xianchang, Shen Hesheng, Li Shenghua (1998). Investigation of adhesion of diamond coatings deposited on tools by the chemical vapor deposition method (in Chinese). J Shanghai Jiatong Univ 32:103–106

6 Wear Resistance of Coated Materials

6.1 General Overview

6.1.1 Terms and Definitions in Wear

Below are given standard terms and definitions of general notions used in science and engineering as applied to friction and wear. To avoid confusion, the authors tried to refrain from using synonymous terms in this chapter.

Wearing The process of destruction and detachment of material from the surface of a solid body and/or build-up of residual deformations under friction.

Wear The result of wearing determined in standard units (length, volume, mass).

Wear rate The ratio of the wear value to the duration of the time interval within which it occurred. Note: there are instantaneous (at a definite moment of time) and average (over a definite time interval) wear rates.

Wear intensity The ratio of the wear value to the path length within which the wear took place.

Wear resistance The property of any material to resist wear under definite friction conditions represented by the reciprocal of the wear rate or wear intensity.

Relative wear resistance The property of any material characterized by the ratio of wear intensity of one material to that of another upon wearing under identical conditions. Usually one of the materials is taken as a reference.

Frictional seizure The condition of local bonding of two solid bodies due to action of molecular forces under friction.

Scuffing The process of occurrence and development of friction surface damages due to seizure and transfer of material. Scuffing can result in blocking of relative movement.

Scoring The damage of the friction surface in the form of wide and deep furrows in the direction of sliding.

Scratching The formation of recesses on the friction surface in the direction of sliding under the action of solid body bulges or solid particles.

Fatigue wear The wear caused by fatigue failure under repeated deformation of microvolumes of surface layer material.

Scaling The detachment of material from the friction surface in the form of scales upon wear.

Chipping The detachment of material particles under fatigue wear which leads to formation of pits on the friction surface.

Run-in The change in friction surface geometry and physicochemical properties of material surface layers during early friction period which usually shows itself as a reduction in frictional force, temperature and wear intensity under constant external conditions.

Friction coefficient The ratio of the frictional force acting at the two bodies to the normal force pressing the bodies to each other.

Some special terms are also defined in the appropriate sections of this chapter when considering specific material wear conditions.

6.1.2 General Principles of Combined Hardening for Wear Resistance Enhancement

Critical parts of machines and mechanisms are required to have low metal consumption, high durability and reliability. Metal consumption is determined by strength indexes of the alloy and can be lowered by volume hardening. Measures for providing durability are determined by the operational conditions of the parts. When the parts work under friction conditions, their durability depends mainly on their wear resistance. Besides other parameters, reliability of parts is characterized by the ability of material to resist initiation and propagation of cracks.

Since materials satisfying the entire complex of rather demanding and contradictory requirements (crack resistance combined with wear resistance) either do not exist or would be very expensive, this problem is solved by designing structurally composite materials [6.81]. Such materials can be produced by modifications in chemical composition, structure and properties of the surface through the use of thermochemical steel treatment techniques, surface alloying, surface hardening, or other kinds of surface treatment. Another way of producing composite materials is deposition of coatings on the parts previously subjected to volume hardening.

Volume-hardened parts are coated using hard-facing, metallization, vacuum deposition, electrolyte, detonation, electric-spark and other techniques. Thus, combined hardening is produced: volume hardening serves to ensure higher reliability, while surface hardening provides wear resistance and, if necessary, special physicochemical properties of the surface.

6.1.3 Peculiarities of Producing Base Metal-Wear Resistant Coating Compositions

In combined hardening it is important to provide an optimum combination of structure and properties of materials forming the core, the surface, and the transition zone between them. The structure of the transition zone often plays a decisive role in the choice of technology and materials for combined hardening. If either the presence of the transition zone or its structure properties are ignored, the material with a wear resistant coating deposited on a crack resistant base may fail over the transition zone, which often accounts for scaling of the coating. Reduction of crack resistance in the contact zone may be caused either by chemical or structural heterogeneity, or by formation of phases of brittle chemical compounds as a result of reactions between base and coating components, as well as by weak base-coating bonding.

The strength of base-coating composition is generally improved by heat treatment of the coating deposited on the working surfaces of parts using various heat energy sources and activation techniques. However, adhesion strength enhancement does not always have a beneficial effect on reliability of the parts. Cracks developing in wear resistant but brittle coatings may propagate into the volume of the part. This phenomenon may initiate propagation of fatigue cracks. A very rigid layer on the surface of the part has an effect on its impact strength, similar to a sharp notch. The base, in its turn, also substantially looses its fatigue resistance because of thermal loading during treatment of the coating.

6.1.4 Hardening of Machine Parts With Wear Resistant Coatings

Thus, in designing and implementing technological processes of combined hardening of machine parts by depositing a wear resistant coating on a volume-hardened substrate, the following three design problems are to be solved:

1. The structure of the deposited coating must ensure the required wear resistance. In most cases the latter is provided by solid hardening particles of carbides, borides and other compounds in the coating composition.
2. The transition zone between the substrate and the coating must be sufficiently plastic and viscous. This can be ensured when a continuous sequence of solid solutions is formed as a result of volume interaction between substrate and coating components. Otherwise, if chemical and structural heterogeneity takes place, it is necessary to take care of reducing the embrittling influence of the transition zone by, for instance, depositing a viscous underlayer.
3. Modifications in the substrate structure, which occur under the action of energy flux during coating deposition, should not deteriorate the structure and, consequently, the mechanical properties of the substrate.

In solving the posed problems it is possible to organize machine operation in such a way that the physically worn working surfaces are periodically restored by depositing wear resistant coatings. If the structure and the properties of the core of the part is retained, its lifetime is then determined solely by the obsolescence of the mechanism.

The choice of materials and technology for combined hardening of machine parts is determined by their service conditions. One should also take into consideration the nature of active loads and the degree of wear, and investigate the peculiarities of coating wear for various types of frictional contact. This chapter deals with the latter direction in solving the problem of improving material wear resistance.

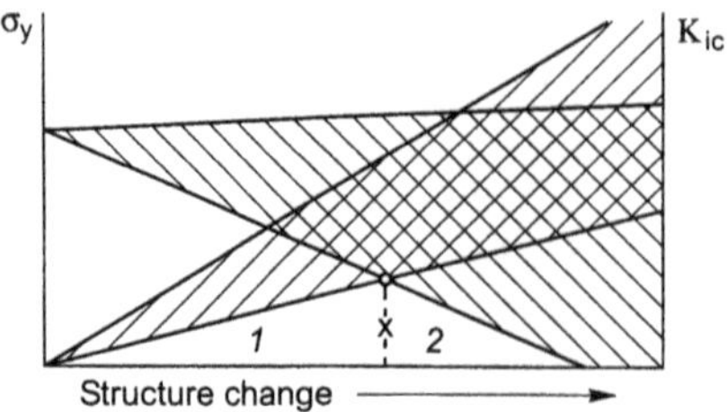

Fig. 6.1. Qualitative diagram of the threefold relation: yield point (σ_y) – fracture toughness (K_{ic}) – structure in hardened alloys. *1* zone of reliable operation of products; *2* zone of possible sudden failures

6.2 Foundations of Structural Theory of Wear Resistance

6.2.1 Dependence of Strength and Fracture Toughness of Materials on Their Dislocation Structure

Let us consider the general principles of producing optimum alloy structures having high yield point, fracture toughness and wear resistance. First we shall consider the dependence of the first two indexes on the dislocation structure of alloys.

It is expedient to evaluate the tendency of industrial alloys to brittleness or plasticity by considering the contradiction between the yield point σ_y and the fracture toughness K_{ic} in obligatory connection with the structural state of an alloy (see Fig. 6.1). Both indexes involved in this plot, the flow stress, or the onset of active and multiple movement of dislocations, and the tendency to brittle crack development from an available concentrator, K_{ic}, are determined experimentally and, consequently, with due account of the actual structural state of an alloy. A conventional point X divides the diagram in two parts, the zone of high reliability *1*, and the zone of increased brittleness *2*. If a real alloy in accordance with its structural state (performed hardening, e.g., by heat treatment) lies in zone *1*, its service is safe since any occasional overloads (external stress growth on ordinate upwards) will be relieved by active plastic deformation due to low values of σ_y and high values of K_{ic}.

For a strongly hardened alloy (i.e. active barriers to moving dislocations have been created), σ_y has significantly grown and K_{ic} decreased. Under these conditions every external load peak first reaches the critical value of K_{ic} thus creating favorable conditions for development of brittle fracture. In order to evaluate the possibility of really effective hardening of metal alloys, i.e. of increasing the yield point without sacrificing fracture toughness (upper limits for curves in Fig. 6.1), it is necessary to consider dislocation mechanisms of increasing the yield point and compare them with possible change in fracture toughness. Within the framework of the modern theory of plastic deformation and fracture (structural strength theory), yield point and fracture toughness depend on the following factors [6.81]:

$$\sigma_{y,} K_{ic} = f\,(\sigma_{p\text{–}n}, \sigma_{disl}, \sigma_{sol}, \sigma_{ph}, \sigma_{grain}), \tag{6.1}$$

where $\sigma_{p\text{–}n}$ is the Pyerls–Nabarro force, i.e. friction stress in crystal lattice, or the stress which a moving dislocation in a lattice free from any barriers must overcome. Experimental determination or calculation of this stress is extremely difficult since it requires information not only on properties of the crystal lattice of the matrix, but on the dislocation (or even on many dislocations) itself, e.g., its width; σ_{disl} is hardening due to increased number of dislocations, i.e. the stress of resistance exerted on a moving dislocation by dislocations placed along its path.

Quantitatively, this stress is determined by the energy of interaction of dislocations in metal, both those located chaotically in the form of the "forest" ($\sigma_{disl\,f}$), and arranged regularly as sub-boundaries of polygons, or cells ($\sigma_{disl\,pc}$):

$$\sigma_{disl} = \sigma_{disl\,f} + \sigma_{disl\,pc}; \tag{6.2}$$

σ_{sol} is hardening due to dissolved atoms.

Initial stress of dislocations or flow stress in solid solutions is higher than in pure metals since alloying creates barriers for displacement of any dislocation as a result of either its elastic interaction with distorted crystal lattice of the matrix ($\sigma_{sol\,mat}$) or blocking of the dislocation by atomic atmospheres of the dissolved element ($\sigma_{sol\,atm}$):

$$\sigma_{sol} = \sigma_{sol\,mat} + \sigma_{sol\,atm}; \tag{6.3}$$

σ_{ph} is hardening by dispersed phases according to three possible models:

$$\sigma_{ph} = \sigma_{phOr} + \sigma_{phH} + \sigma_{phNM}, \tag{6.4}$$

where: σ_{phOr} is hardening due to interaction between moving dislocations and hard non-deformable particles forming dislocation loops by Orovan's mechanism. This contribution may be quantitatively calculated using the known formulas; σ_{phH} is hardening due to formation of prismatic dislocation loops and spirals by Hirsch's mechanism in plastic deformation of an alloy containing hard particles; σ_{phNM} is hardening due to cutting of deformable particles by moving dislocations according to Nicholson-Mott's mechanism; σ_{grain} is hardening by microstructural barriers hindering moving dislocations such as grain boundaries, interfaces, second phase narrow interlayers, i.e. microtranslation stress in a real alloy.

Most calculations of this contribution are based on the known Hull-Petch's equation:

$$\sigma_y = \sigma_0 + Kd^{-1/2}, \tag{6.5}$$

where yield point σ_y is related to the matrix dislocation structure σ_0 and the grain size d, or, generally, for any kind of microstructural barrier:

$$\sigma_y = \sigma_0 + Kd^{-n}. \tag{6.6}$$

6.2.2 Choice of Optimum Dislocation Hardening Mechanisms

Only two of the five dislocation mechanisms satisfy the requirements of structural strength enhancement for industrial steels: substructural hardening $\sigma_{disl\ pc}$ through creation of polygonal-cell dislocation constructions or through comminution of grains (alloy phases) (σ_{grain}).

These dislocation models allow increasing the yield point of an alloy with no danger of its embrittlement. It is these two mechanisms that are effective for refinement of the known and creation of new technological processes of metal hardening treatment such as thermoplastic treatment, controllable rolling, etc. For these technologies special conditions weakening the influence of other unfavorable dislocation mechanisms are usually created.

If a product is designed basing on the principle of a composite material with combined volume plus surface hardening, then new scopes are opened for successful application of every dislocation hardening mechanism: $\sigma_{disl\ pc}$ and σ_{grain} for volume hardening and $\sigma_{disl\ f}$, $\sigma_{p\text{-}n}$, σ_{ph}, σ_{sol} for surface hardening in the course of coating deposition. This new approach to hardening various metal products (i.e. development of a new principle of combined hardening) makes possible a new reconsideration of the coating problem in its entirety. From the new viewpoint the coatings can be applied with great benefit not only in reconstruction of the worn surfaces, but mostly in production of new machine parts working in wear conditions.

6.2.3 Laboratory Methods of Testing Materials for Wear Resistance

The most reliable way to control the correctness of the choice of material or methods for its treatment as applied to certain machine parts is testing the parts in real field conditions (full-scale test). This control method is sometimes used in practice, although it tends to be too labor- and time consuming (long duration of the test, the necessity of disassembling the machine to determine the degree of wear). In this case wear-testing results can be evaluated by changes in the machine service properties, change in part sizes, and also indirectly, e.g., by increased oil leakage.

Due to the difficulty of full-scale testing, laboratory wear tests have become quite popular [6.3, 6.4, 6.25, 6.53–6.55]. Depending on the wear type the testing schemes can be conventionally divided into the following groups:

- friction wear against fixed abrasive particles;
- friction wear against loose abrasives;
- impact abrasive wear;
- wear with particles located in a clearance (abrasive oil interlayer);
- gaseous and hydroabrasive wear;
- fretting corrosion.

The different mechanisms of wear make it necessary to discuss each group of tests separately (see Fig. 6.2).

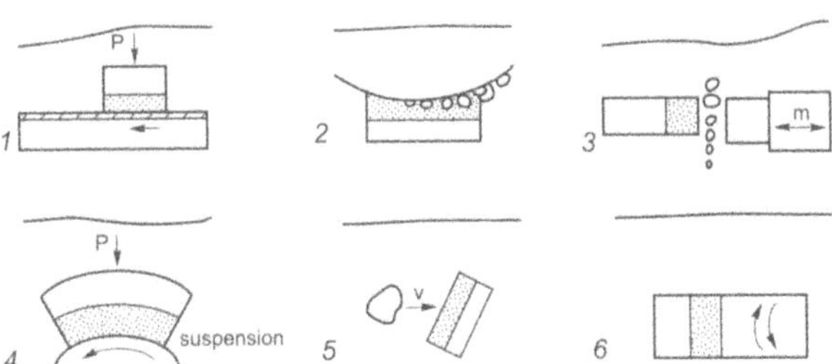

Fig. 6.2. Schemes for wear testing materials in contact with abrasive particles. *1* against fixed abrasive; *2* against non-fixed abrasive; *3* impact-abrasive; *4* with abrasive oil interlayer; *5* gaseous (hydroabrasive); *6* fretting-corrosion

6.3 Wear Against Fixed Abrasive Particles

6.3.1 General Overview

The results of comprehensive laboratory tests of wear resistance of pure metals and structurally heterogeneous non-ferrous alloys and steels under friction against solid grains fixed firmly on a stationary surface in the absence of heating and aggressive environment are given in [6.40]. Wear testing was carried out using an abrasive cloth with characteristics which were maintained constant. It was found that relative wear resistance of pure metals linearly depends on their microhardness. The experimental points are located on the straight line passing through the origin. Testing tin-based, lead-based and tin-lead-based babbits and leaded bronze with structural components of different hardness has not revealed any definite relation between wear resistance and microhardness. Nevertheless, in all cases wear resistance of alloys proved to be lower than that of pure metals with the same hardness. For steels in thermally untreated state the dependence of wear resistance on hardness is the same as for pure metals. Hardened steels with subsequent tempering show linear rise in wear resistance proportional to their hardness growth, though it is less steep as compared to pure metals and thermally untreated steels. Testing has shown that preliminary cold-work hardening does not increase wear resistance of pure metals and steels.

The abrasive particles in the considered technique are fixed (see Fig. 6.3) and cannot roll over the surface, so the material fails solely due to the cutting action of the hard abrasive particles. Khruschov and Babichev [6.40] have shown that under such conditions the degree of wear is determined by the ratio of abrasive and tested material hardnesses H_a/H_m. The wear intensity does not depend on the hardness of abrasive if the latter is at least 1.4 times higher than the hardness of the tested material. The lower the ratio H_a/H_m, the higher is material wear resistance. Therefore, an increase in wear resistance under friction against fixed abrasive particles can be achieved by surface hardening [6.40].

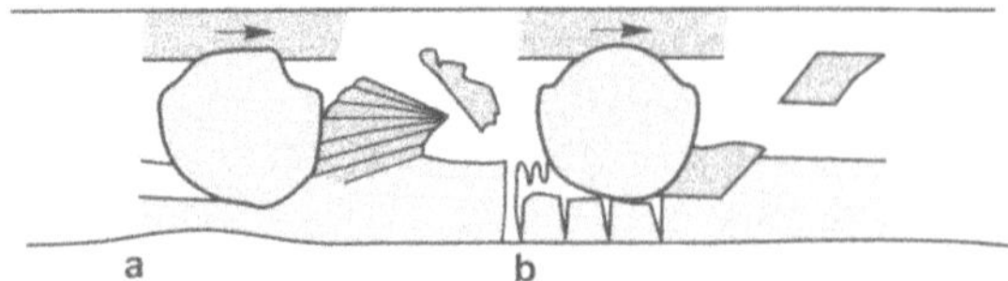

Fig. 6.3. Surface failure with fixed abrasive particles at: **a** high crack resistance; **b** high hardness of material

Wear of different materials with fixed abrasive particles was thoroughly investigated by Zum Gahr [6.92], Horhbogen [6.33, 6.34], Rosenfield [6.65], and Moore [6.57]. They found that the interaction between abrasive grains and tested material has a rather complicated nature. The mechanisms of grain failure change depending on hardness and fracture toughness, and on the loading conditions. In the process of friction the ability of material to deform is completely exhausted. If this is the case, crack propagation energy and fracture toughness of the material play a very important role. When a low-plastic material is used, there is a critical load per abrasive particle, which should not be exceeded, otherwise, wear mechanism will change: microcutting will be replaced by brittle chipping. Wear intensity at the loads below critical is determined basically by the yield point; at the loads over critical it depends, first of all, on the crack resistance index K_{ic}. The value of critical load is a function of material structure and its ability to relax dangerous peak stresses by dissipation of plastic deformation energy.

6.3.2 Specimens

Smooth cylinders with diameter 20.0±0.1 mm and length 15–20 mm are used as specimens; working surface roughness of reference specimens $R_z \leq 0.32$ μm; the material is annealed Steel 45 (0.45% C) with hardness HV185–195. During coating deposition on the end surfaces of specimens special measures to avoid cracking, warping, local overheating, etc. are taken. Requirements to abrasion surface roughness are imposed in each particular case with due account of surface porosity. Specimens with scaled or chipped coatings are rejected. At least three reference and three coated specimens are to be provided for each variant. The hardness of abrasive particles should be at least 1.6 times higher than the hardness of the tested coating.

6.3.3 Equipment

Two types of installations with different shape of wear surface are typically used in this testing scheme. In the first device an abrasive cloth is fixed on the rotating disk (see Fig. 6.4). Specimens to be worn are attached to the lever and pressed to the abrasive by a predetermined load using an axis and a rope passing over a block. Friction moment that tends to divert the lever from the horizontal position is balanced by a counterweight which mass is used for determination of the friction coefficient. The friction unit may be enclosed by a chamber in which pressurized cooling air is supplied.

In the next device (see Fig. 6.5) an abrasive cloth is wound around the rotating cylinder. The specimen is fixed in the holder connected with the carriage through the strain-sensing element used for measuring frictional forces. Rotation of the lead screw translates the specimen along the cylinder. Normal load is set by the weight. Using these or similar schemes a number of devices for testing various coatings (decorative, polymer, etc.) have been designed [6.7].

For wear testing thick (over 0.2 mm) coatings by fixed abrasive particles a device shown in Fig. 6.6 was used. The main units of the device include the disk with abrasive cloth driven by an electric motor at a maximum speed of 1500 rpm and the lead screw with two specimen holders. During testing the specimens are pressed against the disk by the total weight of the holders and additional weights attached to the rod. The specimens move on a spiral with respect to the disk. Radial movement of specimens is provided by lead screw rotation. Movement direction can be altered by reversing the lead screw in automatic mode using switches [6.81].

However, the device is unsuitable for testing thin coatings for the following reasons. Firstly, as described in [6.46], the accuracy of results, even for testing in the standardized scheme (according to State Standard 17367) is rather poor.

Secondly, specimen holders made in the form of grippers do not provide fine alignment of cylindrical specimens. Minor specimen misalignments, which are not very important for investigation of volume-hardened materials, make testing incorrect when studying thin coatings (both the coating and base metal are worn simultaneously).

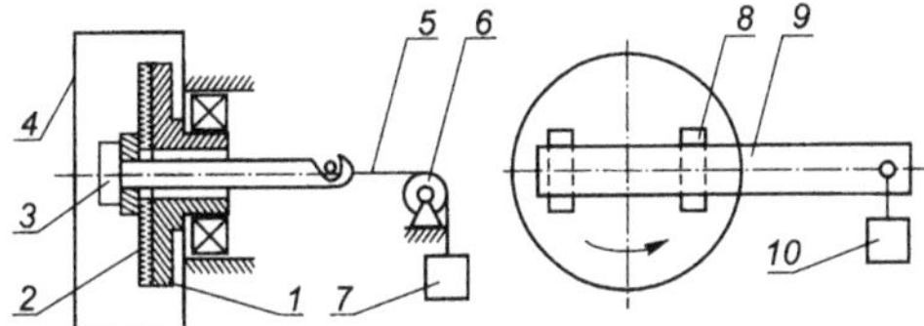

Fig. 6.4. Mechanical diagram of the device with flat wear surface. *1* rotating disk; *2* abrasive cloth; *3* axis; *4* chamber; *5* rope; *6* block; *7* weight; *8* specimen; *9* lever; *10* weight

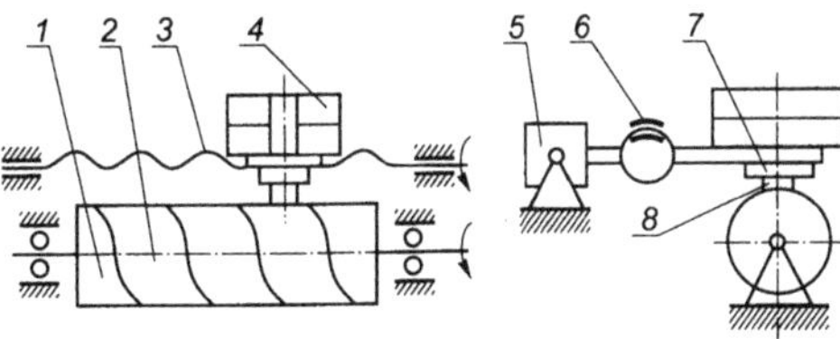

Fig. 6.5. Mechanical diagram of the device with cylindrical wear surface. *1* rotating cylinder; *2* abrasive cloth; *3* lead screw; *4* weight; *5* carriage; *6* strain-sensing element; *7* specimen holder; *8* specimen

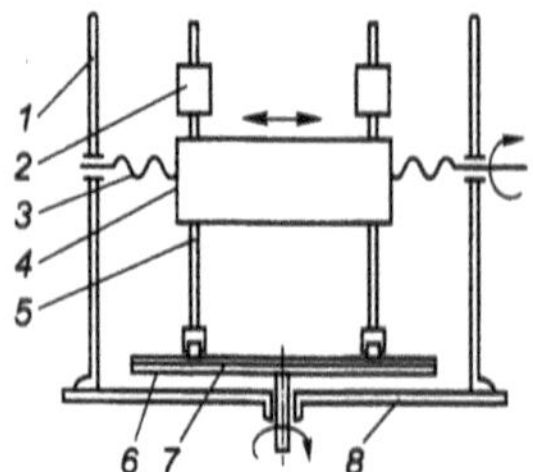

Fig. 6.6. Installation for wear testing with fixed abrasive. *1* post; *2* weight; *3* lead screw; *4* sliding block; *5* rod with specimen holder; *6* specimen; *7* disk with abrasive cloth; *8* base

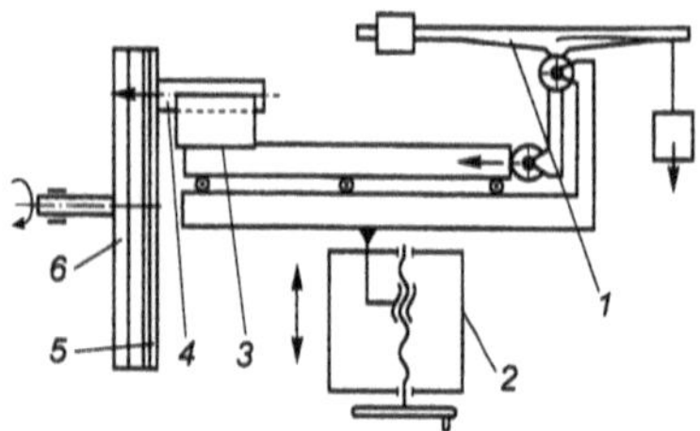

Fig. 6.7. Installation for wear testing thin coatings against fixed abrasive. *1* lever system; *2* support; *3* prismatic holder; *4* specimen; *5* abrasive cloth; *6* spindle

Thirdly, even under the smallest possible loads (without weights) the wear is too large because of the weight of specimens, rods and holders pressing a coating to the abrasive cloth. Fourthly, high speeds of disk rotation and screw movement lead to reduction in time interval between consecutive weighings to several seconds (which means less accurate results), while it is impossible to increase the interval because of small thickness of the coating and the danger of its complete wearing. For these reasons new techniques are being suggested [6.15, 6.38] and new installations are being designed specially for studying thin coatings.

One of the mentioned installations (see Fig. 6.7) is made in the form of a special fixture fitted in the tool-holder of a standard lathe [6.83]. The specimen is secured in the prismatic holder which guards against misalignments of the specimen. The horizontal arrangement of elements completely eliminates the influence of the natural total weight of a specimen and the holder on the force of clamping against abrasive paper. The speed of specimen movement past the abrasive paper is adjusted over a wide range by changing rotation speed of the lathe spindle and cross-feed of the support with the tool-holder. Minimum specimen clamping force corresponds to the weight of 5 g with due account of the lever system.

Recently a similar device was designed in Sweden [6.15]. A rotating metal disk covered with a polishing cloth sprinkled with fine-grained diamond powder was applied to investigation of abrasive wear characteristics of TiN and NbN thin coatings.

In 1970 Khrushchov and Babichev [6.40] proposed another method of wear testing using fixed specimens (see Fig. 6.5), which was subsequently developed into State Standard 17367. This method is quite applicable for testing thick coatings. However, testing thin coatings (under 0.2 mm) has a number of specific features which are not taken into account by the standard technique [6.46, 6.73].

6.3.4 Testing

Prior to testing the density and hardness (microhardness) of the coating and the reference material are measured. Thoroughly cleaned and washed in gasoline and acetone specimens are then dried and either weighed or, otherwise, their linear dimensions are measured. Errors of weighing and measuring should be less than 0.1 mg and 0.01 mm (0.001 mm for thin coatings), respectively. For a better fit to abrasive cloth the specimens should be run in. In the course of testing the reference and tested specimens should pass equal wear path along the fixed abrasive. Relative wear resistance of the coating is then calculated from the following equations:

$$K_c = \frac{g_r \rho_c}{g_c \rho_r}, \tag{6.7}$$

$$K_c = \frac{h_r}{h_c}, \tag{6.8}$$

where g_r and g_c are the average mass losses of the reference and the coating; ρ_r and ρ_c are the densities of the reference and the tested coated material; h_r and h_c are the average linear wear values for the reference and the tested coated material.

Unfortunately, in many papers wear resistance is presented in absolute rather than in relative values, which makes impossible comparison of the results obtained in different laboratories.

6.3.5 Results and Discussion

Testing plasma-sprayed coatings on NiAl, NiTi and self-fluxing alloys carried out using the installation shown in Fig. 6.6 has proved that their wear resistance is directly proportional to their hardness. Fig. 6.8 shows the curve of wear rate plotted for coating PG–SR4 (composed of 3.0% B, 0.8% C, 3.9% Si, 16% Cr, 4.0% Fe, and 72.3% Ni), which was tested with fixed abrasive particles after fusing at several temperatures. The maximum wear resistance of the coating corresponds to the fusing temperatures at which the hardest structure is formed. In these experiments the conditions of intense brittle chipping of surface layers were not achieved. Wear resistance of the tested coatings was 1.5–8 higher than the resistance of Steel 45 (HV190). The lowest wear rate was found for fused self-fluxing coatings containing chromium carbide additives.

Chromium coatings (thickness 20–30 μm) with various content of ball-shaped diamond particles (up to 3 μm in diameter) were studied using the installation shown in Fig. 6.7. The coatings had different hardness depending on the technology of deposition, HV1650 and HV1300 for the first and the second techniques, respectively. The advantages of the first variant compared to the second one are revealed at heavy loads ($P = 240$ g), wear resistance rates being 0.167 mg/s and 0.229 mg/s, respectively (see Fig. 6.9). However, as the load is reduced down to 140 g, the difference between the two variants vanishes. At the load of 20 g the coating produced by the first variant of technology has better resistant qualities, and its wear rate is 2.3 times lower than the rate for the second variant (0.276 mg/s and 0.0686 mg/s, respectively). Probably, as the load exceeds critical point, the mechanism of wearing changes from viscous to brittle wear [6.83].

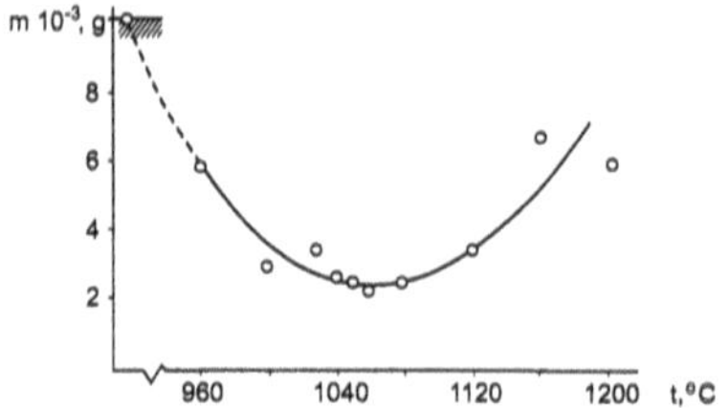

Fig. 6.8. Rate of wear with fixed abrasive particles for coating PG–SR4 fused at different temperatures. *m* mass loss of the specimen over a 50 m friction path; the hatched area corresponds to non-fused coating [courtesy of Bataev A.A.]

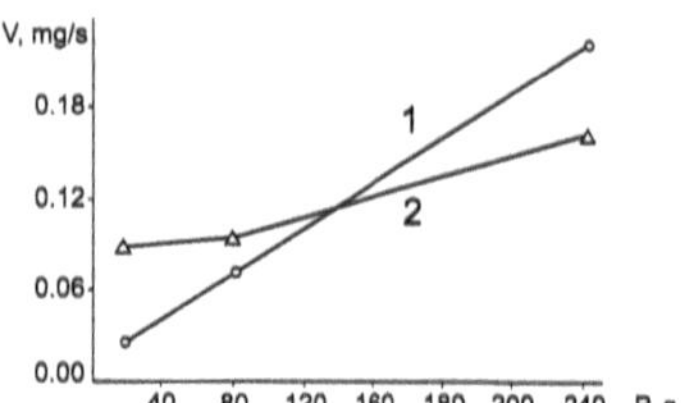

Fig. 6.9. Dependence of wear rate on the applied load for chromium coatings of different hardness. *1* HV1650; *2* HV1300

As was demonstrated in the paper [6.81], at loads on abrasive lower than critical all dislocation hardening mechanisms may be used to enhance wear resistance of coatings. In these conditions finely dispersed grain structures are preferable, and the strength of phases and interphase boundaries is of great importance. More homogeneous microstructures located along the direction of abrasive movement are desirable since they reduce stress concentration at interphase boundaries and, consequently, the probability of local brittle failure. Since at loads over critical surface failure depends on fracture toughness parameters, an increase in wear resistance under these conditions can be achieved either through weakening by σ_{p-n}, σ_{sol}, σ_{ph}, and $\sigma_{disl\,f}$ mechanisms, or hardening due to $\sigma_{disl\,pc}$ and σ_g.

Analyzing the technological aspects of wear resistance enhancement and taking into account dislocation hardening mechanisms for the wear scheme under consideration, Tokarev [6.77] puts forward the following suggestion. For heterophase hard-faced alloys the oriented crystallization of eutectics is the most promising way. A rather finely dispersed carbo-boride phase in NiCr solid solution can be produced by treating self-fluxing alloys in sintering modes. Application of conventional methods of coating deposition such as gas-thermal spraying followed by fusing can deteriorate the mechanical properties of the coating material because of the necessity to heat the powder layer above the melting point. As a result, the hereditary dispersity of the structure and microdistortions of the lattice are generally lost. This phenomenon makes necessary intense development of the techniques which would allow depositing coatings without their complete melting (burning).

Coating burning technology provides considerable reduction in temperature and temporal parameters of the process, taking advantage of various activation methods. One of the simplest activation techniques consists in application of an external force at the different stages of the process. A special feature of self-fluxing powders is their ability to clean the surface from oxide films which ensures chemical activation of the process. At the temperatures 970–1000 °C corresponding to the burning stage, the oxide film is in the liquid state. This provides production of a dense finely dispersed coating with enhanced strength parameters upon activating burning by applying external force. To shorten the duration of the process, it is obviously expedient to apply concentrated energy sources such as various beam heating methods, induction, plasma, and electrocontact techniques [6.77].

6.4 Wear Against Loosely Fixed Abrasive Particles

6.4.1 General Overview

The wear occurs due to movement of the surface of the part through the mass of abrasive particles. This process is most accurately modeled as movement of the part in a reservoir filled with abrasive. In real conditions the processes of wear with fixed and loose abrasive particles accompany each other, and the degree of wear depends on various factors such as composition of the medium, material of the coating, presence of brittle inclusions, etc. [6.5, 6.77, 6.81]. Typical for this process small contact area between an abrasive particle and the working surface causes considerable stresses which depend on the shape and mechanical properties of the particle and on the pressing force. Two situations are possible: when the produced stresses exceed the elastic limit but still remain lower than the yield point, fatigue failure takes place; if the stresses are above the yield point, the wear is accompanied by plastic deformation of microvolumes (microcutting) and post-deformation failure occurs (see Fig. 6.10) [6.77].

Sometimes charging process significantly reducing the wear is observed. The latter may even take on negative values, i.e. the size and mass of the specimen can increase. Apparently, charging may be caused by inevitable impact action of sharp abrasive particles including their shattering, introduction in the surface and some adhesion processes. Charging effect depends on the speed of abrasive mass movement and the relation between abrasive and specimen hardnesses. Probably, it may be observed only in the case of soft, plastic coatings.

Wear resistance strongly depends on the shape of abrasive particles. Thus, when corundum abrasive having fragmentary shape is changed to quartz spherical abrasive, wear resistance of normalized Steel 25 (0.25% of C) is increased by an order of magnitude [6.24]. In wear of metals with loosely fixed corundum abrasive relative wear resistance non-linearly depends on hardness HV. If quartz sand is used as abrasive, then relative wear resistance K_c can be expressed by the following equation:

$$K_c = a(HV)^h, \tag{6.9}$$

where a and b are constant coefficients.

The relation between the material wear (W) and the applied load is also exponential [6.24]:

$$W = kp^n, \tag{6.10}$$

where k and n are constant coefficients, and $n > 1$.

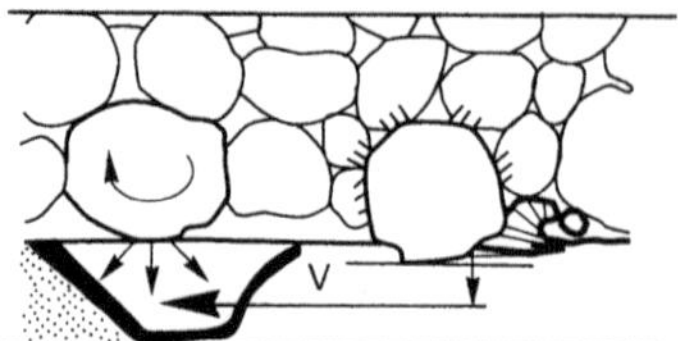

Fig. 6.10. Destruction of surface with loosely fixed abrasive particles. **a** fatigue failure; **b** microcutting

The value of index n indicates the nature of the wearing process. The larger the share of plastic deformation in the process of wearing, the bigger the value of index n, while it tends to 1 with increasing the share of fatigue failure.

High wear resistance in loosely fixed abrasive medium is required for numerous working parts of quarry, building and road machines such as excavator bucket or rotor teeth, bulldozer blades, grader parts, tracks of caterpillar machines, suction-tube dredge buckets, drilling rigs, pipe-lines for transportation of granular bodies, soil cultivation tools (stubble plough disks, cultivator teeth, plough shares, etc.) as well as for many parts of metallurgical equipment.

In the majority of wear testing methods the reference and tested specimens are rubbed against abrasive particles fed to the friction zone and pressed to the specimen in identical conditions. Then the degree of wear of both reference and tested materials is directly measured, and wear resistance of the tested material is determined by comparing its wear with the wear of the reference specimen.

6.4.2 Specimens

The specimens of the reference material are usually prepared in the form of plates with thickness of at least 5 mm. The coating is deposited on the wide face of the plate as described in section 6.3.2. Annealed Steel 45 (0.45% C) with hardness HV190–200 is recommended as the reference material. Repeated wearing of specimens is possible after removing wear traces formed as a result of previous tests by mechanical finishing.

6.4.3 Equipment

All machines for determination of resistance to wear against loosely fixed abrasive using plate specimens can be classified into three types of schemes (see Fig. 6.11).

The most commonly used is the Brinnele's scheme (see Fig. 6.12), in which the tested specimen is pressed with its flat side to the ribbed rubber roll 12 mm wide and 220 mm in diameter. Quartz sand with grains 0.2–0.6 mm in diameter is fed from the hopper through a funnel and a pipe with a calibrated hole. Abrasive is picked up by the roll (roll surface recess radius 4 mm, the number of recesses 20) and dragged along the surface of the specimen at sliding speed 2.29 m/s and load 147 N.

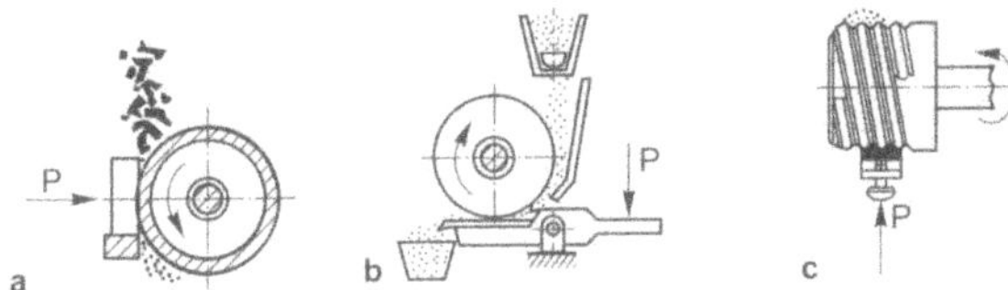

Fig. 6.11. Installation for wear testing of materials with loosely fixed particles: **a** Brinelle's method; **b** Tennebaum's method; **c** screw method

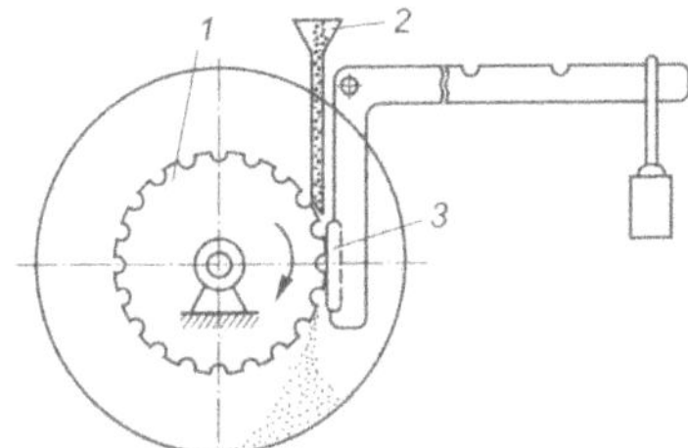

Fig. 6.12. Loading unit in Brinelle's device. *1* roll; *2* feeder; *3* specimen

In his review paper on methods of testing material for abrasive wear, Sorokin, one of the most competent experts in the field, recommends application of an installation with cylindrical specimens [6.68]. The installation (see Fig. 6.13) is designed for abrasive wear testing of materials in the bulk of dry or wet abrasive with varying hardness, grain size and mineralogical properties. It allows simultaneous testing of three specimens, when each specimen is rotated around both its own axis and the carrier's axis, thus providing uniform peripheral wear. Corroding additives can be introduced in the abrasive bowl, thus creating corrosion-mechanical wear conditions. As for practical application of this installation for testing coatings, depositing coatings on the end surfaces of finger-shaped specimens will, probably, cause some difficulty.

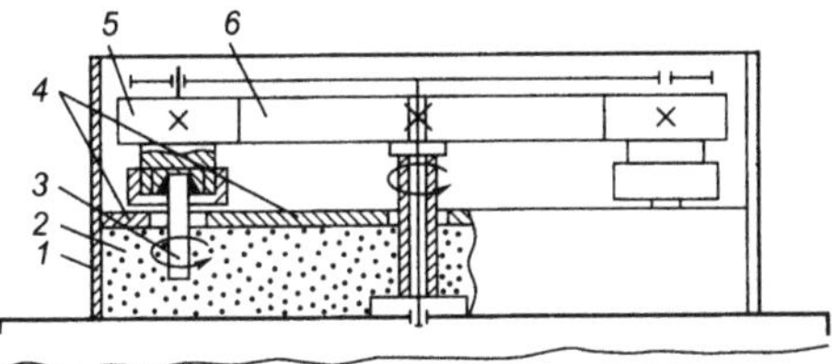

Fig. 6.13. Installation for testing materials in abrasive (by [6.68]). *1* housing; *2* abrasive mass; *3* specimen ($d = 10$ mm, $l = 70$ mm); *4* O-ring seals; *5* spindle gear; *6* stationary wheel

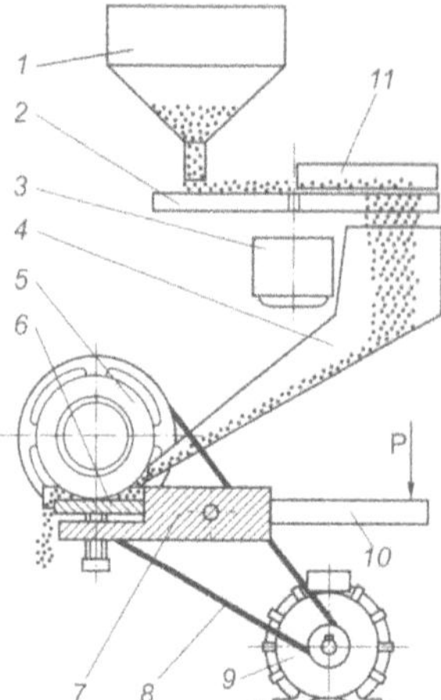

Fig. 6.14. Installation for wear testing with loosely fixed abrasive particles. *1* hopper; *2* metering device disk; *3,9* electric motors; *4* chute; *5* rubber roll; *6* specimen; *7* holder; *8* belt transmission; *10* lever; *11* abrasive cut-plate

The Tennenbaum's scheme is standardized (see State Standard 23.208) as follows: dimensions of the specimen 30x30(50)x5 mm; diameter and width of the rubber roll 48–50 and 15 mm, respectively. The specimen is pressed against the roll by the force of 44.1±0.25 N. The Standard does not cover coatings with hardness higher than HV1400 and porous materials with average pore size larger than 0.1 mm, as well as coatings with hardness varying more than by ±10% at the depth of 0.3 mm. These restrictions set the limits of applicability of the Standard for testing hard thin wear resistant coatings.

Besides, the device for feeding abrasive particles in the zone of frictional contact is not perfect and cannot provide constant and steady feed of abrasive. These drawbacks are partially overcome in the installation shown in Fig. 6.14.

In order to provide steady feed of abrasive to the contact zone the installation is fitted with a metering device consisting of a funnel hopper, lower part of which is placed at a fixed gap above the slowly rotating disk. Abrasive consumption can be adjusted by changing the size of the clearance between the funnel and the disk. A cut-plate located some distance away from the hopper sends abrasive to the chute, lower part of which is near the roll-specimen contact zone.

Another special feature of the installation as compared to standard devices is the possibility to change the force pressing the specimen against the roll. This allows changing the degree of fixing of abrasive particles which is the main parameter distinguishing this type of wear from the wear with fixed abrasive particles. It determines whether a particle performs translational motion, thus causing plastic deformation of surface layers, or rolls causing fatigue failure due to heavy contact loadings.

For comparative testing fused corundum is used as abrasive. When evaluating wear resistance under specific wear conditions it is recommended to use abrasive material corresponding to the material affecting the coating in service but with grain size of 1.0 mm or less.

6.4.4 Preparation for Testing

Prior to testing the rubber roll is run in by rubbing against the surface of abrasive cloth fixed on a flat plate until the plate surface snugly fits the roll surface along its entire length.

Then the specimens are washed, dried, fitted in the specimen holder and pressed against the roll with the desired load. Continuous steady feed of abrasive material to the contact zone is required and can be checked by the presence of particles along the entire roll length.

6.4.5 Testing

State Standard 23.028 defines the following technical characteristics for standard testing: roll rotation speed 60±2 min^{-1}, fixed pressing force 44.1±0.25 N, number of revolutions during testing of the reference plate 600. The tested material is abraded during time interval which depends on the hardness HV of the specimen. For specimen hardness HV lower than 400 the required number of roll revolution is equal to 600; (400–800) – 1800; over 800 – 3600.

It should be noted, that standard conditions are not suitable for testing thin coatings with thickness less than 0.1 mm. In this case the load applied to the roll, the number of revolutions and grain size of the abrasive should be reduced.

The wear of tested and reference specimens is determined by weighing them before and after the testing with error not exceeding 0.1 mg.

Relative wear resistance (K_c) of the tested coatings can be calculated from the next equation:

$$K_c = \frac{g_r \rho_c N_c}{g_c \rho_r N_r}, \tag{6.11}$$

where g_r, g_c are the mass losses of the reference material and the tested coating, respectively, g; ρ_r, ρ_c are the densities of the reference material and the tested coating, respectively, g/cm^3; N_r, N_c are the numbers of roll revolutions during testing the reference material and the tested coating, respectively.

Unfortunately, in many papers absolute rather than relative values of wear resistance are often given, which makes impossible comparison of results obtained in different laboratories.

6.4.6 Results and Discussion

If stresses arising in contact zone during friction exceed the yield point, the wear is accompanied by plastic deformation of microvolumes, cold hardening and failure. Otherwise, fatigue failure due to accumulation of microdefects and cracking takes place.

In the case of gas-thermal coatings rapid cooling of the particles forming the coating may lead to formation of metastable supersaturated solutions in far-from-equilibrium conditions. Because of its low fracture toughness, this structure will have poor resistance to fatigue failure. Alloying matrix content in the structure is reduced by exposure to high temperature (annealing) or by alloy fusing and further crystallization in the conditions close to equilibrium.

Thus, after plasma-spraying the PG–SR4 coating (composed of: 4.0% Fe, 3.0% B, 0.8% C, 3.9% Si, 16.0% Cr and 72.3% Ni), the structure of its particles consists of supersaturated solid solution of the elements in nickel and inclusions of carbides, borides, and chromium carboborides. The size of inclusions is 0.3–1.5 μm, the space between them 0.5–3 μm. The wear with loosely fixed abrasive particles proceeds due to development of cracks and chipping of fragments 10–30 μm in size. Fusing the coating in furnace at 1040 °C followed by crystallization in air leads to growth of hardening crystals (up to 2–6 μm), formation of eutectics and reduced alloying of solid solution which forms the alloy matrix. Microhardness of the solid solution is reduced from 6000–8000 to 2500–3500 MPa. During wearing the size of chipping fragments of such coatings decreases to 3–7 μm, wear resistance exceeds the initial value by 25%. Coarsening of structural components and formation of dendrites of solid solution of the elements in nickel at elevated fusing temperatures (1100–1200 °C) reduce the hardness and wear resistant advantages of the fused coating.

As has already been mentioned, charging is an interesting effect leading to enhanced resistance in wearing with loosely fixed particles. Charging is possible with very rough or porous surfaces (which is characteristic of gas-thermal coatings), when dimensions of irregularities are comparable with the size of abrasive particles. In this case a macroheterogeneous structure is created. Abrasive particles stick in pores and cracks thus hardening the structure and preventing wear to a certain extent. This effect has been found in testing the PN70U30 plasma-sprayed coating (composed of 30% Al and 70% Ni) having high open porosity (more than 20%). Despite the poorer strength properties as compared to the PN85 coating (composed of 15% Al and 85% Ni), its wear resistance is 1.5–2 times higher.

Studies of coatings PN55T45 and PN65T35 with porosity 9% and 24% and composition 45/35% Ti and 55/65% Ni, respectively, have led to the conclusion that in the conditions of friction against loosely fixed abrasive particles the PN55T45 coating is more stable (by about 32%). Apparently, the effect of enhancing coating resistance due to increase in porosity, as found for "nickel–aluminum" coating system, does not reveal itself in this case.

Table 6.1. Wear rate of coatings under friction against loosely fixed abrasive particles (mg/s)

Load applied to the roll [N]	Steel 45	PN55T45	PN65T35	PN70U30	PN85U15
10	0.31	0.14	0.23	0.14	0.21
20	0.70	0.34	0.36	0.38	0.66

The degree of preliminary surface preparation effects wear rate only at the beginning of the testing period. After short run-in wear rate of polished and non-polished coatings is stabilized and remains at the same level. This pattern has been observed when the load applied to the roll changed from 10 to 30 MPa.

The results of comparative testing of some plasma-sprayed Ni–Ti and Ni–Al system coatings are given in Table 6.1. Wear rate of the tested coatings is 1.5–2 lower as compared to Steel 45 (0.45% C).

Fused self-fluxing coatings also wear out at a lower rate than specimens made of carbon steel. The decrease of the wear rate of self-fluxing coatings is inversely proportional to their hardness increase.

Because of their strong structural heterogeneity gas-plasma sprayed coatings generally wear at a higher rate as compared to plasma-sprayed coating. An increase of chromium carbide content in the source powders of self-fluxing alloy leads to reduction in coating wear rate (see Fig. 6.15). Analysis of the reported studies shows that coatings with more than 20% of chromium carbide additives can be effectively utilized, the additives being more efficient when alloys with lower hardness were used.

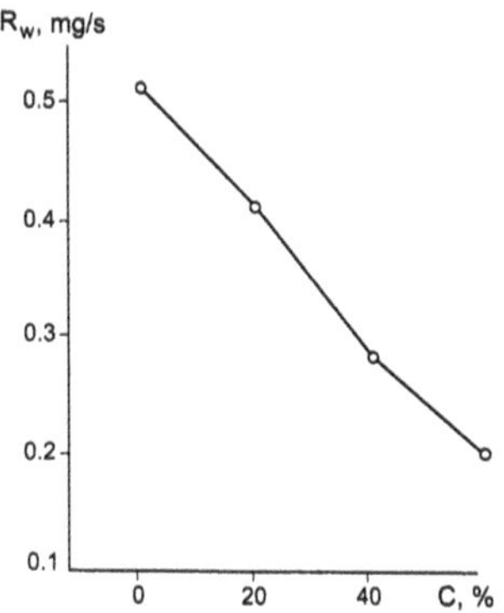

Fig. 6.15. Wear rate R_w of self-fluxing coating PG–SR4 with varying chromium carbide content; C concentration of chromium carbide (load 30 N) [courtesy of Bataev A.A.]

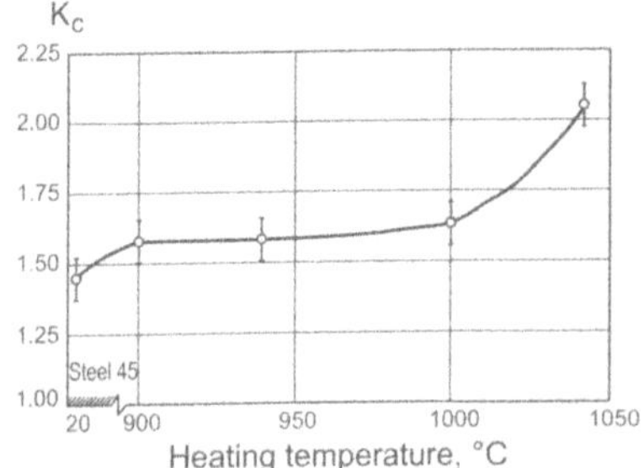

Fig. 6.16. Influence of heating temperature on the relative wear resistance K_c of self-fluxing coating PG–SR4 [courtesy of Poteryaev Yu.P.]

The effect of heating temperature on relative wear resistance of the PG–SR4 coating is illustrated in Fig. 6.16. Studies of the coating surface using Linnik's interferometer after exposure to wear with loose abrasive showed that friction trace profile was different for fused and non-fused coatings. Bottom relief of the mark left by abrasive grain on the non-fused coating surface is an even straight line with rare low "humps" (see Fig. 6.17a). In the case of fused coatings the mark is a strongly bent line with lots of "humps" and "hollows" (see Fig. 6.17b). Comparison of surface interferograms taken with equal magnification after microstructural wear has been completed indicates that carbide phase dimensions probably have the decisive influence on the depth of plastic deformation zones in surface layers in "ploughing through" and shifting with abrasive. For the fused coating the distance between surface interferogram peaks is 20–30 μm which is approximately equal to the distance between carbide phases having the size of 5–10 μm and more. Apparently, structure heterogeneity on such scale level agrees in the order of magnitude with the depth of layers actively deformed by abrasive. Alternations of plasto-elastic deformations of phases with different microhardnesses (from 2360 MPa for Ni-based solid solution to 25000 MPa for chromium carbides) along the zone movement front, and, most important, the presence of large carbide particles hinder displacement of surface layers causing detachment of wear particles, which means enhanced wear resistance.

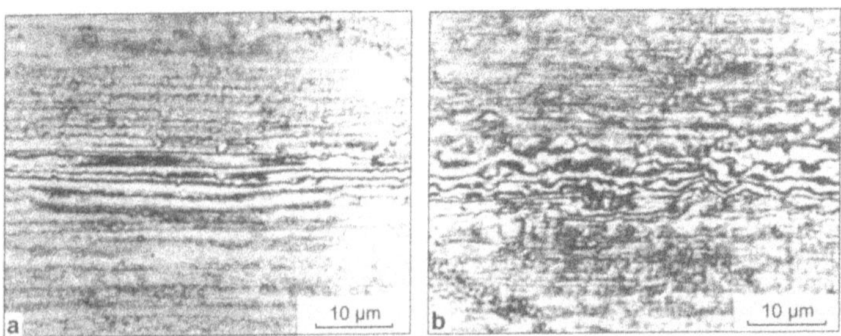

Fig. 6.17. Surface interferogram of the PG–SR4 coating after wearing with loosely fixed abrasive particles. **a** plasma spraying; **b** fusing at 1040 °C for 10 minutes [courtesy of Poteryaev Yu.P.]

In the non-fused coating the structural carbide phase is dispersed with the size up to 1 μm. Probably, fine carbides in Ni-based solid solution matrix do not lead to heterogeneity of mechanical properties over the whole deformation zone under friction and, therefore, do not produce an efficient barrier to abrasive that "plough the coating through". Bottom relief of the mark left by abrasive in the form of straight line on the coating surface is an evidence of uniform plastic flow during wearing of the non-fused coating. Widely spaced interferogram peaks may indicate fall-throughs at the places of particle joints, oxide films chipping, wedging out of microcracks and pores.

In order to enhance wear resistance many researchers tend to avoid plastic material deformation through rise in yield point using one of the dislocation hardening mechanisms. However, thus we get in the fatigue failure area. Now the wear process is controlled by structural fracture toughness (see Chaps. 2 and 8) or by fatigue characteristics (see Chap. 7). Structural studies lead to the conclusion that heterogeneous structure created by the Sharpi's principle (solid inclusions in plastic matrix) is optimum in such conditions. The size of inclusions, the space between them, the mechanical properties of matrix play an important part here. In constructional alloys and coatings the role of the hardening phase is generally played by crystals of carbides, borides or other hard chemical compounds. The degree of matrix hardening should be such as to prevent considerable reduction in fracture toughness. The preferable retardation mechanisms of matrix dislocations are σ_{grain}, $\sigma_{disl\,pc}$, and σ_{ph}. In wear with loose abrasive particles strong solid-solution matrix hardening (σ_{sol}) should be avoided, which is especially important for gas-thermal coatings.

Electric arc surfacing with alloys forming primary carbides and other compounds during crystallization is quite widely used for hardening and restoring machine parts. Among the most popular alloys are carbides of chromium, tungsten, vanadium, molybdenum, titanium, and niobium. High wear resistance combined with fracture toughness can be obtained by surfacing with electrodes forming unstable austenite-carbide and austenite-martensite structures. Laser surfacing is efficient for products of small dimensions and complicated shape as well as for local hardening of the working surfaces of such parts as briquetting presses and moulds, extruder augers and pressing machine screws, parts of mixers and agitators. A welded layer of matrix having austenite-martensite structure with uniformly spread reinforcing carboborides can be produced using highly concentrated energy flux [6.77].

6.5 Impact Abrasive Wear

6.5.1 General Overview

Impact abrasive wear is a mechanical wear produced by dynamic contact of interacting surfaces in the presence of particles with hardness higher than hardnesses of both indenter and coating, located between them. This mechanism of wear is common for working tools of many machines used in oil and mining industry (drilling blast-and bore holes, impact and vibratory impact rock grinding, vibratory loading, etc.), in machine building (riveting, stamping, vibratory impact cleaning, fettling, cutting, etc.), in construction (concrete destruction, ripping up frozen ground, driving piles, etc.), for switching railroad transport.

Impact abrasive wear occurs when hard particles capable of forming hollows at contacting surface which are the traces of their direct dynamic inclusion into the coating hit the surface. A specific surface formed by combination of hollows linked with crosspieces is characteristic of impact dynamic wear. On the surface subjected to impact abrasive wear there is no oriented roughness in the form of scratches and, consequently, no relative movement of solid particles along the surface.

At the heart of the mechanism of impact abrasive wear is direct dynamic inclusion of particles into metal accompanied by deformation which results in failure of metal microvolumes and formation of wear particles (see Fig. 6.18). Being embedded into the wear surface, the solid particle tends to shift the metal of crosspieces by repeated deformation or brittle chipping depending on the hardness. In these conditions of solid particle - wear surface interaction shearing becomes the leading process in the formation of wear particles, and resistance to shearing or breaking off becomes the main criterion for wear resistance [6.77].

In impact abrasive wear mechanism wear resistance of carbon steels depends not only on their hardness, but also on composition and structure of a particular kind of steel. Maximum wear resistance have steels with carbon content about 0.7%. Steels with higher carbon content have poorer wear resistance due to brittle wear. When carbon content is less than 0.7%, steels are subject to plastic deformation and thus wear out at a higher rate. The stronger the impact interaction, the larger effect carbon content has on the wear resistance of steel.

Two trends can be seen in the design of laboratory devices for impact testing. On one hand, a test pattern qualitatively similar to real field conditions is usually strived for. On the other hand, a highly desired feature is unification of the testing procedure so that the obtained results be comparable.

Sorokin [6.67, 6.68] has developed the techniques and designed devices for testing at impact against: loose abrasive layer of definite thickness placed on a metal base; abrasive fixed on a cloth base; monolithic abrasive; abrasive mass; metal surfaces without abrasive. Two of these devices have been adopted as standardized (State Standards 23.207 and 23.212) for impact abrasive wear and impact at low temperatures, respectively.

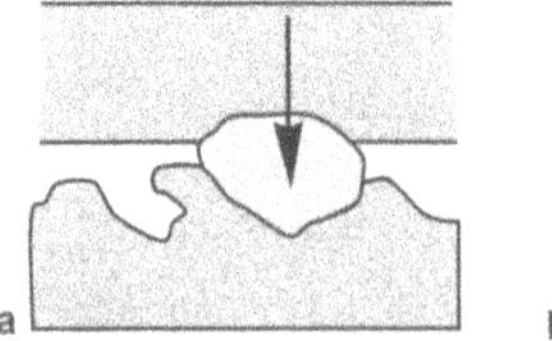

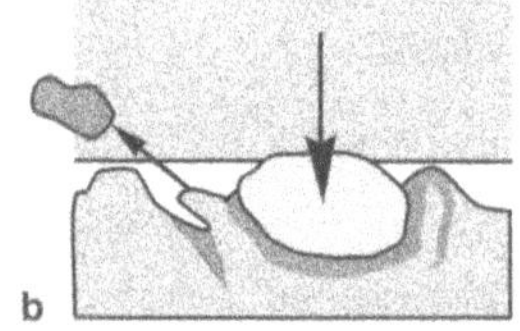

Fig. 6.18. Scheme of impact abrasive wear of materials. **a** plastic; **b** brittle

6.5.2 Specimens

In tests under consideration one of the parts is the specimen, the other is a counter-body (counter-specimen).

For both methods unified cylindrical counter-specimen and reference specimen, both 10 mm in diameter, made of Steel 45 (0.45% C) were chosen.

In impact abrasive wear the separable bush (counter-specimen) passed heat treatment has hardness HV 640–675 (hardening at 840 °C in water, tempering at 100 °C). Hardness of the reference specimens after heat treatment is HV520–622 (hardening at 840 °C in water, tempering at 200 °C).

Counter-specimen for impact wear testing of materials at low temperatures is made of hardened and tempered steel with hardness HV 520–580. For reference specimens annealed steel with hardness HV 190–200 is used.

Standard testing procedures do not cover coatings with hardness less than HV 100, and average pore size more than 0.10 mm, as well as coatings with thickness less than 0.40 mm.

The tested coatings are deposited on the end face of the specimens following the guidelines given in section 6.3.2.

6.5.3 Equipment

In impact abrasive wear (see Fig. 6.19) a cylindrical specimen with a coating deposited on one of its end faces is fixed in the striker. Impact speed can be adjusted by accessory weights, while impact energy is regulated by separable torsions. From the hopper with open valve abrasive material is fed through the box-shaped directing channel to the impact zone between the specimen and the separable bush fixed on the anvil. Impact speed should be in the range 0.5–5.0 m/s, collision frequency 20–400 hits per minute. The stationary anvil weighs 50 kg or more.

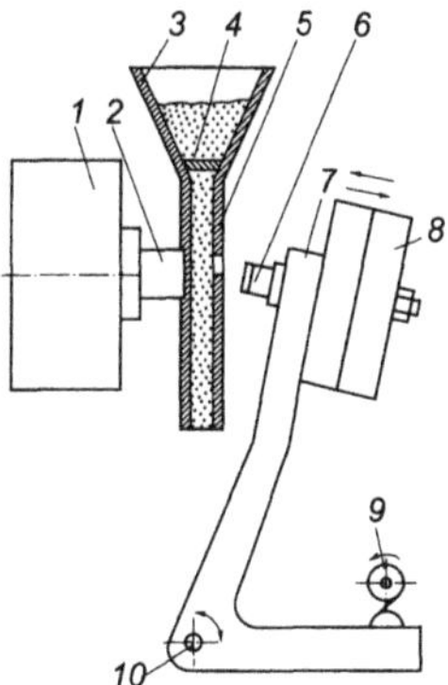

Fig. 6.19. Simplified diagram of installation for impact abrasive wear testing. *1* anvil; *2* separable bush; *3* hopper; *4* valve; *5* directing channel; *6* specimen; *7* striker; *8* removable weights; *9* torsion bar with cam; *10* axis

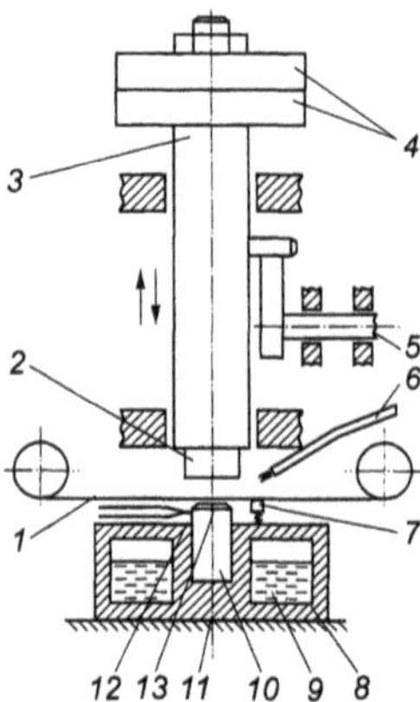

Fig. 6.20. Simplified diagram for impact wear testing at low temperatures. *1* abrasive band; *2* counter-specimen; *3* striker; *4* accessory weights; *5* drive; *6* mechanism for removing wear products; *7* gear for swinging the abrasive band off the specimen; *8* cooling chamber; *9* cooling medium; *10* specimen; *11* frame; *12* thermal element

The device for impact testing at low temperatures (see Fig. 6.20) is capable of wear testing metal coatings subjected to repeated impacts against an abrasive or metal coating at low temperatures determined by particular operating conditions. Counter-specimen lifted at 25±5 mm with a driver strikes the specimen cooled to the desired temperature. Impact energy is adjusted with accessory weights placed on the striker. The abrasive band is periodically moved, a special gear swings the band off the specimen during its movement, and wear products are removed by a compressed air jet dispensed from the nozzle. The temperature of the cooled specimen is adjusted by the signal coming from the thermal element through the automatic temperature control unit which determines the volume of cooling medium dispensed from the cooling chamber fitted on the frame. Collision frequency is 60 hits per minute, specimen temperature from 173 to 293 K.

6.5.4 Preparation of Tests and Testing

The density, hardness, porosity and thickness of the coating is determined prior to testing using specimens-witnesses. The obtained values of these parameters should meet the requirements described above.

Both tested and reference specimens are run-in before testing.

The following testing conditions are common for both methods: fixing of the specimens, maintenance of stable dynamic loadings, testing duration schedule.

Specimens should be fixed so that both their axes and the axes of the counter specimens coincided with the direction of impact and were perpendicular to the surface of the coating. Impact energy is picked from the following series depending on the specific wearing conditions: 0.25, 0.49, 0.60, 1.50, 2.50 Joules. The number of strikes is 1000, 2000, or 4000 depending on operating conditions of the coating.

The relative wear resistance of the coating is calculated from equation 6.11, where N_r and N_c are the number of strikes for the reference and coated specimens, respectively.

6.5.5 Results and Discussion

Sorokin has demonstrated [6.67, 6.68] that the mechanism of failure in impact abrasive wear, besides resistance to shearing of the material, is determined by a great number of various factors such as impact energy, physic-mechanical properties of the abrasive, composition of the tested material, degree fixture of abrasive particles, etc. The conventional characteristics of strength and plasticity (yield point, breaking point, hardness, percent elongation, reduction of area, impact strength) have ambiguous effect on wear resistance in impact abrasive wear. Enhancement of strength and plasticity is favorable only up to a certain threshold level. Further enhancement of these parameters leads to wear increase, though the reasons of wear resistance enhancement vary. If strength enhancement is accompanied by viscous-brittle transition rising, then the degree of wear grows because of intensification of brittle chipping. Considerable plasticity enhancement causes reduction in wear resistance because of active plastic flow and accompanying cold hardening. Apparently, the maximum wear resistance is attained for alloys at the boundary between brittle and viscous failure.

Conventional methods of steel surface hardening (carbonization and nitration) are not applicable for the parts operating in the conditions of impact abrasive wear. Thin surface hardened layers having high brittleness are rapidly chipped off under impact loading [6.63], the cracks being initiated not only on the surface but also along the boundary separating the hardened layer from the base metal. The advantages of hard alloys which have shown a good performance in abrasive wear have not been revealed. In certain loading conditions their wear resistance turned out to be lower than that of Steel 45 (0.45% C) [6.67].

The effect of Fe–C–Cr–W–Ti coating-surfacing on impact abrasive wear resistance was studied at impact energies 5–10 Joules [6.67]. A layer of hard alloys with thickness 7–8 mm and hardness from 35 up to 62 HRC was deposited on the end faces of cylindrical specimens. Silicon carbide was used as abrasive. Coating wear resistance was evaluated by the weighing method taking into account different specific densities of the tested materials.

The results of testing numerous surfacings of various composition and properties have shown that wear rate of all surfacings except two ones was higher than wear rate of the reference specimen made of Steel 45 (0.45% C) after hardening and low tempering. It was found that the effect of hardness of the surfacing on wear resistance is not straightforward, i.e. wear resistance is maximum at HV 4500 MPa, whereas further hardness increase is accompanied by reduction in wear resistance. The pattern of failure also changes: while at low values of hardness detachment of particles takes place after considerable plastic flow (the zone of viscous failure), at values above HV 4500 MPa brittle chipping without noticeable plastic deformation (the zone of brittle failure) has been observed.

Consideration of general wear patterns combined with simultaneous analysis of structural changes has led to a definite conclusion: material resistance to direct action of abrasive grains under impact conditions shows obvious dependence only on shearing resistance, and the growth of the latter causes decrease in wear.

Analysis of published data and our own results demonstrate low efficiency of material strength enhancement in impact abrasive wear using retardation of dislocations by σ_{pn} and σ_{sol} mechanisms, which increase brittleness. Doping of alloy may be used to alter the width of the packing defect. Hardening by σ_{grain} and $\sigma_{disl\,pc}$ mechanisms as well as moderate hardening by σ_{ph} that do not aggravate the contradiction between the yield point and the fracture toughness is quite useful. Heterophase micro-scale structures (by Sharpy's principle) cannot compete with single-phase structures, since local action of abrasive particles causes strong but brittle phase to fail, though the limit of energy dissipation through plastic deformation is not reached in the surrounding viscous phase. Hard but brittle phases can exist only in the form of finely dispersed, preferably globular, inclusions. The main disadvantages of gas-thermal coatings restricting their application for enhancement of impact abrasive wear resistance of alloys are low plasticity and low bond strength between particles caused by defects (by the presence of oxide films, pores, cracks). Testing at room temperature has shown that impact abrasive wear of plasma-sprayed coatings made of PN85U15 and PG–SR4 powders was accompanied by brittle failure of particles into fragments or by detachment of the whole particles. Wear of carbon and chromium-nickel austenite steels was less intense as compared to plasma-sprayed coatings. Resistance of fused PG–SR4 and PN85U15 coatings was close to that of steels despite the fact that the coating also failed by brittle chipping from the surface.

When using instable austenite–carbide and austenite–martensite alloys as a coating material high crack resistance is provided by the presence of austenite as a tough structural component, while high wear resistance is achieved due to ability of the material to undergo transition to martensite under mechanical impact. [6.77].

High resistance of wear resistant high-manganese 110G3-type steel (composed of 1.10% C and 13% Mn) to impact abrasive wear is caused by crushing of the austenite grains into very small-sized blocks and blocking slip planes by carbides. Ability of the 110G3 steel to phase transformation as a result of impact was also noted. All steels described above consist of dispersed solid particles in viscous solid solution. However, alloys with relatively rough initial crystals of solid compounds are also applied to harden parts working in the conditions of impact abrasive wear. Specially designed electrodes that provide a surfacing structure containing chromium boride crystals carburized by eutectics were successfully applied to harden hammer mill beaters, bucket teeth of stripping excavating machines and other heavily loaded parts. Good results have also been obtained in hardening working surfaces of rock roller bits, blast furnace loading devices, etc. using grains of relite, melted tungsten carbide. The size of grains reaches 200–300 μm.

The base for the hard surfacing layer is provided by self-fluxing alloys or low-carbon steel [6.77].

6.6 Wear With Abrasive Particles Located in the Gap Between Adjoint Parts

6.6.1 General Overview

Abrasive particles find their way into the gap of friction pairs together with oil and fuel, or because of unsatisfactory sealing.

Having got into the gap, abrasive particles take up a part of the load applied to the parts and depending on particular conditions may be pressed into the friction surface, crushed into finer fragments, slip or roll over the surface subjected to wear, causing its elastic or plastic deformation (see Fig. 6.21) [6.77]. The mechanism of surface failure is similar to the processes of wear with fixed and loose abrasive particles under heavy loads. The main cause of the failure is fatigue because of contact loading through forcing abrasive into the material followed by displacement of the latter by scratching.

The wear of the friction pair is increased as the hardness of abrasive particles grows. The harder the abrasive particle, the deeper it is embedded into the material subjected to wear and, consequently, the higher can be wear rate.

The effect of specific load P on the state of the particle in this wear type is not straightforward. The grain is destroyed for $P > P_{cr}$. The critical load P_{cr} can be calculated from the following equation [6.24]:

$$P_{cr} = yd^x, \tag{6.12}$$

where y and x are constants, d is the size of the grain.

Crushing of abrasive grain intensifies wear process causing a sort of micro-strike that leads to formation of crater-shaped recesses into which the fragments of abrasive are stuck.

As hardness of metal surfaces subjected to wear grows, their wear resistance generally increases. In the absence of charging effect the harder material is less wearable. However, when one surface is charged, the wear rate of another one will increase, even if this surface is harder [6.24].

The parts working in the conditions of sliding or rolling friction are subjected to wear with abrasive oil interlayer. This class of parts includes open plain bearings, reduction gears with clogged lubricant, machine guides, open transmission gears, cylinders and pistons of internal combustion engines operated in polluted air.

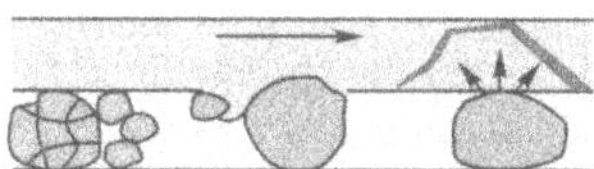

Fig. 6.21. Abrasive particles located in the gap between adjoint parts. **a** crushing into finer fragments; **b** sliding; **c** rolling over

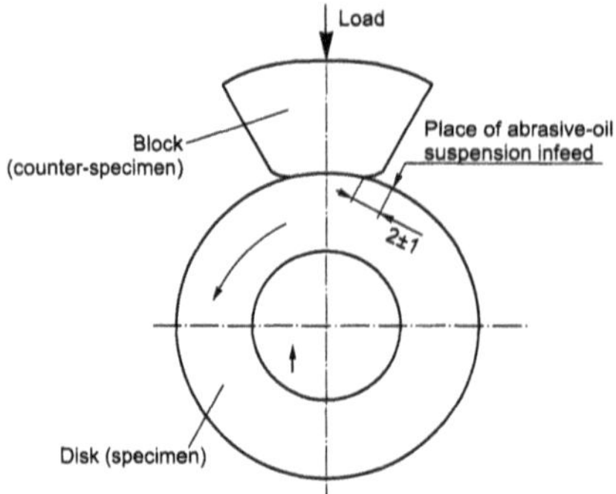

Fig. 6.22. Scheme of wear testing with abrasive-oil interlayer

The essence of testing consists in wearing tested and reference materials against counter-specimen with simultaneous metered feeding of abrasive-oil suspension to friction zone (see Fig. 6.22).

6.6.2 Specimens and Equipment

Reference specimen (disk) is made of annealed steel 45 (0.45% C) with hardness HV 190–220. Diameter and width of the disk are 40 mm and 12 mm, respectively. Counter-specimen (block) is made of thermally treated Steel 45 with hardness HV 520–580. Outer diameter of the block is 88 mm, its width is 10 mm.

Quartz sand is usually used as abrasive material.

This method does not cover materials with hardness less than HV 20, porous materials with average pore size lower than 0.1 mm, and coatings with thickness 0.05 mm and smaller.

When testing specimens with thickness of coatings as low as 0.15 mm, specimens and counter-specimens should be carefully chosen to provide the required adjacency tightness.

It is recommended to use commercial friction machines fitted with an automatically adjusted metering device. The metering device (see Fig. 6.23) is designed so that, in order to provide a measured injection of abrasive-oil suspension, a current pulse is applied to the coil on the moving part of electromagnet. The keeper overcomes the springs and, being attracted to the stationary magnet part, raises valves thus opening the holes at the expansion piece to feed the suspension.

The length of the current pulse and, consequently, one-time feed of abrasive-oil suspension is set using time relay. In the absence of current the springs force the valves to block the holes of the expansion piece.

Operation of a simplified metering device feeding suspension using a similar scheme should be clear from Fig. 6.24.

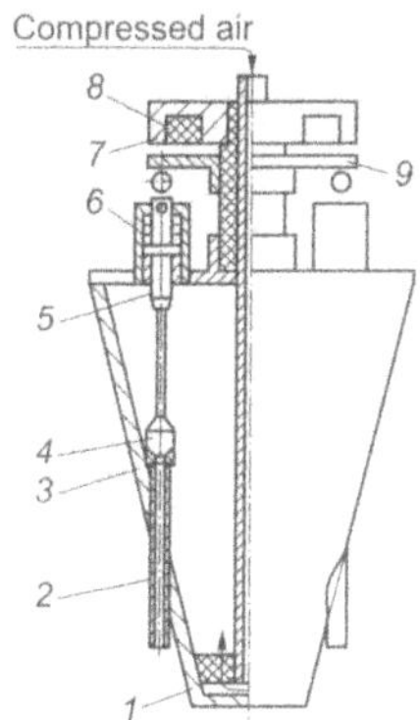

Fig. 6.23. Automatic measuring device for feeding abrasive-oil suspension to the friction zone. *1* housing; *2* pipe; *3* extension piece; *4* valve; *5* pusher; *6* spring; *7* stationary part of electromagnet; *8* current coil; *9* mobile keeper

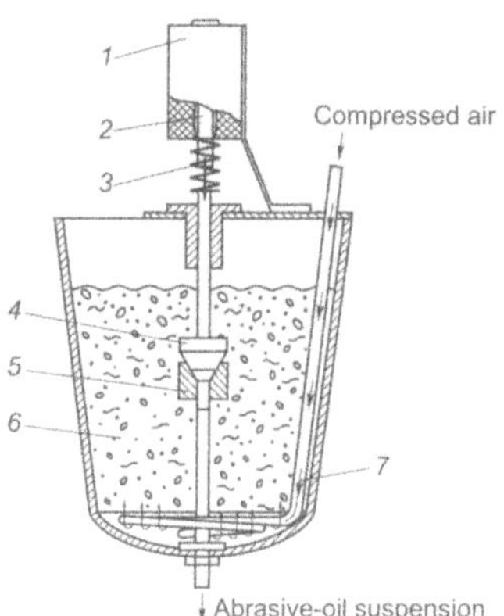

Fig. 6.24. Metering device for wear testing with abrasive-oil interlayer. *1* electromagnet; *2* core of electromagnet; *3* opposing spring; *4* valve; *5* expansion piece; *6* abrasive-oil suspension; *7* pipe

The abrasive-oil mixture can also be dispensed to the friction zone using another device (see Fig. 6.25). Abrasive-oil suspension with the given concentration of particles is placed in reservoir connected to the sucking and pressure lines of the closed-looped main. A pump connected in series transfers suspension along the main, the pressure part of which is fitted with an electrically driven valve. When electric voltage is applied to the winding, the valve opens and passes a portion of suspension to the friction zone of specimens fixed in the testing machine.

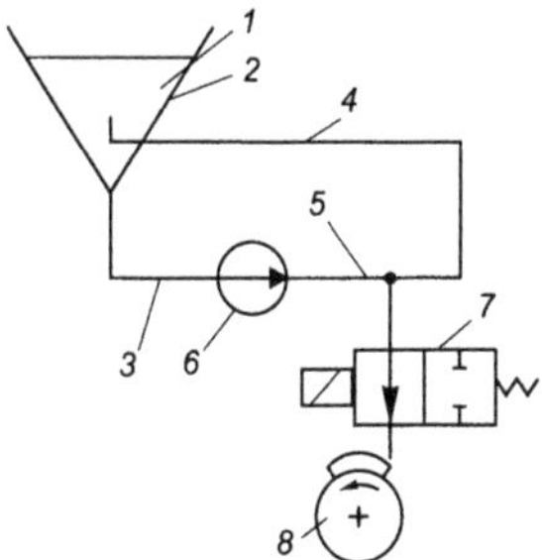

Fig. 6.25. Hydraulic device for producing abrasive-oil suspension and dispensing it to the friction zone. *1* abrasive-oil suspension; *2* reservoir; *3* sucking line; *4* pressure line; *5* main; *6* pump; *7* electrically driven valve; *8* specimens

6.6.3 Preparation of Tests and Testing

Specimens and counter-specimens are run-in, washed, and their initial mass is determined. The tests are carried out for time interval corresponding to friction path 2500 m for materials with hardness HV 800 and lower, and to the path 5000 m for materials with hardness above HV 800 (according to regulation RD50–339). Coating continuity over the friction surface should be preserved after the test has been completed.

Relative wear resistance (K_c) of the tested coating can be calculated from the following equation:

$$K_c = \frac{g_r \cdot \rho_c \cdot S_c}{g_c \cdot \rho_r \cdot S_r} \tag{6.13}$$

where g_r, g_c are the mass losses of the reference material and the coating, respectively, g; ρ_r, ρ_c are the densities of reference material and the coating, respectively, g/cm^3; S_c, S_r are the friction paths for the reference material and the coating, respectively, m.

6.6.4 Results and Discussion

Testing fused and non-fused coatings of self-fluxing PG–SR4 alloys (composed of 3.0% B, 0.8% C, 3.9% Si, 16% Cr, and 72.3% Ni) using special equipment has proved that their resistance to wear with abrasive-oil interlayer over a wide range of loads and speeds is lower than the resistance of Steel 45 (0.45% C). This can be probably accounted for by the following two circumstances:

- low fracture toughness K_{ic} of the coating; the main hardening mechanisms providing high yield point and hardness of the phases making up the coating are σ_{pn} and σ_{sol} that strongly reduce index K_{ic};
- defectiveness of the coating, i.e. the presence of pores, microcracks, oxide films, unstable bonds between particles. Fusing PG–SR4 coating reduces its failure intensity by a factor of 1.2–1.8 (see Fig. 6.26).

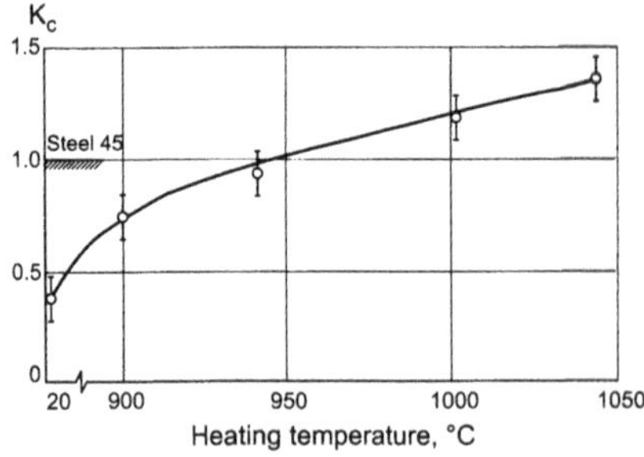

Fig. 6.26. Dependence of relative wear resistance of self-fluxing PG–SR4 coating on heating temperature in wear with abrasive-oil interlayer [courtesy of Poteryaev Yu.P.]

Like in many other types of abrasive wear, in this case high wear resistance is achieved by optimal combination of hardening which hinders plastic deformation, and sufficient fracture toughness preventing development of brittle cracking in the surface layers. Therefore, the most preferable mechanisms of dislocation hardening are σ_{grain}, $\sigma_{disl\,pc}$, or σ_{ph} that do not aggravate the existing contradiction between the yield point and the crack resistance of the alloy.

Abrasive wear resistance of "shaft"-type parts working in the conditions of boundary friction with lubricant is enhanced when the distance between inclusions of the dispersed phase rules out penetration of abrasive particles into the matrix. Taking into account possible dimensions of abrasive particles produced as a result of cold hardening, ash particles formed in fuel combustion, etc., the distribution of hardening solid phase in the matrix of a wear resistant coating should be rather dispersed. When reconstructing and hardening parts with powder self-fluxing coatings, crushing of the solid phase (i.e. carboborides and thin eutectics zones) is efficiently achieved by superimposing ultrasonic oscillations during plasma spraying. Extremely dispersed structure can be produced using continuous wave CO_2–laser for thermal treatment of the deposited gas-thermal coating. [6.77].

When describing parts working in a friction pair, the parts embracing "shafts" are collectively referred to as "bushes". This term covers actual bushes of various types and purposes, plain bearings and other kinds of supports, and also guiding elements of different gears, frames and similar mechanisms, cylindrical bushes of internal combustion engines, extruder sleeves, valve seats, current collector brushes. When lubricants are used, the main condition determining wear resistance of "bushes" is the presence of oil trapping microrelief on the working surface. This can be a structure produced by Sharpy's principle, i.e. hard inclusions in a plastic matrix.

In the production of guides, gears and machines these requirements are met for cast iron structure including initial carbides and graphite. Such surfaces are restored and hardened by arc surfacing using stellite type materials, as well as stellite-like alloys on the basis of Fe-Ni solid solution (sormite type) and boride surfacing alloys. For cylindrical bushes of internal combustion engines the structure must include cementite plates uniformly distributed in pearlite, and uniformly distributed lamellar graphite. Structurally free ferrite and cementite must be avoided. Powder metallurgy provides ample technological possibilities for production of composite antifriction matrix-filled metal-based materials with regularly alternating wear resistant and antifriction regions. The indicated technological potentials of powder metallurgy can also be realized using gas-thermal deposition of powder alloys with or without fusing of the produced coatings [6.77].

6.7 Gas Abrasive Wear

6.7.1 General Overview

Gas abrasive wear (erosion) is a result of surface interaction surface with solid particles suspended in gas and moving past a body subjected to wear. Gas abrasive wear results in failure of gas turbine blades, parts of electric power station equipment working on solid fuel, sand slinger barrels, many parts of gas pipeline and gas pumping equipment.

Wear with abrasive particles in gas flow is a complicated process which depends not only on physicomechanical properties of materials, but also on the angle of incidence, speed of abrasive particles at the instant of impact against the part surface, ratio of the material and abrasive hardness values, concentration of abrasive particles in the medium, etc. [6.16, 6.45, 6.50].

It is convenient to distinguish two cases of interacting between the flow of abrasive particles and material: straight impact (angle of incidence $\alpha = 90°$) and oblique impact (angle of incidence $0 < \alpha < 90°$) (see Fig. 6.27).

The first case results in such types of wear as elastic deformation, plastic deformation, wear hardening with cracking and detachment of separate scales, and brittle failure depending on physicomechanical properties of materials and intensity of impact [6.90]. Since the impact energy of each abrasive particle is small, high wear resistance at $\alpha = 90°$ is typical for elastic materials (e.g. rubber) capable of reducing stress in the contact spot and damping the impact energy.

For angles of incidence in the range 0°–90° the pattern of surface damage is strongly influenced by the tangential component of the impulse and material resistance to the action of stresses tangential to its surface. As the angle of incidence decreases, cutting with abrasive particles becomes dominant. The failure proceeds similar to wear with loosely fixed abrasive particles. The dynamic nature of loading increases the probability of either brittle or fatigue material failure.

Gas-abrasive wear is substantially more complicated process as compared to wear against fixed or loose abrasive. First of all, the angle of incidence plays an important role [6.75]. Sorokin argues that none of the known mechanical characteristics can be considered a criterion for wear resistance when the angle of incidence and flight speed of the particles change [6.68].

At present there is no accepted opinion regarding the influence of the size and mass of abrasive particles on wear rate. One of the reasons of the data discrepancy is that during testing a considerable lag of peripheral particles behind the base flow is not taken into account [6.24]. Particles lag behind the flow in proportion to their masses and dimensions. Increasing the concentration of abrasive in the flow leads to monotonous increase of the wear rate up to a certain limit, and at larger concentrations wear rate falls off because of the shielding action of rebounding particles. Additions of surfactants generally weaken both abrasive and tested material which results in increased wear rate.

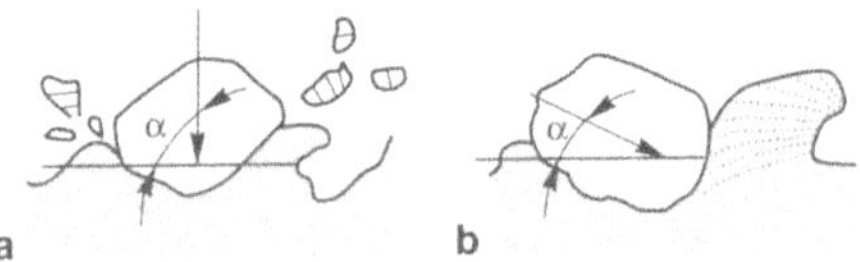

Fig. 6.27. Interaction between abrasive particles and the surface at different angles of incidence α. **a** right angle; **b** oblique angle

Metal erosion has been addressed in a sufficient number of works, while only few investigations are devoted to studying wear resistance of coatings. However, it appears to be impossible to evaluate wear resistance of a hardened surface in abrasive flow without conducting appropriate experiments. Therefore, it is recommended to use a standardized testing method based on simultaneous effect of gas-abrasive flow on tested and reference specimens in fixed testing modes (State Standard 23.201).

6.7.2 Specimens

Reference specimens in the form of plates with dimensions 20x15x4mm with working surface roughness Rz = 0.16–0.32 μm without sharp edges are made from annealed Steel 45 with hardness HV 185–195.

The coating is deposited on the wide face of the specimen following the guidelines given in section 6.3.2.

Standard testing procedures do not cover materials and coatings with hardness less than HV 20, porous materials with average pore size more than 0.02 mm, as well as coatings with thickness less than 0.30 mm.

6.7.3 Equipment

Equipment for gas-abrasive wear testing materials can be divided into three groups: employing free fall of abrasive particles; acceleration of free particles with gas flow; mechanical acceleration of abrasive material (see Fig. 6.28).

Devices using free fall of abrasive in the air medium find only limited application, since they are applicable only for low collision speeds. In vacuum installation collision speed up to 1.9–8.5 m/s can be obtained [6.67].

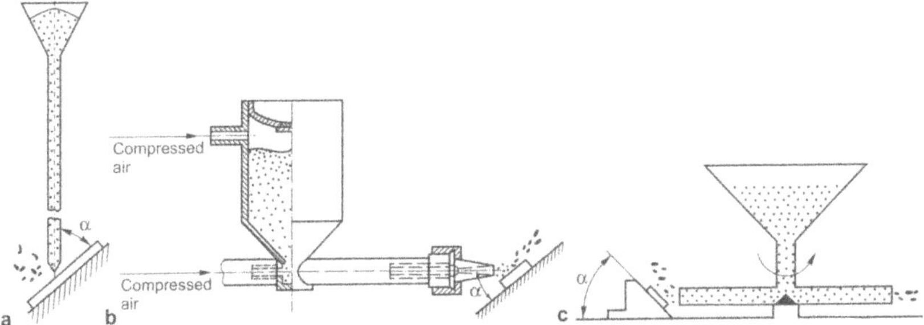

Fig. 6.28. Installations for gas abrasive testing of materials (α is angle of incidence). **a** with free fall of abrasive; **b** with acceleration of abrasive particles by gas flow; **c** with mechanical acceleration of abrasive

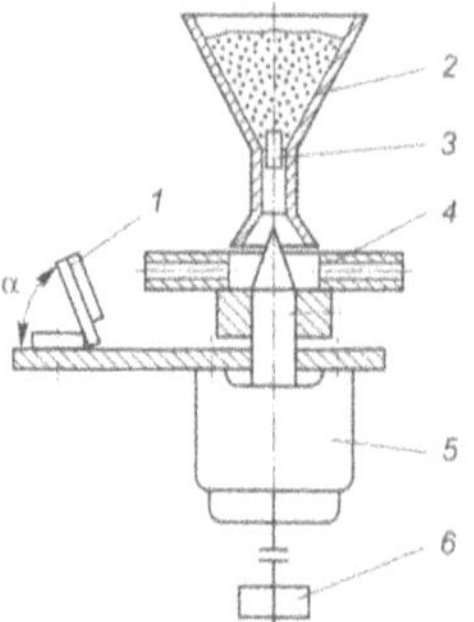

Fig. 6.29. Installation for gas abrasive wear. *1* specimen; *2* metering hopper; *3* valve; *4* rotor; *5* electric motor; *6* rotation speed sensor

Devices using pneumatic feed can be injection- or discharge-type (the latter is shown in the diagram). Injection-type devices cannot provide high speeds of abrasive particles and therefore are seldom applied. With discharge-type devices, besides the effect of abrasive particles, the influence of aggressive medium and temperature on wear resistance can also be evaluated.

In the devices with mechanical acceleration abrasive grains are accelerated by centrifugal forces. Either gas or vacuum can be the medium where particles move. Such devices are applied fairly often. Five modifications of this type of installation have been created in Talinn Polytechnical Institute. The devices differ by the rotor diameter (40–600 mm), range of rotor rotation speeds (15–350 m/s), type of adjusting rotation speeds (stepwise, continuous), possibility of evacuation [6.20].

The installation with centrifugal acceleration of the jet is typical of this group. Here the conditions of collision between abrasive mixture and a specimen can be controlled over a wide range which makes possible simultaneous testing of several specimens at various angles of incidence. The device is easy to assemble and operate.

The device (see Fig. 6.29) consists of a measuring unit hopper containing abrasive mass and a rotor fitted with four channels that serve as abrasive jet outlets. From the open valve of the hopper abrasive mass is dispensed to the channels of the rotor revolving at the given angular velocity using an electric motor connected with a rotational speed sensor. Under the action of centrifugal forces the jet of abrasive emerges from the rotor channel and hits tested and reference specimens. The specimens are fixed in the specimen holders at the angle of incidence α adjusted by inclining the working surface with respect to horizontal within 15–90°. At least six specimens (three tested and three reference) are tested simultaneously at equal angles of incidence. Adjustment of rotational speed provides collision speed of abrasive flow in the range 38–76 m/s. Spent abrasive is collected into a hopper from which it is drawn off with a vacuum cleaner. Continuous adjustment of rotational speed is also provided. The device is standardized [State Standard 23.201]. Quartz sand with grain size 0.5–0.9 mm and maximum relative moisture content 0.15% is used as abrasive material. The standard allows utilizing abrasive material that corresponds to the abrasive found in field conditions, but with maximum grain size 1.2 mm.

Testing conditions in a standard installation are given below:

Specimen dimensions [mm]	20x15x4
Rotor rotational speed [min^{-1}]	3000 and 6000
Maximum number of tested specimens	15
Space between the center of the working surface of a specimen and rotor axis [mm]	118±1
Channel cross-section [mm]	3x7
Channel length [mm]	82±0.4
Diameter of hopper output channel [mm]	6±0.2
Hopper taper angle [degree]	90
Angles of incidence [degree]	15, 30, 60, 90

Standard techniques of gas-abrasive wear testing have the following disadvantages. Firstly, the amount of abrasive interacting with the surface subjected to wear must be calculated and cannot be determined directly. This introduces a systematic error in measured values of relative wear resistance and wear rate. Secondly, an abrasive jet is non-homogeneous over its cross-section, which aggravates measurement error, especially in wear of particularly thin coatings.

Vinnitskaya et al. [6.87] have designed and built an installation in which these drawbacks have been overcome. To ensure homogeneity of the wearing field the construction is fitted with an additional unit which provides rotor scanning along the rotation axis. A diaphragm mounted between the rotor and the specimen extract the part of the jet with the most uniform flow of particles. Special collectors are added to determine the amount of abrasive actually interacted with the specimen.

A principally new type of testing installation was suggested in [6.28]. Plasma- and gas-plasma sprayed coatings were tested in humid atmosphere inside a chamber filled with particles of Al_2O_3. The specimens were moved along a circle with radius 188 mm at a rotational speed 10 rpm. Simultaneously the specimens were spun around their axes at a speed 57 rpm. Maximum linear speed of the specimens reached 24.47 m/s.

In the majority of works the tests were conducted at room temperature, however, experiments carried out at low (125, 136K) [6.77] and high temperatures have also been reported. Thus, Eaton and Novak [6.26] investigated gas-abrasive wear of plasma-sprayed coatings with the following flow characteristics: Al_2O_3 abrasive 27 μm in size, temperature 1560 K, angle of incidence 15°, particle flow speed 244 m/s. In another work [6.62] high-temperature corrosion resistance test was conducted in the conditions close to typical found in operation of gas-turbine engine compressor, namely: air-abrasive jet speed 250 m/s, jet temperature 370 °C, angle of incidence on the coating surface (5–10)°, abrasive concentration up to 5 g/m^3. Rutile powder with particles 10–15 μm in size was used as abrasive.

A device for high-temperature erosion tests designed in E.O. Paton Electric Welding Institute (Kiev) proved to be very efficient [6.84]. Compressed air is heated up to 830 °C by natural gas. Gas jet carries along abrasive dispensed from a metering unit. Abrasive particles are accelerated, heated and after passing through a nozzle hit the specimen. Up to six specimens tested in turn can be placed inside the working chamber.

In certain cases gas flow not only has an elevated temperature but also contains aggressive components which worsens wear conditions. Among the installations for testing in such aggressive conditions a device made in the Central Scientific Research Institute of Heavy Industry (Russia) is worth paying close attention. The testing device reproduces the working conditions of the gas-turbine blade material including temperature, stressed state and wear with hot gas flow carrying abrasive particles. Blades having different lengths and thus different values of stretching stresses are used as specimens [6.84].

To study high-temperature erosion of coatings deposited on such a specific object as gun barrel surface, Schlett [6.66] applied two types of explosive substances differing by ther thermochemical properties.

It can be concluded that modern testing equipment is capable of satisfactory modeling wear processes in gas-abrasive flow at different temperatures [6.64, 6.69]. Testing conditions range as wide as 50–675 μm for particle size, 2200–4000 kg/m^3 for density of abrasive, 10–109 m/s for collision speed, and 4600–11700 MPa for hardness [6.85].

6.7.4 Preparation for Testing

In standard tests using centrifugal accelerator the specimens are run-in before the test. Mounting of the specimen in the holder must leave an open area of working surface 12±0.1 mm wide for interaction with abrasive material.

Rotor rotation speed (n) is set to either 3000 or 6000 rpm providing flow speed of abrasive particles V equal to 38 and 76 m/s, respectively. The two parameters are related by the following equation: $n = 78.2V$.

Relative wear resistance (K_c) of the tested coating can be calculated from equation (6.11), where N_r and N_c are numbers of rotor revolutions for the reference material and for the coating, respectively.

6.7.5 Results and Discussion

Analysis of the results of tests conducted using the installation described in [6.82] (see Fig. 6.29) showed that resistance of plasma-sprayed coatings in the gas abrasive conditions is generally low, and sometimes even lower as compared to resistance of carbon steels 45 (0.45% C) and Y8 (0.8% C). Intermetallic PN85U15 and PN70U30 plasma-sprayed coatings (composed of 15(30)% Al and 85(70)% Ni, respectively); PN55T45 and PN65T35 intermetallic plasma-sprayed coatings (composed of 45(35)% Ti and 55(65)% Ni, respectively); PG–SR4 self-fluxing coating (composed of 3% B, 0.8% C, 3.9% Si, 16.0% Cr, 4.0% Fe, 72.3% Ni) were studied. River sand with average size of particles 0.7 mm was used as abrasive. Abrasive consumption during testing was $4.2 \cdot 10^{-3}$ kg/s, angle of incidence of abrasive particles was 30, 60 or 90°, and rotational speed of the rotor was 3000 rpm.

In testing nickel-titanium coatings at the angles of incidence equal to 30, 60 or 90° the effect of preliminary surface finishing is seen only during the run-in period. Henceforth the wear rates of polished and non-polished surfaces are nearly equal despite the net of cracks developing in the course of polishing (see Fig. 6.30). Apparently, impact energy of each separate abrasive particle is not sufficient to detach large pieces into which the coating is cut by the net of cracks.

Self-fluxing coatings tend to wear out less actively than intermetallic coatings. In certain testing conditions the wear rate of self-fluxing coatings is half of the rate for Steel 45. Introduction of chromium carbide Cr_3C_2 in self-fluxing coating powders prior to spraying in some cases leads to increased wear rate (see Table 6.2). This phenomenon can be accounted for by the fact that the brittle component of the alloys is susceptible to chipping under the action of abrasive particles. Increased wear rate growth also favors a coarser structure of self-fluxing alloys that occurs as a result of material overheating in fusing. Wear rate of plasma-sprayed coatings is basically 15–25% higher than the rate for gas-plasma sprayed coatings.

Failure mechanisms of Steel 45 (0.45% C) and plasma-sprayed coatings in identical conditions of gas abrasive wear are substantially different. This can be explained by peculiarities of internal structure of sprayed coatings, in particular, by the presence of pores and cavities.

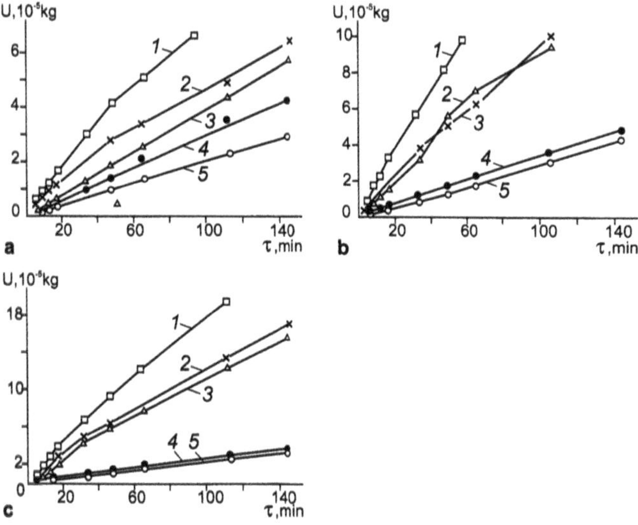

Fig. 6.30. Wear curves of PN55T45, PN65T35 plasma sprayed coatings and carbon steels in gas abrasive medium. Angle of incidence: **a** α = 30°; **b** 60°; **c** 90°. *1* non-polished PN65T35 coating; *2* non-polished PN55T45 coating; *3* polished PN55T45 coating; *4* Steel 45; *5* Steel Y8 [courtesy of Bataev A.A.]

Table 6.2. Wear rate of self-fluxing PG–SR4 coating in gas abrasive medium, $\cdot 10^{-8}$ kg/s

Deposition technique	Angle of incidence [degrees]	Amount of chromium carbides [weighing %]			
		0	20	40	60
Gas-plasma spraying	90	0.38	0.29	0.42	0.52
	60	0.38	0.42	0.40	0.43
	30	0.22	0.21	0.23	0.31
Plasma spraying	90	0.33	0.30	0.30	0.38
	60	0.32	0.33	0.28	0.32
	30	0.18	0.16	0.16	0.19

Porosity of the PN70U30, PN85U15, PN55T45, PN65T35 coatings measured by metallographic method is equal to 11±3%, 20±3%, 9±2%, and 24±5%, respectively. The pores in materials are located uniformly over the entire deposited layer. In such coatings as PN70U30, PN55T45 and PN65T35 the majority of pores has the size of 1–10 μm, while in PN85U15 it is equal to 5–60 μm. The influence of the defects on the wear pattern can be considered from two viewpoints [6.82].

Firstly, pores and cavities have the ability to considerably deteriorate strength properties of a coating by decreasing the "live" cross-section of a sprayed layer and acting as concentrators of local stresses.

Secondly, microrelief features caused by the presence of large surface defects change the wear conditions of the specimen. In each specific point the angle of incidence of abrasive is determined by the sizes and shapes of pores and cavities. As a rule, coating wear conditions in the vicinity of defect zones are harder as compared to flat areas without defects. These factors make possible intense pricking out of large pieces of coating which have considerable volume share of large pores and cavities.

Maximal wear rate of specimens made of steel 45 (0.45%C) with hardness HRB 91 has been obtained in tests with angle of incidence equal to 60° (see Table 6.3). At the same time the tested coatings wear out more actively when a flow of particles is incident on the specimen at an angle equal to 90°.

It should be noted that changing the angle of incidence from 60° to 30° causes reduction in wear rate of Steel 45 by 10%, while wear rate of coatings is reduced by a factor of 1.5–3. These results demonstrate that at the angles of incidence less than 60° coating can be used with great efficiency.

Table 6.3. Wear rate of various materials in gas abrasive medium, 10^{-8}, kg/s

Coating material	Angle of incidence [degrees]		
	90	60	30
Steel 45	0.43±0.03	0.55±0.03	0.50±0.02
PN55T45	1.78±0.12	1.53±0.06	0.63±0.03
PN65T35	2.63±0.18	2.47±0.20	0.88±0.05
PN85U15	11.62±1.12	11.47±1.01	4.27±0.31
PN70U30	0.92±0.08	0.92±0.05	0.65±0.04
PG-SR4	0.55±0.02	0.53±0.04	0.30±0.02

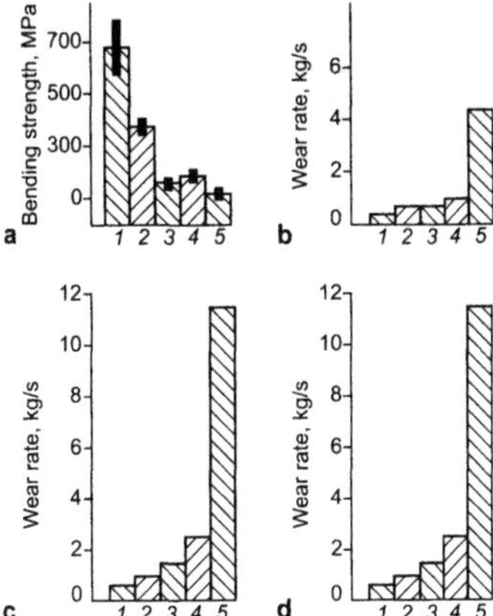

Fig. 6.31. Relation of coating strength properties at bending (**a**) to the rate of wear in gas abrasive medium at angles of incidence $\alpha = 30°$(**b**), 60°(**c**), 90°(**d**): *1* PG–SR4; *2* PN30U30; *3* PN55T45; *4* PN65T35; *5* PN85U15 (dark zones are confidence intervals)

The observed wear patterns are explained by higher hardnesses of the tested coatings as compared to Steel 45. In testing coatings at large angles of incidence harder materials, which the coatings are, need less time as compared to steel for reaching the limit of plasticity and creating a cold-hardened layer prone to failing (i.e. for realization of the basic wear mechanism in these testing conditions). At small angles, when microcutting becomes the basic wear mechanism, hard coatings may turn to be more resistant than steel. However, high porosity and low cohesion strength result in a 10 times higher wear rate of the PN85U15 coating as compared to Steel 45 even in testing at angle of incidence equal to 30°.

The studies have led to the conclusion that wear rate of plasma-sprayed coatings varies inversely with breaking stress at bending, i.e. as breaking stress goes up, wear rate decreases (see. Fig. 6.31). This inference was verified by testing self-fluxing PG–SR4 coating. The strength properties of coatings are in turn determined by various factors, among which the structure of materials should be highlighted.

As has already been noted, reduction in cohesion strength of plasma-sprayed coatings is assisted by presence of pores, non-homogeneous structure of materials, in particular, the presence of boundaries between separate layers and equiaxial volumes formed during spraying of powders. PN70U30 coating is less porous and more homogeneous than the PN85U15 one, and, consequently, the breaking stress at bending for PN70U30-coated specimens is 3.4 times higher, and the wear rate in testing at 60° angle of incidence is 12.5 times lower than the corresponding values for PN85U15-coated specimens.

The presence of boundaries between separate layers in the coating, as well as the presence of pores, obviously, favors a deterioration of mechanical properties of the coating, which has been indirectly confirmed by the results of vacuum etching of the PN70U30 coating. The boundaries between separate layers are etched more actively than the coating itself, which points to substantial porosity and poor cohesion properties of certain areas of the boundaries [6.82]. To reveal the role of structure (the presence of oxide net) in wearing, Kingswell [6.43] has deposited tungsten plasma coatings by the following two technologies. In the first case the coating was sprayed in open air, in the second one it was deposited in vacuum. In both cases the coatings had equal thickness (about 450 μm). Both coatings show similar behavior of wear characteristics as functions of angle of incidence, speed and size of particles, but steady-state erosion rates in "air" coatings are an order of magnitude higher than the rates in "vacuum" coatings. This can be explained by chipping of some particles in brittle failure caused by the presence of oxide layers at the boundaries of the particles. As for the "vacuum" coatings, their failure mechanism involves plastic deformation, i.e. abrasive microcutting of the surface of tungsten particles not separated by oxide nets.

Eaton and Novak [6.26] have determined the following three stages in the gas abrasive wear of zirconium-yttrium plasma-sprayed coating in the flow of Al_2O_3 particles: single scratches; failure and chipping of separate particles; formation of worn coating areas. They also verified that erosion resistance of coatings can be enhanced by improving cohesion resistance and reducing porosity.

Good resistance to gas abrasive wear was demonstrated for multi-layer combined coatings [6.29]. Oswald and Husemann [6.59] additionally deposited TiC or TiC/TiN films from vapor on an annealed plasma-sprayed layer. Abrasive jet (SiC with the size of particles <20 μm) having speed 12–130 m/s was directed from a revolving disk at the specimens at angles of incidence 15–90°. Al_2O_3 and TiO_2 coatings have shown the worst erosion resistance, while the best resistant properties were demonstrated by the two-layer coating produced by additional deposition of TiC. For all seven tested coatings wear rate was maximum at the angle of incidence $\alpha = 30°$.

For coatings made of Ni and Cr–B–Si alloys Fukumoto et al. [6.28] have found that erosion rate W depends on the testing period τ as $\lg W \approx \mathrm{R} - n\lg\tau$, where R and n are constants. On the plot of $\lg W$ vs. $\lg\tau$ two areas with different slopes can be clearly seen which characterize initial and steady stages of erosion.

Among original works the investigations which proved that inner surface of cannon barrels subjected to erosion wear can be protected by Fe–Cr-based coatings instead of conventionally applied chromium coatings should be mentioned [6.66]. Oxides with high protective properties are formed on Cr- and Fe–Cr-based coatings with high chromium content. Erosion of coatings is low, although sometimes cracks leading to failure still develop. It has been noted, that even though application of Fe–Cr-based coatings is limited by high temperatures in the barrels, cheaper coatings containing 25–30% Cr may be of interest for covering cannon barrel canals.

Sophisticated high-temperature erosion studies of various coatings in a jet of SiC and Al_2O_3 particles at speeds 70 and 150 m/s were conducted by Levy Alan [6.48]. The following systems were investigated at temperatures 25, 500 and 538°: Cr_3C_2, WC–Ni, FeB, TiN, CrB, WC–Co and some others. Coatings were produced by various techniques: thermal diffusion saturation, plasma spraying, nitration, physical precipitation from vapor, etc. The author demonstrated that small grain size, reduced porosity, the absence of cracking are the main factors leading to enhanced wear resistance of the coating. Hardness, composition and uniformity of the distribution of the second phase with higher hardness are all factors having relatively small effect on erosion resistance of various coatings. Coatings with high hardness and brittleness such as Cr_3C_2, WC–NiCrB are characterized by increased wear rate [6.48].

Serious investigations of gas abrasive wear resistance of various coatings were carried out by Kulu [6.47]. His work generalizes observations drawn from studying fourteen gas plasma, plasma and detonation coatings and eleven surfaced self-fluxing coatings. Of all sprayed coatings the detonation coating BK9 (91% WC; 9% Co) had the best wear resistance. Wear resistance of gas plasma- and plasma sprayed coatings (in the first instance, of oxide coatings) is low because of their high porosity and brittleness. Surfaced powder coatings made of Ni-based self-fluxing alloys have somewhat better wear resistance at the angle of incidence 30° than Steel 45 (0.45% C). Hardness has different effect on wear resistance of different coatings. In identical testing conditions powder coatings deposited using different techniques but having equal hardness may have wear resistance differing by an order of magnitude. For surfaced and sprayed detonation coatings at small and average angles of incidence of abrasive particles ($\alpha \leq 45°$) as the hardness goes up, the wear resistance tends to enhance. On the contrary, at large angles ($\alpha > 45°$) increased hardness leads to reduction in the wear resistance of the coating [6.47].

Blasting of eight metal gas-thermal coatings with abrasive particles of various grain composition and at different angles of incidence (15–90°) has not revealed any satisfactory correlation between volume wear and hardness by Vikkers either [6.37]. The authors indicate that wear resistance can be enhanced using heat treatment (600 °C, 1 hour) leading to stress-relief and structure stabilization.

Later the same journal has published a paper presenting further results of the same authors in studying plasma-sprayed ceramic coatings by blasting air jet containing Al_2O_3 particles at pressure 539 kPa and at the same angles of incidence [6.36]. For Al_2O_3, Cr_2O_3, and WC–12%Co coatings with 80%Ni–20%Al underlayer no relation between volume erosion rate of the specimen hardness was found. In providing high erosion resistance not only hardness, but also crack resistance of the material plays an important role. Besides, the failure pattern depends on the angle of incidence.

Despite the difference in coating deposition techniques and feasibility of producing surface layers with substantially different properties, general requirements for their resistance enhancement can still be formulated. The most important requirement is, probably, concerned with the necessity to improve the level of the cohesion strength of the coating material. This characteristic in turn depends on chemical composition of the powder, proportions of structural components in the coating, residual stress level and other properties.

In the adopted gas abrasive wear testing conditions surface layer failure is preceded by high specific dynamic contact loads. In these conditions enhancing material hardness helps only up to some critical point. As a consequence of yield point increase and fracture toughness drop the share of brittle failure in the wear will increase [6.91]. Therefore, the most homogenous or ultradispersed structure is required.

Particle impacts should be taken up by uniformly distributed strengthening phases. The matrix of viscous solid solution should experience plastic deformation under the action of abrasive particles, thus preventing failure of the solid phase.

In this case it is expedient to harden materials applying self-fluxing alloys having the structure of nickel-chromium solutions with ground uniformly distributed compound carbides. Unstable austenite-carbide and austenite-martensite alloys also provide high wear resistance. At small angles of incidence the requirements for wear resistant surface structure coincide with the requirements for alloys working in the fixed abrasive medium. Enhanced wear resistance of self-fluxing alloys is achieved by their hardening using high-melting dispersed additives [6.77].

6.8 Hydroabrasive Wear

6.8.1 General Overview

Hydroabrasive wear occurs under the action of solid particles suspended in liquid and moving past a body exposed to wear. Hydroabrasive wear is typical for diaphragms and working wheels of pumping and pipe-line equipment.

Results of investigations conducted by Tenenbaum [6.76] indicate that hydroabrasive resistance is a complicated process depending, first of all, on such characteristics as the angle of incidence, speed of abrasive particles at the point of impact against the surface of a part, the ratio of the hardness values of abrasive and material subjected to wear (hardness coefficient), and concentration of abrasive particles in liquid. Hydroabrasive wear is determined not only by the action of abrasive particles, but also by physicochemical reactions between the material and liquid. In certain conditions interaction with liquid may be so active that hydroabrasive wear is suppressed either by cavitation or corrosion. Hydroabrasive wear is usually preceded by plastic deformation, microfatigue or microcutting processes, on which hydraulic impacts of collapsing cavitation bubbles and adsorption-corrosion reactions are superimposed.

In actual hydorabrasive wear the most essential variable factor is the angle of incidence. From numerous materials (more than 80) exposed to various kinds of hardening only rigid vynil plastic has shown low sensitivity to the angle of incidence. For steels and cast irons ambiguous effect of the angle of incidence has been reported, namely, changing the angle of incidence from 10° to 70° causes a wear rate increase by a factor of 1.5–8 [6.76].

The next significant factor is hydroabrasive flow rate. Wear degree and impact speed of abrasive particles are related by a power-law type expression, where the index of power for speed changes from 1.5 to 4 and is determined by physicomechanical properties of the material exposed to wear and the angle of incidence [6.24].

Similar to some other kinds of abrasive wear, hardness has dual effect on wear rate. Hardness growth is accompanied by increase in wear resistance in tough failure and its decrease in brittle failure. The dependence of wear on the time of hydroabrasive jet action is linear in most cases, which allows to evaluate wear as a function of time for any duration of tests.

Up to certain values the wear also linearly depends on the concentration of abrasive particles. As the concentration grows further (above 25 g/l), the wear slows down [6.24].

In many respects hydroabrasive wear is similar to gas abrasive, which has already been described in detail. So only the peculiarities of testing in hydroabrasive flow as compared to gas abrasive will be given below.

6.8.2 Equipment

Since hydroabrasive wear is determined by a large number of various factors and there is no failure theory for this complicated process at present, wear evaluation of real materials and resource tests should be carried out in the conditions simulating actual operation. Tenenbaum [6.76] has designed two versions of a device that meets these requirements. A machine for specimen wear in a constant volume of suspension is intended for performing tests in chemically active liquids. Through electrically driven wire spiral suspension from the tank is dispensed to the branch pipes terminated by calibrated hard alloy nozzles. The suspension is continuously thrown from the nozzles and interacts with the tested specimen attached to the rotor at a specified angle of incidence. The specimen mounting tank is lowered using a screw mechanism. Tank volume is 9 l; the specimens have the shape of plates with dimensions 30x20x4 mm.

Another version of this machine may be used for studying specimens in flowing liquid using renewed abrasive. The tank is separated into two parts. Both abrasive and liquid are fed through a metering unit and a cock. After being ejected from the nozzle, the liquid runs down the partition and is then removed.

Among other installation most closely emulating the operating conditions in the tests, the following should be noted.

An original scheme for testing erosion of the hard surface of helicopter propellers caused by rain drop was chosen by Maier [6.51]. The specimens for erosion test had elliptical shape typical of propeller blade edge and were blasted by water drops 2 mm in diameter at a speed of 225 m/s for 30 hours, which had approximately the same effect as 2000 hours of operation.

Bihat et al. [6.8] studied cavitation-erosion failure of Fe–Al–Mo plasma-sprayed coatings using a special test bench. Cavitation was excited by a striker vibrating in water with amplitude 50 μm and frequency 20 kHz.

Golubev et al. [6.31] studied hydroabrasive resistance using a machine simulating operation of hydraulic turbines (concentration of abrasive particles up to 30 g/l, flow rate reached 25 m/s).

6.8.3 Results and Discussion

Fillers, first of all, high-chromium cast irons, demonstrated good wear resistance to hydroabrasive medium. Testing surfaced specimens, built into the elbow of a suction-tube dredge pump driving water with 10–30% sand content, has proved that cast irons composed of 1.5–5% C, ≤6% Mn, 25–35% Cr, 4% Ni and 1.5% W with hardness HRC 45 has the best wear resistance, being 10–60 times higher as compared to carbon steels [6.24].

Hard sintered tungsten alloys also have high wear resistance which enhances as the cobalt content goes down. Alloys with identical composition show better wear resistance for finer grain size. Alloys with different nickel content may have equal wear resistance due to favorable combination of hardness and strength. For alloys containing 10% and 30% Ni and having equal hardness, the alloy with higher nickel content has 2 to 3 times better wear resistance [6.18].

In [6.31] the influence of Cr, Mn, B, Ti, Si, Cr–Mn, and Cr–Si-type diffusion coatings on failure and resistance of steels 30 (0.3% C) and 20X13 (0.2% C; 13% Cr) was reported. For weak hydroabrasive impact (at small angles of incidence) all coatings provide approximately equal enhancement of the resistance of steel (1.5-2.5 times). At the angle of incidence 15° the most wear resistant are boron- and chromium-based coatings, increasing the resistance by a factor of 6–18. Relatively soft Mn-, Cr–Mn-, and Cr–Si-based coatings are unable to provide any resistance enhancement at these incident angles.

Boride and chromium coatings are distinguished by their high corrosion resistance, large layer thickness and high coupling strength. Probably, these properties have the decisive effect on their wear resistance as well. It was noted that the pattern of failure is approximately similar for all types of coating. Detachment of particles starts from occurrence of pits and their further growth in the area of the maximum effect of the hydroabrasive jet [6.31].

Unfortunately, there is little information on studying wear processes of gas thermal coatings. Three most interesting works are considered below.

Two gas thermal NiCr–Cr_3C_2 and CoCr–W_2C coatings sprayed on helicopter propellers made of aluminum alloy were studied in [6.51]. Both coatings had layered structure but contained separate crack-like defects. The cracks were initiated at carbide boundaries. At early stages of blasting by water drops no damage was observed, however, at later times scaling starts over the interface. Takeda [6.71] also demonstrated that only coatings having optimal structure with no defects are capable of noticeable wear resistance enhancement.

Plasma-sprayed Fe–Al–Mo coatings were tested for hydroabrasive wear after deposition and heat treatment (annealing for 12 hours at 600 and 300 °C and for 6 hours at 800 °C) [6.8]. The length of incubation period before the onset of coating failure increases proportionally to the duration of heat treatment (from 5 minutes without heat treatment up to 16 minutes after annealing at 600 °C for 12 hours). Cavitation-erosion resistance enhancement after annealing has been explained by increased coupling strength between particles, porosity reduction, and by favorable changes in coating microstructure.

Analyzing an alloy reinforced with high melting dispersed particles, Tokarev [6.77] demonstrated that under hydroabrasive impact at the angle of incidence equal to 30° the wear occurs by microcutting mechanism. During the process wear resistance changes along a curve with the peak point depending on the amount of additives. At first, wear resistance increases in proportion to increase in the number of hard grains. However, when the number of high melting particles exceeds the optimum point, toughness reserve of the base becomes insufficient. Consequently, brittle cracking and chipping of hard particles takes place. As the angle of incidence increases up to 75°, wear resistance reduces monotonously with increasing the fraction of the solid phase. Besides fatigue failure, hard inclusions may be chipped off solely because of the poor resistance of the surrounding base to the impact of abrasive particles [6.77].

The described peculiarities of abrasive wear and the methods for producing alloys with optimal structures did not consider the following two essential circumstances. Firstly, in many cases the damage to the real parts is caused by a combination of two or more types of abrasive-surface contact rather than by single-type action, which complicates revealing and analysis of the predominating physicomechanical processes leading to surface failure. Secondly, quite often, especially in mining and petroleum chemical industries, the parts come in contact with abrasive in aggressive liquid or gas medium. As a result, the requirements of corrosion-resistance and passivity towards the environment are imposed on the materials in use. Thus, the range of alloys suitable for usage is significantly reduced.

All the above shows that abrasive wear resistance of alloys can be actually enhanced only through carrying out comprehensive investigations in all trends of tribonics.

6.9 Main Conclusions

Now we can formulate some conclusions regarding wear resistance in abrasive media.

1. Wear with abrasive particles is characterized by considerable local stresses arising in the surface, short duration and repeated recurrence of the impact.
2. Relation between yield point, hardness and fracture toughness of an alloy is extremely important for abrasive wear resistance. Abrasive wear rate of alloys is reduced for all wear types when plastic flow is impeded (i.e. when σ_y increases). However, excessive hardening leads to embrittlement of the material, and the failure is accompanied by rapid pricking out of the surface layers (low K_{Ic}).
3. The tendency to increase yield point and hardness still avoiding strong brittle surface failure is represented by the following rule: for equal values of σ_y a material with higher K_{Ic} will have better abrasive resistance; for equal values of cracking resistance a material with higher yield point will wear out less actively.

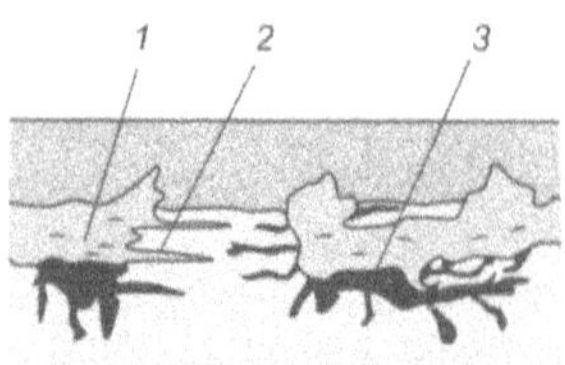

Fig. 6.32. Fretting corrosion scheme. *1* oxide; *2* microcracks; *3* cavity

4. Granular structures taking maximum advantage of hardening by σ_{grain}, $\sigma_{disl\,pc}$, and σ_{ph} mechanisms, which do not aggravate the discrepancy between σ_{disl} and K_{Ic}, are preferable in abrasive wear resistance enhancement. Matrix-type structures with dispersed globular hardening phase inclusions are also efficient.
5. Gas thermal coatings produce maximal effect (1.5–8-fold resistance enhancement) in wear with fixed and loosely fixed abrasive particles.
6. Application of many gas thermal coatings to resistance enhancement in the conditions of gas abrasive and impact abrasive wear, as well as under the action of abrasive particles located in the friction pair gap is not expedient because of their low resistance to brittle failure. Fusing the coatings brings their resistance closer to the resistance of constructional steels in the same conditions.
7. Abrasive wear resistance enhancement of gas thermal coatings is assisted by reducing the number of structural microdefects (reducing the number of pores, microcracks, unstable oxide films, enhancing the strength of coupling between particles) and, consequently, by increasing the uniformity and decreasing the stress concentration at these defects.
8. Abrasive wear resistance of gas thermal coatings is proportional to their cohesion strength as determined by bending tests.

6.10 Fretting Corrosion

6.10.1 General Overview

Fretting corrosion is the type of wear observed when one surface performs oscillatory motion relative to another one in a corroding medium. Slipping between the adjoint surfaces is a necessary condition for occurrence of fretting corrosion. Fretting corrosion considerably deteriorates the quality of the surface layer (increases roughness, cavities, undersurface microcracks) (see Fig. 6.32), which leads to reduction in fatigue strength of parts and increased wear [6.6, 6.9]. Fatigue damage caused immediately by the development of fretting corrosion is called fretting fatigue.

Fretting corrosion is a special form of wear and has the following distinctions by both the conditions of its occurrence and the way it shows up as compared to common wear at unidirectional slipping [6.30, 6.89]:

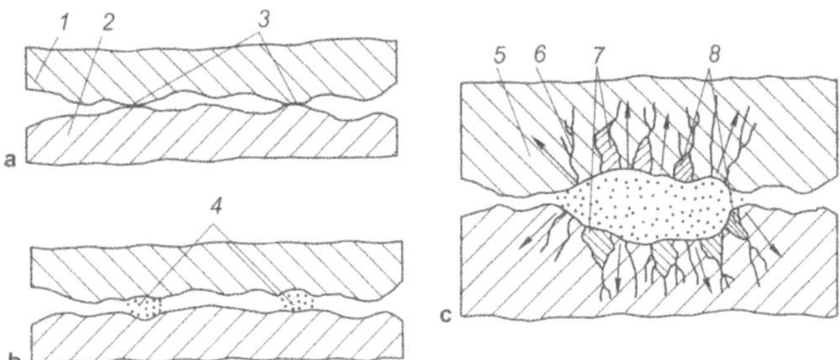

Fig. 6.33. Kinetics of fretting corrosion wear of metal surfaces (adopted from [6.30]). *1,2* contacting parts; *3* points of contact between surfaces; *4* small incipient cavities; *5* big shared cavity; *6* cracks; *7* split off metal volumes; *8* split off solid oxides

1. Due to the small displacement amplitudes the removal of wear products from the friction zone is hindered and fretting corrosion damage is confined to the actual contact areas.
2. Relative movement speed of contacting zones is much lower than conventional speeds of unidirectional sliding friction. Thus, at sliding amplitude of 0.025 mm and vibration frequency of 30 Hz the maximum speed is only 4.7 mm/s, while average speed is only 3 mm/s.
3. While in conventional sliding friction the presence of oxygen may decrease wear rate, in the conditions of fretting corrosion oxygen makes damage even worse. Typical products of metal fretting corrosion are their oxides.

A simplified mechanism of wear in fretting corrosion is illustrated in Fig. 6.33. The parts initially contact in separate surface points (see Fig. 6.33a). Upon vibration oxide films in the actual contact zone are destroyed forming small cavities filled with oxide films (see Fig. 6.33b), which gradually grow in size coalescing into one large cavity (see Fig. 6.33c).

The last stage of fretting corrosion is the ultimate destruction of damageable zones previously loosened by fatigue and corrosion processes. Since the possibility of electrochemical processes should also be taken into account, this stage can be called the stage of corrosion fatigue failure. During this period metal surface layers, which have been subjected to cyclic deformation for a long time, become so weakened that their resistance properties are completely lost resulting in their gradual detachment leading to increase in wear rate.

Fretting corrosion is accompanied by dislocation changes in metal surface layers. Depending on the distance from the surface, three different dislocation zones can be distinguished. In the volumes immediately adjacent to the contact zone high dislocation density can be observed, the metal is grain-oriented and cold-hardened. The second zone is characterized by a large number of twins and rapid fragmentation with high degree of block desorientation. In the third zone, which is the most remote from the surface, a usual dislocation pattern can be observed. It is believed that in the conditions of fretting corrosion formation of nonequilibrium structures approaching to amorphous state is possible on the contact surface [6.30].

Fretting corrosion occurs in shafts, threaded connections, rolling bearings, couplings and other parts in movable contact. Considerable wear rate and wear intensity of contacting part surfaces, which are typical for aircraft engines, force to discard large quantities of expensive products. Corrosion of this type, being one of the most dangerous processes of machine parts failure, may take place with many materials both in the conditions of lubrication and dry friction in the wide temperature range.

Surface failure due to fretting corrosion reveals itself as rubs, sticking, and presence of cavities or tear-outs filled with products of wear. The first diagnostic indication of fretting corrosion is occurrence of stained spots containing deformed oxides on the friction surface.

Besides changing the geometry of the part and spoiling its appearance, fretting damages at early stages lead to more harmful consequences related to violation of dimensional accuracy. Three following situations are possible in this case. If there is some egress of fretting corrosion products out of contact zone, then it leads to the loss of initial fitting and loosening of tension between the contacting parts. On the other hand, if fretting corrosion products have no free egress out of contact zone, jamming and even seizure may occur in the system, especially as the volume of forming corrosion products generally exceeds the volume of metal converted into the oxide state. This situation becomes especially dangerous when contacting parts are to be disengaged from time to time during work (e.g. in safety valves and regulators).

6.10.2 Specimens and Equipment

The modern state of the theory of fretting corrosion does not allow calculating wear rate of materials using exclusively analytical methods. As for the existing experimental techniques, they differ, first of all, by the contact scheme.

According to this criterion all installations used in the studies of fretting corrosion can be divided into the following two types: the devices of the first type are based on point or linear contact, while the devices of the second type produce vibrations in the contact zone of flat surfaces. The installations used to study point (sphere-plane) or linear (cylinder-plane) contact have a number of disadvantages: wear products are removed out of the contact zone, which usually is not the case in real conditions; pressure distribution over the contact surface is non-uniform; as the load changes, the nominal contact area changes too.

The above drawbacks can be avoided when the contact occurs between sufficiently extended surfaces. Therefore, face friction of contacting specimens such as a bushing (movable counter-specimen) and a cylinder (stationary specimen) can be considered the most suitable scheme (see Fig. 6.34). Contact area in this case is 0.5 cm^2.

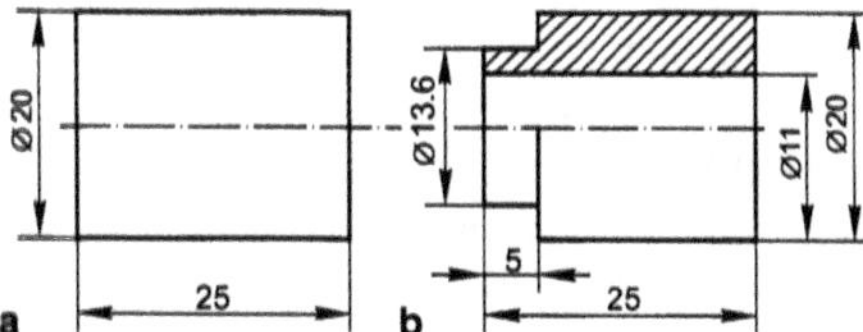

Fig. 6.34. Specimens for fretting corrosion testing. **a** stationary; **b** moving

Counter-specimen is made of hardened and tempered Steel 45 (0.45% C) with hardness HV 550–600. When depositing coatings on the end faces of the specimen, special care is paid to avoid cracking, warping, local overheating, etc. The specimens are polished, specimens with scaled or chipped coatings are rejected. It is expedient to deposit the coatings and polish specimens using a special gear with openings for fixing specimens. After polishing the thickness of the coating should be at least 0.1 mm.

The method of testing for fretting corrosion wear in the "bushing-cylinder" scheme developed by Golego [6.30] is standardized (State Standard 23.211). The standard defines the procedure of testing metal and non-metal coatings for fretting corrosion wear with or without lubrication. The tests should be performed using a special installation (see Fig. 6.35). The face of the counter-specimen comes in contact with a stationary cylindrical specimen fixed in a self-orienting collet. The specimens are loaded through the shaft of a moving head. An electric motor transfers rotary motion to eccentric with adjustable eccentricity. The eccentric is connected with the drive shaft link providing reciprocating motion of the counter-specimen through a connecting rod. Amplitude of the counter-specimen movement is varied using the eccentric and an adjusting device.

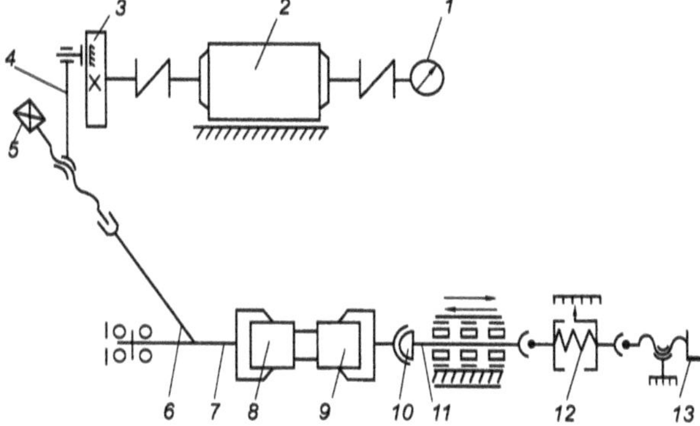

Fig. 6.35. Mechanical diagram of experimental installation for fretting corrosion testing of coated specimens (adopted from [6.30]) *1* revolution counter; *2* electric motor; *3* eccentric; *4* connecting rod; *5* adjusting device; *6* link; *7* shaft; *8* counter specimen; *9* tested specimen; *10* collet; *11* moving head; *12*, *13* loading devices

An installation on the basis of optical microscope was applied to observation of kinetics of surface failure due to fretting corrosion (formation of sliding strips, development of cracks, surface oxidation) in [6.44]. The installation allows to continuously monitor accumulation of damages on microsections.

Most reported studies of the coatings were conducted at room temperature, although in some works high-temperature testing was also used. Thus, plasma-sprayed alumina coatings were studied over temperature range 233–1073 K [6.80], and certain gas-plasma and detonation coatings – in the range 223–673 K [6.35].

6.10.3 Testing

Working surfaces of specimens are thoroughly cleaned, washed in benzine and acetone and dried. If necessary, either a tank for liquid lubrication medium is mounted on a testing machine, or specimens are smeared with plastic lubricants.

The specimen is fixed so that longitudinal axes of the clamps, specimen and counter-specimen (see Fig. 6.35) coincide and be perpendicular to the working surface. The following standard values of test parameters are set: load 500±25 N; amplitude 50±5 μm; frequency 30±3 Hz; number of loading cycles $5x10^5$±50. After completion of the test oxides are removed from the friction surface both mechanically and by rinsing. In most cases oxides can be removed using the following solution: 4 g of hydroquinone, 22 ml of concentrated orthophosphoric acid, and 20 ml of alcohol per 100 ml of water solution [6.30].

Then circular friction tracks are examined using optical microscope. The character of failure can be defined using the results of metallographic analysis. Fatigue cracks, oxidation products, exfoliated scales, secondary phases, seizing traces, plastic deformation areas, tear-outs, scores and scratches are all noted.

Mass loss or linear wear (or similar values referred to friction path) may be used as criteria for fretting resistance. Average linear wear is determined by profilographing of a friction track.

The above methods only provide a measure of the overall damage, so in some specific cases they may be insufficiently objective. Thus, for example, deep pits located on a small area can be more dangerous than shallow pits of the same volume. Damage determination by mass loss is inefficient in the case of small wear, especially as there is a possibility of material transfer from one surface onto another and vice versa.

The results of testing can also be represented as plot of the measured linear wear or wear intensity versus applied load (compression stress), temperature (when the temperature is varied), or other variable external parameter.

6.10.4 Results and Discussion

Despite the fact that fretting corrosion can lead to fatal failure of critical parts, coatings that can enhance wear resistance are still not applied as widely as they deserve. This is probably related to the fact that peculiarities of work and failure of coatings have not yet been sufficiently investigated, and at present there are no data on tribomechanical characteristics for oscillatory displacements of contacting bodies.

The results of structural changes in the surface layer and their effect on wear for nine types of steel subjected to laser treatment were reported in [6.2]. A complicated character of laser radiation influence on the structure of the surface layer and fretting damage depth was demonstrated. The work proves that laser surface hardening allows creation of a favorable structure and enhancement of wear resistance by a factor of 1.5–3 depending on carbon content in steel and testing parameters.

The work [6.44] is devoted to investigation of the effect of ion implantation of six different elements into surface layers of Steel 45 (0.45% C) and its influence on tribotechnical characteristics in fretting. By ion implantation the authors basically mean technology which allows production of a coating film, or peculiar surface "alloy" with stepwise composition, gradually transiting into base metal. Tests for fretting corrosion wear have proved that implantation reduces wear rate of specimens. Thus, implantation of barium ions increases fretting fatigue resistance for 10^7–10^8 cycles by more than 30%. This effect can be explained as follows: firstly, dense, strong and plastic $BaTiO_3$ film is formed on the surface of the specimen; secondly, no coalescence takes place, and thirdly, rather high compression stresses are induced in surface layers. Deposited films reduce friction coefficient by 10–17% and maintain it constant for extended periods of time during the test, the part wearing out being the non-hardened counter-specimen.

In the work conducted by specialists of Kiev Civil Aircraft Institute [6.35] the role of gas-plasma and detonation-sprayed coatings in enhancement of reliability and durability of materials subjected to fretting corrosion was evaluated. For detonation-sprayed BK15 (85% WC, 15% Co) coatings optimal and limiting operating conditions including specific loading, relative displacement amplitude and temperature were determined. Fretting corrosion wear was evaluated by linear wear and well as by a special parameter characterizing non-uniformity of the friction track relief. Combined analysis of the results of wear at room temperature and metallographic data has allowed to identify three zones (A, B, C) differing by the pattern of damageability accumulation and failure mechanisms.

The combination of parameters in the A zone (temperature, load, amplitude of vibration) provides reliable and stable work of coatings. Wear shows up as plastic deformation and surface microroughness leveling.

The B zone is characterized by wear process associated with detachment of small particles, which is typical of fretting corrosion, while failure in this case is caused by fatigue and abrasive mechanisms.

The C zone is characterized by the most unfavorable combination of variable factors, which makes it a zone of inadmissible operating conditions. Metallographic analysis reveals cracks, chips, and deep furrows on the friction surface. Wear rate and wear intensity are sharply increased.

As for the effect of temperature on the performance of coatings, wear degree is practically constant in the range 223–673 K. Temperature increase speeds up coalescence and oxidation processes accompanied by sharp wear increase [6.35].

However, a different point of view has also been suggested. Thus, in his thorough work Waterhouse has demonstrated that failure of some metal coatings at high-temperature fretting corrosion can be much lower than at room temperature due to formation of indestructible protective oxide films on the surface [6.89].

Fretting corrosion wear resistance of ion-plasma sprayed coatings was investigated with due account of strength characteristics [6.17]. Electron-fractographic studies of the deformation relief of titanium nitride-based coating demonstrated that the leading process of mechanical damage is fatigue failure, which is realized when accompanying development of special forms of abrasive wear occurs. When $H^c/H^a < 0.6$ (where H^c and H^a are microhardnesses of the coating and abrasive particles, respectively), abrasive wear intensity is determined by the value of coating microhardness. At the same time, fatigue failure damageability is controlled by fracture toughness K_{ic}.

Chipping of coatings is strongly influenced by defects which are already present in the coating, initiating crack development both in brittle and fatigue failure [6.74].

6.11 Wear in Friction Pair

6.11.1 General Overview of Testing Coatings

The curve of wear intensity of parts working in friction pair (see Fig. 6.36) can be split into the following three stages: 1 – run-in; 2 – steady wear; 3 – accelerated wear.

The first stage is characterized by increased wear intensity, which can be explained by small area of surface contact owing to macro- and microroughness resulting in significant contact loads. By the end of the run-in stage surface roughness reaches the steady state. Simultaneously structural conversions in the surface layer forming secondary structures occur. During the steady wear stage the intensity of wear is low and constant. Upon worsening of the working conditions the third stage, i.e. accelerated wear, can be observed. Not all three stages are actually observed in real operating conditions, so the presence of all stages should not be sought after in testing either.

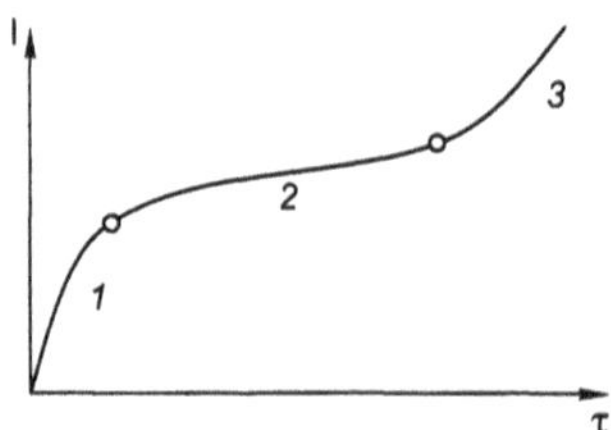

Fig. 6.36. General form of the wear curve for an arbitrary surface including coatings. *1* run-in stage; *2* steady wear stage; *3* accelerated wear stage; *I* wear intensity; τ time

In standard loading regimes or in the conditions that simulate real operation, laboratory friction testing machines allow: timely comparative evaluation of coating deposition regimes, optimization of powder chemical composition, control of stability of material properties with friction, obtaining original data required for tribotechnical calculations, determination of limiting field conditions for tested wear resistant coatings.

The results of comparative laboratory tests may be used in bench or service tests of materials. Laboratory tests using friction machines are regarded as the first stage in the series of successive tests "laboratory–bench–service". The results of each preceding stage serve as input data for evaluation of wear resistance at a subsequent stage of testing.

The factors determining the processes of friction and coating wear in the tests are divided into three groups: technological, constructional and operational.

The first group of factors includes structure (dispersity of oxides, inclusions, amorphous compositions, etc.); physical properties (heat conductivity, porosity, temperature coefficient of linear expansion); mechanical properties (coating-base metal coupling strength, fatigue characteristics, cohesion strength, fracture toughness, hardness) [6.10, 6.23, 6.32, 6.60].

Constructional factors include contact schemes (kinematic types of connection) [6.11, 6.72, 6.79]; parameters of roughness (first of all, R_a, R_z, t_p, as well as orientation of irregularities); shape and dimensions of working surfaces [6.19, 6.21, 6.22, 6.49]. The data on relief and surface microgeometry are required, in particular, for determination of the actual contact area (which is the sum of separate contact surface zones of both contacting bodies) and the actual specific load, i.e. the normal load per unit actual contact area of the contacting surfaces.

The group of operational factors consists of temperature, force, speed and energy parameters of loading [6.27, 6.39, 6.88]. The integrated effect of three groups of factors determines structural evolution of the friction surfaces.

In surface contacting layers both macro- and microlevel changes are possible. Macrochanges include plastic deformation of surface volumes, scratching, scaling and chipping of material in friction contact zone, as well as coalescence, seizing and scoring. At the microscopic level occur the processes of phase, structural and substructural conversions, dislocation rearrangement, cleavage of existing and formation of new chemical bonds, modification of electron structure of the surface.

The choice of a particular technique for testing coating wear is first of all determined by the purpose of investigation. The study may be aimed at solving the following problems: analysis of the failure process in order to figure out the general rules of coating wear; determination of the influence of the technological parameters of deposition on wear resistance of coatings; evaluation of the influence of coating structure and properties on its wear resistance in the given external conditions. In the last two cases the correct choice of testing technique is especially important. The scheme of testing, the values of variable parameters of loading and the working medium should be chosen as close to real operating conditions as possible. In designing testing equipment for evaluation of wear resistance the following principle should be observed: on one hand, the main characteristics of the experimental installation must provide as good modeling of the real operating conditions as possible; on the other hand, it is desirable to intensify wearing process in order to speed up testing.

6.11.2 Specimens

Specimens for tests must all have identical size and shape, surfaces prepared by the same method, equal roughness parameters and coatings deposited using the same technology. At least 3–5 specimens must be tested.

The shape of the specimens is chosen depending on the kinematic type of connection and on the available friction machine. The size of specimens is limited by technical characteristics of each specific testing device and the necessity to simulate the required levels of pressure, speed and frictional torque. Practically the specimens have size from tens of millimeters to several millimeters, i.e. the size can differ by an order of magnitude. Fig. 6.37 gives an example of specimens tested using a modern friction machine SMT–1.

Specimens made of the base metal are often used as references.

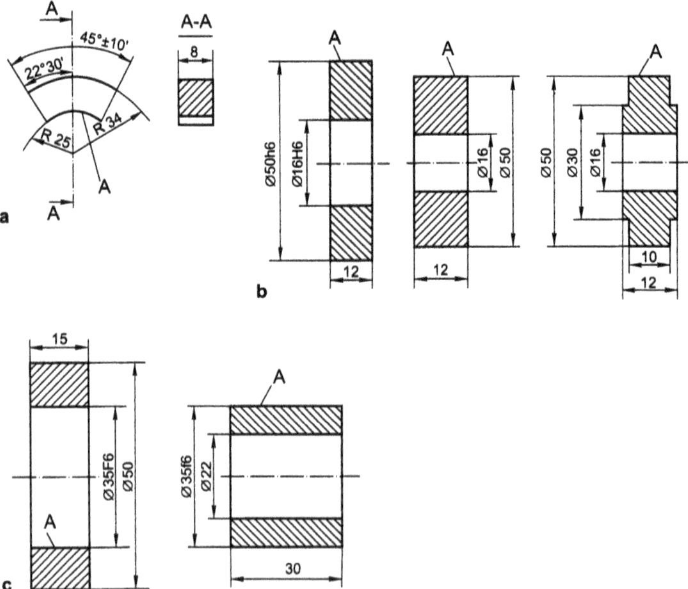

Fig. 6.37. Types of specimen tested using SMT–1 machine. **a** "disk–block" friction pair; **b** "disk–disk" friction pair; **c** "shaft–bushing" friction pair; roughness of surfaces $R_z = 20$; A – the surface to be coated

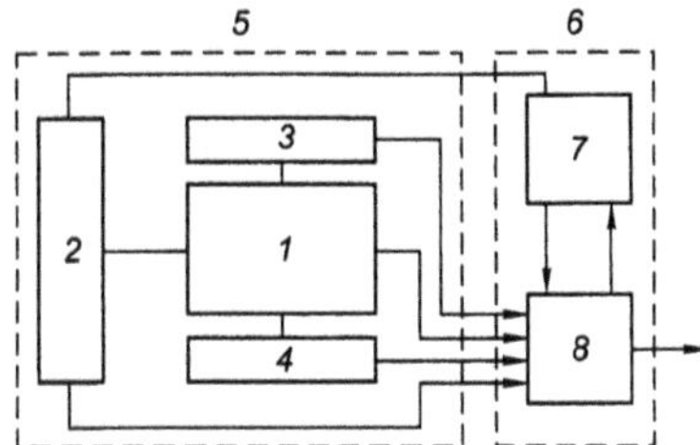

Fig. 6.38. Block scheme of commercial friction machines (adopted from [6.13]). *1* testing unit; *2* drive; *3* loading assembly; *4* chamber; *5* friction machine; *6* control panel; *7* control unit; *8* measuring unit

6.11.3 Equipment

Modern commercial friction machines generally have the following block scheme (see Fig. 6.38). The main machine part, which usually provides several testing schemes, is the testing unit. A drive and a loading assembly provide movement speed and specific load in sufficiently wide ranges. The chamber is used for hermetic encapsulation and feed of lubricants into the friction zone. The friction machine is controlled from a console fitted with control and measuring units. The machines are capable of measuring and displaying the parameters required to obtain information on the friction process, including: frictional torque (force), normal pressure, movement speed (number of revolutions of the specimen or a counter-body), temperature in the given zone, and, if necessary, other values.

Nine of the fourteen surface contact schemes realized in friction machines and described in [6.12, 6.14] have been used for testing coatings. The scheme can be divided into two groups: planar and linear friction surface contacts.

The first group includes the following schemes (see Fig. 6.40) [6.14]:

A. Contact between the end faces of rotating and stationary bushings. Used in the commercial IM–58 friction machines and tribotechnical complexes (UMP–1, 2168UMT).

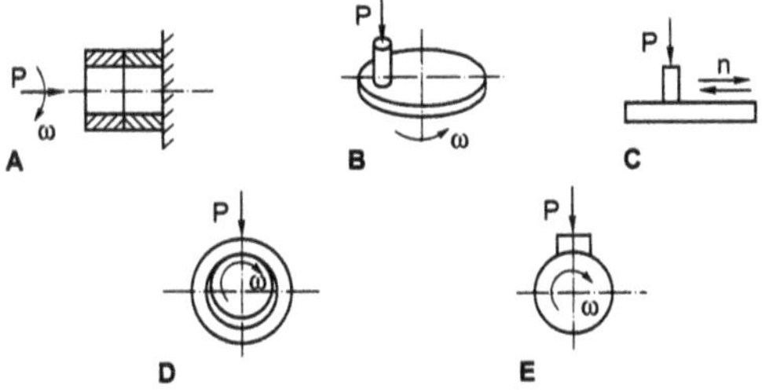

Fig. 6.39. Schemes of planar friction surface contacts. *P* load at the friction assembly; *ω* angular velocity; *n* movement frequency (adopted from [6.14])

B. Contact between the end face of a stationary cylindrical specimen and a rotating disk. Used in the 2168UMT complex and in the commercial T–01M machine produced in Poland. A similar scheme is also employed in machines used for wear testing against fixed abrasive.
C. Contact between two flat specimens, one of which performs reciprocating movement. Used in the 77MT and 2168UMT friction machines. This is one of the friction schemes implemented in the friction machines manufactured by the "Optimole" company (Germany).
D. Shaft–bushing friction scheme. Used in the UMT–1, 2168UMT, SMC–2 and some other commercial friction machines.
E. Shaft–partial bushing (block) friction scheme. Used in the SMT–1, SMC–1, MI–1M friction machines and in the friction machine produced by the "Amsler" company.

The second group (see Fig. 6.40) includes the following friction schemes [6.14]:

A. Rotating shaft–rectangular bar. Used in the friction machines made by "Timken" company as well as in the MI–1M and Schkod–Savin friction machines.
B. Rotating shaft–two bushings (blocks). Used in the Almen–Viland machine with stepwise increase of the load up to the point of seizure.
C. Rotating sphere–ring. Contact is performed over a 0.2 mm wide circular zone at the surface of the sphere.
D. Rotating roller is pressed from two sides by the end faces of two cylinders with a large diameter and V-shaped cuts at the angle of 90°. This scheme is used in the friction machine produced by "Falex".
E. Two rotating cylinders. The scheme is used in such machines as SMT–1, SMC–2, Amsler, MI–1M. Both sliding and rolling motion with slipping are possible depending on the direction and magnitude of the angular velocities ω_1 and ω_2.

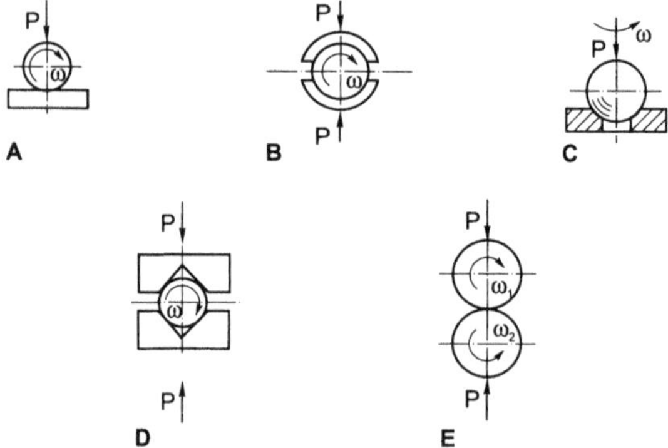

Fig. 6.40. Schemes of linear friction surface contacts. *P* load; ω angular velocity (adopted from [6.14])

Let us consider the schemes and potentials of the above mentioned friction machines in more detail [6.12–6.14].

MI–1M is a multipurpose machine for wear testing in the conditions of sliding friction, rolling friction, rolling friction with slipping with or without lubrication. The machine comes with a special gear for feeding abrasive to friction contact zone. SMC has the same purpose as MI–1M but provides a broader range of speeds and loads. MFT–1 is mostly applied to evaluation of frictional heat resistance of materials. Using ring specimens pressed to each other with a pneumatic membrane mechanism, MDT–1 uses the basic "disk-finger" testing scheme. At low speed the machine provides sufficiently high frictional torque. Its main drawback is low vibration resistance.

UMT–1 and SMT–1 are among the most versatile machines. A universal friction machine UMT–1 (see Fig. 6.41) is used for testing and modeling working conditions of basic frictional pairs used in machine building. The base version of the machine is designed for testing "disk-finger" type specimens. Electric motor through V-belt transmission rotates regulating spindle 10 connected with faceplate 8, on which disk specimen 7 is mounted. Disk 6 used to secure three specimens 6 is fitted on the shaft 12 of the head 11. Disk 5 through a flexible link 4 is connected with a hollow shaft interacting with elastic element of the force measuring unit 2. The specimens are loaded by membrane pneumatic mechanism 1. The following four testing schemes are provided:

1. End faces of three pin specimens with diameter 5 or 10 mm come in contact with the rotating disk; the specimens are arranged at the angle of 120°; friction radius can be up to 150 mm;
2. Two ring specimens with external diameter 28 mm come in contact with each other;
3. "Shaft–bushing" friction pair is tested in a chamber with liquid lubricant (friction surface diameter from 30 to 40 mm);
4. Wear resistance is determined using sign-variable friction in the "shaft–bushing" pair (the scheme simulates the work of a hinge performing swinging motion).

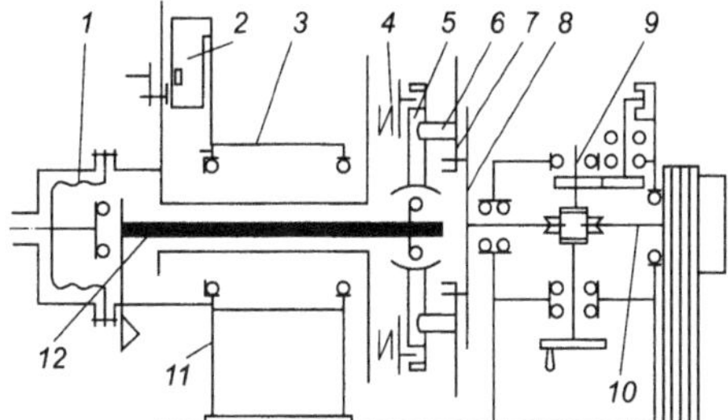

Fig. 6.41. Universal friction machine UMT–1. *1* loading mechanism; *2* force measuring device; *3* shaft; *4* flexible link; *5* disk; *6* finger specimen; *7* disk specimen; *8* faceplate; *9* reduction gear shaft; *10* reduction gear spindle; *11* head; *12* head shaft

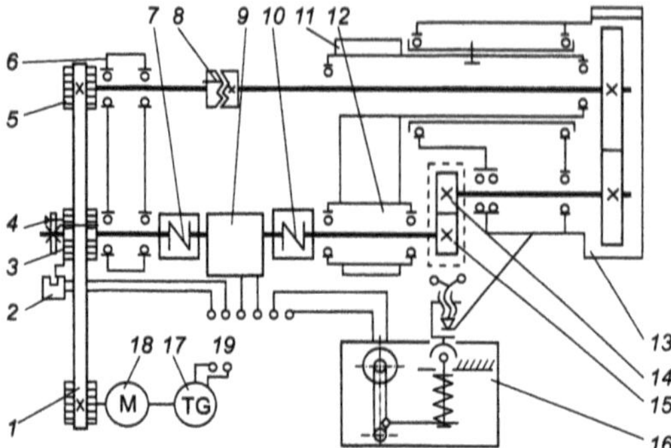

Fig. 6.42. Kinematic scheme of the SMT–1 machine. *1* gear belt; *2* non-contact sensor; *3,5* pulleys; *4* safety pin; *6* rear bearing assembly; *7,10* couplings; *8* movable coupling; *9* induction friction torque sensor; *11* front bearing assembly; *12* front head; *13* carriage; *14,15* specimens; *16* loading mechanism; *17* tachogenerator; *18* electric motor; *19* control panel

In the latter case the pair is driven using a crank mounted on the output shaft of a reduction gear. In all schemes the console indicates the following parameters: friction torque, temperature in the friction zone, rotation speed and pressing force of the specimens.

In the commercial SMT–1 installation electric motor 18 (see Fig. 6.42) through the toothed belt rotates pulleys 3 and 5. Pulley 3 through safety pin 4, shaft, coupling 7, frictional torque sensor 9, and coupling 10 drives head shaft 12, onto which the specimen 14 is mounted. Specimens 14 and 15 are pressed against each other by the spring of the loading mechanism 16. The value of the load is adjusted by axis-screw, which transmits the load to specimens through the pivot, bracket and the body of the carriage 13. Drive rotation speed is adjusted using the tachogenerator 17, the number of the revolutions of the specimen 15 is regulated by the non-contact sensor 2. The magnitude of the friction torque is measured by the sensor 9 and continuously registered by recording potentiometer.

The machine supports three testing schemes (see Fig. 6.43). A special hermetically sealed chamber provides lubrication of contacting surfaces. Because of continuous recording of the friction torque values, the installation allows observation of the wear process and registration of changes in friction coefficient immediately during operation. The machine, besides wear testing, is capable of modeling the operation of gear transmissions, wheels and rails of railway transport, as well as the operation of rolling bearing, i.e. it can provide contact-fatigue loading.

During last years several companies ("Tochpribor", Russia; "Shimadzu", Japan; "Wolper-Amsler", Germany, etc) have announced new types of universal friction machines. Constructional approaches to machine design offered by these companies ensure application of 3 to 8 optional testing schemes and multi-purpose measuring and recording systems [6.14].

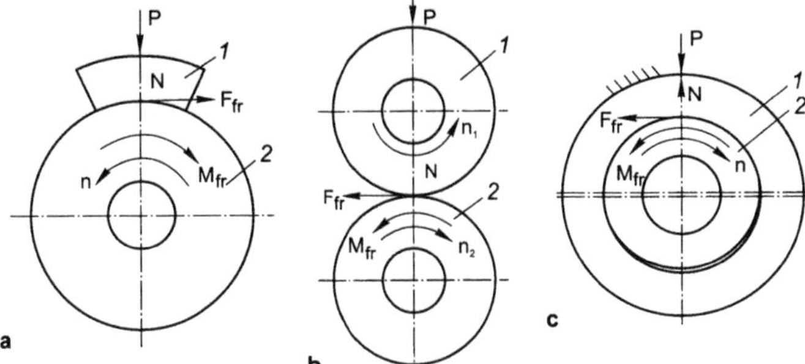

Fig. 6.43. Scheme for wear testing of coated materials with determination of the friction coefficient in the conditions of: **a** sliding friction ("disk–block" friction pair); **b** sliding friction with slipping ("disk–disk" friction pair); **c** sliding friction ("shaft–bushing" friction pair); *1,2* specimens

A recently designed new universal SI machine for fatigue wear testing sets a new standard for testing equipment [6.52]. Its unique capabilities include: 13 supported testing schemes, stepwise and continuous friction speed adjustment in the range 0–22 m/s, adjustment of the specimen rotation speed in the range 50–5000 rpm, arbitrary time profiles of contact loading in a wide range.

Specially for the SI machine a modern PC-based measuring-control system with 27 measuring and 16 controlling built-in channels has been created.

6.11.4 Preparation for Testing

The operations are carried out in the following order: a required contact surface (which amounts to at least 20% of the nominal calculated contact surface) is ensured; idle surfaces are marked (indicating their mutual orientation required for mounting on a testing machine); artificial bases are formed (if necessary, depending on the selected method of wear determination); working surfaces are washed (first in benzine and then in acetone) and dried; feed of lubricant is provided (if necessary).

It is recommended to perform metallographic analysis for one specimen from the batch in order to determine its hardness, microhardness of separate phases, and microhardness of the surface layer.

6.11.5 Testing

The test is performed in the following sequence: the specimens are set into the testing machine; the movable specimen is set in motion at a rotation speed providing the desired sliding speed; the contact between the specimens is established; the specified load is applied. The specimens are then run-in so that the surface of their mutual contact amounts to at least 90% of the nominal calculated contact surface. The control of the contact quality is performed visually by the contact spot.

The tests are carried out according to selected regimes for periods of time at least equal to the period of stabilization of the friction force moment and the temperature of the stationary specimen. In the course of testing the steady-state values of the friction force moment and temperature are recorded. Upon completion of the test the degree of wear and the roughness of the worn surface are determined.

Contact and non-contact devices are used for this purpose. The operation of commonly used roughness indicators and profilographs is based on probing the surface with a diamond tip.

A profilograph records the coordinates of the surface profile and scales them up in the form of a profilogram. During the test the probing prod moves over profile irregularities thus turning an attached mirror of the optical system, which serves for reflection of the light beam coming through the lens onto a uniformly moving paper or photographic paper. Vertical and horizontal magnification factors can be set to desired values.

A combined profilograph-roughness indicator plots profilograms in orthogonal coordinates and also determines the R_a and R_z values from the device readings.

The non-contact roughness measurement equipment is represented by such devices as scanning measuring microscopes, light section sets, interference microscopes, and some others.

Laser method, as one of the most state-of-the-art techniques, is suitable for probing at practically any level of surface microrelief.

Scanning electronic microscopy is extremely widely used for analysis of the worn surfaces [6.42, 6.52, 6.58, 6.86].

6.11.6 Main Methods of Measuring Wear

The method of wear evaluation is chosen depending on the particular testing conditions, shape and material of the specimen.

The main methods of wear measurement are defined by the State Standard 23.224 (see Table 6.4).

Micrometry

Micrometry is used when wear is accompanied by considerable changes in the size of a part. The value of the linear wear of the part is determined as the difference between its dimensions measured before and after testing. Such devices as end gages, optical instrumental microscopes, micrometers and some others are used as measuring tools. Devices measuring part dimensions to an accuracy of 1 μm make possible evaluation of linear wear accurate to 5 μm. A larger error comes from the presence of deformations, inaccuracy of setting the measuring device itself, instability of the measurement temperature, and so on. Micrometry technique allows determination of the final wear degree only rather than its continuous measurement in the course of testing.

Table 6.4. Main methods of measuring wear

№	Name of the method	Measured parameter	Conditions of applicability
1.	Micrometry	Part dimensions	Wear size significantly exceeds the limiting error of the measuring tool
2.	The method of artificial bases, method of cut indentations	Indentation size, distance from the base surface	No filling up, corrosion, plastic deformation of the base (indentation)
3.	The method of recording the closing of the specimens	Displacement of the free surface of a stationary specimen relative to the base of the moving specimen	The absence or compensation of the heat and vibration effects on the measuring system and the measured parameter
4.	Profilographing	Difference between surface profiles measured before and after wear	No surface deformation under the prod of the profilograph
5.	Weighing	Mass and contour area of the specimen	Available data on density and porosity of the specimen, complete removal of the lubricating oil
6.	Analysis of wear products in lubricating oil	Volume or weight of wear products in lubricating oil	Available data on density and porosity of the wear products, their complete washing out of the friction zone by lubricating oil
7.	The method of activation surfaces	The number of radioactive radiation impulses within the given time interval	Available sufficiently accurate "radioactivity-wear" calibrating diagrams for the materials under investigation
8.	The method of residual coating thickness	Coating thickness	Sufficiently thick coating
9.	The method of the joint gap	The flow or flux (air, electron, etc.) passing through the gap	Constructional feasibility of passing the measuring flow or flux

Method of Artificial Bases

In the method of artificial bases the linear wear of the surface is determined by measuring specially prepared markers of well-defined shapes and dimensions before and after testing. The method is commonly applied to wear evaluation of the guides of metal-cutting machine tools, textile machine parts, cylinders of aircraft and tractor engines, piston rings, etc. Usually the bottom of a recess (indentation), from which a distance to the worn surface is measured, serves as an artificial base. An indentation may be produced by pressing a tetrahedral pyramid using a Vickers hardness tester or a microhardness tester into a specimen. For a pyramid with square base the linear wear size in mm is calculated from the following equation:

$$\Delta b = b - b_1 = \frac{c - c_1}{z}, \tag{6.14}$$

where b, b_1 are the depths of the indentation produced with a microhardness or Vickers hardness tester and measured before and after a certain stage of wear, respectively, mm; c, c_1 are the lengths of the indentation diagonal measured before and after a certain stage of wear, respectively, mm; z is the proportionality factor (for a pyramid with square base and vertex angle between two opposite sides 136° $z = 7$).

When using a microhardness tester to produce an indentation, wear measurement accuracy can be as high as a fraction of micrometer (≈0.3 μm). Pressing a pyramid using hardness tester is accompanied by material swelling along the edges of the indentation, which reduces the accuracy of the method. Besides, small indentations are difficult to detect after intense wear. The dimensions of an indentation produced on a very hard coating may be reduced noticeably due to elastic recovery after removing the load. When studying plastic coatings working in the conditions of heavy contact loads, filling up, swelling and smearing of the indentations can take place.

Local porosity of coatings and anisotropy of their properties, which can influence the relationship between the depth and the length of the indentation and thus increase testing error, should also be taken into account.

It should be noted, that the method of artificial bases can be applied only to coatings with minor surface roughness. Extension of the method to rough-machined surfaces leads to corruption of the results.

The method of artificial bases was successfully applied to evaluation of the wear resistance of TiN-based coatings on the parts of an automatic loom. Working surface of the coated part was indented with a diamond pyramid, and, after certain periods of operation, the required indentation measurements were performed (see Fig. 6.44).

Method of Cut Indentations

The method of cut indentation provides higher accuracy as compared to the method of artificial base. Special commercial equipment is available for testing flat, convex and concave cylindrical surfaces. Indentations are cut out using a diamond cutter with working part in the shape of a regular trihedral pyramid.

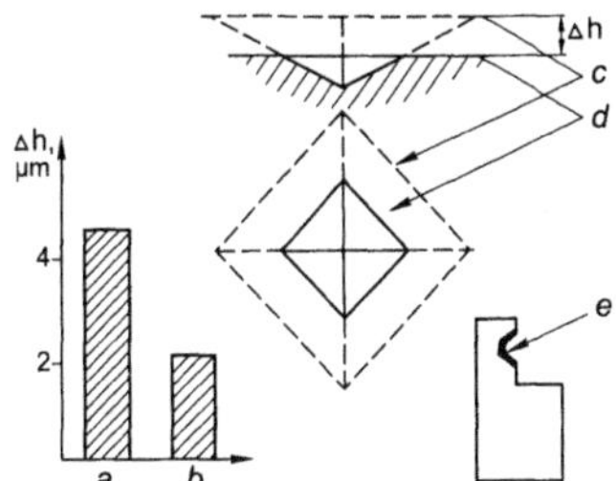

Fig. 6.44. Example of application of the artificial base method to evaluation of the efficiency of TiN-based coatings on a loom part. Δh is the wear of non-coated (*a*) and coated (*b*) surface; *c*,*d* wear surface before and after the test, respectively; *e* tested surfaces of the "block" part

Firstly, the locations and depths of the indentations are chosen based on the expected wear size and the purposes of the study. The indentations are cut out at thoroughly degreased and dried surface, using cutter feed 0.002–0.003 mm/rev and the radius of its top rotation 11±0.1 mm (see Fig. 6.45). The length of the indentation is measured with a microscope, first immediately after cutting and then upon completion of the test (State Standard 17534).

Linear wear of flat (see Fig. 6.46) and cylindrical surfaces with the indentation located along the cylinder element (Δh) is calculated from the equation:

$$\Delta h = h - h_1 = 0.125\left(l^2 - l_1^2\right)\frac{l}{r}, \tag{6.15}$$

where h, h_1 are the indentation depths before and after a certain stage of wear, respectively, mm; l, l_1 are the indentation lengths before and after a certain stage of wear, respectively, mm; r is the rotation radius of the cutter top, mm.

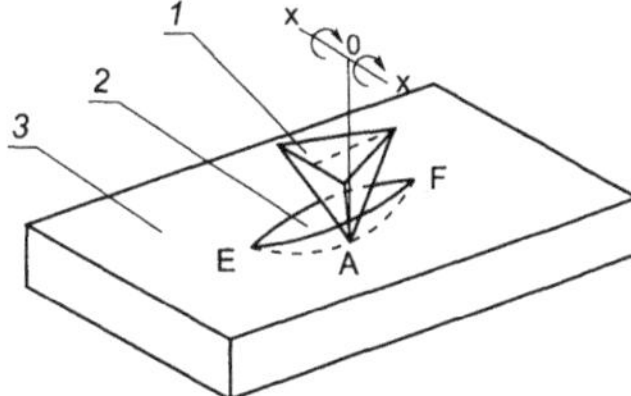

Fig. 6.45. Scheme of cutting an indentation. *1* trihedral diamond cutter; *2* indentation; *3* friction part surface; *X* rotation axis of the cutter; *EF* indentation length; *OA* rotation axis of the cutter top

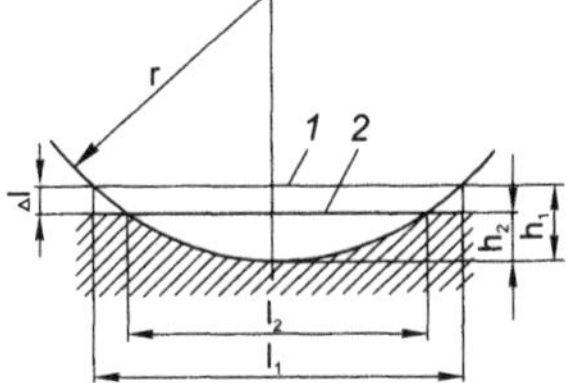

Fig. 6.46. Scheme of wear measurement using the method of cut indentation. *1*,*2* upper border of the friction surface before and after wear, respectively; h,h_1 indentation depths before and after a certain stage of wear, respectively, mm; l,l_1 indentation lengths before and after a certain stage of wear, respectively, mm; Δh linear wear; *r* rotation radius of the cutter top

Linear wear of cylindrical surfaces with the indentation located perpendicularly to the cylinder element (Δh) is calculated from the equation:

$$\Delta h = h - h_1 = 0.125\left(l^2 - l_1^2\right)\left(\frac{1}{r} \pm \frac{1}{R}\right), \tag{6.16}$$

where R is the radius of surface curvature at the place of indentation, mm.

The signs "plus" and "minus" are taken for convex and concave surfaces, respectively.

Method of Recording the Closing of the Specimens

Wear measurement by recording the closing of the specimens is performed using impedance transformers, either inductive (contact and non-contact versions) or capacitive. The main advantages of the method is continuous wear recording, high measurement accuracy and sensitivity of measuring devices. When a strain gauge is employed as a signal transducer, its ohmic resistance is determined by elastic deformation of the gauge, which depends on the displacement of the free surface of the stationary specimen relative to the base of the movable specimen [6.7].

The device (see Fig. 6.47) consists of the housing 3 attached to the free surface of the stationary specimen with angle pieces 5; elastic beam 2 made of flat bronze spring; strain gauges 4, glued onto the spring beam; prod 6 contacting with the movable specimen; spring 1 for preloading through the spring beam.

Wear evaluation using inductive contact transformer (a mechanical prod) takes advantage of the change of inductivity caused by increasing air gap.

Capacitive transformers are the most suitable devices for tribosystems with high wear rate. In this case the movement of the free surface of the stationary specimen relative to the base of the movable specimen is transformed into the change in the gap between the plates of the measuring capacitor.

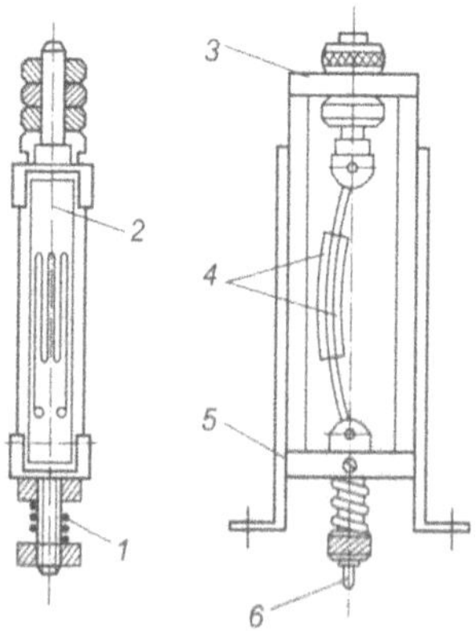

Fig. 6.47. Device for wear measurement using impedance transformer. *1* spring; *2* elastic beam; *3* housing; *4* strain gauges; *5* angle pieces; *6* prod

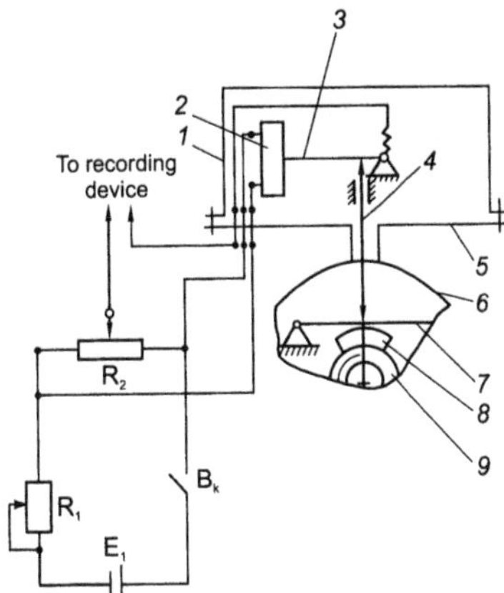

Fig. 6.48. Arrangement of wear measurement using pressure transducer. *1* housing; *2* potentiometer; *3* slide block; *4* pin; *5* housing; *6* hermetic chamber; *7* lever; *8* movable specimen; *9* stationary specimen

The device shown in Fig. 6.48 is made on the basis of an electric pressure sensor [6.7]. The potentiometer 2 with slide block 3 driven by pin 4 is placed inside the housing 5 connected with the hermetic chamber 6. Reduction of the total size of the loaded specimens 8 and 9 in the course of testing lowers the pin 4. The specimens are loaded through the lever 7.

Linear wear can be recorded using any high-sensitive device with high input impedance. Wear of specimens results in lowering the pin 4 and, consequently, in displacement of the potentiometer slide block. This disturbs the balance of the bridge, and the produced signal directly proportional to the specimen size change (the total linear wear) is applied to the recording device. The device is calibrated using any clockwork-type indicator.

Profilographing

The profiligraphing method provides sufficiently accurate wear measurement of flat cylindrical surfaces. Two versions of this method have been suggested: repeated profilographing with base application (indentation), and single profilographing. In the former version the surface of the tested specimen is indented twice using Vickers hardness tester, so that wear surface be located between the indentations. The indentations have equal depth exceeding the expected wear size not more than two times. The profilographing path should pass through the lowest points of the indentation, which are taken as the base. Wear size is determined by the distance between the average lines of surface profiles formed at each subsequent stage of wear.

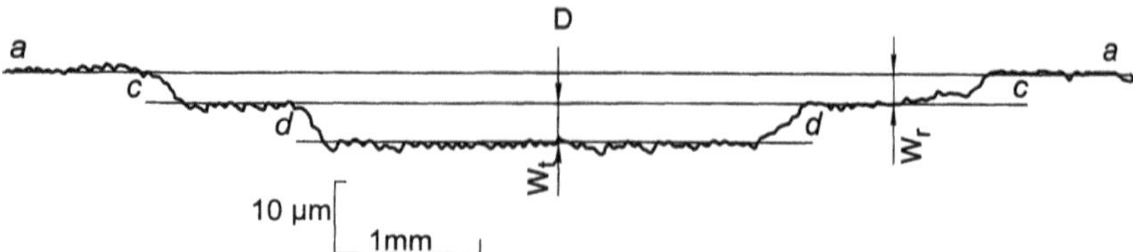

Fig. 6.49. Scheme of wear measurement by single profilographing. *a–a* base surface; *c-c* run-in surface; *d–d* surface formed as a result of wear; W_r run-in wear; W_t testing stage wear

If wear measurements at each stage of testing are required in single profilographing, it is recommended to reduce at least one dimension of the tested specimen surface area after each stage of wear (State Standard 23.224). In this case wear size is determined using the areas of worn surfaces formed at each of the stages and preserved on the specimen after all wear stages have been completed. A typical form of the profilogram obtained using this technique is shown in Fig. 6.49. Worn surface areas preserved on the specimens were formed at various stages of wear: a–a – at the lapping stage; c–c – at the stage of run-in testing; d–d – at the stage of wear resistance testing. Thus, the profilogram yields the sizes of the run-in wear W_r and steady-state wear W_t as the distances between the average lines of surface profiles formed at the appropriate wear stages.

Weighing of Wear Products and Specimens

Two ways of evaluating wear by weighing have been suggested: determination of the mass of wear products, and determination of the mass of the worn specimen. In the former case testing error depends on how thoroughly the wear particles have been collected. In the conditions of dry friction of the specimens wear products can be collected using forced air-cooling of the contacting surfaces. For friction in the presence of a lubricant the mass of forming wear particles is determined by the difference between the results of weighing of finely porous membrane filters through which the lubricant is passed during the test. Impossibility of separating wear products of the specimen and counter-specimen is the main drawback of the first weighing method. The second method can be applied only to small specimens, when wear mass loss is comparable with the total mass of the specimen. This method, however, may be completely unacceptable for evaluation of the wear of porous coatings working with lubrication because of the presence of hardly-removable oil and wear products in the pores. When specimens with significantly different chemical composition and porosity are tested, the difference between coating densities must be taken into account, and the results should be represented in relative units as compared to the reference specimen.

Both variants do not allow taking into account transfer of material in friction when determining mass loss.

Wear Evaluation of Metal Coatings by Suppression of Scintillations (Glow) of the Lubricating Oil

In order to produce a scintillator special additives are introduced in the lubricating oil. Testing is carried out on a device fitted with a system providing complete washing of wear products from the friction zone into the oil. During the test equal volumes of the scintillator are periodically sampled into measuring sleeves with subsequent radiometric analysis using a radiometric device (radiometer) and a special gamma-ray source located under the sleeve containing the sampled probe. As the mass of wear products grows, the intensity of oil scintillations (the number of pulses per unit time) goes down from sample to sample. Wear sizes are determined from previously prepared calibrating curve giving the number of pulses from the sample of the lubricating oil as a function of the weight content of mechanical particles in the oil.

Method of Activation Surfaces

Wear of metal coatings both in the course of testing and during operation can be measured by the surface activation method. Selected areas of the tested surface are activated by direct beam of charged particles, e.g. protons, accelerated up to the specified energy using a cyclotron. Activated areas of the specimen (part) surface and the measuring transducer are fixed at the minimal distance relative to each other, which remains constant within the entire wear period. Wear size may be estimated by reduction in gamma-ray emission intensity, since the thickness of the previously activated area decreases during wear. The exact values of average linear wear are determined from calibrating diagrams giving the number of pulses (which are corrected for background) versus the thickness of the removed layer.

Methods of Wear Determination by Residual Thickness of Coatings and Joint Gap

In the method of the residual thickness of the coating wear is measured using a suitable thickness gauge (see section 3.9).

The joint gap is measured using pneumatics principles. Compressed air is supplied to the gap from a nozzle, and the exiting airflow that depends on the gap size, i.e. on the wear, is determined using a pneumatic transducer.

6.11.7 Determination of Wear Intensity

Wear intensity in friction pair (I) is calculated from the equation:

$$I = \frac{W}{L}, \tag{6.17}$$

where W is linear wear of a specimen, m; L is friction path of a specimen, m.

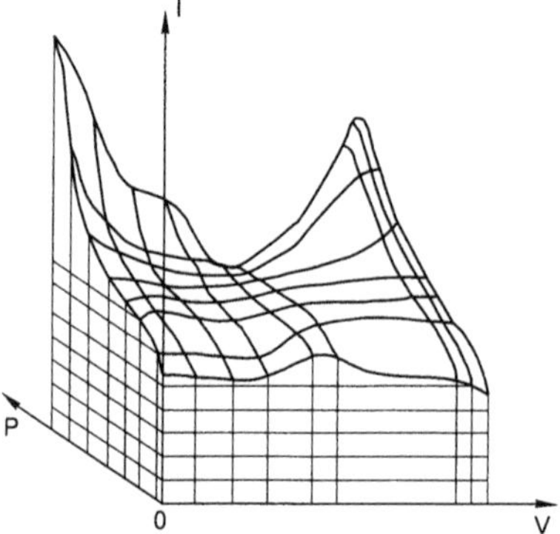

Fig. 6.50. Diagram of wear intensity (*I*) versus sliding speed (*V*) and specific pressure (*P*)

It is convenient to plot the wear intensity curve as a function of sliding speed and specific pressure on three-dimensional Cartesian coordinates (Fig. 6.50).

These curves may serve as initial data for solving resource prediction problems using a computer. The computer can provide the picture of the changes in the state of the tested friction unit from run-in starting point till attainment of ultimate wear. In a very short machine time a computer-based system is able to solve problems of resource prediction and wear of various friction units with due account of part deformation due to wear, non-uniform distribution of specific load over the contact surface, and heat generated by friction. Specialized resource prediction program applications for several kinematic types of connection are available. For certain products (working members of cylinder-piston engine group, automobile brake straps and drums, pumping plant parts) such characteristics as residual resource, residual resource variance, and probability of survival can be calculated from the measured values of wear intensity [6.83].

6.11.8 Evaluation of Frictional Compatibility of Materials in Friction Pair

Friction machines can also be used to evaluate frictional compatibility of metal coatings deposited on the movable specimen and material of the stationary specimen (State Standard 23.222). In this case the extreme temperature, force and speed operating conditions in the faulty lubrication mode are determined. Specimens of the "disk-block" type are recommended. A thermoelectric temperature transducer head is caulked in the stationary specimen at the distance 2 mm or less from the friction surface. Once the specimens have been run in and friction torque has stabilized at the frequency (300±10) min^{-1}, the desired load is applied. The steady-state values of friction force moment and specimen temperature are recorded at the fixed temperatures of 60, 80, 100, 120, 140, 160, and 180°C (the oil is heated up in a bath by a heater).

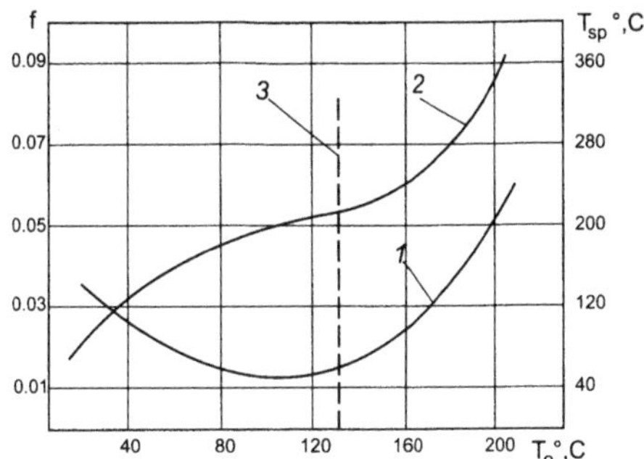

Fig. 6.51. Example of the dependence of friction coefficient and temperature of the specimen on the temperature of oil. *1* friction coefficient; *2* temperature of the stationary specimen; *3* critical oil temperature

When extreme temperature, force and speed operating conditions are determined in the faulty lubrication mode, average friction coefficient values for the given temperature (f) are first calculated from the following equation for each testing run:

$$f = \frac{M}{RP}, \tag{6.18}$$

where M is the average value of friction force moment, N·m; R is the radius of the movable specimen, m; P is the applied load, N.

Then the critical oil temperature, at which monotonous growth of friction coefficient begins (see Fig. 6.51), is found from the plot of specimen friction coefficient and temperature versus oil temperature.

For comprehensive compatibility evaluation the compatibility criterion (C), calculated from the following equation, is used:

$$C = \frac{1}{f_{cr}} \cdot \frac{T_{o.cr}}{T_{sp.cr}}, \tag{6.19}$$

where f_{cr} is the friction coefficient calculated from the equation (6.18) at the critical oil temperature, °C; $T_{o.cr}$ is the critical oil temperature, °C; $T_{sp.cr}$ is the specimen temperature attained at the critical oil temperature, °C.

Materials with higher C values have better compatibility.

6.11.9 Study of Friction Surfaces

Numerical results of tests are usually complemented by the data of macro-, micro- and submicroanalyses of friction surfaces, since it is structural transformations occurring in the zone of friction contact that determine wear and failure mechanisms. Besides structural studies, macrolevel analysis provides data on changes in the friction surface geometry. For instance, basing on two-dimensional profilorgaph measuring system and having the results of several tens of measurements, isometric representations of the area before and after wear can be plotted. At micro- and submicrolevel phase structural, substructural and dislocation changes, chemical transformations, and electronic surface structure are evaluated.

Friction surface macroanalysis is performed on longitudinal, transversal and slanted sections. The latter type of section is preferable for studying thickness and structure of the forming films, and also for determination of microhardness of the near-surface layer. For determination of the profile of mated surfaces before and after testing it is expedient to use transversal sections. Longitudinal sections yield data on surface profile, as well as on qualitative and quantitative microstructural changes (after etching). Among the disadvantages of longitudinal sections is difficulty and, at times, even impossibility of studying rough relieves because of low focus depth of light microscopes.

The increase in the number of friction surface investigations on micro- and submicrolevels is, to a great extent, caused by application of new techniques of studying very thin films [6.1, 6.56, 6.70, 6.78]. The main distinguishing feature of these methods is high depth resolution, allowing to obtain information from a layer one to some tens of interatomic distances deep. Their broad introduction in laboratory practice is caused, first of all, by commercial availability of ultravacuum devices, required for preparing clear surfaces, and further refinement of measuring equipment.

All submicroanalysis methods are based on interaction between primary radiation and surface material, as well as on studying scattered or secondary radiation. To date several tens of such methods have been suggested. Some of them were considered in some detail in the first chapter.

6.12 Abrading Ability of Coated Surfaces

6.12.1 General Overview

Abrading is removal of the layers of material due to wear in the conditions of sliding friction.

Values of the height parameters R_a and R_z, which are usually defined to normalize the roughness of coatings, generally cannot characterize the abrading ability of the coating. The latter depends on the values of step parameters and relative support length of profile, as well as on the orientation of surface irregularities. Besides, the requirements for roughness are defined without reference to defects, which are always present on the surface of a coating (scratches, punctures, large surface pores, dents, exfoliations, swells).

According to Khruschov and Pruzhansky [6.41], for surfaces with the equal roughness index R_a the values of abrading ability indexes may differ as much as six times.

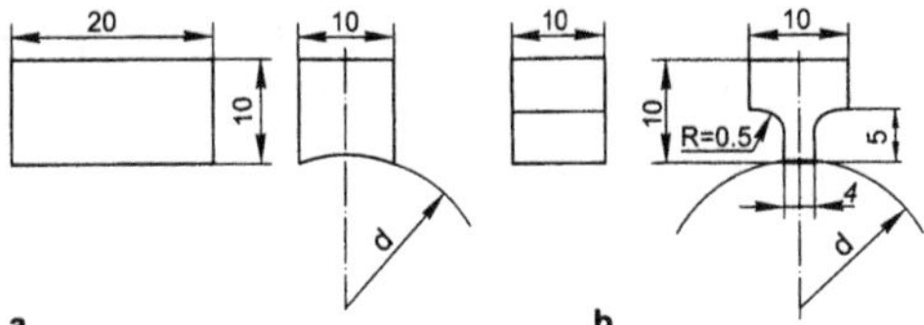

Fig. 6.52. Specimens for testing. **a** supporting specimen; **b** working specimen

Abrading ability of the coating should be determined concurrently with its wear resistance evaluation. A coating with sufficient wear resistance cannot be regarded as satisfactory if it actively abrades the mated counter-body.

To evaluate the abrading ability of a coating the surface of a specimen made of reference material is worn out by the surface of a shaft covered with the tested coating. Abrading ability of the coating is judged by the linear wear of specimens over a certain friction path.

The method is capable of comprehensive and conclusive evaluation of shaft surface abrading ability immediately at the depositing post or circular grinder.

The standard (State Standard 23.220) regulating application of this method covers surfaces of shafts with diameter 10–100 mm with minimal roughness R_a 0.16 μm and hardness HV 30 and higher.

6.12.2 Specimens

The following requirements are specified for the reference specimen: frictional compatibility with the tested coating; high wear rate not accompanied by changes in the coating surface relief; no chemical interaction with the coating. Extruded fluoroplastic with density 2.2 g/cm^3 meets these requirements [6.61]. The roughness R_a of the working surface of the specimens (see Fig. 6.52) is 0.6 μm or less.

6.12.3 Equipment

In the testing diagram (see Fig. 6.53) the working specimens made of reference material with initially flat friction surfaces and a supporting specimen made of the same material and previously run in over the shaft are located in the holders attached to the cramp. The specimens are pressed to the coated surface with a fixed force using springs. Oil from the lubricator via a tube is supplied to the friction contact zone.

6.12.4 Preparation for Testing

Polished surface of the coated shaft and radial surfaces of the specimen are washed in benzine and acetone and dried up. Depending on the selected way of wear evaluation either height or mass of specimens are measured. Rotation speed (n) in min^{-1} and time of testing (t) in seconds are calculated from the equations:

$$n = \frac{19 \cdot 10^3}{d} V, \tag{6.20}$$

$$t = \frac{9.55 \cdot 10^6}{nd}, \tag{6.21}$$

where V is the sliding speed, m·s^{-1}; d is the shaft diameter, mm.

Testing duration t corresponds to 1000 m friction path.

6.12.5 Testing

Specimens are pressed to the shaft by the equal load of 98.2±1.96 N and run in until cylindrical recesses are formed all over the working surface and reliable contact in the mating zone is provided. Upon completion of the testing at the specified rotational speed n for time period t, the specimens are taken from the holders, rinsed and dried. Linear wear is determined using micrometry or weighing.

6.12.6 Processing of Test Results

Wear intensity (I) is calculated from the equation:

$$I = \frac{\Delta h}{5 \cdot 10^5}, \tag{6.22}$$

where Δh is the linear wear, mm.

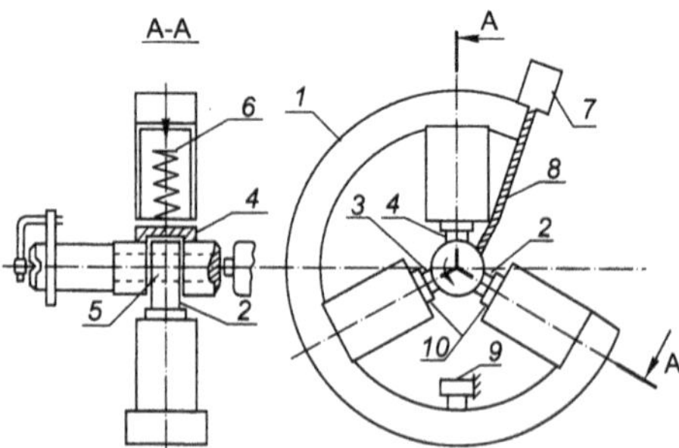

Fig. 6.53. Installation for evaluating the abrading ability of recovered shafts (adopted from [6.61]). *1* cramp; *2,3* specimens; *4* supporting specimen; *5* shaft; *6* spring; *7* lubricator; *8* tube; *9* stop; *10* holder

The obtained value is characteristic of the relative abrading ability of the coated shaft surface. The method is recommended for comparative tests of coatings deposited on shaft surfaces after finishing mechanical treatment.

References

6.1 Ahn HS, Kwon OK (1993) Wear behaviour of plasma-sprayed partially stabilized zirconia on a steel substrate. Wear 162–164:162–164

6.2 Alyabiev AYa, Kovalevski VV, Melnikov VV (1983) Effect of laser treatment of steels with different carbon content on wear resistance in the conditions of fretting (in Russian). Trenie i Iznos 4:508–513

6.3 Axen N, Jakobson S, Hogmark S (1994) Influence of hardness of the counterbody in three-body abrasive wear on overlooked hardness effect. Tribology 27:233–241

6.4 Bani SY, Raman N, Sechan S (1996) Effect of structural features on erosion resistance of cast irons. Tribology Letters 2:99–111

6.5 Barbezat G, Nicoll AR, Sickinger A (1993) Abrasion, erosion and scuffing resistance of carbide and oxide ceramic thermal sprayed coatings for different applications. Wear 162–164:529–537

6.6 Batchelor AW, Stachowiak GW, Stachowiak GB, Leech PW, O Reinhold (1992) Control of fretting friction and wear of roping wire by laser surface alloying and physical vapor deposition coatings. Wear 152:127–150

6.7 Batuev GS, Golubkov VS, Smushkovich BA (1982) Methods of wear and friction testing of materials. In: Klyuev VV (ed) Testing techniques (in Russian). Mashinostroenie, Moscow, pp 190–240

6.8 Bihat H, Herman H, Smith MF (1985) Cavitation-erosion of plasma-sprayed alloys. 2nd Nat Conf Therm Spray, Long Beach, Calif, 31 Oct–2 Nov 1984. Metals Park, Onto, 1985, 101–104

6.9 Blanpain B, Celts JP, Roos JR, Ebberink J, Smeet SJ (1993) A comparative study of the fretting wear of hard carbon coatings. Thin Solid Films 223:65–71

6.10 Borisov Yu, Borisova A, Dyadko E (1995) Composite Ti–Si–C system coatings. Mater and Manuf Processes 10:831–835

6.11 Boving HJ, Hintermann HE, Julia C (1992) Characterization of CVD and PVD coatings. Adv Mater Technol Int: 81–84

6.12 Braun ED, Buyakovski IA (1997) Trends in the development of tribological testing techniques (a review) (in Russian). Zavodskaya Laboratoria 1:3–8

6.13 Braun ED, Smushkovich BL (1984) Modern commercial friction machines (in Russian). Trenie i Iznos 5:94–99

6.14 Braun ED, Buyakovski IA, Smushkovich BL (1997) Techniques of tribological testing (a review) (in Russian). Zavodskaya Laboratoria 10:29–37

6.15 Bromark Michael (1996) A novel abrasive wear test for thin hard coatings. Tribology 15:29–40

6.16 Bromark Michael, Larsson Mats, Hedengvist Per, Olsson Mikael, Hogmark Sture (1992) Influence of substrate surface topography on the erosion resistance of TiN-coated tool steels. Tribology 11:153–160

6.17 Byakova AV, Vasiliev AI, Vlasov AA (1992) On the wear resistance of titanium nitride coatings in the conditions of fretting corrosion (in Russian). Trenie i Iznos 13:674–682

6.18 Callas PK, Pirso YuYu, Valdma LE (1980) Hydroabrasive wear of sintered chromium carbide alloys (in Russian). Trenie i Iznos 1:859–863

6.19 Cartier M, McDonnell L, Cashell EM (1991) Friction of tungsten carbide–cobalt coatings obtained by means of plasma spraying. Surface and Coating Technology 48:241–248

6.20 Clace IR, Waymace CC (1986) Wear resistance of the parts of crushing machines (in Russian). Mashinostroenie, Moscow

6.21 De Bruyn K, Celts JP, Roos JR, Stals LM, Van Stappen M (1993) Coating thickness and surface roughness of TiN-coated high speed steel in relation to coating. Wear 166:127–129

6.22 Dennis JK, Sagoo KS (1991) Wear behavior of engineering coatings and surface treatments. Metal Finish 89:111–122

6.23 Diao DP, Sawaki Y (1995) Fracture mechanisms of ceramic coatings during wear. Thin Solid Films 270:362–366

6.24 Dobrovolski AG, Koshelenko PI (1993) Abrasive wear resistance of materials. A reference book (in Russian). Tekhnika, Kiev

6.25 Dowson D (1985) Wears on where? Wear 103:189–203

6.26 Eaton HE, Novak RC (1987) Particulate erosion of plasma-sprayed porous ceramic. Surface and Coating Technology 30:41–50

6.27 Erdemir A, Switala M, Wei R Wilbur P(1991) A tribological investigation of the graphite-to-diamond-like behavior of amorphous carbon films ion beam deposited on ceramic substrates. Surface and Coating Technology 50:17–23

6.28 Fukumoto Masahiro, Nakaoko Makoto, Okane Isao (1987) Erosion resistance of Ni and Cr–B–Si coatings (in Japanese). J Jap Weld Soc 5:385–389

6.29 Gard D, Dyer PN (1993) Erosive wear behavior of chemical vapor deposited multilayer tungsten carbide coating. Wear 162–164:552–557

6.30 Golego NL, Alyabiev AYa, Shevelya VV (1974) Fretting corrosion of metals (in Russian). Tekhnika, Kiev

6.31 Golubev VI, Sytov VP, Dovgal YuG, Sorokin VG, Bournyshev IN (1978) Hydroabrasive resistance of metals and alloys with diffusion coatings (in Russian). Zaschitnye Pokrytia na Metallah 12:66–68

6.32 Govindaraju Madhav Rao, Molian PA (1994) Enhancement of wear and corrosion resistance of metal-matrix composites by laser coatings. J Mater Sci 29:3274–3280

6.33 Hornbogen E (1975) Der Einfluss der Bruchzaehigkeit auf den Verschleiss metallischer Werkstoffe (in German). Zeitschrift fuer Metallkunde 66:507–511

6.34 Hornbogen E (1980) Werkstoffeigenschaften and verschleiss (in German). Metallwissenschaft und Technik 34:1079–1086

6.35 Ilinski II, Dukhota AN, Sergeev VV (1981) Optimal and limiting conditions of fretting resistance of detonation coatings on the basis of tungsten carbide (in Russian). Trenie i Iznos 2:850–855

6.36 Itoh Y, Takahashi M, Saitoh M (1991) Dust erosion characteristics of sintered ceramics and sprayed coatings (in Japanese). Quart J Jap Weld Soc 9:74–79

6.37 Itoh Yoshiyasu, Saitoh Masahiro, Kashiways Hideo, Asai Satoru (1991) Dust erosion resistance of thermally sprayed metallic coatings (in Japanese). Quart J Jap Weld Soc 8:419–426

6.38 Kassman Asa, Jacobson Staffan, Erickson Lynn, Hedengvist Per, Olsson Mikael (1991) A new test method for the intrinsic abrasion resistance of thin coatings. Surface and Coating Technology 50:75–84

6.39 Kerkush IR, Melnik PI (1997) Properties improvement of plasma-sprayed iron coatings by chromium plating. Poroshkovaya Metallurgia 40:230

6.40 Khrushchov MM, Babichev MA (1970) Abrasive wear (in Russian). Mashinostroenie, Moscow

6.41 Khrushchov MM, Pruzhansky LYu (1973) A method and device for evaluating abrading ability of treated shaft surface (in Russian). Mashinostroenie 4:112–115

6.42 Kiiski AA, Ruuskanen PR, Rubin JB (1996) Wear resistant coatings produced by shock-wave compaction of powders. Met and Mater Trans A 27:2297–2304

6.43 Kingswell R, Rickerby DS, Bull SJ, Scott KT (1991) Erosive wear of thermally sprayed tungsten coatings. Thin Solid Films 198:139–148

6.44 Kovalevski VV, Melnikov VV, Bolshunov VV (1983) Study of wear resistance in the conditions of implanted ion films fretting (in Russian). Trenie i Iznos 4:344–347

6.45 Krai MV, Davidson JL, Wert JJ (1993) Erosion resistance of diamond coatings. Wear 166:7–16

6.46 Kroha VA, Planida VE (1988) Evaluation of the accuracy of standard metal abrasive resistance tests (in Russian). Trenie i Iznos 36:1128–1133

6.47 Kulu P (1988) Wear resistance of powder materials and coatings (in Russian). Valgus, Talinn

6.48 Levy Alan V (1988) Erosion of hard material coating systems. Wear 121:325–346

6.49 Lin Jen Fin, Li Tzuen Ren (19993) Analysis of the friction and wear mechanisms of multilayered plasma-sprayed ceramic coatings. Wear 160:201–212

6.50 Ma Xiang-Dong, Lin Fu-Yan, Shao He-Sheng (1993) Study of antiwear adhesive coating under different erosion conditions. Wear 162–164:569–573

6.51 Maier HJ, Christ H–J, Hammer J, Schindler R (1994) Erosion of hard coatings by rain-impact. EUROMAT'94 Top: 15th Con, Mater, Test Metall, Balatonszeplak, 30 May–1 June 1994: Conf Proc Vol 2- [Balatonszeplak] 1994- pp 482–487

6.52 Makhutov NA, Bogdanovich AV, Andropov PV, Marchenko AV, Tyurin AS, Sosnovski LA (1995) Methods of fatigue wear testing and their realization in the SI machine (in Russian). Zavodskaya Laboratoria 6:37–41

6.53 Misra A, Finnie I (1980) A classification of three-body abrasive wear and design of a new tester. Wear 60:111–121

6.54 Misra A, Finnie I (1981) Correlations between two-body and three-body abrasion and erosion of metals. Wear 68:33–39

6.55 Misra A, Finnie I (1981) On the size effect in abrasive and erosive wear. Wear 68:41–56

6.56 Moonir-Vaghefi SM, Saatchi A, Hedjazi J (1997) Tribological behavior of electroless Ni–P–MoS_2 composite coatings. Z MetallK 88:498–501

6.57 Moore MA (1974) The relationship between the abrasive wear resistance hardness and microstructure of ferritic material. Wear 28:59–68

6.58 Nicoletto G, Tucci A, Esposito L (1993) Comparative dry wear behavior of hard coatings. Wear 162–164:925–929

6.59 Oswald A, Husemann K (1992) Verschleissverhalten plasmagespritzter Schichten bei Strahlbeanspmchung (in German). Schweiss und Schneid 8:432–436

6.60 Peteres John A (1996) New approaches to surface engineering. Metallurgia 63:408–409

6.61 Pruzhanski LYu (1978) Analysis of wear testing techniques (in Russian). Nauka, Moscow

6.62 Revo SL, Borisova AL, Dashevski NN (1992) Effect of thermoload impact on erosion resistance of coatings. In: Borisenko AI (ed) Corrosion resistant coatings (in Russian). Nauka, Leningrad, pp 111–114

6.63 Rezakhanlou R, Guerout F, Zbinden M, Lina A, Clemendot F, Leiait L, Van Duysen JC (1992) Correlative study of scratch and wear behavior of several surface treatments for potential uses in nuclear industries. Vide, Couches Minces 48:358–360

6.64 Rogers PM, Hutchings IM, Little JA (1992) TiN coatings for protection against combined wear and oxidation. Surface Eng 8:48–54

6.65 Rosenfield AR (1990) Wear and fracture mechanics: are they related? Scripta Metallurgica 24:811–814

6.66 Schlett V, Stuke H, Weiss H (1986) Morphology and erosive wear of Fe–Cr coatings for gun bores deposited by physical vapor deposition. J Vac Sci and Technol A 4:3032–3037

6.67 Sorokin GM (1976) Techniques of impact wear testing (in Russian). Vestnik Mashinostroeniya 4:11–14

6.68 Sorokin GM (1989) Development of techniques for abrasive wear testing of materials (in Russian). Zavodskaya Laboratoria 9:74–78

6.69 Srinivasan V (1994) High-temperature corrosion and erosion in gas turbine engines – where do we stand? J Miner, Metals and Mater Soc 46:34-35

6.70 Su YL, Lin JS (1993) Friction and wear behavior of a number of ceramic-coated steels matched as sliding pairs to various surface-treated steels. Wear 166:27–35

6.71 Takeda Koichi, Ito Mitihisa, Takeuchi Sunao, Sudo Katuji, Koda Masamichi, Kazama Koichi (1993) Erosion resistant coating by low-pressure plasma spraying. ISIJ International 33:976–981

6.72 Tamura Motonori, Fukuda Kanao (1993) Properties and tribological behavior of Ti(C,N) coatings deposited by reactive ion plating. ISIJ International 33:949–956

6.73 Tarasov VV (1993) New methods for determination of wear resistance of coatings (in Russian). Trenie i Iznos 14:1087–1091

6.74 Taylor DE, Waterhouse RB (1972) Spayed molybdenum coatings as a protection against fretting fatigue. Wear 20:401–407

6.75 Taylor KA, Emrick AJ (1992) Comparison of stress deposited thick coatings of TiB_2+Ni on polymeric composites. J Vac Sci and Technol A 10:1734–1739

6.76 Tenenbaum MM (1982) Hydroabrasive wear resistance of materials (in Russian). Trenie i Iznos 3:76–82

6.77 Tokarev AO (2000) Hardening machine parts by wear resistant coatings (in Russian). Novosibirsk State Academy of Water Transport Press, Novosibirsk

6.78 Tomlinson WJ, Bransden AS (1994) Sliding wear of laser alloyed coating on Al–12%Si. J Mater Sci Lett 13:1086–1088

6.79 Tsai C, Nelson J, Gerberich WW, Heberlein J, Pfender E (1992) Metal reinforced thermal plasma diamond coatings. J Mater Res 7:1967–1969

6.80 Tsidulko AG, Dokukina IA, Rusakov VM, Kulagin NS, Iljinski II (1991) Wear resistance of gas-thermal coatings on the basis of alumina at elevated temperatures (in Russian). Poroshkovaya Metallurgia 5:27–29

6.81 Tushinski LI (1990) Theory and technology of metal alloy hardening (in Russian). Nauka, Novosibirsk

6.82 Tushinski LI, Bataev AA, Bataev VA, Geltman IS (1988) Wear of protective coatings under the action of gas abrasive medium (in Russian). Problemy Prochnosti 5:108–110

6.83 Tushinski LI, Plokhov AV, Sindeev VI (1996) Protective properties of surface layers after high-energy impacts (in Russian). Novosibirsk State Technical University Press, Novosibirsk

6.84 Tylkin MA (1981) A handbook of thermist of a service station (in Russian). Metallurgia, Moscow

6.85 Urbanovich LI, Kramchenkov EM, Chukosob YuN (1994) Gas abrasive wear of metals and alloys (in Russian). Trenie i Iznos 15:389–393

6.86 Vancoille E, Celts JP, Roos JR (1993) Dry sliding wear of TiN-based ternary PVD coatings. Wear 165:41–49

6.87 Vinitskaya GI, Gorpinchenko SD, Trahtenberg IS (1993) Installation for wearing materials and coatings in the flow of abrasive particles (in Russian). Zavodskaya Laboratoria 3:53–54

6.88 Wang Yinglong (1993) Friction and wear performances of detonation-, gun- and plasma-sprayed ceramic and cermet hard coatings under dry friction. Wear 161:69–78

6.89 Waterhouse RB (1984) Fretting wear. Wear 100:107–118

6.90 Yi Maozhong, Zhang Xianlong, Ji Gengshun, Zheng Jihong, He Jiavven (1997) Erosion wear of AlSi–graphite and Ni/graphite abradable seal coatings. Trans Nonferrous Metals Soc China 7:99–102

6.91 Yushchenko K, Borisov Yu, Murashov A, Kolomytsev M, Lugscheider E, Iokiel P, Feldhege M (1995) Thermal sprayed aluminium–silicon alloy coatings. Mater and Manuf Processes 10:837–841

6.92 Zum Gahr K-H (1987) Microstructure and wear of materials. Elsevier, Amsterdam

7 Fatigue Failure of the Base Metal–Coating Composition

7.1 High- Low-Cycle Endurance of Surface-Hardened Metals

7.1.1 General Overview

In operation many machine parts are subjected to loads changing in value and direction. Such repeatedly varying stresses entail gradual fatigue failure of metal.

"Fatigue" is interpreted as gradual accumulation of damages leading to a change in properties, formation and development of cracks and failure. Failure occurs without visible signs of plastic deformation and takes place at stresses not only lower than the ultimate strength and yield point, but even below the limit of elasticity as well.

Failure is preceded by appearance and development of typical microcracks caused by structural features of metal and coating.

At repeated action of varying loads applied to locations of drastic increase in stresses due to holes, turnings, cavities, inclusions, and cracks, etc., sudden fractures of fatigue failure occur. A place of fatigue fracture usually coincides with the zone of stress concentration caused by a drastic change in a cross-section.

Let us consider the main terms, definitions and designations related to fatigue resistance.

The stress cycle is a set of successive values of varying stresses per one period of their alteration. Stresses σ, τ of a cycle may be plotted as diagrams (Fig. 7.1), and defined from the equations [7.1, 7.9].

$$\sigma = \sigma_m + \sigma_a f(t), \tag{7.1}$$

$$\tau = \tau_m + \tau_a f(t), \tag{7.2}$$

where: σ_m, τ_m are the average stresses over a cycle; σ , τ are cycle amplitudes; $f(t)$ is the continuous periodic function characterizing the cycle mode.

Maximal stresses of a cycle are the highest algebraic values of cycle stresses:

$$\sigma_{\max} = \sigma_m + \sigma_a, \tag{7.3}$$

$$\tau_{\max} = \tau_m + \tau_a. \tag{7.4}$$

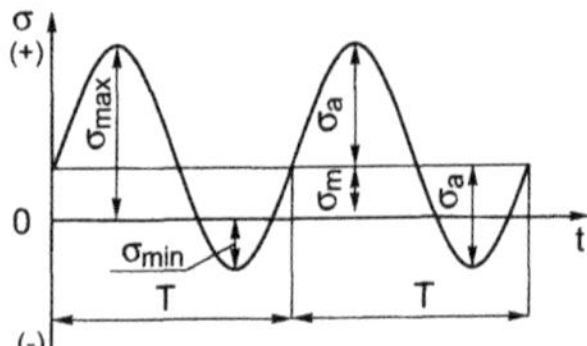

Fig. 7.1. Stress cycle in fatigue test

Minimal stresses of a cycle are the lowest algebraic values of cycle stresses, equal to the difference of the average cycle stress and amplitude:

$$\sigma_{\min} = \sigma_m - \sigma_a , \tag{7.5}$$

$$\tau_{\min} = \tau_m - \tau_a . \tag{7.6}$$

Average stresses are determined from the equations:

$$\sigma_m = \frac{\sigma_{\max} + \sigma_{\min}}{2} , \tag{7.7}$$

$$\tau_{\min} = \frac{\tau_{\max} + \tau_{\min}}{2} . \tag{7.8}$$

Amplitudes of cycle stresses are the highest (positive) values of the variable component of a stress cycle, determined from the equations:

$$\sigma_a = \frac{\sigma_{\max} - \sigma_{\min}}{2} , \tag{7.9}$$

$$\tau_a = \frac{\tau_{\max} - \tau_{\min}}{2} . \tag{7.10}$$

At high-cycle fatigue the material fails or damages in the area of elastic deformation, and at low-cycle fatigue it fails in the elastic-plastic area (a conventional boundary of high and low-cycle fatigue amounts to $5 \cdot 10^4$ cycles).

The fatigue life is interpreted as a fatigue resistance characteristic of material, which is defined by the number of loading cycles, withstood by a specimen before failure at a given stress.

Asymmetry parameter *R*, equal to algebraic ratio of the minimal cycle stress to the maximal one, is an important characteristic of a cycle:

$$R = \frac{\sigma_{\min}}{\sigma_{\max}} . \tag{7.11}$$

If the maximal cycle stress is equal to the minimal one in magnitude, the cycle is called symmetrical, and $R = -1$.

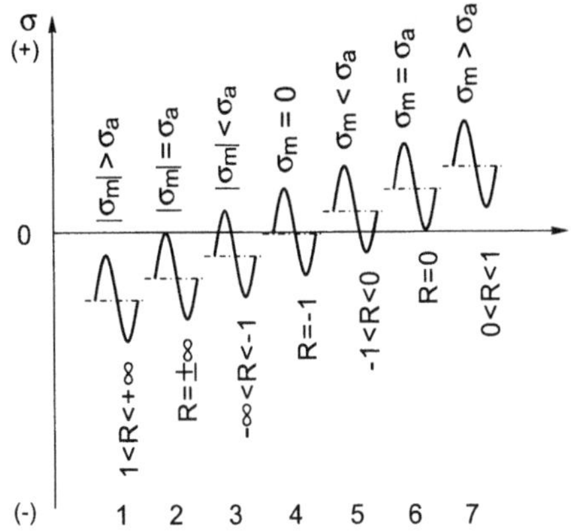

Fig. 7.2. Varieties of stress cycles and corresponding values of asymmetry coefficients

When testing specimens with constant R, the fatigue limit is defined as the highest value of the cycle stress maximal in magnitude, whose action does not lead to fatigue failure after unlimited number of cycles. In testing with constant σ_m, the fatigue limit is the highest amplitude of cycle stresses, whose action does not limit the service life of specimens. The fatigue limit is designated by symbol σ_R with asymmetry coefficient R shown as the index. Thus, in case of a symmetrical cycle the fatigue limit is represented by σ_{-1}.

All diversity of cycle loading situations can be reduced to seven main cases depending on relationships between the average stress and cycle amplitude (Fig. 7.2). Here 1, 2, 6, 7 are asymmetrical cycles with a constant sign (cycles 2 and 6 are also called pulsating or zero-to-compression stress cycles: $\sigma_{\max}$ or $\sigma_{\min}$ are equal to zero); 3 and 5 are sign alternating asymmetric cycles; 4 is the symmetrical cycle (the symmetrical cycle is always sign alternating).

Fatigue failure is an urgent problem, and technological progress mostly depends on solution to this problem. Of particular importance is the development of hypotheses on propagation of fatigue cracks within solid bodies depending on material structure, loading cycle parameters, physical-mechanical properties of medium, etc. [7.8].

7.1.2 Test Schemes and Specimens

The following structural schemes of fatigue tests for metal specimens with coatings can be recognized (Fig. 7.3).

The main task in studying coated articles at cyclic loading is to reveal the contribution of coatings to fatigue resistance.

Tests are carried out according to the methods established for metal specimens. The standard specimens are coated, and then the effect of the coating on fatigue resistance is estimated.

Table 7.1. Sizes of smooth specimens (by State Standard # 25.502)

Specimen type	Sizes [mm]							
I	d	5.0	7.5	10	12	15	20	25
	R			$\geq 5d$			≥ 90	
II	d	5.0	7.5	10	12	15	20	25
	$l=5d$	25	37.5	50	60	75	100	125
	R	5.0	7.5	10	12	15	20	25
III[a]	h		≤ 3.0				> 3.0–10	
	b		$10h$				15–30	
	R				$> 2b$			
III[b]	h				3.0–20.0			
	b				$0.5h$–$2h$			
	R				$> 5h$			
IV	h		≤ 3.0				> 3.0–10	
	b		$10h$				15-30	
	l				$5.6\sqrt{bh}$			
	R				$\geq 2b$			

[a]Bending in the plane of size h.
[b]Bending in the plane of size b.

Smooth specimens with round cross-sections of types I and II as well as specimens with rectangular cross-sections of types III and IV are used (Fig. 7.4, Table 7.1). The roughness of the working part of metal specimens is $Ra = 0.32$–0.16 μm.

Special procedures are performed before coating deposition (ultrasonic cleaning, sand blasting, chemical etching).

Measures excluding significant structural and physical-chemical changes in the base metal should be taken in coating deposition. There should not be local overheating, cracks, warping, etc. Within the limits of the planned experimental sets specimens are made by the same technology and in the same deposition regimes. Specimen working surfaces are usually ground after coating deposition. In every particular case demands to the roughness of ground surface are determined with consideration of the coating porosity. For gas-thermal coatings $Ra < 2.5$–1.25 μm is recommended. When analyzing the obtained results, difference in roughness should be taken into account.

The coating thickness should be uniform along the whole length of the working surface. Specimens with exfoliated and flaked coating are rejected.

In order to clarify the role of a coating in the specimen behavior under specified loading conditions, it is necessary to set up experiments to study the effect of preliminary preparation of specimen surfaces before spraying. The experiment should allow differentiation of the effects of preparation procedures from those of the coating presence. Therefore, two types of specimens are recommended as the controls: standard specimens and specimens, whose surface was treated in a way similar to the treatment before spraying.

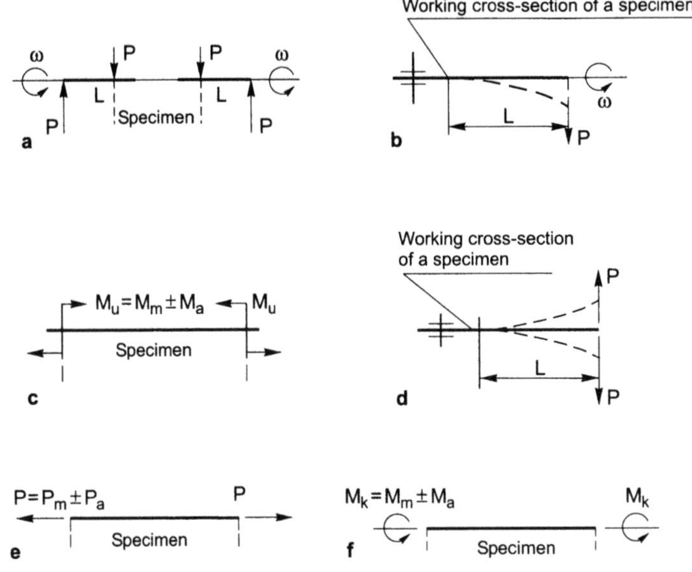

Fig. 7.3. Schemes of specimen loading in fatigue testing. **a** pure bending with rotation; **b** lateral bending with rotation under cantilever loading; **c** pure bending in one plane; **d** lateral bending in one plane under cantilever loading; **e** repeatedly variable extension-compression; **f** repeatedly variable torsion

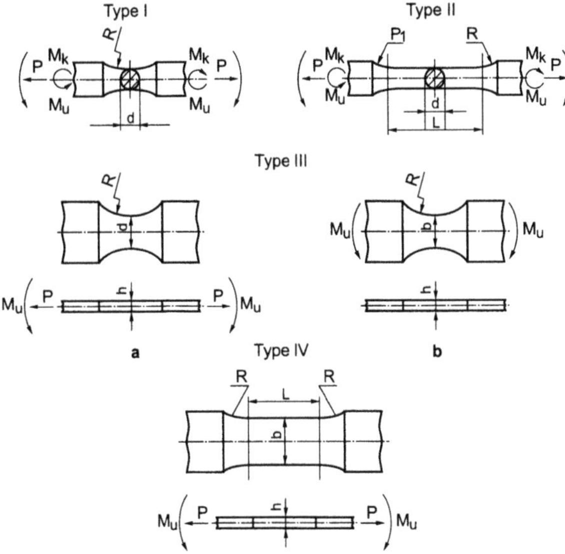

Fig. 7.4. Working parts of specimens of I–IV types

Not less than fifteen specimens should be used for tests in the high-cycle area, and five specimens are required for experiments in the low-cycle field.

Depending on loading schemes (see Fig. 7.3) the following types of specimens are used (see Fig. 7.4): pure bending with rotation – I, II; lateral bending with rotation under cantilever loading – I, II; pure bending in one plane – I–IV; lateral bending in one plane under cantilever loading – I–IV; repeatedly variable extension-compression – I–IV; repeatedly variable torsion – I, II.

7.1.3 Equipment

A great number of possible loading schemes have resulted in creation of numerous versions of test methods and designs of testing machines.

Many countries produce standard testing machines providing every loading scheme considered above. Installations and loading schemes are selected with consideration of operational demands to the base metal–coating composition. Specimens can be tested at hard and soft loading. For the first case alternating deformation is assigned, while for the second one the stress is variable. Machines for fatigue tests are divided into the following types: mechanical, electromechanical (with electromagnetic resonance excitation) and hydraulic. The main controlled parameters are as follows: maximal static and dynamic loads, maximal bending moment and frequency.

Mechanical machines for bending tests of rotating specimens at pure and cantilever bending are loaded by compound weights and springs. During the whole test the acting force is constant.

Operation principle of electromechanical machines is based on excitation of specimen dynamic load through a mechanical system connected to an electromagnet fed by alternating current of an appropriate frequency. The application of electromagnetic converters allows testing in a frequency range from 50 to 600 Hz. In the case of hydraulic machines a mineral oil is injected into the loading cylinder. Hydraulic methods provide cyclic loads of hundreds of tons over a wide range of loading frequencies from fractions of Hertz to 100 Hz.

Let us consider the most widespread machines and methods for fatigue tests.

Machines for bending test with rotation are the most popular. Two main loading techniques are used for bending with rotation (Fig. 7.5): symmetrical circular bending of a specimen fixed from one side, or pure bending with rotation of a specimen fixed from both sides. In the first case (Fig. 7.5a) the specimen has only one head, which is centered and then fixed in a chuck by screws, while the bending moment is applied through the bearing to the other end.

When the specimen is rotated, a stress in a surface point changes its sign twice per one revolution. If the rotational velocity is constant, the sinusoidal law of stress alteration with peak value is realized in a specified critical cross-section. This causes the main disadvantage of the test method with one-side fixing: the presence of one maximally loaded cross-section, which does not always characterize material endurance along the whole length of a specimen.

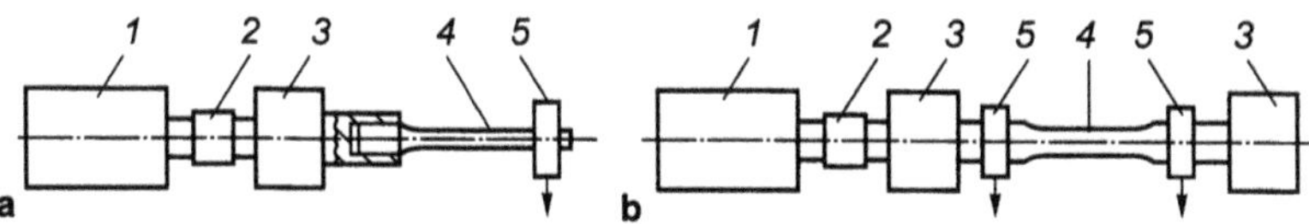

Fig. 7.5. Testing machines for bending with rotation: **a** single-point loading; **b** two-point loading. *1* engine; *2* elastic clutch; *3* basic shaft; *4* specimen; *5* loading bearing

Despite this disadvantage, the method is widely used for experiments due to a simple design of its associated machines and their high productivity. Quantity-produced machines of the UKI–type in Russia as well as Schenk's machines and their versions abroad are based on this principle.

The UKI–10M machine is intended for fatigue tests at cantilever bending of a rotating specimen due to a constant load applied to its free end at the temperature of 20±10 °C. This is a two-section machine (Fig. 7.6) with independently controlled sections. A two-speed engine allows machine tuning for two different regimes: 3000 and 6000 rpm.

The method of pure bending test is even more distributed (Fig. 7.5b). This method allows determination of material endurance along the whole working length of the specimen because all cross-sections are uniformly loaded. This test method is the most unbiased also because the loading scheme to a greater extent corresponds to the practical usage of axes, shafts and similar parts. Some machines with programmable alteration of load use the principle of pure bending [7.1].

All machines for bending tests with rotation are equipped with cycle counters, which automatically switch off at the moment of complete specimen failure. They are relatively productive devices because the construction of bearing units allows specimen to reach the rate of 6000 rpm and more.

Independently on loading condition, the common disadvantage of the bending test with rotation is impossibility to obtain the asymmetrical stress cycle.

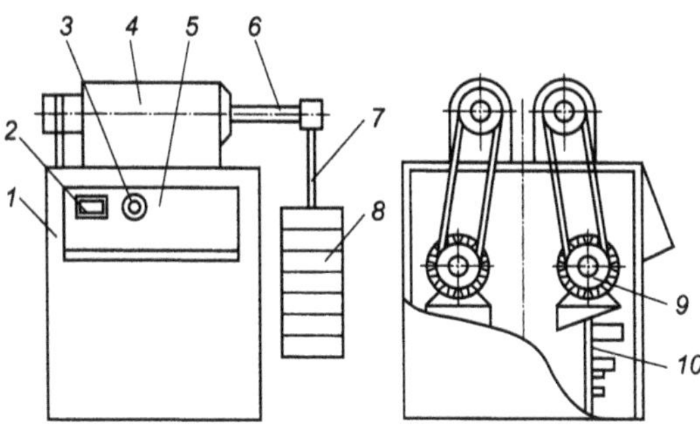

Fig. 7.6. Diagram of an UKI–10M-type machine. *1* stand; *2* loading cycle counter; *3* starting button; *4* spindle head; *5* control panel; *6* specimen; *7* load hanger; *8* loads; *9* device for specimen rotation; *10* electric equipment

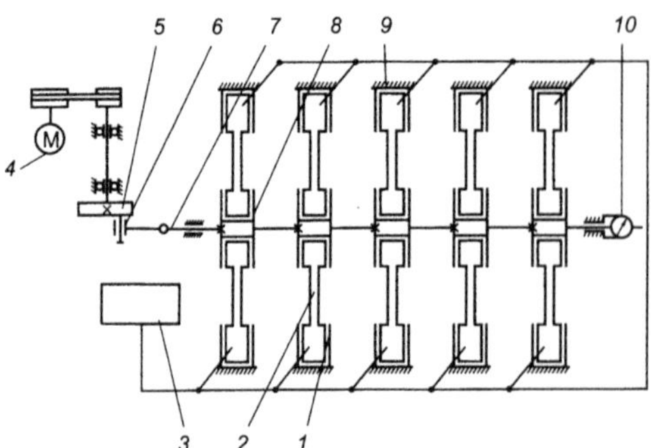

Fig. 7.7. Diagram of a machine for lateral bending tests of rectangular specimens in one plane at cantilever loading. *1* stationary blocks (holders) of specimens; *2* tested specimens with coatings; *3* automated unit for machine shutdown at failure of a specimen; *4* electric motor; *5*, *6* rod-crank mechanism; *7* movable rod; *8* movable blocks (holders) of specimens; *9* base of stationary holders; *10* indicator of deformation amplitude

The simple bending test is less spread, however, it is required for determination of sheet material endurance. One of the possible test variations is shown in Fig. 7.7.

A machine providing simultaneous loading of ten rectangular specimens 2 is convenient to carry out lateral bending tests of specimens in one plane at cantilever loading. Bending amplitude is measured using the rod shift indicator or a measuring microscope. Cycle asymmetry may be changed by displacement of the rod blocks or screw adjustment. The installation has an automatic unit 3. All specimens are successively connected into a low-voltage electric circuit. Breakage of any specimen opens the circuit. If this is the case, the electric motor and electric clock are shut off via the relay system and an alarm is actuated.

If necessary, this machine can be used to immediately observe the stages of fatigue failure of specimens. A narrow surface of a specimen in its working cross-section should be prepared as a microsection. This allows determination of the moment of fatigue crack initiation for preliminary found microstructure, estimation of crack propagation rate and observation of growth kinetics.

Usually the machines for extension-compression testing have larger dimensions and more complicated construction. Nevertheless, this type of testing is widely used because it provides the required cycle mode and, moreover, uniform stressed state of specimens. Machines with axial loading can have the mechanical drive, but more often they use the hydraulic or electromechanical ones. The diagram of this machine is shown in Fig. 7.8. Holders fix specimen 8, and static load is applied by rotation of flywheel 4 through the static loading spring 10. Cyclic loading is performed using the spring 9 and vibrator 11 controlled by the electromechanical drive 12. The microscope measures specimen deformation.

In repeatedly variable torsion testing material endurance is determined at the action of varying shearing stress. The results of these tests are especially important to estimate properties of the base metal and coating used for fabrication of such parts as torsion bars and springs. The mechanical and electromechanical machines are the most common.

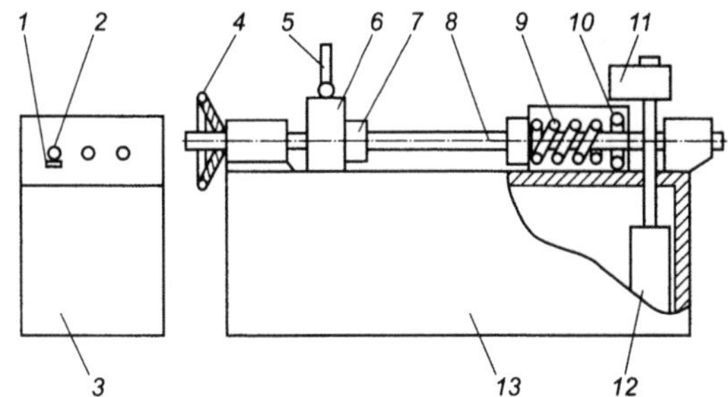

Fig. 7.8. Diagram of a machine with electromagnetic resonance excitation. *1* electric power handles; *2* switch-on of electromechanical drive; *3* control panel; *4* flywheel of the loading system; *5* microscope; *6* dynamometer; *7* holder; *8* specimen; *9* springs of dynamic loading; *10* spring of static loading; *11* vibrator; *12* electromechanical drive; *13* stand

The most famous firms that produce machines for fatigue tests are the following: "Schenk" and "Losenhausen" (Germany); "MTS", "Baldwin" and "Richlilos Bull" (USA); "Dowty Rotol", "Instron" and "Avery" (UK); "Amsler" (Switzerland); "Toyo Seiki" and "Shimadzu" (Japan).

7.1.4 Preparation for Testing

There are two methods of testing: at constant coefficient of cycle asymmetry R with specimen-to-specimen variation of the maximal cycle stress (Fig. 7.9), or at constant average cycle stress σ_m with varying amplitude (Fig. 7.10).

In any case, the relationship between the number of cycles N before specimen failure and the stress σ will be obtained. This relationship is called the Weller curve or fatigue diagram. In this diagram stress values are plotted on the ordinate axis, while logarithms of cycle number ln$N$ are laid off as abscissa.

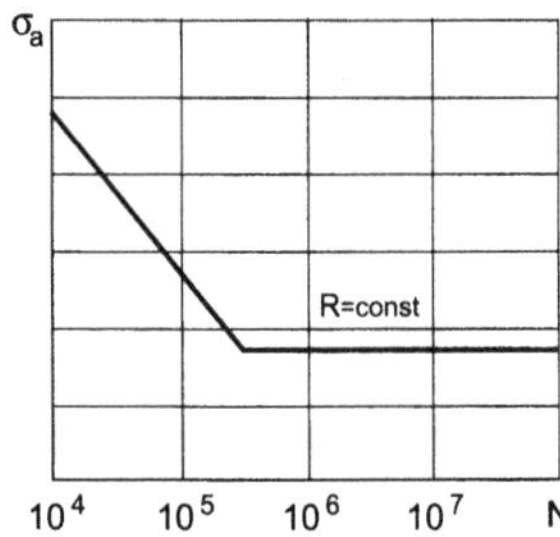

Fig. 7.9. Fatigue curve plotted from the coefficient of the stress cycle asymmetry

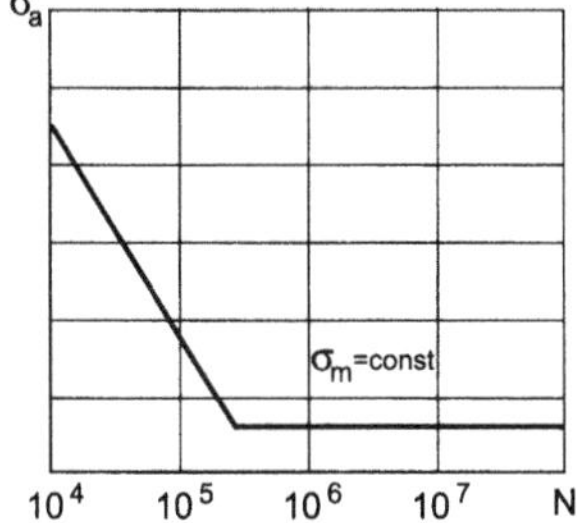

Fig. 7.10. Fatigue curve plotted from the parameter of the average cycle stress

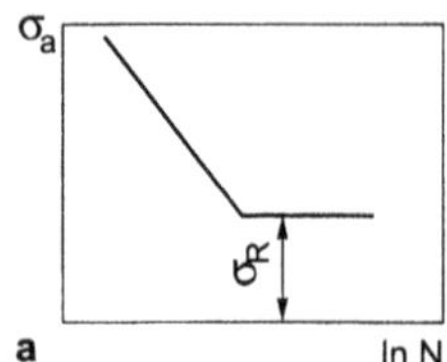

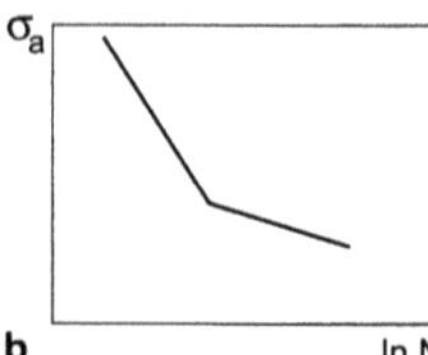

Fig. 7.11. Form of fatigue diagrams for materials with (**a**) and without (**b**) physical limit of endurance

There are two types of fatigue diagrams (Fig. 7.11): the first one (Fig. 7.11a) is typical of iron and its alloys as well as some titanium and magnesium alloys, the second type (Fig. 7.11b) is typical of other non-ferrous metals and alloys and non-metal materials. It is clear from the figure that for all materials the number of cycles before failure increases with decreasing acting stress. For iron and steels failure does not occur at stresses below some certain value independently on test duration, and the Weller curve has sharp bend and becomes parallel to the abscissa axis. Usually, the bend point for steels falls into the range from $5 \cdot 10^6$ to $10 \cdot 10^6$ cycles. Due to this fact, the number of cycles for such materials is assumed to be 10^7. If the specimen endures the higher number of basic cycles, the test is halted prior to specimen failure because there is no sense in the following testing. For materials with the fatigue diagram of this type, the physical limit of endurance can be estimated.

For materials with fatigue diagram of the second type (Fig. 7.11b) the failure stress decreases constantly with the increase in a number of loading cycles. However, when this diagram is plotted in logarithmic or semi-logarithmic scale, the bend point can be often found [7.1].

7.1.5 Testing

As already mentioned before, the test basis in the high-cycle area is assumed to be 10^7 cycles. The resistance to low-cycle fatigue is estimated at durability of up to $5 \cdot 10^4$ cycles. The main loading scheme therewith is "extension–compression".

The accelerated fatigue tests are permitted. Time shortening can be obtained due to an increase in test frequency or by exclusion of small not-damaging amplitudes. To reduce the test time for the estimate of fatigue limit, the method of step-wise load growth (Locatti method) is used. A specimen or part is loaded by the initial stress and tested during a certain number of cycles. The stress is increased up to some level without intermediate pauses, and then the test is continued during the same number of cycles as at the first test stage. With a step-by-step increase in stress, specimens are brought to failure. Using special monograms, the fatigue limit can be calculated. The Locatti method makes the time of testing 30–70 times shorter.

Fatigue damages (coating flaking and exfoliation, and appearance of cracks of a certain length) and fatigue failure (specimen splitting into two parts) are recognized as criteria for defectiveness. Tests are continued at the specified stresses to coating damage, specimen failure or to the basic number of cycles. As this takes place, the number of cycles is registered.

To register the fatigue damages or failure specimens are examined in certain periods of time. Damaged or failed specimens are rejected.

Not less than three specimens should be tested at the level of fatigue limit. The absence of coating damages and integrity of a specimen is considered the measure of fatigue resistance.

Table 7.2. Calculation of stress amplitude σ_a (τ_a) and average cycle stress σ_m (τ_m) for various loading schemes

Loading scheme	$\sigma_a(\tau_a)$, $\sigma_m(\tau_m)$	Auxiliary formulae
Pure bending with rotation	$\sigma_a = \frac{M_b}{W_{ax}}$	$W_{ax} = \frac{\pi d^3}{32}$
Lateral bending with rotation at cantilever loading	$\sigma_a = \frac{M_b}{W_{ax}}$	$W_{ax} = \frac{\pi d^3}{32}$
Pure bending in one plane	$\sigma_m = \frac{M_m}{W_{ax}}$	$W_{ax} = \frac{\pi d^3}{32}$
	$\sigma_a = \frac{M_a}{W_{ax}}$	$W_{ax} = \frac{bh^2}{6}$
Lateral bending in one plane at cantilever loading	$\sigma_a = \frac{M_b}{W_{ax}}$	$W_{ax} = \frac{\pi d^3}{32}$
		$W_{ax} = \frac{bh^2}{6}$
Repeatedly varying extension-compression	$\sigma_m = \frac{P_m}{F}$	$F = \frac{\pi d^2}{4}$
	$\sigma_a = \frac{P_a}{F}$	$F = bh$
Repeatedly varying torsion	$\tau_m = \frac{M_m}{W_P}$	$W_P = \frac{\pi d^3}{16}$
	$\tau_a = \frac{M_a}{W_P}$	

W_{ax} is the axial moment of resistance of calculated cross-section of a specimen; W_p is the polar moment of resistance of calculated cross-section of a specimen; F is the area of the working area cross-section; b, h, d are width and thickness of a rectangular specimen and diameter of a round specimen respectively

Test results allow: construction of the fatigue curve and fatigue curves in the low-cycle area, determination of the fatigue limit, and calculation of influence coefficients of coating and preliminary preparation on fatigue strength.

The amplitude of cycle stresses and average stresses are calculated from the formulae of theoretical mechanics (Table 7.2).

When calculating $\sigma_a(\tau_a)$ and $\sigma_m(\tau_m)$ for specimens with coatings, the thickness of the latter should be taken into account in calculation formulas.

The fatigue curve is plotted on semi-logarithmic or logarithmic coordinates from the data obtained.

The role of a coating can be expressed through the coefficient of coating effect on the fatigue strength

$$K_l = \frac{\sigma_{Rcoat}(\tau_{Rcoat})}{\sigma_R(\tau_R)}, \tag{7.12}$$

where $\sigma_{Rcoat}(\tau_{Rcoat})$ is the fatigue limit for the specimen with coating, MPa; $\sigma_R(\tau_R)$ is the fatigue limit of a standard specimen, MPa.

If necessary, the coefficient of preliminary treatment effect on the fatigue strength can be determined as

$$K_{sp} = \frac{\sigma_{Rsp}(\tau_{Rsp})}{\sigma_R(\tau_R)}, \tag{7.13}$$

where $\sigma_{Rsp}(\tau_{Rsp})$ is the fatigue limit of a specimen after preliminary treatment, MPa.

The fatigue curve is plotted from results of low-cycle fatigue tests on double logarithmic coordinates: "amplitude of complete deformation ε_a – number of cycles before the fatigue damage of a coating N_n or specimen failure N" (for instance, Fig. 7.12).

When estimating the development aspects of fatigue testing methods, it should be noted that the agreement between testing and operational conditions is not always achievable for coated specimens, because the level and distribution of residual stresses, correspondence of scales, structure modifications from the coating surface to base metal, etc. in a specimen can be hardly established the same as those in a part. Besides, there is a considerable scatter in the data for specimens with coatings. Therefore, the most reliable and precise characteristics of fatigue resistance can be obtained in immediate testing a coated part, when experimental conditions are close to operational.

There are several reasons why at present fatigue resistance characteristics of particular coated parts cannot be calculated using standard techniques without experimental data. First of all, this is caused by anisotropy of mechanical properties in the cross-section and residual stresses in near-surface layers of the base metal and coating. With this the pattern of stressed state in elastic and plastic-elastic areas becomes more complex, and the calculation problem for the fatigue resistance characteristics of relatively simple parts such as a shaft or a plate becomes stubborn.

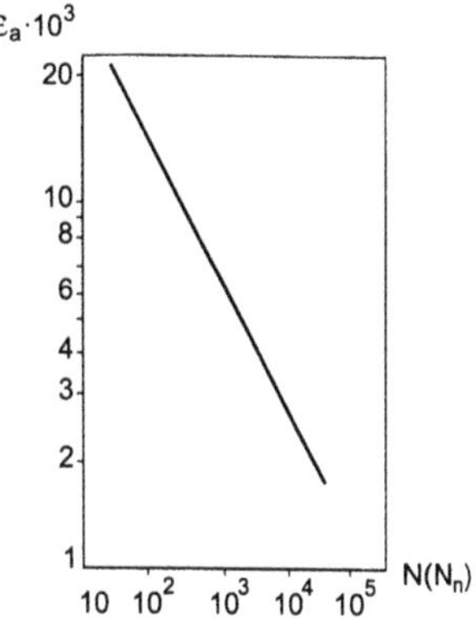

Fig. 7.12. Fatigue curve for plastic-elastic low-cycle area on $\lg\varepsilon_a$–$\lg N(N_n)$ coordinates

The aforesaid does not exclude the development of calculation methods for fatigue resistance of coated articles without experiments. It can be assumed that the unified universal calculation methods (without experiments) will be put forward soon. This is proved by the fact that the standard calculation methods have been already developed for coefficients of surface strengthening influence on fatigue strength after hardening by high-frequency currents, chemical-thermal treatment (nitration, carbonization, cyanidation), surface cold-hardening on roller treatment and shot blasting.

7.1.6 Analysis of Fatigue Fracture

The study of fracture surface provides important information on specific reasons for failure of a part with coating and makes for understanding of fatigue process features. Some versions of fatigue fracture of the base metal are schematically shown in Fig. 7.13. The fracture consists of two main zones. One of them corresponds to the area, where a fatigue crack develops, and has a typically smooth appearance. Concentric contours of the crack propagation front, which meet in the fracture center, are often seen on the surface of this zone. In the base metal–coating composition there can be several fracture centers, especially in the area of low-cycle fatigue.

The second one is the fracture zone (shaded), which appears due to the fast final failure within one or several last cycles. It can have brittle, viscous or mixed structure depending on test conditions.

The detailed study of fracture surfaces by means of microfractography provides required information on the mechanism and kinetics of crack development and character of the stressed state near its top point at the stage of slow and accelerated growth and at the stage of unstable propagation.

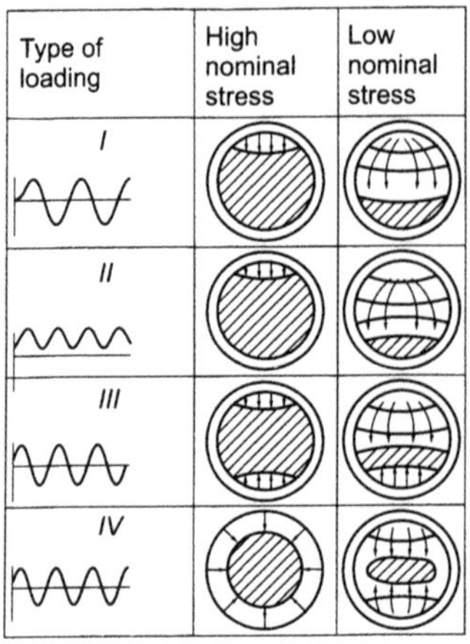

Type of loading	High nominal stress	Low nominal stress
I		
II		
III		
IV		

Fig. 7.13. Schemes of fatigue fractures at some types of loading. *I* repeatedly varying extension-compression; *II* lateral bending in one plane at cantilever loading for sign-alternating asymmetric cycle; *III* lateral symmetrical bending in one plane at cantilever loading; *IV* pure bending with rotation

7.1.7 Results and Discussion

Fatigue characteristics depend on a great many variable factors like none of the other criteria for structural strength. These factors are so diversified that can hardly be enumerated. The main of them are the structures of coating and base metal [7.4, 7.7, 7.12, 7.22] (structure, porosity, roughness, coating thickness; the bond character at the "base metal–coating" boundary; the presence or absence of a transition zone near this boundary, etc.); fabrication technology and specimen geometry [7.1, 7.11] (size and shape, the presence or absence of stress concentrators); method of coating [7.5, 7.10, 7.21, 7.26] (value, distribution and sign of residual stresses, preliminary surface preparation before coating, etc.); loading conditions [7.17] (frequency, period, maximal stress, amplitude, asymmetry factor of the stress cycle, etc.); test medium [7.12, 7.16, 7.23] (the presence of aggressive atmosphere, increased or decreased temperatures, etc.).

Some factors exerting an effect on the fatigue resistance (chemical composition, coating thickness, loading conditions) are partially analyzed, the rest should be studied and examined.

Analysis of the coating forming conditions and accompanying physical-chemical processes allows assumption that the properties of coating-substrate combination are mainly effected by the composition and structure of the sprayed coating and transition zone. In general case, the transition layer formed in the process of spraying or after homogenizing, may play both negative and positive roles depending on experimental conditions. Composition, structure and their operational stability tell mainly on the stressed state of the contact zone, which is determined, along with other factors, by residual stresses.

The transition layer is a barrier to dislocations, which tend to reach the surface.

Maximovich et al. [7.12] studied kinetics of initiation and development of fatigue cracks. The coating impedes plastic deformation and serves as a barrier, in front of which dislocations heap up and local stresses increase. Under the action of tangential stress the source of dislocations *S* starts generating dislocations (Fig. 7.14). Once the cross-point of the diffusion layer with the grain boundary is reached, the head dislocation stops. Other dislocations heap up behind this one, which results in occurrence of local extension stresses σ_s in front of the head dislocation, thus promoting cracking of relatively brittle intermediate layer. So a crack may nucleate in a brittle transition (solid solution, intermetallic compound) layer with subsequent propagation deep into material.

A sprayed layer with numerous defects in the form of pores, where a destructive crack may be initiated, can also serve as a source of crack formation in gas-thermal coatings. In this case the residual stresses play negative role because even if they do not lead to immediate cracking of coatings, they augment the stressed state in the contact zone and precipitate the moment of crack generation. First of all, this is attributed to residual stresses appeared in spraying of tungsten and molybdenum with strong chemical bonds, whereas nickel spraying leads to coating flaking, not effecting the strength.

The coating deposited on the base metal has a double effect. On the one hand, this technological process augments the surface defectiveness and makes for formation of extension stresses; as a result, a new weakly deformable layer constrains plastic deformation and increases sensitivity to overloading. In the context of construction strength, these factors are negative for the longevity of the "base metal–coating" composition.

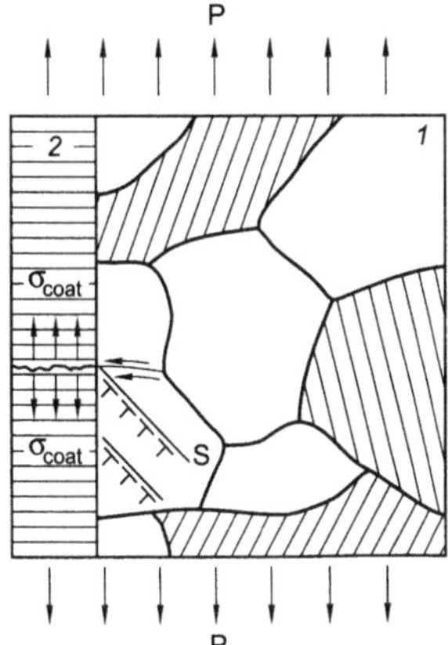

Fig. 7.14. Scheme of initiation mechanism for fatigue cracks in material (*1*) with a sprayed coating (*2*), (by [7.12])

Sometimes, "healing" of defects in the base metal are observed; some coatings deposited by the optimal technology create residual compression stresses in surface layers, i.e., the coating plays the positive role in fatigue resistance. The effect of these two groups of factors is usually shown by reduction in the fatigue limit, i.e., factors of the first group prevail. The fatigue limit decreases when the coating is more brittle than the base metal. The thicker the coating and the lesser the working area of the base metal, the greater decrease in fatigue limit.

Rakitsky et al. proved the double role of coatings at fatigue loading [7.17]. Fatigue tests of medium-carbon steel with a gas-flame coating of powder with particles of 40-100 μm in size and percent content of Ni – 75.5; Cr – 12.5; C – 0.6; Fe – 7.4; Si – 4.0 revealed the role of pores in crack initiation. The failure of coated specimens was brittle, and cracks developed from surface discontinuities. Fatigue tests were carried out under conditions of torsion (sinusoidal loading, frequency of 2 GHz) and by the scheme of impact with extension at the load of ±50 kN. In the first case the fatigue strength of specimens with coatings was lower, while in the second case it was slightly higher than that of uncoated steel [7.17].

Data on the increase in fatigue resistance of steel after coating is rather the exclusion than the rule. An increase in endurance and improvement of protection properties due to coating is usually accompanied by reduction in fatigue strength by 15–60% and more [7.18, 7.22].

According to Pokhmursky et al. [7.16], metal coatings deposited by the electrolytic method also decrease fatigue characteristics, and the reduction coefficient of the fatigue limit is proportional to the coating thickness.

Kalmutsky [7.10] has developed the methods for statistical calculation of strength and reliability of parts with electrolytic coatings at cyclic loading, determined statistical regularities of damage and failure, and checked the advanced proposals in operation. The reason for reduction in endurance of specimens with electrolytic metal coatings is the residual extension stresses at the "base metal–coating" boundary, which can reach 100–960 MPa. These stresses have a negative effect on the crack resistance of smooth cylindrical specimens at asymmetrical loading cycle and cause the special character of deformation and failure. The fatigue limit may decrease by 50%.

Maximovich et al. [7.12] noted the discrepant character of coating effect on specimen failure. A plasma installation was used to deposit the coatings based on tungsten, molybdenum and nickel. The pure bending was chosen as a loading scheme. Some specimens with coatings were subjected to homogenizing, and as a result a maximal decrease in low-cycle strength was observed. This is explained by formation of brittle transition layers. Low-cycle endurance of specimens with thin plasma coatings (without annealing) is almost indistinguishable from that of control specimens (without coating). According to microscopic investigations of cross-sections, fatigue failure always starts from a specimen surface or transition zone. Microcracks nucleate in the near-surface coating layers and then spread into the base metal. When testing specimens with nickel coatings, a different pattern is observed. In the range of slow bending stresses a slight increase in fatigue strength is pointed out, however, local exfoliation of some coating areas takes place. The authors explain this phenomenon by the different character of bonds between the nickel coating and base metal: mechanical bonds are generated, while no chemical interaction exists.

Residual internal stresses generated in formation of coatings play an ambiguous part in the context of initiation and propagation of fatigue cracks. If compression residual stresses exist in coating and near-surface layers of the base metal, they increase longevity by inhibition of fatigue crack initiation and propagation. When extension stresses (which are more frequent) unfavorable for structural strength are generated, specimen failure is accelerated due to intensification of the stressed state and initiation of crack formation.

Umansky et al. [7.28] have obtained results, which show the connection between longevity of plasma coatings made of titanium carbide and distribution of residual stresses over the coating thickness. In the near-surface coating layer significant extension stresses lead to coating exfoliation even at the first stage of loading. Residual compression stresses of up to 450 MPa found in the layer center change the character of crack formation, moving the center of crack nucleation under the coating.

Mora-Marquer and Lira-Olivares [7.14] examined initiation and propagation of cracks in the self-fluxing coating of the following composition (in %): Cr – 15; Fe – 4; C – 0.25; B – 3.25; Si – 4.5; Ni – the rest by the method of acoustical emission. It was determined that slug inclusions are the reason for initiation of cracks, which lead to coating failure.

Sometimes fatigue loading of specimens is accompanied by coating peeling. The failure may have adhesion, cohesion or mixed character. In some instances the fatigue strength is limited by the bond strength of the coating and base metal or cohesion strength of the coating.

A positive effect of nickel-aluminum and nickel-titanium plasma coatings was found by Pokhmursky et al. [7.16] when testing coatings in a corrosion medium (3% water solution of NaCl). The limit of corrosion resistance of steel with NiAl coating increases almost twice, whereas with NiTi coating it increases by 25%. As a result of fractographic analysis by means of a scanning electron microscope, two chip facets were found on specimens with NiTi coatings, i.e., the failure of brittle type was determined. The fracture surface of specimens with NiAl coating is different: the typical cup-shaped structure is observed, i.e., failure has a tough nature. Microscopic studies of microsections cut out in the zone of fatigue failure demonstrated many cracks normal to the boundary with the base metal. Usually, the cracks pass through the boundaries of particles and stop near the separating boundaries. This feature is typical of both brittle and plastic coatings.

At cyclic loading, preliminary treatment of a surface before spraying has a considerable effect upon behavior of coated products. Khasui [7.11] considered the effect of preliminary surface preparation on relative change in fatigue limit for various steels and coatings. The influence of the following preparatory procedures was estimated: blasting by steel crumbs, defective thread, turning of grooves, knurling, and electric-spark treatment. It was shown that only steel shot blasting exerts a positive influence on the fatigue strength of coated specimens. All the remaining kinds of preparatory treatment (especially turning of grooves and defective threading) significantly reduce the fatigue limit.

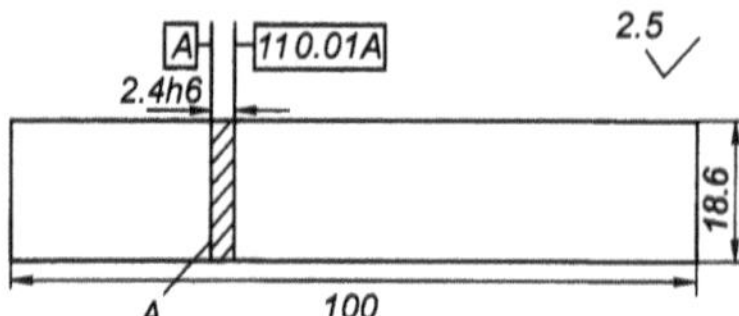

Fig. 7.15. Specimen for contact fatigue test. Limit tolerance of dimensions (not indicated) is h_{12}; *A* coated surface

7.2 Contact Fatigue

7.2.1 Specimens

Four types of specimens – with cylindrical, spherical, toroidal and flat working surfaces are tested. A flat specimen is the most convenient for the "pulsing contact" scheme since a coating can be easily deposited on its surface (Fig. 7.15). A specimen is fixed in a mandrel to provide convenience of testing and prevent its displacement under load (Fig. 7.16).

Technological process of coating deposition should be stable enough. Slight differences in structure, porosity, and coating hardness of specimens from the same lot may cause a wide scatter in contact fatigue characteristics. After deposition the working surfaces are ground. Special attention must be given to quality of the coating surface after grinding. There should not be any cracks, cleavages, exfoliation, burns and other defects. Significant non-uniformity of the coating thickness is a rejection criterion as well. Metal specimens can serve as controls. Their number is determined by the number of chosen stress levels.

While testing flat specimens according to "pulsing contact" scheme, the same specimen may be used repeatedly by successive shifting of a contact application point. Number of specimens used to plot contact fatigue curve and determine contact fatigue strength is less by far than that for standard tests, i.e. four to six instead of thirty to forty.

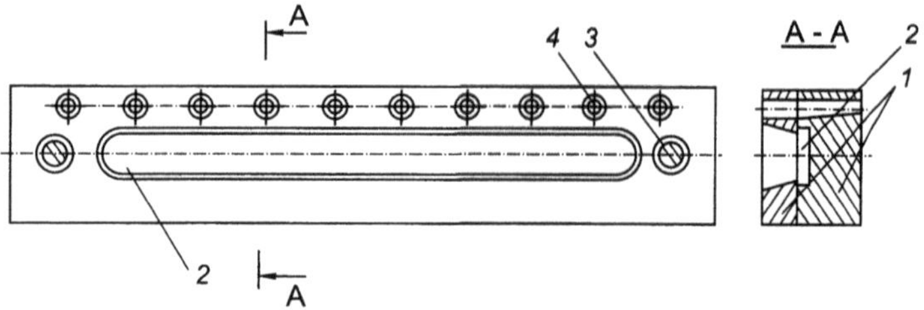

Fig. 7.16. Mandrel to fix specimen. *1* clamps; *2* specimen; *3* clamping screws; *4* cone orifice

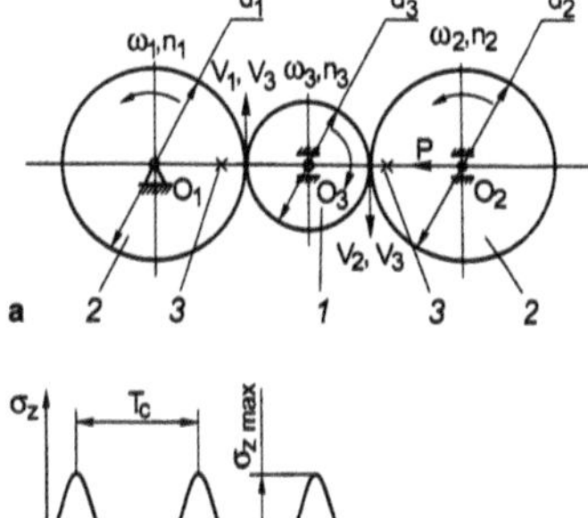

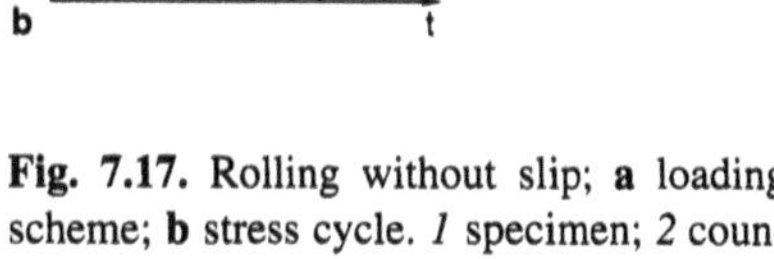

Fig. 7.17. Rolling without slip; **a** loading scheme; **b** stress cycle. *1* specimen; *2* counter-bodies; *3* drive elements; T_C cycle period

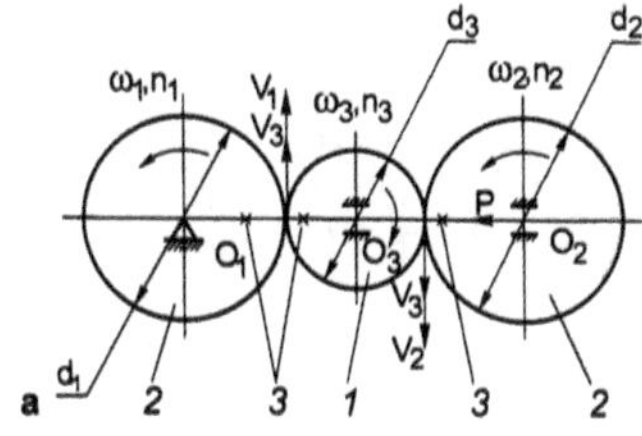
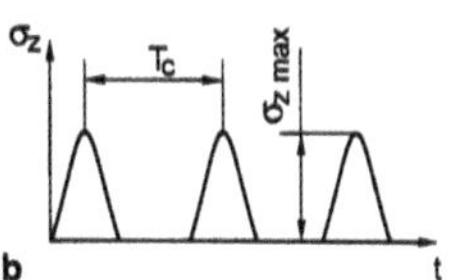

Fig. 7.18. Slip rolling; **a** loading scheme; **b** stress cycle. *1* specimen; *2* counter-bodies; *3* drive elements; T_C cycle period

7.2.2 Equipment

The next four loading schemes are usually realized according to the Russian State Standard: rolling without slip (Fig. 7.17) (shakedown of a specimen between two counter-bodies), slip rolling (Fig. 7.18), rolling with an outer tangential load (Fig. 7.19), pulsing contact (Fig. 7.20) [7.15].

For the first scheme:

$$\lambda_{C0} = \frac{v_3 - v_1}{v_3} = \frac{v_3 - v_2}{v_3} = 0 \ ^{(1)}; \tag{7.14}$$

$$d_1 = d_2\,;\ n_1 = n_2\,;\ \omega_1 = \omega_2\,;\ v_1 = v_2 = v_3 = \frac{\pi d_1 n_1}{1000 \cdot 60}\,; \tag{7.15}$$

$$\omega_1 = \frac{\pi n_1}{30}\,;\ \omega_2 = \frac{\pi n_2}{30}\,;\ \omega_3 = \frac{\pi n_3}{30}\,; \tag{7.16–7.18}$$

where d_1, d_2, d_3 are working part diameters of counter-bodies and a specimen, mm; n_1, n_2, n_3 – rotational velocities of counter-bodies and a specimen, 1/min; ω_1, ω_2, ω_3 – angular velocities of counter-bodies and a specimen, radian/s; v_1, v_2, v_3, – circumferential velocities of counter-bodies and a specimen, m/s.

Drive of counter-bodies is carried out through a differential device.

[1] relative value of slipping

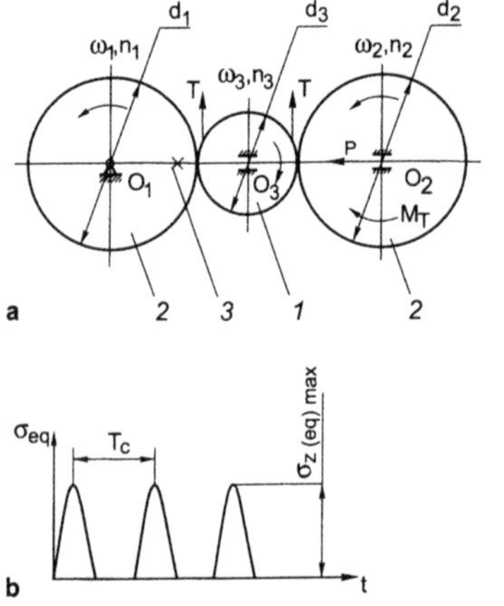

Fig. 7.19. Rolling with outer tangential load; **a** loading scheme; **b** stress cycle. *1* specimen; *2* counter-bodies; *3* drive elements; T_C cycle period. d_1=d_2; $M_T = T(d_2/2)$ braking torque; T tangential load

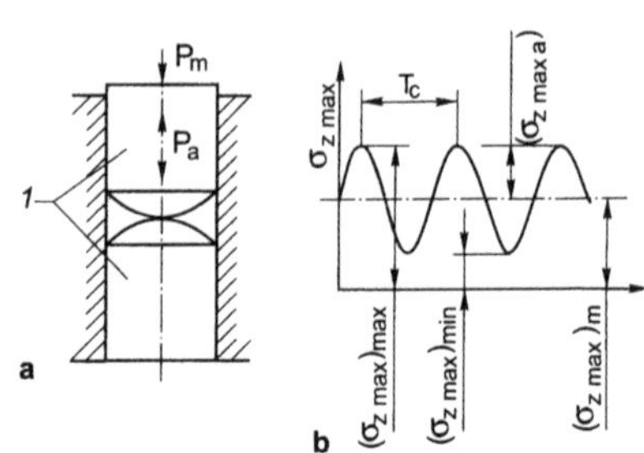

Fig. 7.20. Pulsing contact; **a** loading scheme; **b** stress cycle. *1* specimens; T_C cycle period

For the first scheme:

$$\lambda_{C3} = \frac{v_3 - v_1}{v_3} = \frac{v_3 - v_2}{v_3} \neq 0 \tag{7.19}$$

$$d_1 = d_2\,;\; n_1 = n_2\,;\; \omega_1 = \omega_2\,;\; v_1 = v_2\,;\; \omega_3 = \frac{\omega_1 d_1}{(1-\lambda_{C3})d_3} = \frac{\omega 2 d_2}{(1-\lambda_{C3})d_3}\,; \tag{7.20}$$

$$n_3 = \frac{30\omega_2 d_2}{\pi(1-\lambda_{C3})d_3}\,; \tag{7.21}$$

$$v_3 = \frac{\omega_3 d_3}{2000} \tag{7.22}$$

To realize the first and third schemes a base test bench of "MKV–KM" type (see Fig. 7.21) has been installed. This equipment enables sufficiently high-frequency loading of a specimen, automatic registration of its breaking moment (appearance of pittings of a certain length) with simultaneous shutdown of the machine. An AC electric motor transmits rotation through the V-belt transmission and a driving counter-body to a specimen and driven counter-body on the principle of friction transmission. Normal P and tangential T loads are set by a spring mechanism and through braking of the driven counter-body controlled by a DC electric motor, which operates in a dynamic braking mode, respectively [7.15].

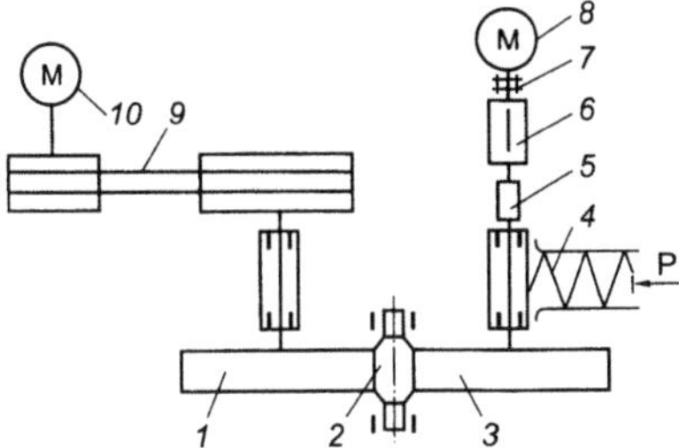

Fig. 7.21. Test machine of MKV–KM type: *1* driving counter-body; *2* specimen; *3* driven counter-body; *4* spring mechanism; *5* tensometric beam; *6* current collector; *7* elastic clutch; *8* DC electric motor; *9* V-belt transmission; *10* AC electric motor

Measurement of electric current in excitation winding allows variation of breaking torque that is transmitted through an elastic clutch to the shaft of the driven counter-body. Breaking torque is measured using tensometric beam with glued foil strain sensors and a current collector with a mercury contact. Transmission of braking torque results in strain of the tensometric beam that generates an electric signal proportional to the tangential load. Maximal values of normal and tangential loads set by means of the test bench are equal to 2 kN and 50 N respectively. Many of the existing friction machines may be employed for slip rolling tests.

For "pulsing contact" tests hydro-pulsing and resonance fatigue machines are suitable. A setup providing pulsing contact loading of specimens with "sphere-surface" working surfaces was successfully applied to "base metal–coating" composition testing (Fig. 7.22).

A kinematical scheme driven by the electric motor *1* provides periodical specimen loading using the indenter *13*. Once a specimen is removed, the force of the spring *8* is measured by means of the dynamometer *14*. A number of loading cycles is determined by the register.

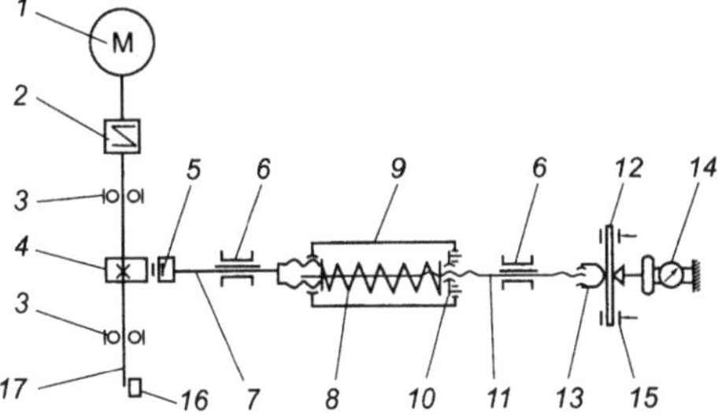

Fig. 7.22. Setup for tests of "pulsing contact" with working surfaces of a specimen of the "counter-body surface–sphere type". *1* electric motor; *2* clutch; *3* rolling contact bearing; *4* eccentric; *5* push bur; *6* slipping contact bearings; *7* rod; *8* spring; *9* barrel; *10* screw nut; *11* rod; *12* specimen; *13* indenter; *14* dynamometer; *15* mandrel; *16* register; *17* shaft [courtesy of Potapov V.M.]

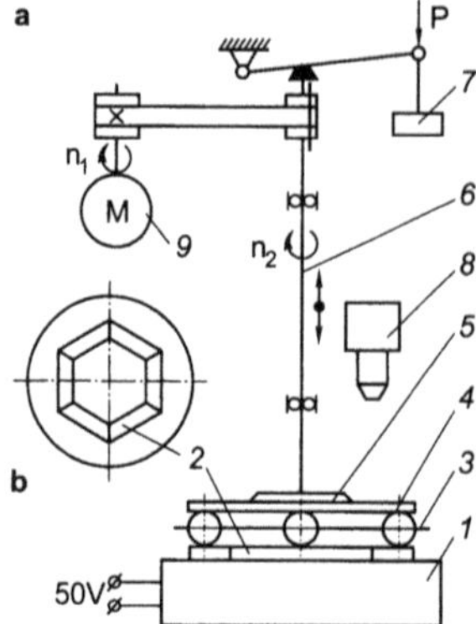

Fig. 7.23. Setup for estimation of contact endurance at rolling out by spheres (**a**) and arrangement of specimens on the magnetic plate (**b**). *1* electromagnetic plate; *2* specimen; *3* separator; *4* sphere; *5* thrust bearing band; *6* spindle; *7* weight; *8* microscope; *9* electric motor; *P* loading force; n_1 rotation frequency of the electric motor shaft; n_2 spindle rotating frequency

Detailed information on kinetics and failure mechanism of specimens at contact loading can be obtained through the use of a testing setup shown in Fig. 7.23. The essence of the test consists in rolling out of a closed contour formed by six specimens through the use of hardened steel spheres. The specimens *2* are installed in a hexagon shape on an annual clearance of the magnetic plate *1* and fixed additionally by mechanical stops in order to prevent shear. Spheres revolve over the specimen surfaces. Required contact pressure is provided by the weight *7*. Visualization, photographing and measurement of parameters of the specimens' contact zone after a required number of loading cycles is carried out using the microscope *8*.

7.2.3 Preparation for Testing

A loading scheme is selected to provide the most realistic testing conditions of a coated product. A test basis and load are specified. The test basis should be 2–3 times greater than abscises of the inflection point of a contact fatigue curve and lie within the range of 10^7–$5 \cdot 10^8$ cycles. The number of loading cycles depends on the coating hardness and availability of horizontal part on the contact fatigue curve.

The choice of load when testing coatings is made difficult by the fact that there are no relationships between contact fatigue strength and coating properties including mechanical characteristics. On the one hand, the load should be much below its limiting value in order to prevent significant decrease in contact strain due to accumulation of plastic deformations during the test. On the other hand, if load is below a certain level at which accumulation of plastic deformations is revealed the contact-fatigue lifetime will be too long.

7.2.4 Testing

When using rolling schemes, each specimen should be tested on a new track of a rolling cylindrical counter-body. Specimens are lubricated by oil of the same brand, which is tested periodically for compliance to standards or technical requirements.

Frequency of load cycles for a planned testing set should vary within the range from 30 to 1000 Hz. If at a given loading frequency the temperature in the contact zone causes change in structure and physical and mechanical properties of material, the loading frequency should be reduced.

Installation, fixing and loading sequence should be the same for each specimen. The specimens should be loaded only when a specified frequency of cycles is achieved. Frequent start-ups and stops of equipment have an influence on experimental results. Appearance of flaking pits on a working surface may be considered a criterion for rolling schemes failure. For "pulsing contact" scheme the moment of failure is identified by appearance of fatigue cracks near the profile of a contact spot. Flaking pits and cracks are registered either by a special gear or by means of Brinell magnifier.

Usually it is hard to determine the first microcrack. Therefore, in research tests it is allowed to use various failure criterions. For "pulsing contact" scheme it is recommended to plot a graph (Fig. 7.24) on D_S–N coordinates; where D_S is the diameter of a contact spot, μm; N – a number of loading cycles. At the moment of pitting formation speed-up (beginning of the third stage of failure development) a rapid increase of the contact spot is observed. This implies the beginning of fatigue failure at a given stress.

In testing flat specimens according to the "pulsing contact" scheme it is expedient to take structure photographs of transverse microsections to reveal kinetics of contact-fatigue failure.

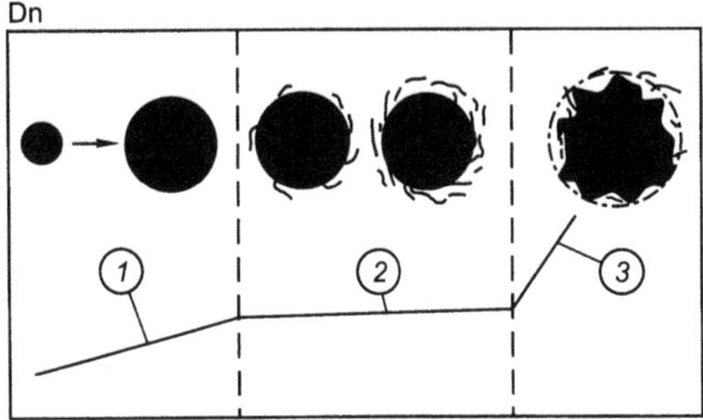

Fig. 7.24. Diagram of failure development for testing by "pulsing contact" scheme using flat specimens: *1* elastic-plastic strain; *2* accumulation of fatigue damage; *3* contact-fatigue failure [courtesy of Potapov V.M.]

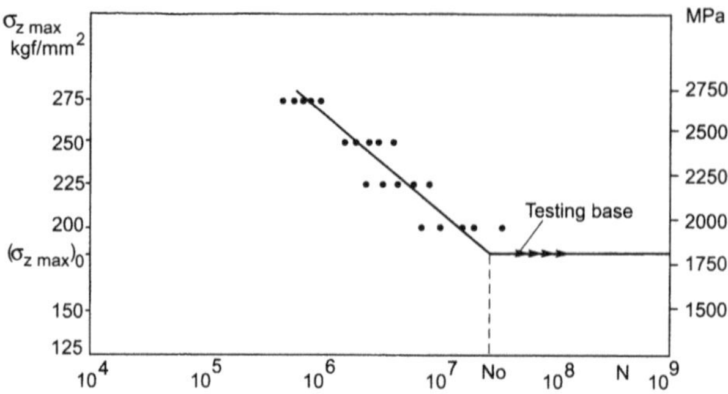

Fig. 7.25. Contact fatigue curve

When testing according to schemes I–III, a contact fatigue curve is plotted on semi-logarithmic "σ_{max}–lg$N$" (Fig. 7.25) or double-logarithmic "lgσ_{max}–lg$N$" co-ordinates based on experimental data obtained. Durability N is expressed either by a number of cycles preceding the appearance of coating surface-fatigue damages or by the basic number of cycles.

Contact endurance limit (σ_R) in MPa is determined in the contact fatigue curve depending on disposition of its horizontal branch.

The part played by the coating is expressed through the coefficient of coating influence on the contact-fatigue strength K_C

$$K_C = \frac{\sigma_{Rcoat}}{\sigma_R} \tag{7.23}$$

where σ_{Rcoat} is a contact endurance limit of coated specimens, MPa; σ_R – that of a standard metal specimen.

7.2.6 Results and Discussion

Stress concentrators are of great importance in surface failure process. Surface pores, cracks formed during deposition or grinding, deep hairlines, and scratches are qualified as stress concentrators, which facilitate failure. Boundaries between layers, oxide films, pore channels, "coating–base metal" interface are very dangerous surface stress concentrators. Especially unfavorable situation develops in case of coincidence of coating subsurface concentrators with the effective area of maximal shearing stresses.

Contact endurance of "base metal–coating" composition depends on mechanical properties of the substrate (ultimate strength and yield stress) and coating (hardness, ultimate strength) [7.24]; the value, sign and distribution of residual stresses [7.26]; scale factor and size of a contact area [7.15]; coating composition and structure [7.27]; availability or lack of lubrication [7.29]; etc. According to Kalmutsky [7.10], for electrolytic coatings a number of facts influencing contact endurance limit runs to fifteen. As for gas-thermal coatings, a number of such variable factors is, probably, considerably more. Kalmutsky suggests solving the problem of contact strength enhancement for coated metals by taking into account the probabilistic statistical character of their actual fabrication and loading conditions. Optimization of fabrication conditions in subsequent processing of some electrolytic coatings allowed increase in their lifetime at a contact loading up to 15–20%.

When depositing a coating, its thickness value should be such that the maximal shearing stress area does not superimpose on "base metal–coating" interface. Middleton with coworkers [7.13] revealed fourfold increase in bearing steel lifetime with TiN coating 1 μm thick as compared with that for coating of 0.25 μm in thickness. Base metal should be hardened enough. Otherwise, at considerable contact loads it is quite possible that the coating impresses into substrate.

Low specific loads in wear-resistant plasma coatings composed of Ni – 85% and Al – 15% usually stipulate easy initiation of cracks with subsequent formation of indents. If tests are carried out at high contact loads, the coating may exfoliate from base metal in the vicinity of the contact spot. This phenomenon is accompanied by bulging and intensive flaking of the coating along the contact spot perimeter.

Research results of self-fluxing coatings composed of (in %): B – 3.2; Si – 4.1; C – 1.0; Cr – 17.0; Fe – 4.0, Ni – the rest, differ from those obtained when testing plasma coatings PN85U15. Despite high hardness (HRC53), coating 0.6 mm thick undergoes considerable plastic strain without formation of large cracks. In testing under large contact pressures (load of 900 N, indenter diameter of 2.5 mm) impression of coating material into base metal was observed. After two million loading cycles metallographic investigations carried out at a depth of 0.2–0.5 mm, revealed availability of microcracks 0.1–0.7 mm long situated in parallel to "base metal–coating" interface. No cracks were displayed on the boundary. Increasing contact spot diameter is accompanied by formation of tangential and radial microcracks on the coating surface. Coalescence of separate microcracks along the spot perimeter results in formation of indents (Fig. 7.26).

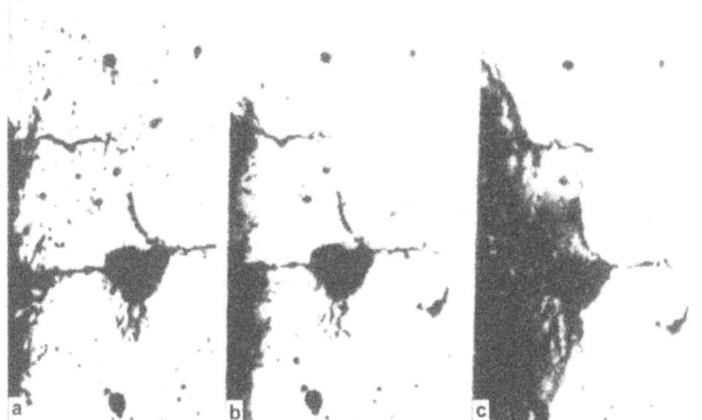

Fig. 7.26. Contact fatigue failure kinetics of a self-fluxing coating (load of 900 N) x100. Number of cycles: **a** $1{,}1 \cdot 10^5$; **b** $1{,}4 \cdot 10^5$; **c** $1{,}5 \cdot 10^5$ [courtesy of Potapov V.M.]

Due to peculiarities of coatings microstructure, it is not always possible to display the three stages of contact fatigue failure typical of bulk-hardened base metal.

7.3 Thermal Fatigue

7.3.1 General Overview

Thermal fatigue is a particular case of thermomechanical fatigue with a loading caused by constraint of thermal deformations during a cyclic heating-and-cooling process.

Various factors that influence the fatigue resistance of coated specimens can be classified into external and internal. The first class of factors influencing the thermal fatigue involves the minimal and maximal temperatures over a cycle, frequency and number of cycles of thermal loading. The additional external factors during thermomechanical loading are the law of mechanical loading (a function describing temporal changes in loading); frequency, period, maximal, average and minimal stresses (deformations) over a cycle, etc.

The internal factors (structure and characteristics of the coating material and base metal) are determined by the modulus and limit of elasticity, porosity, the magnitudes of thermal coefficients of linear expansion and their ratio, coating thickness, its bond strength with the base metal and others.

Thermomechanical and thermal fatigue can damage some elements of power plants such as gas pipelines, parts of rotors and turbine bodies, superheater chambers, and other important units and parts. Cracks initiated on the surface of parts of different metallurgical equipment result from either thermal cyclic variations or joint effect of mechanical loading and temperature. That is why the cutting elements of hot-cutting saws, dies for hot closed-die forging, mill rollers and guides, arms of charging machines often fail to operate. After crack formation, the conditions of heat exchange between the part and medium (deformed metal, air, cooling water, furnace atmosphere) become different. The amount of heat accumulated by the part goes up, the number and sizes of cracks also increase. The cracks can intersect making up a grid. As a result, some areas of the operating surface of these parts are crumbled out.

7.3.2 Thermal Fatigue Testing Techniques

The techniques for thermal fatigue testing can be divided into qualitative, quantitative and full-scale tests.

For qualitative testing specimens are exposed to heat cycles without estimation of the resulting deformation and stresses. Only a number of heat cycles until the onset of failure is counted.

For full-scale testing a special technique is utilized in every specific case to simulate actual operation conditions of various materials such as pipeline materials, turbine blades, or hot forging dies. The studies using test benches allow estimation of working capacity of a construction, selection of a proper material, and design of thermal treatment technology aimed to a higher thermal stability. However, these tests are of a particular character; they do not allow establishment of the pattern of different effects and correlation with other types of studies.

Conversely, the quantitative techniques create the opportunity for studying thermal fatigue failure with account for thermal stresses and deformations.

Getsov et al. [7.6] considered the most typical and widespread constructional shapes of tested specimens (Fig. 7.27). Specimens 3–7 fit for modeling stressed states and failure conditions of parts with similar shapes and sizes. However, thermal stresses in such specimens vary in time not only due to changing temperature, but also depending on the specimen cross-sections. A complex pattern of static stresses causing the creep and cyclic stresses producing fatigue failure makes the calculation of stresses and deformations a rather complex task and brings uncertainty into final results. Therefore, those specimens cannot be used for plotting thermal fatigue curves.

In specimens 1–2 with ends fixed toughly, the stress induced by thermal cycling is the same for the entire working zone at a certain instant of time. In hollow specimens of type 1, due to a small wall thickness, even a fast heating does not bring any non-uniformity of the thermal field and, consequently, any differences in the stress level. For thin specimens a material testing method is often used which implies the external fixing of the heads and internal heating of the material by electric current (Coffin's setup).

7.3.3 Thermal Fatigue Testing of Materials Using the Coffin's Setup

Coffin has established the relationship between elastic deformation and a number of cycles before failure

$$\Delta\varepsilon_{el} = N_m D, \tag{7.24}$$

where $\Delta\varepsilon_{el}$ is a change of elastic deformation per cycle; N is a number of cycles until the onset of failure; m and D are constants.

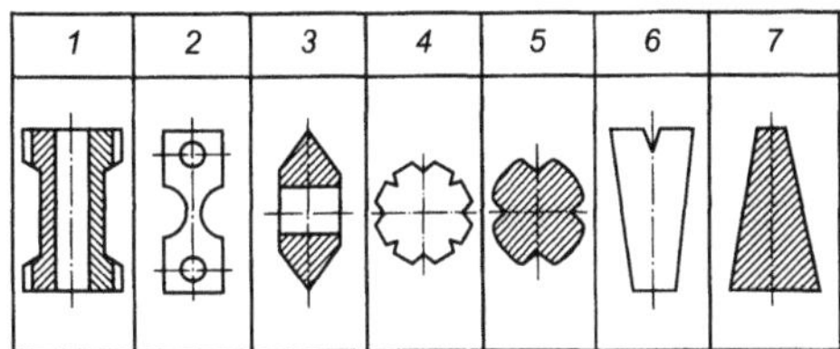

Fig. 7.27. Types of specimens for thermal fatigue tests (by [7.6]): *1* Coffin's specimen (tubular); *2* flat corset specimen; *3* disk-shaped specimen; *4* disk with concentrators; *5* cylinder with grooves; *6* flat with concentrator; *7* wedge-shaped specimen

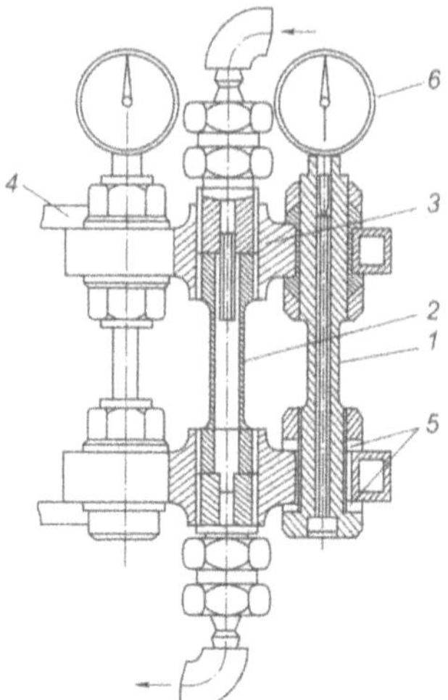

Fig. 7.28. Coffin's setup diagram for thermal fatigue testing of tubular specimens: *1* supporting columns; *2* coated specimen; *3* fixing flanges; *4* lead wire; *5* insulators; *6* clock-type indicators

However, it was demonstrated later that the deformation conditions are dependent on the number of cycles and that some other parameters influence the deformation magnitude. Therefore, the magnitude of deformation has to be recorded directly during testing procedure.

In Coffin's setup (Fig. 7.28), specimens with the wall thickness about 0.2–1.0 mm are fixed in flanges. Supporting columns provides the total inflexibility. The specimens are heated by electric current through the lead bars and cooled by cold air blown into the specimen's hole. The heating and cooling are alternated with a specified periodicity. The clock-type indicators or glued strain gauges display the deformation level of supporting columns. A thermocouple welded in the middle part of a specimen controls the maximal and minimal temperatures. The number of heat cycles N that caused a formation of a through crack in the working zone of the specimen is to be determined.

The test results are plotted in the form of a fatigue curve on $\lg\Delta\varepsilon_{max} - \lg N$ coordinates, where $\Delta\varepsilon_{max}$ is the maximal change in the plastic deformation.

Bockstain et al [7.3] studied the effect of a coating on operational characteristics of a heat-resistant alloy applied to manufacturing gas-turbine blades using the Coffin's setup. To reveal the nature of failure the ruptures were analyzed and metallographic analyses of longitudinal microsections were performed. Multicomponent CoCrAlY, NiCrAlY, NiCoCrAlY-based coatings were deposited using the electron-beam technique at a condensation rate of about 2 μm/min.

A complex relation between the loading and reliability was demonstrated, and the role of coating in failure of "base metal–coating" structure was cleared out. For $\Delta\varepsilon_{max} > 0.8$–$1.0\%$, coated specimens stands the amount of heat cycles by 30–

50% less than control specimens without coating; however, for $\Delta\varepsilon_{max} < 0.6\%$ the critical number of cycles is almost the same.

In all instances the main center of specimen failure coincides with the crack in the coating; but the patterns of failure in the coating and base metal can be different and determined by the coating properties. In a more plastic NiCoCrAlY-based coating a crack is characterized by higher degree of branching and slowed-up development, thus providing greater number of cycles until the onset of failure. The cracks in CoCrAlY-based coating spread out in radial directions and penetrate easily into the base metal, which leads to considerable reduction in the number of cycles before failure.

The Coffin's setup has several advantages, which results in a wide usage of this method; however, it does not allow tracking down structural changes occurring in material during thermal fatigue loading, or controlling development of microcracks during testing.

7.3.4 Thermal Fatigue Testing Using Vacuum Metallography

Rybnikov et al. [7.19, 7.20, 7.25] have developed a method of continuous microstructural study of damages in metals caused by thermal fatigue. A distinguishing advantage of the vacuum metallography method is a possibility of direct observation of changes in microstructure and kinetics of crack formation in the "base metal–coating" composition.

Tests are conducted using installations for high-temperature study of material structure and properties (IMASH series, Russia). These devices provide simultaneous microstructure analysis and micromechanic testing. The microstructure of the specimen surface can be observed and photographed for several types of loading (stretching, cyclic loading), and for heating and cooling in vacuum.

In IMASH-series devices a specimen can be heated from the room temperature up to 900 °C in 5 s. As the temperature goes down from 900 to 500 °C and from 600 °C to 150 °C, the maximal time of specimen cooling in vacuum is 5 s and 15 s respectively, but below 150 °C the cooling rate drastically decreases. In order to constrain thermal deformations during heating and cooling cycle Getsev et al. [7.6] proposed a special facility providing a high rigidity of head fixing. The size of this facility allows putting it into the working chamber of IMASH-series installation. The shape and size of commonly used flat corset-type specimens (Fig. 7.29) provide rigid fixing without pre-test deformation. Fabrication of the working zone (corset) and preparation of a microsection are simple. It was shown that the temperature in the central part of a specimen 4 mm in length is constant and this length is recognized as the research zone. The plain surface of the corset is fabricated in the form of a microsection. The general testing scheme is shown in Fig. 7.30.

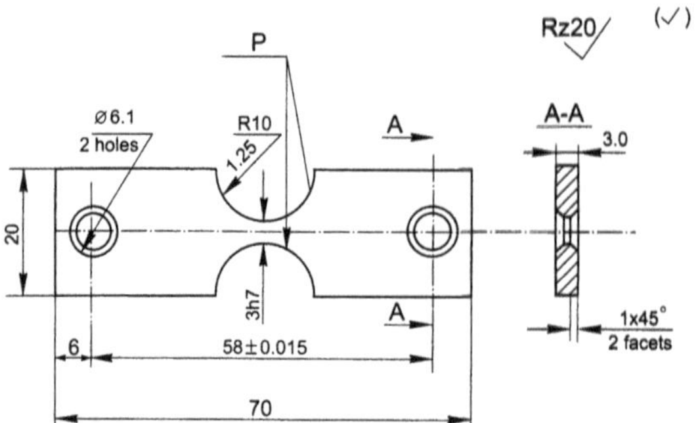

Fig. 7.29. Flat corset-type specimen for thermal fatigue testing with vacuum metallography. Non-specified deviations of longitudinal sizes belong to class H12, h12, (JT12)/2, c stands for a coated surface

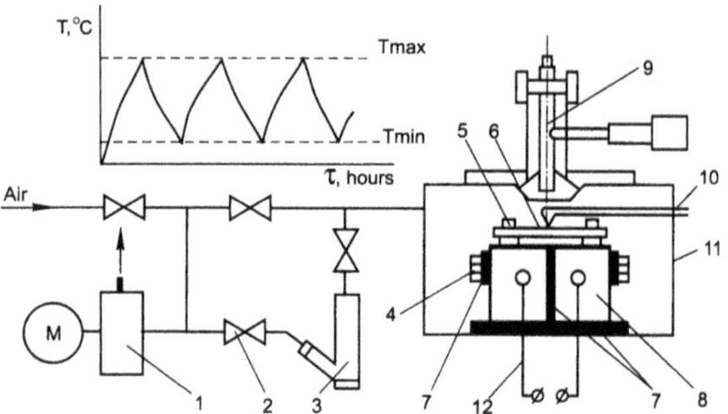

Fig. 7.30. The general testing scheme using an IMASH device. *1* preliminary evacuation pump; *2* vacuum valves; *3* diffusion pump; *4*, *5* bolts; *6* specimen; *7* gasket; *8* rams; *9* microscope; *10* thermal transducer; *11* vacuum chamber; *12* temperature cycling device; T_{max}, T_{min} maximal and minimal temperatures over a cycle; τ testing time

Prior to testing, indentations (bench marks) are made in the research zone using a microhardness meter, so the loading rigidity can be calculated from these distances during thermal cycling. First, vacuum is created in the IMASH device chamber, after which the specimen is heated by electric current.

The correct results can be obtained at the maximal temperature of the thermal cycle about 900–950 °C. The parameters of testing are the maximal and minimal temperatures, cycle period, and loading rigidity over a stable cycle. During thermal cycling a microscope is used for observation of a polished research zone. The moment of crack formation is recorded not only for the central zone, but also for remote areas. A number of cycles to crack initiation in the coating and base metal as well as a number of cycles until the specimen is disrupted into two pieces should be counted.

There are two scenarios of the observed failure of "coating–base metal" composition. In the first variant a crack emerges in the base metal and then spreads into the coating; in the second one (at a low deformability of the coating) cracks first develop in the coating, and only then the failure of the base metal takes place. It is also possible to follow the change in deformation profile and to measure the crack spreading velocity in the coating and base metal. The moment of crack initiation in the coating can be detected easily with devices based on the effect of acoustic emission [7.2].

The span of the total cycle deformation ($2\varepsilon_a$) in the central part of the specimen is determined by the formula developed by Getsov at al. [7.6]:

$$2\varepsilon_a = (\alpha_1 T_{\max} - \alpha_2 T_{\min})\varphi \tag{7.25}$$

where $T_{\max}$, $T_{\min}$ are the maximal and minimal temperatures over a cycle; α_1 α_2 are the temperature coefficient of linear expansion of the base metal for the ranges 20–$T_{\max}$, 20–$T_{\min}$; φ is the loading rigidity in a stable cycle, $\varphi = 1-\Delta K/\Delta L$; ΔK is the displacement of indentations (bench marks) in the central part of a specimen with the length of L=2.0 mm; ΔL is a free expansion of an area L after heating from $T_{\min}$ up to $T_{\max}$.

The ratios between numbers of cycles *Nc*, *Nbm*, *Nd*, which stand for crack formation in the coating and base metal and specimen disintegration into two pieces respectively, characterize the coating contribution into reliability and durability of the "base metal–coating" composition. The coating is still serviceable after thermal cyclic loading if $Nc > Nbm$, but it reduces the service life of the specimen if $Nc < Nbm$. If the ratio Nc/Nd remains almost the same for different temperature spans ΔT, then the coating serviceability can be estimated from short-run testing with a higher ΔT.

Rybnikov et al. [7.20] studied thermal fatigue of a nickel alloy with Co–20Cr–8Al–Y, Co–20Cr–11Al–Y, Co–25Ni–20Cr–11Al–Y electron-beam coatings, and for a two-layer metal-ceramic coating in the thermal cycling 150↔850 °C and 500↔850 °C. The number of cycles until the onset of failure of the coating and base metal was recorded. The authors established that a higher plasticity of the coating is related with increase in the lowest temperature over a cycle. The higher content of β-phase (depending of aluminum content of the coating) and formation of γ-phase (related with a high chrome content of the base metal) makes the surface layer more brittle. The Co–25Ni–20Cr–11Al–Y coating is apt to lamination. The use of a plastic layer between the base metal and coating increases the service life of the composition [7.20].

Later the same authors and Malashenko [7.25] studied thermal fatigue of nickel alloys with CoCrAlY/ZrO_2+8%Y_2O_3 coatings having a different thickness of the ceramic layer. A coating was deposited by physical precipitation from vapor using an electron beam. The relationship between the service life and the thickness of a three-layer coating can be represented by a curve with the maximum corresponding to the thickness of 100 microns [7.25].

References

7.1 Bernstain ML, Zaimovsky VA (1979) Mechanical Properties of Metals (in Russian). Metallurgia, Moscow

7.2 Billi F, Santulli C (1996) Thermal fatigue ceramic coated cutting tips by acoustic emission. Bull Cercle Etud Metaux 16:14–17

7.3 Bokstain SZ, Bychkov NG, Dulnev RA (1980) Thermal fatigue resistance of superalloys with protective coatings (in Russian). Problemy Prochnosti 4:59–63

7.4 Bomas H, Mayr P, Kurth B (1997) Fatigue properties of steel coated with TIN by a PVD process. Mater and Manuf Processes 12:17–27

7.5 Cholvy G, Cuntz JM (1992) Evaluation of aluminum coatings and steel used in aeronautical industry. Vide, Couches Minces 48:17–24

7.6 Getsov LB, Rybnikov AI, Olshanskaya EL (1982) The use of vacuum metallograpy methods to estimate the efficiency of materials with protective coatings (in Russian). Zavodskaya Laboratoria 7:74–77

7.7 Ibrahim A, Berndt CC (1998) The effect of high-velocity oxygen fuel, thermally sprayed WC–Co coatings on the high-cycle fatigue of aluminum alloy and steel. J Mater Sci 33:3095–3100

7.8 Ivanova VS, Shanyavsky AA (1998) Quantitative fractography. fatigue failure (in Russian). Metallurgia, Moscow

7.9 Ivanova VS, Terentiev VF (1975) The Nature of metal fatigue (in Russian). Metallurgia, Moscow

7.10 Kalmutsky VS (1983) Criteria for fatigue failure of parts with coatings (in Russian). Problemy Prochnosti 12:7–10

7.11 Khasui A (1975) Deposition techniques (in Russian). Mashinostroenie, Moscow

7.12 Maximovich GG, Shatinsky VF, Kopylov VI (1983) Physical-chemical processes in plasma spraying and failure of materials with coatings (in Russian). Nauchnaya Mysl, Kiev

7.13 Middleton RM, Huang PJ, Wells MGH, Kant RA (1991) Effect of coatings on rolling contact fatigue behavior of M50 bearing steel. Surface Eng 7:319–326

7.14 Mora-Marquez JG, Lira-Olivares J (1987) A study of crack initiation and propagation in Ni–Cr thermally sprayed coatings using acoustic emission techniques. Thin Solid Films 153:243–252

7.15 Orlov AV, Chermensky ON, Nesterov VM (1980) Testing of Construction Materials on Contact fatigue (in Russian). Mashinostroenie, Moscow

7.16 Pokhmursky VI, Borisov YuS, Kalichak TN (1980) Studies on the influence of Ni–Al and Ni–Ti plasma-sprayed coatings on fatigue resistance and corrosion resistance of medium-carbon steel (in Russian). Fizika I Khimia Obrabotki aterialov 14:22–25

7.17 Rakitsky AA, De los Rios ER, Miller KJ (1994) Fatigue resistance of medium carbon steel with a wear resistant thermal sprayed coating. Fatigue and Fract Eng Mater and Struct 17:563–570

7.18 Rhys-Jones TN, Cunningham TP (1990) The influence of surface coatings on the fatigue behavior of aero engine materials. Surface and Coating Technology 42:13–19

7.19 Rybnikov AI, Getsov LB, Malashenko IS (1995) Thermal cyclic response of EB PVD yttria-stabilized zirconia /CoCrAlY coatings. Thin Solid Films 270:247–252

7.20 Rybnikov AI, Ogurtzov AP, Tchizhik AA, Getzov LB (1991) Thermal fatigue resistance of gas turbine blades protective coatings. Vide, Couches Minces 47:138–140

7.21 Schueider , Grunling HW (1983) Mechanical aspects of high temperature coatings. Thin Solid Films 4:395–416

7.22 Smith T (1994) The effect of plasma-sprayed coatings on the fatigue of titanium alloy implants. JOM: J Miner Metals and Mater Soc 46:54–56

7.23 Sonobe Masaru, Shizawa Kazuaki, Motobayashi Kou (1997) Corrosion resistance and corrosion fatigue strength of carbon steel coated with chromium nitride by multistage. JSME Int J A 40:436–444

7.24 Sproul William D (1994) Multilayer, multicomponent, and multiphase physical vapor deposition coatings for enhanced performance. J Vac Sci and Technol A 12:1595–1601

7.25 Tchizhik AA, Getsov LB, Rybnikov AI, Malashenko IS (1995) Creep studies of the EB PVD coatings with a ceramic layer. Thin Solid Films 270:243–246

7.26 Tushinsky LI, Plokhov AV (1986) The Study of Structure and Physical-Mechanical Properties of Coatings (in Russian). Nauka, Novosibirsk

7.27 Tushinsky LI, Plokhov AV, Stolbov AA, Sindeev VI (1996) Structural strength of the base metal–coating composition (in Russian). Nauka, Novosibirsk

7.28 Umansky ES, Afonin NI, Borisov YuS (1977) The effect of plasma-sprayed coating on fatigue strength of steels. Problemy Prochnosti 10:112–113

7.29 Wei Ronghua, Wilbur Paul J, Liston Mary-Jo, Lux Gayle (1993) Rolling-contact-fatigue wear characteristics of diamond-like hydrocarbon coatings on steel. Wear 162–164:558–568

8 Breaking Strength and Stressed State of Base Metal–Coating Composition

8.1 Static Crack Resistance (Fracture Toughness)

8.1.1 General Overview

Fracture toughness is the ability of a material to resist crack propagation under mechanical and other impacts.

Today evaluation of metal quality only by indexes of strength and plasticity is not sufficient for determination of metal reliability in practical use. A general trend of the majority of strengthening treatments is that an improvement in material strength is inevitably accompanied by deterioration of crack resistance. Therefore, these important indexes of structural strength contradict each other: when the road of strength increase is taken, the operation reliability of the material suffers; on the other hand, when choosing the road of enhanced reliability (thus reducing material strength), the specific material consumption of a given construction is considerably increased. Therefore, determination of the optimal levels of yield point and crack resistance is one of the most difficult problems in the present-day physical metallurgy. Its solution becomes substantially more complicated for products covered with corrosion and wear resistant coatings.

During the last three decades the progress in studying strength physics and especially fracture mechanics has led to elaboration of special methods for evaluation of failure resistance. New characteristics of mechanical properties obtained on the basis of these methods are used in the certification of alloys, calculation of carrying capacity and survivability of products, in the diagnostics of failure [8.5, 8.17, 8.25, 8.33, 8.45].

Any process of material failure can be divided into three stages. At the first stage incipient cracks appear. The second and the third stages are the phases of subcritical and then critical crack propagation. In accordance with this pattern of failure process, there are various integral methods for evaluation of alloy failure resistance, as well as methods for evaluation of crack propagation resistance at subcritical and critical stages. The latter methods are well developed and substantiated on the basis of fracture mechanics. As for the first stage, there are no validated engineering methods for evaluation of crack initiation resistance owing to physical complexity of the phenomena that govern this stage [8.34, 8.35].

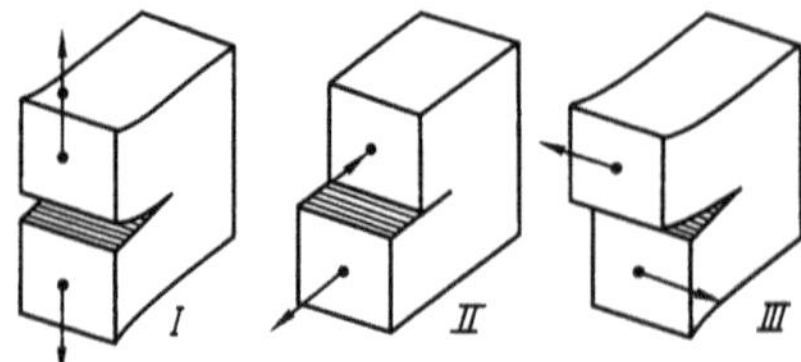

Fig. 8.1. Three main techniques for deforming a plate with concentrator having the shape of a crack: *I* break-off; *II* lateral shift; *III* longitudinal shift

Fracture toughness tests under static loading are among the most approved and theoretically substantiated methods that are now gaining wide acceptance in practical metal science [8.27, 8.40].

The principal characteristics of material crack resistance include: force parameters – critical coefficients of stress intensity K_c or K_{ic}; deformation parameters – critical crack corner opening σ_c; energy parameters – critical values of J-integral J_c or J_{ic}.

The theoretical basis for the methods of evaluating crack resistance of materials was found in the linear mechanics of fracture put forward in the works of Griffits. Based on the energy approach, he was the first to demonstrate that the striking discrepancy between the actual and theoretical strengths of solids may be caused by the presence of small defects which act as stress concentrators, thus creating local stresses reaching the theoretical strength limit. Developing the ideas of Griffits, Irvin has proved that in the most general case of break-off loading of a cracked body these local stresses are proportional to the so called stress intensity coefficient K, which can be written as

$$K = \sigma\sqrt{\alpha\pi l}\ , \tag{8.1}$$

where σ are the nominal stresses evaluated without considering the crack in the body; l is the length of crack; α is a factor depending on the shape of the specimen and geometry of the crack.

There are three main schemes for loading of a cracked body (break-off, lateral shift, and longitudinal shift, see Fig. 8.1), which are called I_{st}, II_{nd}, and III_d loading techniques, respectively. The first type of crack edges displacement is realized in case of the most dangerous brittle failure; therefore, all engineering methods for evaluation of fracture toughness are concerned with this type.

When formulating his criteria for material crack resistance, Irvin proceeded from the fact that once instable spontaneous crack growth is reached, the stress intensity coefficient K reaches its critical value K_c, which had been considered the constant of a material. However, it has been found out that the value of this characteristic depends on the thickness of the product under test, and as the latter goes up, K drops due to change-over from the planar stressed state at the corner of the crack to the most dangerous in terms of brittle failure planar deformed state, until it finally reaches its minimal stable value K_{ic}. When the conditions for the correct determination of crack resistance characteristics are satisfied, K_{ic} is the main quantity that characterizes material properties.

Crack resistance characteristics are very important for practical calculation of machine parts and details, since they allow to determine critical crack dimensions which give rise to brittle failure from the known values of working stress and, vice versa, having determined crack dimensions and shapes by some flaw detection methods, the breaking stress value can be determined [8.34].

8.1.2 Specimens

According to State Standard 25.506, the following specimen types are recommended for determination of the crack resistance characteristics: type I is a flat rectangular specimen with a central crack intended for axial stretching test (see Fig. 8.2a); type II is a cylindrical specimen with a ring-shaped crack intended for axial stretching test (see Fig. 8.2b); type III is a rectangular compact specimen with an edge crack intended for eccentric stretching test (see Fig. 8.2c); type IV is a flat rectangular specimen with an edge crack intended for three-point bending test (see Fig. 8.2d).

For steels with yield point below 800 MPa the assumptions of linear fracture mechanics (*vide infra*) are satisfied only when large-scale specimens are used. Their weight can reach 1.5 t, with thickness up to 300 mm. So, owing to methodical difficulty crack resistance evaluation using K_{ic} is complicated for commonly used low-carbon steels.

The value of K_{ic} for low-carbon steels necessary to evaluate crack resistance of a large product cannot be determined using compact laboratory specimens. When a structure is used at low temperatures, the appropriate tests allow determination of K_{ic} using smaller specimens, since in this case as the testing temperature goes down, yield point goes up, while K_{ic} decreases.

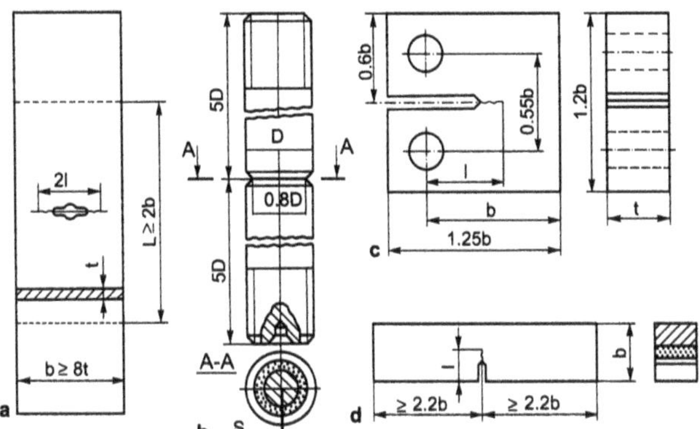

Fig. 8.2. Types of specimens for K_{ic} determination: **a** for stretching tests on sheet materials; **b** cylindrical specimen for stretching tests; **c** for eccentric stretching tests; **d** prismatic specimen for bending tests

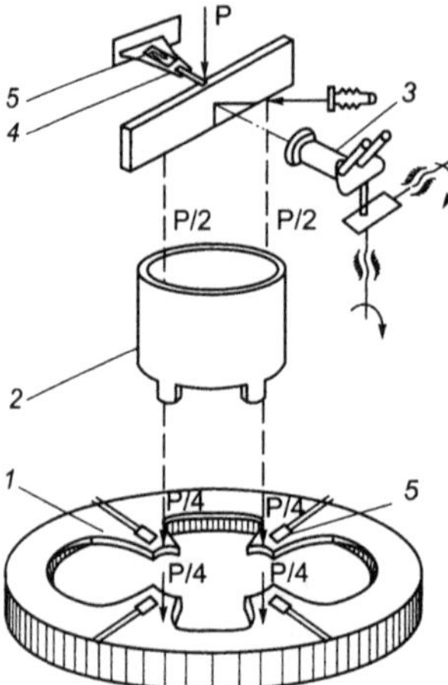

Fig. 8.3. Scheme of testing fracture toughness for notched and cracked flat specimens. *1* elastic force element; *2* intermediate ring; *3* microscope; *4* elastic deflection element; *5* strain gauge; *P* applied load

8.1.3 Equipment

The prepared specimens with an induced crack are tested for failure using standard (but necessarily rigid) testing machines, at loading rates in the range 0.02–0.2 mm/s.

A scheme of loading flat prismatic specimens at a testing machine is shown in Fig. 8.3. The specimens are loaded using the three-point bending scheme. The load applied to the specimen is transmitted through the ring to the four-tab elastic element and is converted to electric signal by the strain gauges. The signal is then fed to the "Y" input of an XY-recorder. Readings of the load strain gauge are calibrated using a compression spring dynamometer. Specimen deflection at the point of load application is measured by the strain gauge glued onto the elastic plate. The gauge is calibrated using a micrometer depth indicator with ±0.01 mm accuracy. A microscope provides visual monitoring of the failure process.

8.1.4 Preparation for Testing

Preparation of specimens for testing consists in depositing coatings on side faces, finishing the faces, cutting of stress concentrators and inducing fatigue cracks. To facilitate length control of fatigue cracks during their growing the side faces of the specimens should be polished. For specimens with thickness larger than 25 mm it is recommended to use chevron-shaped notches as stress concentrators. A notch produced using an electric-spark discharge machine with 0.1–0.15 mm diameter wire is an efficient concentrator facilitating fatigue crack initiation. Special machines are used to grow fatigue cracks from the mouths of notches.

When producing fatigue cracks in specimens, nominal stresses σ_0 should not exceed $0.5\sigma_{0.2}$ for the material at maximum load of the cycle, while the number of loading cycles should be at least $5 \cdot 10^4$.

8.1.5 Testing

During the test the failure diagram is automatically registered by XY-recorder in the coordinates applied load P – displacement of crack edges V. An appropriate crack opening sensor mounted on the specimen provides displacement readings.

Fig. 8.4 shows four different characteristic types of failure diagram P–V. For diagram types I and II the maximum load in the sector cut off by a 5% secant line (α_5 angle) is taken as the calculated load P_Q to determine fracture toughness. For diagram types III and IV P_c is taken as P_Q. The sought value of K_Q is calculated form the P_Q values using the following equations (Y_1, Y_2', Y_2'', Y_3, Y_4 are correction functions):

1. For specimens of type I (see Fig. 8.2a):

$$K_Q = \frac{P_Q Y_1}{t\sqrt{b}}, \tag{8.2}$$

where $Y_1 = 0.380[1 + 2.308(2l/b) + 2.439(2l/b)^2]$, for $0.3b \le 2l \le 0.5b$.

2. For specimens of type II (see Fig. 8.2b):

$$K_Q = P_Q\left(Y_2' + Y_2''\right)\sqrt{D}, \tag{8.3}$$

where $Y_2' = 6.53[1 - 1.8167(d/D + 0.9167(d/D)^2]$, $Y_2'' = 3.1(2S/d)$, for $0.6D \le d \le 0.7D$ and $2S < 0.08d$.

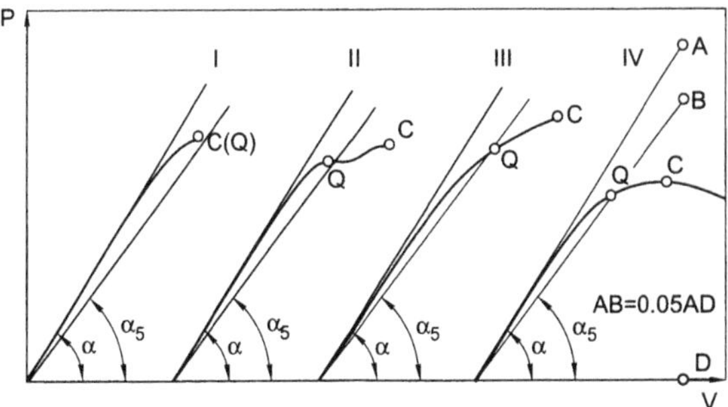

Fig. 8.4. Characteristic load–displacement diagram types (*I–IV*) and their processing

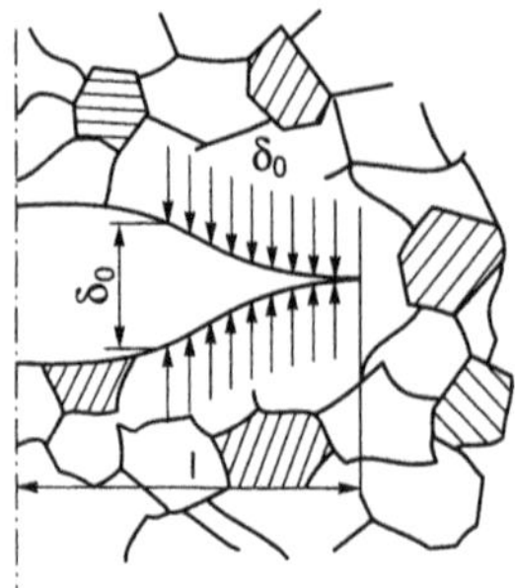

Fig. 8.5. Distribution of stresses near crack edges and critical opening value δ_c

3. For specimens of type III (see Fig. 8.2c):

$$K_Q = \frac{P_Q Y_3}{t\sqrt{b}}, \tag{8.4}$$

where $Y_3 = 13.74[1 - 3.380(l/b) + 5.572(l/b)^2]$, for $0.45b \leq l \leq 0.55b$.

4. For specimens of type IV (see Fig. 8.2d):

$$K_Q = \frac{P_Q L Y_4}{t\sqrt{b}}, \tag{8.5}$$

where $Y_4 = 3.494[1 - 3.396(l/b) + 5.839(l/b)^2]$; L is the distance between supports, $L = 4b$; for $0.45 \leq l \leq 0.55b$.

The calculated K_Q values are taken to be equal to K_{ic}, if the following requirements on the size of the cross-section of specimens are satisfied:

- $t_k \geq 2.5(K_Q/\sigma_{0.2})^2$ for flat specimens and
- $D_k \geq 2.3(K_Q/\sigma_{0.2})^2$ or $d_k \geq 1.6(K_Q/\sigma_{0.2})^2$ for cylindrical specimens.

As has already been mentioned, rather demanding geometric requirements needed for the realization of the planar deformed state failure substantially complicate evaluation of material crack resistance using the parameter K_{ic}. K_{ic} can be readily evaluated for high-strength steels, whereas its application to medium- and especially low-strength steels, which are most commonly used, requires considerable increase in size (thickness) of the tested specimens and in the power of testing devices. As a result, crack resistance evaluation of constructional steels using K_{ic} becomes either impossible simply because the material with the required thickness is not available, or rather difficult when testing large-size specimens [8.35].

This difficulty in evaluation of crack resistance of low-strength steels may be overcome if the so-called critical crack opening δ_c is used as a measure of material crack resistance. The measurement of δ_c corresponding to crack opening at the onset of its spontaneous growth (see Fig. 8.5) can be performed using a special device, although more often it is evaluated by displacement of crack edges determined from P–V diagrams (see Fig. 8.4).

According to State Standard 25.506, the value of δ_c is calculated for point C of P–V diagrams, types I–IV, from the following equations corresponding to different specimen types.

1. For specimens of types I and II:

$$\delta_C = \frac{K_C^{*}\left(1-\mu^2\right)}{2\sigma_{0.2}E}. \tag{8.6}$$

2. For specimens of type III:

$$\delta_C = \frac{K_C^{*}\left(1-\mu^2\right)}{2\sigma_{0.2}E} + \frac{(b-l)}{0.4b+0.6l+7}. \tag{8.7}$$

3. for specimens of type IV:

$$\delta_C = \frac{K_C^{*}\left(1-\mu^2\right)}{2\sigma_{0.2}E} + \frac{0.4(b+l)}{0.4b+0.6l+7}. \tag{8.8}$$

K^{*}_c in these equations is the value of the stress intensity coefficient corresponding to point C of the P–V diagram.

Last years have brought successful attempts to create a new universal method for evaluation of fracture toughness of low- and high-strength alloys in terms of so-called J-integral, which is the change in potential energy in elastoplastic continuum in the course of crack propagation.

The values of J-integral are determined from the diagram of load P versus specimen deflection at the load application point f. Critical values of J-integral are evaluated at the onset of cracking.

There are a number of laboratory techniques for J-integral evaluation. According to State Standard 25.506, J values are calculated from the following equation:

$$J = \frac{K^2(1-\mu)}{E} + \frac{A_p}{(b-l)t}\frac{\chi}{k}, \tag{8.9}$$

where A_p corresponds to plastic part of the area under P–f diagram limited by the load relief point; χ and k are coefficients depending on the geometry of the specimen and length of crack.

Testing a batch of specimens loaded up to various levels f and, accordingly, to various initial crack gains Δl yields a series of J and Δl values which are used to plot a curve of crack growth resistance (see Fig. 8.6). Extrapolation of the curve to zero crack gain allows determination of the critical value J_{ic} of the integral.

Application of J-integral considerably extends the possibilities of determination of crack resistance in metals and alloys, since it accommodates testing low-strength materials using small-size specimens and determination of the K_{ic} value via J_{ic} from the following equation:

$$K_{ic} = \sqrt{\frac{J_{ic}E}{1-\mu^2}} \, . \tag{8.10}$$

8.1.6 Results and Discussion

The results of research of the coating structure and the base metal-coating interface show that near-surface volumes of material practically always contain preformed failure centers of various sizes and shapes. Pores, non-uniformities on the base metal boundary, incipient cracks developing in the process of deposition, loose boundaries between layers, etc. play in the coating the role of stress concentrators. If a coating is deposited at sufficiently high temperatures, domains with increased density of dislocations and vacancies are formed in the diffusion zone.

Redistribution of excess vacancies and their sink at certain points lead to formation of micropores. The areas of stretching and compression arising in the diffusion zone promote microplastic deformation of the base metal and transformation of micropores into cracks. Thus, deposition of coatings is generally accompanied by an increase in defectiveness of base metal surface layers; the higher the level of coating hardening, i.e. the higher its susceptibility to brittle failure, the more dangerous any discontinuity flaws and pores are [8.35].

Cherepanov is his paper reviewing the modern problems of fracture mechanics pays special attention to behavior of coated materials in the presence of cracks [8.8]. He has formulated the necessary conditions that a protective coating must satisfy. Let a fatigue corrosion crack be developing from a certain point O that is one of the corners of an edge crack at the base metal-coating interface. In this case three ways of crack development are possible: along the old direction, but now inside the substrate at speed V_g, along the boundary and in the layer material near the boundary, at speeds V_{lg} and V_l, respectively.

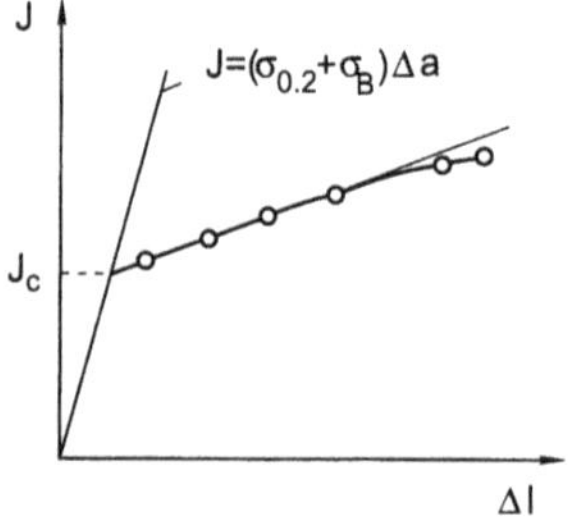

Fig. 8.6. A curve of crack growth resistance

The three speeds are determined experimentally and independently from each other from failure diagrams using estimated coefficient of stress intensity. It is natural to assume that the main crack will propagate from the point O in the direction of maximal speed of its potential growth. Possible crack branches developing in other directions fall in the load relief area and do not grow. Therefore, if $V_g < \max(V_l, V_{lg})$, then the coating will protect from fatigue corrosion cracks. Cherepanov specially points out the most unfavorable case, when a crack goes from the coating directly to the base metal [8.8].

Pokhmursky with co-authors [8.31] made an interesting attempt to create an analytical model of a two-layer coating with a crack perpendicular to the line of layer demarcation in the external layer. Such a model can describe, in particular, coatings with barrier layers, i.e. a composition material consisting of three interconnected strips with different elastic and strength properties, with the upper strip containing a crack normal to the line of layer demarcation.

As was demonstrated by simulations, a five-fold increase in hardness of the intermediate layer as compared to hardness of the coating practically provides good retardation of cracks in the composition, with sufficient thickness of the intermediate layer being equal to thickness of the coating or 1.5–2 times larger.

The authors argue [8.31] that the obtained results allow formulation of a criterion for failure and of optimum conditions for metal-protective coating compositions, after crack resistance of the whole composition and its components have been determined experimentally.

There are few experimental works concerning the influence of gas-thermal coatings on reliability of specimens with crack-type defects. We consider to be the most interesting a paper by Iliinsky et al. [8.12] describing laborious tests that led to definitive conclusions on how crack resistance of a critical construction depends on its specific coating.

The effect of coating on crack resistance of steel used in the manufacturing of load bearing parts of reactor housing was determined. Hull steel with composition 0.15% C, 0.7% V, 1.9% Cr, 1.2% Ni, 1.5% Mo was used as a base material. Standard specimens for eccentric static stretching testing were specially produced. The side faces of the specimens were covered with Austenite steel coating with thickness 7–9 mm using welding deposition. 16 specimens with thickness 50–150 mm, including 6 reference specimens without coating, were tested at room and negative temperatures. The obtained results show that the (welded) coating does not reduce fracture toughness of the base metal over the whole range of the checked temperatures (from –80 to +20 °C). The values of K_c (for temperatures of 20 °C and above) and K_{ic} (for temperatures below 20 °C) of homogeneous and coated specimens agree with each other within the common scatter of experimental data.

In [8.12] the authors considered the influence of the specimen thickness on fracture toughness at room temperature. A principal difference was found between the experimentally obtained and expected results, namely, the values of K_{ic} were found to increase with increasing thickness of the specimen. Earlier Krasovsky et al. [8.21] reported a similar dependence when evaluating crack resistance of the same steel: for specimens with thickness 300 mm K_{ic} was 3 times larger than for specimens with thickness 18 mm. The opposite actual relationship between the parameters t and K_{ic} in comparison with theoretical expectations (see Fig. 2.20) can possibly be explained by some uncertainty in the interpretation of K_{ic} rather than by the presence of coatings on side faces of specimens.

In view of the above mentioned facts, Ilinsky et al. [8.12] recommend evaluation of hull steel crack resistance using J_{ic} index that is not so sensitive to the geometry of the specimen. Analyzing the role of coatings in enhancing constructional strength of the reactor housing base components, they arrive at the conclusion that welding of an austenite steel coating at the housing steel provides such level of plasticity and toughness in the coating and thermal influence zone that, despite the presence of high residual stresses, crack resistance of the composition is not reduced.

In the work [8.46] the influence of plasma- and ion-plasma sprayed coatings on crack resistance of hardened carbon steel (0.8% C) under static loading was evaluated using flat specimens with edge crack in the three-point bending scheme.

The mechanical properties of the base metal as determined after ion-plasma spraying of titanium nitride coatings do not significantly depend on the duration of heating during spraying ($\sigma_{0.2} = 1150$, $\sigma_2 = 1400$; $\Psi = 36\%$). The results of metallographic studies have shown that even on non-etched sections the coating-base interface is relatively sharp; the coating follows metal relief, and the thickness of the coating in the areas normal to the direction of arrival of sprayed particles is larger than in other parts of the surface. The coating surface is irregular, with dips and bumps. Larger dips having the shape of truncated cones generally have fused bottoms and smooth edges. Probably, in this case chemical interaction between coating and base materials takes place. The results of crack resistance determination in the three-point bending scheme are summarized in Table 8.1. The control treatment consists in heating a specimen up to the ion cleaning temperature with subsequent slow cooling.

The analysis of obtained results shows that the influence of surface roughening and restraining of plastic deformation seems to be compensated by the effect of favorable compressing stresses that arise during deposition and can be sufficiently high. Deposition of ion-plasma sprayed coatings does not reduce crack resistance, and J_c values determined for different treatments are equal (i.e. within the common scatter in experimental data).

An ambiguous influence of the plasma-sprayed self-fluxing coating (composition: 3.1% B, 0.8% C, 4.3% Si, 16.0% Cr, 4.0% Fe, Ni – the rest) on crack resistance of non-hardened hot-rolled steel (0.8% C) has also been reported. A 0.3 mm thick coating was deposited on one of the wider faces of a flat compact specimen after shot blasting treatment of its surface.

Table 8.1. Crack resistance of ion-plasma-sprayed coatings based on titanium nitride

Thickness of coating [μm]	Duration of spraying [min]	J_c [kJ/m^2]
Without coating	–	29.0
5	30	29.0
9	45	29.5
16	60	29.6

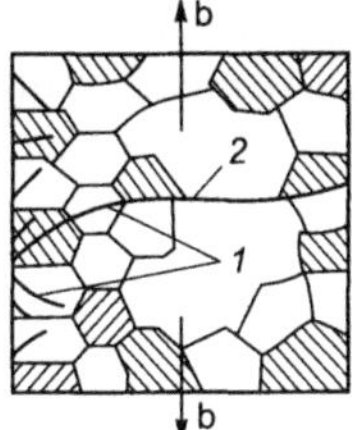

Fig. 8.7. Scheme of fatigue crack development. *1* initial stage (microlevel); *2* main crack (macrolevel)

Laminated structure and the presence of non-deformed particles of the source powder (about 12% by volume) are typical for this type of coating type. The total porosity amounts to 8–10%, the pores are located mainly along the boundaries of non-deformed particles. Fracture toughness of coated specimens was found to be slightly higher than of reference specimens, though the scatter of values is much larger (127±28 and 120±11 kJ/m^2, respectively).

8.2 Cyclic Crack Resistance

8.2.1 General Overview

The time of fatigue crack propagation, or period of survivability, may cover from 10 to 90 per cent of the total durability of a specimen or detail [8.37]. Gradually developing fatigue cracks initiated under cyclic loads create conditions for brittle failure. A scheme of fatigue crack development is shown in Fig. 8.7.

The growth rate of fatigue cracks is an important characteristic of a material. It depends on loading conditions, such as the level of applied stresses, asymmetry and frequency of cycling, the pattern of stressed state of the specimen, the value of average stresses over a cycle. Constructional factors and environment in which the testing is carried out are also of great importance [8.34].

As has already been mentioned, the process of fatigue crack growth in metals can be divided into three periods. During the first period the crack develops along sliding systems located in the zone of maximum tangential stresses. The length and growth rate of the crack are small, and there are no furrows at the fracture. During the second period the crack grows normal to applied stresses. This is a period of stationary crack growth, and its rate is proportional to the swing of the stress intensity coefficient. The third and final period is catastrophic crack development, which ends with failure.

Evaluation of the crack growth rate at the second stage is of primary importance for practical application.

Evaluation of resistance to crack propagation under cyclic loading can be reduced to construction of the kinetic diagram of fatigue failure (KDFF). The diagram sets the correspondence between the crack growth rate V and the stress intensity coefficient K at the corner of the crack (its swing ΔK or maximal value K_{max} with due account of cycle asymmetry). The details of the procedure of material testing for cyclic crack resistance and the drawings of specimens are given in methodical recommendations PD50-345.

8.2.2 Specimens

Flat specimens used for evaluation of static crack resistance (see Fig. 8.2) can also be used for cyclic testing. During cyclic bending or stretching tests of such specimens, ΔK growth is generally observed with increasing the crack length. At the same time [8.35], specimens providing constant ΔK for constant load level over a certain range of crack length are worth special attention. This type of specimens includes, first of all, disk specimens with a central crack intended for cyclic stretching testing (see Fig. 8.8).

The recommended thickness of the specimen is equal to thickness of the coated product, as this parameter has considerable effect on the crack growth rate. "L"-type specimens (see Fig. 8.2) cut out so that a crack could develop parallel with the direction of the base metal rolling are also commonly used. An "L"-type specimen is tested on supports having diameter equal to the width of the specimen set at a distance four times the thickness of the specimen apart.

A scheme of such a specimen with applied coating is shown in Fig. 8.9.

8.2.3 Equipment

Cyclic loading of specimens can be created using various machines for general and special fatigue testing. Most often pulsed stretching and flat bending are used.

Fig. 8.10 shows a scheme of installation for flat bending tests of specimens with constant force amplitude over a cycle. An electric motor rotates the shaft through the flexing coupling. The unbalanced mass creates force during shaft rotation that is transferred through the bearing race first to coupling rod and then to the specimen.

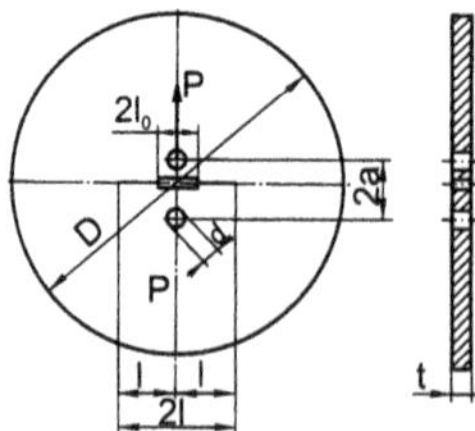

Fig. 8.8. Disk specimen for cyclic crack resistance evaluation

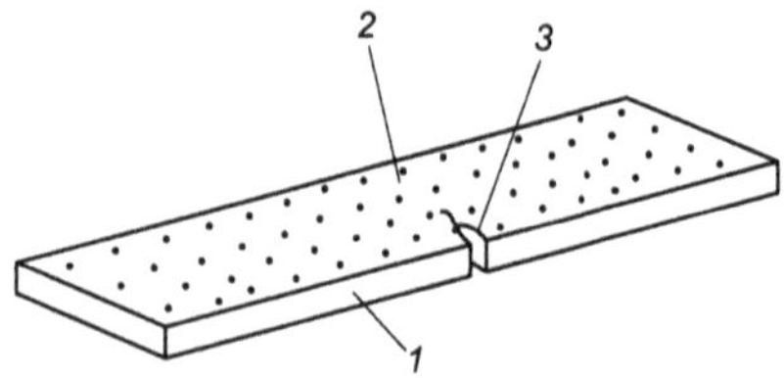

Fig. 8.9. Scheme of a specimen for evaluation of crack resistance. *1* base metal; *2* coating; *3* fatigue crack

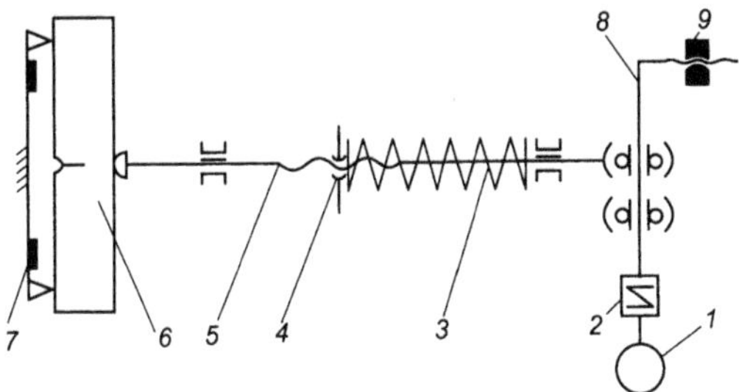

Fig. 8.10. Installation for cyclic crack resistance testing. *1* electric motor; *2* flexing coupling; *3* spring; *4* adjusting nut; *5* coupling rod; *6* specimen; *7* elastic element with a strain gauge; *8* shaft; *9* unbalanced mass

Electric signal proportional to the load applied to the specimen comes from the strain gauge to strain amplifier and then to an oscilloscope. The spring is used for adjusting cycle asymmetry coefficient. A direct current motor provides testing over a wide frequency range (6–20 Hz).

8.2.4 Preparation for Testing

The specimens are prepared by depositing the coating on the side faces and inducing fatigue cracks. These operations are similar to preparation of specimens for static tests and are described in more details in Sect. 8.1.4. It should be noted, that the loads should be chosen in such a way so as to avoid development of considerable plastic deformation zones, which could influence the kinetics of crack propagation.

8.2.5 Testing

Depending on their basic principles the methods for fatigue failure analysis and crack length registration can be divided into three groups: direct observation methods (including visual or optical inspection, fiber optics, and photoelastic coatings); physical methods (including magnetic, ultrasonic, acoustic, and eddy current methods); methods based on registration of change in material properties (such as electric resistance, deformation strengthening, internal friction, magnetic properties).

Measurements of crack growth are usually started after a fatigue crack $0.5L$ longh (L is the thickness of a specimen) has been formed near the end of the notch.

Stress intensity coefficients K are determined for different specimen types (see Fig. 8.2) from the following equations [8.35]:

1. Type I (see Fig. 8.2a):

$$K = Y_1 P\sqrt{l}/tb\,, \tag{8.11}$$

where

$$Y_1 = 1.77 + 0.227(2l/b) - 0.5l(2l/b)^2 + 2.7(l/b)^3. \tag{8.12}$$

2. Type III (see Fig. 8.2c):

$$K = Y_3 P\sqrt{l}/tb\,, \tag{8.13}$$

where

$$Y_3 = 29.6 - 18.5(l/b) + 655.7(l/b)^2 - 1017(l/b)^3 + 638.9(l/b)^4. \tag{8.14}$$

3. Type IV (see Fig. 8.2d):

$$K = Y_4 P\sqrt{l}/tb\,, \tag{8.15}$$

where

$$Y_4 = 1.93 - 3.07(l/b) + 14.53(l/b)^2 - 25.11(l/b)^3 + 25.8(l/b)^4. \tag{8.16}$$

Functions $Y=f(l/b)$ for the above specimen types are given in Fig. 8.11. The distinguishing feature of the specimen with a central through crack (see Fig. 8.8) as compared to all others is that the value of K does not depend on crack length over sufficiently wide length range and is determined only by the value of the applied load.

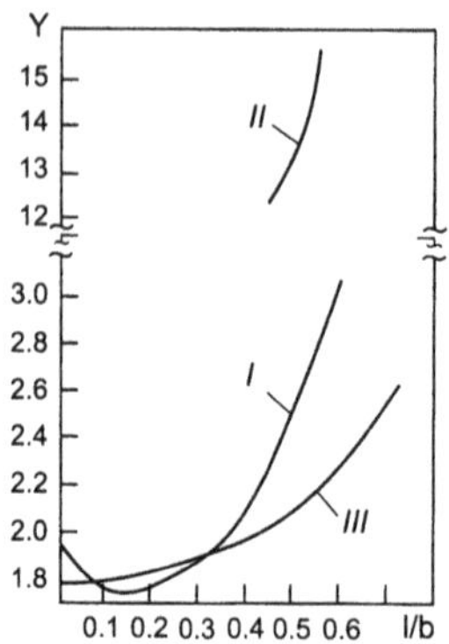

Fig. 8.11. Y as a function of l/b for the specimens type I, II, and III

8.2.6 Results and Discussion

Fig. 8.12 shows a typical KDFF plotted in logarithmic coordinates. Three areas are generally differentiated in the diagram: I – for low (below $v = 10^{-8}$ m/cycle), II – for medium ($10^{-8} \geq v \geq 10^{-6}$ m/cycle), and III ($v > 10^{-6}$ m/cycle) – for high crack growth rates. The main characteristics of the material cyclic crack resistance are as follows:

1. Threshold stress intensity coefficient K_{th}, sometimes called fatigue threshold, corresponding to the maximum value K_{max} at which the crack still does not develop. In most cases $K = K_{th}$ corresponds to crack growth rate 10^{-10} m/cycle or less;
2. Critical stress intensity coefficient (cyclic fracture toughness) K_{fc} corresponding to spontaneous failure of the specimen;
3. Parameters K^*, C and n of the functions

$$v = 10^{-7}(K_{max}/K^*)^n \tag{8.17}$$

and

$$v = Ck_{max}{}^n, \tag{8.18}$$

characterizing material behavior at the second (medium-amplitude) KDFF area.

The character of KDFF strongly depends on cycle asymmetry. The latter is commonly expressed by means of the asymmetry coefficient (see Fig. 8.12):

$$R = K_{min}/K_{max}. \tag{8.19}$$

Characteristics of material cyclic crack resistance are determined by computerized statistical processing of the arrays of experimental values $v - \Delta K$ forming KDFF as the parameters of the equation of the fatigue crack growth rate. For the middle KDFF area the equations (8.17) and (8.18) are used. A large number of empirical equations has been proposed for description of the entire KDFF curve, but most commonly various particular cases of the following general expression [8.34] are used:

$$v = A\Delta K^s \frac{\left(\Delta K^m - \Delta K_{th}^m\right)^q}{\left[(1-R)K_{fc}^n - \Delta K^n\right]^r}. \tag{8.20}$$

For example, when $s = 0$, $m = n = 1$ and $q = r$, we obtain the following equation that is widely used and recommended by methodical instructions PD50–345:

$$v = v_0 \left[\frac{\Delta K - \Delta K_{th}}{(1-R)K_{fc} - \Delta K}\right]^q. \tag{8.21}$$

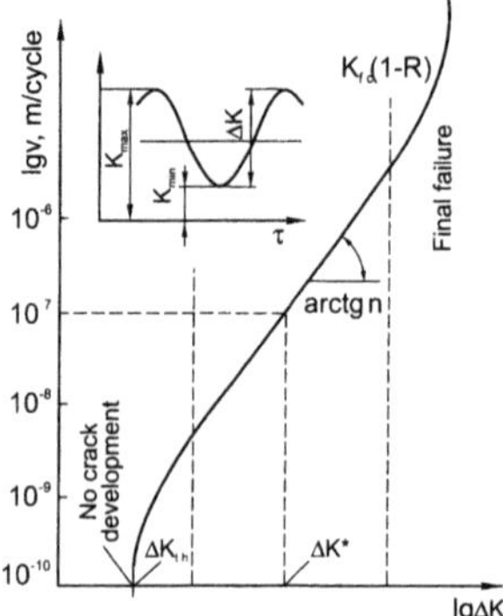

Fig. 8.12. Kinetic diagram of fatigue failure (KDFF)

The influence of self-fluxing PG–SR4 coating (composition: 3.1% B, 0.9% C, 4.1% Si, 16% Cr, 4.0% Fe, Ni – the rest) on crack resistance of the U8 steel (0.8% C) under cyclic loading was evaluated according to PD50–345 recommendations using prismatic specimens (see Fig. 8.9). The swing of stress intensity threshold coefficient ΔK_{th} was determined on the basis of 10^7 sinusoidal cycles at loading frequency 20 Hz. The tested specimens were either plasma sprayed or fused (fusing at 1040 °C for 10 minutes followed by normalization at 800 °C). The negative influence of the coating on cyclic crack resistance of steel found (see Fig. 8.13). Thus, deposition of plasma-sprayed coating reduced the threshold value of stress intensity coefficient of specimens from 9.72 to 7.42 MPa·m$^{1/2}$. Subsequent fusing of the coating caused further drop of the fatigue threshold down to 5.28 MPa·m$^{1/2}$. At the medium amplitude part of KDFF crack propagation rates of specimens with plasma-sprayed coatings and without coatings were the same. Specimens with fused coatings failed more readily. The drop of the fatigue resistance of steel probably occurs mostly as a result of a reduction in the efficiency of boundary layers of the base metal that change during the coating deposition, as well as due to embrittling effect of the coating. Studies of the fracture surface of specimens with fused coatings having high bond strength have verified this conjecture: near the base metal-coating interface at the substrate depth up to 0.8–1.0 mm the front of crack propagation leaves behind the failure of the rest bulk of metal.

The results of structural research of fractures in specimens demonstrate that the coating acts as an initiator of the base metal-coating composition failure. When the boundary has a large number of microdefects (pores, discontinuity flaws), a crack from the coating goes along the interphase boundary and not into the base metal, i.e. it has no effect on steel failure. This phenomenon, probably, explains the similar behavior patterns of non-coated specimens and specimens with plasma-sprayed coatings in the medium amplitude part of the fatigue failure kinetic diagram and in static tests for determination of J_c (which corresponds to the third part of the diagram, where K_{fc}~K_{ic}).

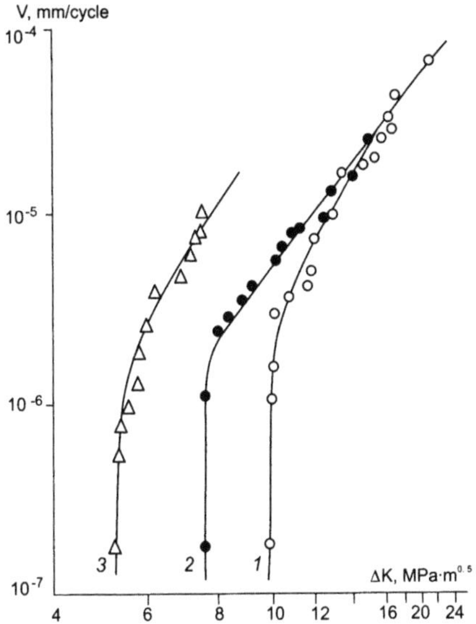

Fig. 8.13. Cyclic crack resistance of steel U8 coated with self-fluxing alloy PG–SR4. *1* without coating; *2* with plasma sprayed PG–SR4 coating; *3* with fused PG–SR4 coating

8.2.7 Cyclic Crack Resistance Tests of Coated Details

Due to a large number of factors that determine strength in cyclic tests, testing of real coated details is very important. Coated details and construction elements may have the gradient of changes in their properties and the base metal structure, the pattern of internal stress, the state of the surface layer and the concentration of stress substantially different from those of the specimens.

Testing of details under cyclic loading should be performed using a scheme and conditions of loading which approximate as close as possible the actual operating conditions of the detail in terms of stresses, environment, temperature, frequency and parameters of the loading cycle [8.37].

The criterion for the correct choice of the loading scheme and conditions of loading is reproduction of the actual failure behavior of the detail in the same section where the operational damages occur. If the actual failure behavior is not known, a load application scheme and loading conditions should be chosen in such a way so as to provide a failure in the weakest section.

Reliable data on crack resistance can be obtained by testing cracked details in the actual operation loading modes using single-step and random loading in multicycle region for simulation of the field operation conditions.

In single-step tests fatigue limits and fatigue curve parameters are determined. In order to optimize the construction of the detail and its production technology, including selection of the base material and the technology of coating deposition, along with single-step testing a random-type test with digital modeling of the sequence extrema is also used [8.37].

8.3 Prospects of the Development of Methods for Evaluation of Crack Resistance of the Base Metal–Coating Composition

When estimating future prospects of the development of static methods for determination of the crack resistance of coated specimens, the following can be noted:

1. Characteristics obtained by testing metal specimens serve, first of all, to calculate strength of products taking into account the presence of crack-type defects. Using the general ideas of linear fracture mechanics, we can establish what critical crack size will cause brittle failure, or evaluate the level of failure stresses at a given defect size. As for the results obtained for coated specimens, at present they are hardly applicable to similar calculations, since a comprehensive approach to strength calculation of composite materials, which include coated base metals, has not yet been developed.
2. The experience gleaned from testing uniformly strengthened specimens can be applied to development of standardized techniques for determination of crack resistance of coated specimens.
3. The influence of coatings on the change in crack resistance of the substrate must be evaluated. On one hand, all circumstances encountered in the formation of a coating, such as plastic deformation constraint, increase in concentration of surface defects, formation of residual tension stresses, creation of additional obstacles for dislocations on the base metal boundary, would decrease the level of fracture toughness. On the other hand, certain positive factors (favorable compression stresses, recovery of base metal surface defects, etc.) can enhance crack resistance of the material. To understand the results of competitive influence of the opposite factors on the fracture toughness, systematical tests of specimens with coatings deposited in various technological regimes are required.

The importance of adhering to the following principle should be especially emphasized: the chosen deposition technique must not reduce fracture toughness factors of the base metal-coating composition. If, for example, in order to enhance wear resistance U the chemical formulation of a composite coating has been changed, i.e. tungsten carbide content has been increased which led to an abrupt drop in crack resistance, then it is hardly expedient to use such a detail in a critical construction, since the probability of catastrophic brittle failure has thus been increased with the corresponding reduction in constructional strength (see Fig. 8.14).

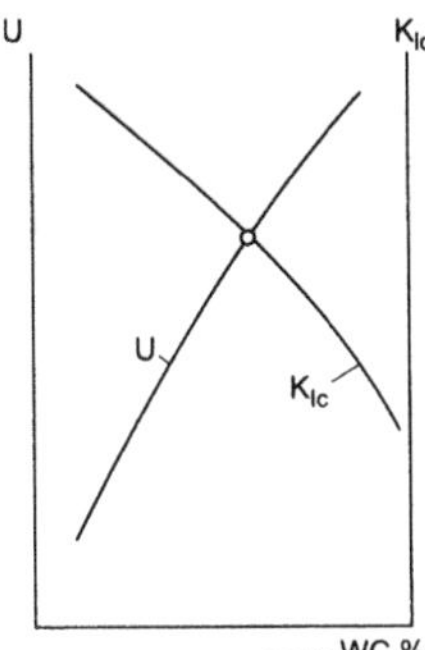

Fig. 8.14. Constructional strength of a coated material

8.4 Internal Stresses (IS) in the Base Metal–Gas Thermal Coating Composition

8.4.1 General Overview

According to Davidenkov, there are three types of internal stresses (I, II, III). IS (internal stresses) of the first type are zonal stresses arising between separate zones of a section and between different parts of the base metal-coating composition. The larger the temperature gradient over the cross-section developing in the course of thermal treatment between various portions of the detail and depending on the rate and uniformity of cooling, size of the detail and a number of other factors, the higher levels of internal stresses of the first type can be produced.

IS of the second type build up either inside a grain or between adjacent particles. They develop between different phases because of their different coefficients of linear expansion or due to the formation of new phases with unequal volumes. Internal stresses of the second type do not depend on the factors causing stresses of the first type.

IS of the third type develop inside a volume of the order of several elementary cells of the crystal lattice. An example of the stress of the third type is a creation of elastic distortions of the crystal lattice around a foreign atom in a solid solution.

Irrespective of their type, all internal stresses eventually cause the same effect – elastic deformations and distortions of the crystal lattice.

IS of the first type, which are the most significant of all stresses since it is them that cause buckling of the detail and crack initiation, depend not only on external factors, but also on the properties of the base metal-coating composition. If a metal has low plasticity, the created IS cannot be relieved by plastic deformation, and when stress values exceed the strength limit, a crack will be produced.

In the course of heating or cooling the base metal-coating composition IS are redistributed and changed. For example, surface layers of a coating undergo compression stresses during heating, since their tendency to expansion is hampered by the colder metal layers of the central part. And vice versa, during cooling the surface layers that have lower temperature as compared to the central portion of the detail experience stretching stresses, while the center is subjected to compression.

The stresses preserved in a detail a long time after their sources have been removed are called residual.

8.4.2 Development of Internal Stresses and Their Influence on Constructional Strength

IS of all three types can occur both in coatings and in the substrate, and their values vary in wide limits depending on the technique of deposition, thickness of the coating, properties of deposited material, preparation of the surface, technological regime of the deposition, cooling conditions and so on. With time residual stresses which can have opposite signs, attain considerable values, and have non-uniform distribution in the deposited layer and the base metal are developed. The presence of residual stresses is typical for all coatings.

The development of internal stresses in coatings is caused by various reasons, such as: different values of linear expansion coefficients of the coating and the base metal (thermal residual stresses); variations in specific volume of structural phases; non-uniformity of the distribution of powder material in plasma or gas flow; constructional properties of the sprayed product, etc. Cooling also promotes normal and tangential to the coating-base metal interface boundary stresses in the coating creating deformations that may ultimately lead to brittle failure [8.43, 8.46].

The mechanism of development of internal stresses in gas thermal coatings has been addressed in a number of works [8.1, 8.10, 8.13, 8.15, 8.22].

The largest effect IS have on the strength of the base metal-coating bonding: at high values of IS spontaneous scaling or crack initiation in a coating are possible [8.20, 8.28]. If the detail is not rigid enough, internal stresses can distort its shape or warp it. The level, sign and distribution pattern of IS determine constructional strength of details and to a certain degree determine chemical, mechanical and physical properties of the coating [8.11, 8.14, 8.30, 8.44, 8.51].

Although the effects of internal stresses on wear resistance of machine details were investigated many times, the obtained results often contradict each other. Some researchers claim that internal stresses, independently of their sign, do not have any effect on wear resistance of details. As for the initiation and propagation of fatigue cracks, compressing IS are favorable for crack resistance irrespective of their origin (thermal, mechanical).

8.4.3 Control of the Stressed State

The following technological approaches provide a means for controlling the internal stresses in the base metal-coating composition:

- matching of properties of the coating and base metal materials, first of all, their coefficients of linear temperature expansion [8.1, 8.22, 8.36];
- adjustment of thermal impact upon the base and the particles by changing the thermal power distribution over the heated spot, as well as by variation in spraying distance or the displacement speed of the laser, plasmotron or burner [8.22, 8.52];
- reduction of the elasticity modulus of the coating material [8.2, 8.52];
- introduction of underlayers between the coating and the substrate to provide smooth property transition from the substrate material to the coating material [8.29, 8.42];
- variation in the coating thickness and deposition of multilayer coatings with alternating layers made of different materials [8.4, 8.23];
- reinforcement of the coating with continuous or discrete fibers and wires [8.22];
- changing the shape of the sprayed surface [8.7].

It should be noted once again that compression stresses in coatings are the safest for details in operation and should thus be preferred.

8.4.4 Methods of Measuring Internal Stresses

Methods of IS measurement can be divided into destructive and non-destructive.

Among the non-destructive techniques X-ray method of measuring IS is most commonly used [8.3, 8.9, 8.16, 8.24]. Wide application of tilted photography allowed application of X-ray strain measurements for determination of internal stresses. The method is based on measurement of microdeformations in the crystal lattice of the material caused by residual stresses. The main advantages of the X-ray method are: determination of stresses without destruction of the object under study; elimination of foreign factors that can influence the object during measurement; the possibility of repeated measurements on the same portion of the specimen; locality; suitability for stress measurement in the details with complex configuration; the possibility to differentiate the measured stresses by the direction of their action; high accuracy and performance.

The main disadvantage of the X-ray method as applied to determination of internal stresses is its unusability in the presence of considerable concentration gradients in the surface layer of material.

Among the destructive methods the so-called mechanical method, based on measurements of specimen deformation caused by load relief, is most often used [8.1, 8.26, 8.48]. Internal stresses inside a body are internally compensated, so that their resultant and moment vanish at any section. If a portion of the stressed body is removed, the balance of stresses in the remainder of the body is disturbed thus causing elastic deformations, which can be measured to determine internal stresses. Load relief is performed by removing the stressed layers using etching, turning and cutting. Thus, determination of internal stresses requires: removing a portion of the stressed body or breaking its bonds; measuring deformation occurring when a portion of the stressed body is removed or its bonds are broken; calculating the stresses from the measured deformations.

Devices allowing automatic recording of sag variations of the tested specimen upon continuous removal of stressed layers using electrochemical or chemical etching are widely used in practical studies.

Disadvantages of the mechanical method are: destruction of the detail in the course of stress measurement; difficulty of stress determination in small objects and separate portions of a body with concentrated stresses; influence of additional factors on the results of measurements for body failure under the action of internal stresses.

Let us consider the two methods, one destructive and one non-destructive, in more detail.

X-ray Method of IS Determination

The results of X-ray photographic determination of I and II type stresses in various coatings carried out by Shmyreva and Vorobiev [8.38, 8.39] give reasons to consider the suggested X-ray photography technique convenient and accessible for research laboratories. Calculation of microstresses using the standard procedure is complicated by at least two factors: phase transformations and changes in chemical composition during deposition. In perpendicular photography the calculation of microstresses by relative change of the crystal lattice of the coating material is not accurate enough, since variations in chemical composition of the coating are not taken into account.

The authors [8.39] suggest a method of tilted photography with subsequent evaluation of microstresses from the equation:

$$\sigma_r = \left(\frac{1}{K} + \frac{1}{L} \sin^2 \Psi \frac{a_\perp}{a_\Psi - a_\perp} \right)^{-1}, \tag{8.22}$$

where $K = -E/\mu$ (E is the stretching modulus of elasticity, μ is the Poisson coefficient); $L = 2E/(I + \mu)$; Ψ is the angle of specimen rotation relative to standard (reference) focusing; $a_\perp$ is the lattice parameter calculated from X-ray photograph taken in the standard focussing of the specimen (perpendicular photograph); a_Ψ is the lattice parameter calculated from X-ray photograph taken after the specimen has been rotated by the angle Ψ relative to standard (reference) focusing.

The technique is capable of sensing the residual stresses without inflicting damage on the specimen. The equation allows determination of microstresses developed during the formation of solid solution without resorting to reference (tabulated) value of the lattice parameter of the coating material that is usually rather hard to find.

The method of tilted photography has been successfully used for determination of microstresses in detonation coatings from nickel powder and solid alloy [8.39]. In the course of depositing these materials both chemical composition and lattice parameters change. X-ray photographs were taken in iron radiation with the specimen set at angles Ψ equal to 90, 90 + 30, 90 + 45, 90 + 65°. The tests have discovered the complicated distribution pattern of residual microstresses over the depth of detonation coatings [8.39].

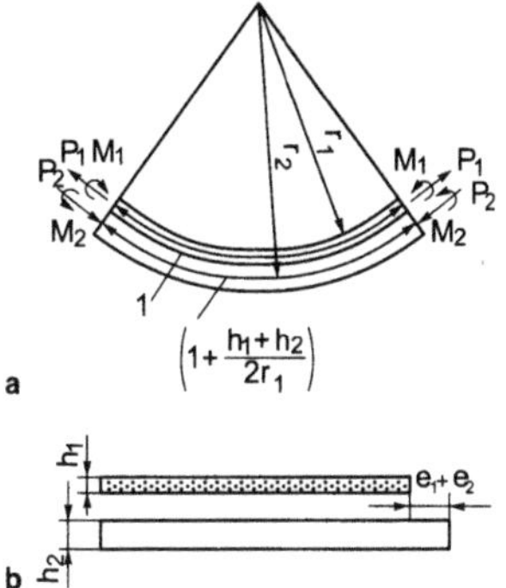

Fig. 8.15. Specimen for determination of residual stress values (**a**) before and (**b**) after separating the coating from the substrate (adopted from [8.15])

Mechanical Method of IS Determination

A coating is deposited on a long, narrow and thin plate. Under the action of the developing internal stresses the plate is deformed into the shape of an arc of a circle. Measuring the radius of the deformed plate yields the values of stress in the coating. Although generally the stresses are distributed non-uniformly over a section of the coating, in determination of internal stresses their distribution is assumed to be uniform, i.e. only average values are evaluated [8.15].

The length of the central axis of the coating bonded with the substrate is taken to be equal to unit. It is assumed that after the coating has been separated from the substrate and relieved from residual stresses, the deformations of the coating and substrate are equal to ε_1 and ε_2, respectively (see Fig. 8.15) [8.15].

The following symbols are adopted in the forthcoming discussion:

E_1, E_2 are the longitudinal moduli of elasticity of the coating and substrate, respectively; E_1J_1, E_2J_2 are the bending toughnesses of the coating and substrate, respectively; h_1, h_2 are the thicknesses of the coating and substrate, respectively; b is the width of the substrate; ρ_1, ρ_2 are the radii of curvature of the coating and substrate, respectively; P_1, P_2 are the axially directed forces of the coating and substrate; M_1, M_2 are the bending torques of the coating and substrate; μ is the Poisson coefficient of the coating and substrate; J_1, J_2 are the axial moments of inertia of the coating and substrate.

The total deformation of the specimen after the coating has been separated from the substrate and residual stresses have been relieved is equal to $\varepsilon_1 + \varepsilon_2$. In this case the deformation of the coating can be written as:

$$\varepsilon_1 = (\varepsilon_1 + \varepsilon_2)\frac{h_2 E_2}{h_1 E_1 + h_2 E_2}. \tag{8.23}$$

Thus, the average residual stress in a coating σ_r may be represented in the following form:

$$\sigma_r = (\varepsilon_1 + \varepsilon_2)\frac{h_2 E_2}{h_1 E_1 + h_2 E_2} E_1 . \tag{8.24}$$

Once $\varepsilon_1 + \varepsilon_2$ are calculated [8.15], the following equation can be obtained:

$$\sigma_r = \left[\frac{h_1^3 E_1 + h_2^3 E_2}{6\rho h_1 (h_1 + h_2)} + \frac{E_1 \left(h_1^3 E_1 + h_2^3 E_2\right)}{12\rho^2 (h_2 E_1 + h_1 E_2)} + \frac{E_1 E_2 h_2 (h_1 + h_2)}{2\rho (h_1 E_1 + h_2 E_2)}\right]\frac{1}{1-\mu^2}, \tag{8.25}$$

where E_1 and E_2 are substituted for $E_1/\left(1-\mu^2\right)$ and $E_2/\left(1-\mu^2\right)$, respectively.

Assuming that Poisson coefficient μ is equal to 0.3 for both the coating and substrate, and experimentally determining longitudinal moduli of elasticity E_1, E_2, thicknesses of the coating and substrate h_1, h_2, and radius of curvature of the specimen, we can calculate the average value of residual stresses in the coating using equation (8.25) [8.15].

8.5 Internal Stresses in the Base Metal–Galvanic Coating Composition

8.5.1 General Overview

IS developing during electrocrystallization are one of the most important characteristics of galvanic coatings. In most cases IS lead to cracking, increase in porosity, and reduction in protective ability and crack resistance of coatings and their scaling off the substrate. On the other hand, certain enhancements of hardness, luster, and wear resistance of some coatings with high IS have been reported.

The nature of IS in galvanic coatings was extensively studied in numerous investigations by Russian and foreign researchers, and the basic hypotheses concerning the origin of IS had been developed approximately by 1970. However, as was stressed in a recent critical review [8.32], most of them explain only individual aspects of the phenomena, while others are imperfect from the modern point of view.

In the last two decades the problems of the theory of IS practically fell out of consideration, excluding rare occasional publications. In this situation the views of different researchers on the development of IS do not coincide, and what probably is even more important, there is currently no settled viewpoint about the practical consequences following from the mechanism of IS influence on the properties of coating. The prospects of manufacturing coatings with controllable IS and high operating characteristics obviously require a model of IS formation developed on the basis of present-day knowledge and adequately representing the effects of various factors of the electrodeposition process on the structure and properties of galvanic coatings.

In view of this, the development of reliable methods of IS determination is of great importance. On one hand, IS determine operating characteristics of the produced coatings. On the other hand, owing to their high sensitivity to conditions of electrodeposition, IS allow detection of fine effects in the process of electrocrystallization.

The first part of the present section summarizes the existing concepts on the nature of IS in galvanic coatings. An atomic-vacancy model of IS formation based on the analysis of structural data and authors' own studies carried out using high-resolution physical techniques is formulated. Within the framework of this model the various phenomena accompanying the processes of production and operation of coatings are explained, and the methods of IS control are demonstrated. The second part presents schemes and structures of various installations for determination of IS in electrodeposited coating, including precision machines employing laser interferometry effects, and provides examples of the application of new techniques for solution of certain scientific and applied problems.

8.5.2 Nature of Internal Stresses

Let us formulate the principal postulates of the existing hypotheses about IS origin [8.19, 8.32].

The hypothesis of Kolschutter accounts for formation of stretching IS by coarsening of the structure in the postelectrolytic period and the tendency of the coating to reduce volume. This hypothesis assumes the recrystallization at room temperature in coating of high-melting metals (such as iron, nickel, cobalt, etc). However, their homologous temperature of the onset of recrystallization, even taking into consideration their high dispersity, is considerably higher than room temperature. In low-melting metals (such as zinc, lead, indium), where recrystallization in postelectrolytic period is possible, contrary to the Kolschutter hypothesis the compression IS develop which cannot be accounted for by the hypothesis at all.

The hydrogen hypothesis, developed by several authors, attributes IS to one or another mechanism of the effect of hydrogen released during electrolysis on the structure of the coating. The following mechanisms have been called upon: capture of adsorbed hydrogen by layers of the coating; formation of hydrides of the deposited metal with their subsequent capture by layers of the coating; formation of hydrides inside the coating due to hydrogen captured by the coating; occlusion of the co-deposited hydrogen by the formed layers of the coating; migration of hydrogen atoms in the crystal lattice or along inter-grain boundaries; decomposition of hydrides; removal of gaseous hydrogen from the coating, etc. Not denying the role which hydrogen may play in IS formation, it is necessary to note that on many occasions electrodeposition is performed practically in the absence of hydrogen, for example, in production of low-melting coatings crystallizing with compression IS.

The energetic concept put forward by Soderberg and Gram relates IS in coatings to the overstress of metal release, which causes high level of free energy. In the opinion of Soderberg and Gram, it is equivalent to deposition of metal at elevated temperatures. Stretching IS then develop because of the tendency of the coating to compression upon transition to the equilibrium state. However, the mechanism of the formation of compression IS, the change of the stress signs and some other facts are not still explained.

The hypotheses of Houre and Arrowsmith include macro- and microscopic variants. Macroscopic variant of the concept is close to Joffe's considerations and does not provide an explanation of the development of compression IS either. In the microscopic variant an attempt was made to employ the dislocation theory for the explanation of the nature of IS. According to the ideas of Houre and Arrowsmith, accumulation of linear dislocations forming in the coating perpendicularly to the growth surface, leads to bending of the coating together with the cathode in the direction of the coating and to the development of stretching IS. When low-melting metals are deposited, compression IS develop and bend the coating in the opposite direction due to absorption of oxides or hydroxides of these metals which promote formation of the dislocations of the opposite sign. However, this concept cannot be accepted either because of the static character of the model, which ignores movement of dislocations, and ambiguity of a number of postulates such as, in particular, the mechanism of formation of dislocations during the development of compression IS.

A large body of experimental data can be explained in the framework of the dislocation-sorption model put forward by Popereka [8.32]. The dislocation model of stretching IS formation is based on the assumption that dislocations move spontaneously toward peripheral areas and annihilate on the boundaries of newly formed crystallites. Reduction in dislocation density requires a decrease in the volume of the crystal, which is prevented by adhesion bonds with the substrate; as a result, a planar stressed state characterized by stretching macrostress is formed.

The sorption model of compression IS formation is based on the assumption that foreign particles captured by the coating tend to increase the volume of the inter-crystallite voids, since these particles escape from the influence of the double-layer electric field. The tendency to increase volume is resisted by metal-base adhesion bonds, which results in a planar stressed state characterized by compression macrostress.

And still, the dislocation-sorption model of IS formation is not perfect, since from the viewpoint of modern knowledge its explanation of many phenomena (such as the mechanism of formation of the compression IS in galvanic coatings, the kinetics of stretching IS formation) are not entirely convincing, and certain effects (postelectrolytic changes of IS in coatings, the influence of detergents, impurities, and atoms of alloying elements on IS) cannot be explained at all.

Thus, despite numerous studies devoted to the problem of IS formation in galvanic coatings, their nature still remains unclear.

Before proceeding to consideration of the nature of IS, it should be noted that all electrodeposited metals (under stationary electrolysis conditions and with no organic additives in the solution) can be divided into three groups according to criterion K_p (see Table 8.2).

Table 8.2. Structure and IS of electrodeposited metals

Group	Kp value	Metals	IS [MPa]	Grain size, [cm]	Prevailing point defect
1	7–40	Re, Ru, Rh, Cr, Pt, Pd, Fe, Co, Ni, Mn	+(100–1100)	10^{-4}–10^{-6}	Vacancies
2	2–11	Cu, Au, Ag, Sb, Ga	±(10–200)	10^{-3}–10^{-5}	Vacancies or inter-nodal atoms
3	1–5	Zn, Pb, Cd, Bi, Sn, Tl, In	–(1–80)	10^{-2}–10^{-3}	Inter-nodal atoms

The combined criterion for the sign of internal stresses determines whether a metal is deposited with stretching or compression stresses and reflects the influence of both electrocrystallization conditions and physical nature of the metal:

$$K_p = \sqrt{\frac{\Delta U z E \rho}{A}}, \frac{J^{1/2} \cdot V^{1/2}}{cm^3}, \tag{8.26}$$

where ΔU is melting heat, z is valency, A is atomic mass, ρ is density, E is the deposition potential.

The metals of the first group, characterized by the strong bonds between atoms and crystallizing at high overvoltages of the cathode, are deposited with stretching IS and have $K_p \sim 7$–40. Conversely, the metals of the third group having weak interatomic bonds and crystallizing at low overvoltages, are deposited with compression IS and have $K_p \sim 1$–5. As for the metals of the second group demonstrating the stresses of both signs, they have $K_p \sim 2$–11. These metals either have strong inter-atomic bonds, but deposit at low overvoltages, or, otherwise, have weak inter-atomic bonds and deposit at high overvoltages, or, finally, are characterized by intermediate parameter values as compared to the metals of the first and the second groups.

Analysis of the data given in Table 8.2 shows the existence of certain correlations between the mechanisms of formation of the structure and properties of the coatings and the conditions of electrolysis for metals of different groups. First of all, it may be noted that the coatings from different groups have different dispersity.

The table clearly demonstrates that the average size of the crystallites for metals of the first group is of the order of 10^{-5}–10^{-6} cm, while for the third group this value is about 10^{-2} cm, i.e. 3–4 orders of magnitude greater. Such a substantial difference determines the details of the fine structure of crystallites of electrodeposited metals from different groups.

It is known that as the size of a crystal metal particles decreases, the number of vacancies inside it abruptly increases due to dimensional vacancy effect. The method of positron annihilation completely verifies the dominating influence of this type of defects, showing vacancy concentration of $\sim 10^{-2}$ after completion of electrodeposition, which considerably exceeds the number of vacancies in thermodynamically equilibrium state. Formation of vacancies leads to local distortions of crystal lattice due to displacement of adjacent atoms from their stable locations. The relative distance of approach for atoms of different metals in the first coordinate sphere varies from 2–3% for close-packed lattices (face-centered cubic, face-centered closely packed) to 6–7% for more open ones (body-centered cubic, simple cubic); relaxation slowly decays in magnitude covering 4–6 coordinate spheres or more. Though atomic displacement is non-monotonous, as a whole a vacancy tends to stretch out the lattice. As demonstrated by the results of computer simulations, for such long-range action of a vacancy and vacancy concentration of $\sim 10^{-2}$–10^{-3} after the electrodeposition is completed, all atoms in the crystal lattice of the coating are found in the stress fields generated by these point defects.

Due to the macrocrystalline structure of the coatings for metals of the third group, the assumption that the presence of vacancies can be the main cause of IS has no ground. However, in these metals the prevailing effect of the point defects with the "reverse sign" with respect to vacancies can be observed, i.e. the effect of inter-nodal atoms that show non-equilibrium concentration after electric deposition. Inter-nodal atoms as well as foreign interstitial atoms always present in electrolytes cause local distortions in the crystal lattice of the coating within at least 5–6 coordinate spheres, since atoms located in the nodal points of the crystal lattice are displaced from their stable positions. Thus, the size of atomic displacement in the first coordinate sphere varies from 12% to 20% depending on the metal. The relaxation displacement is non-monotonous, though as a whole an inter-nodal atom causes the lattice to contract. Taking into account the long-range action of an inter-nodal atom and their concentration $\sim 10^{-3}$ in the coating, all atoms located in the nodal points of the lattice are found in the stress fields generated by these point defects.

Let us make a quantitative estimation of the influence that concentration of inter-nodal atoms and vacancies has on IS in electrodeposited metals.

To determine the value of compression IS, let us use the method of "elastic balls" when an inserted (inter-nodal) atom is considered to be an elastic ball with radius R_0 (see Fig. 8.16). The inter-node cavity is considered as a sphere of equivalent volume with radius R_{eq}. When an atom hits an internode, both spheres joint at some value of radius R.

The final size of the inserted atom is determined from the condition of minimal work requierd for deformation of the atom and the medium A_m:

$$A = A_{at} + A_m \rightarrow \min, \tag{8.27}$$

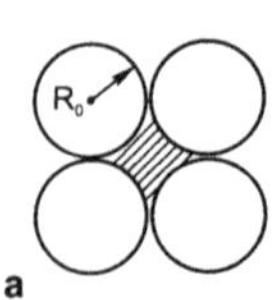

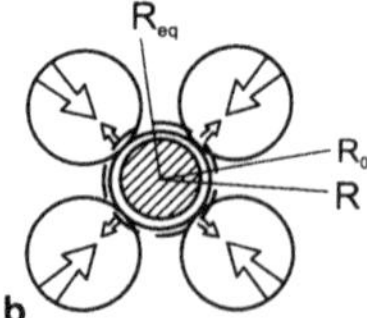

Fig. 8.16. Fragment of a lattice in equilibrium state (**a**) and with an inter-nodal atom (**b**)

$$A_{at} = 6\pi D R_0 (R - R_0)^2, \tag{8.28}$$

$$A_m = 8\pi G R_{eq} (R - R_{eq})^2, \tag{8.29}$$

where D is the modulus of volume elasticity; G is the shear modulus.

It is known that

$$R_{eq} = R_0 (1/\lambda - 1)^{1/3}, \tag{8.30}$$

where $\lambda = V_0/V$ is the relative volume density ($V_0 = 4\pi R_0^3/3$ is atomic volume; V is the volume occupied by one atom in the given structural packing). Thus, for the densest packing $\lambda = (\sqrt{2}/8)(4\pi/3) = 0.74$ and $R_{eq} = 0.706 R_0$. The radius R is found from the condition $\mathrm{d}A/\mathrm{d}R = 0$, giving

$$R = \frac{3D + 4G(1/\lambda - 1)^{2/3}}{3D + 4G(1/\lambda - 1)^{1/3}} R_0 . \tag{8.31}$$

Thermodynamics gives the following expression for the work required to compress an atom:

$$A_{at} = \sigma_0 \frac{4}{3} \pi \left(R_0^3 - R^3\right), \tag{8.32}$$

from which

$$\sigma_0 = \frac{3A_{at}}{4\pi \left(R_0^3 - R^3\right)}, \tag{8.33}$$

or, in view of the condition (8.27)

$$\sigma_0 = \frac{4.5 D R_0 (R_0 - R)^2}{\left(R_0^3 - R^3\right)}, \tag{8.34}$$

where σ_0 is the atom compressive stress created by elastic medium.

In order to calculate the value of stretching IS let us represent a vacancy as a "void atom" with the radius R_0 equal to the radii of other atoms (see Fig. 8.17).

Unlike inter-nodal atoms, the "void atom" tends to draw its neighbors inside, so that the latter by trying to occupy stable positions cause local stretching stresses in the crystal lattice.

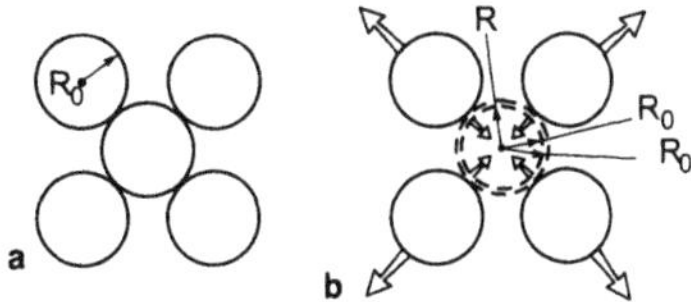

Fig. 8.17. Fragment of a lattice in equilibrium state (**a**) and after formation of a vacancy (**b**)

Deforming the lattice, a "void atom" delivers work that according to the condition (8.29) is equal to

$$A_m = 8\pi G R_0 (R - R_0)^2, \tag{8.35}$$

where G is the shear modulus, $R = R_0(1 - \delta)$ is the radius of "void atom" after deformation (δ is the coefficient of atomic displacement in the formation of a vacancy, equal to 0.02–0.03 for close-packed lattices and to 0.06–0.10 for more open lattices).

At the same time, the work expended on deforming medium is determined as follows:

$$A_m = \sigma_0 \frac{4}{3}\pi\left(R_0^3 - R^3\right) \tag{8.36}$$

from which

$$\sigma_0 = \frac{3A_m}{4\pi\left(R_0^3 - R^3\right)}, \tag{8.37}$$

or in the view of the condition (8.35)

$$\sigma_0 = \frac{6GR_0\left(R_0 - R\right)^2}{\left(R_0^3 - R^3\right)} \tag{8.38}$$

where σ_0 is the elastic medium stretching stress created by a void atom.

Knowing local stretching and compression IS from single point defects and the concentration of inter-nodal atoms and vacancies in electrodeposited metals, IS in the coating can be readily calculated.

For each single point defect the stresses σ_i at the distance R_i from the defect will be equal to:

$$\sigma_i = \sigma_0 \left(\frac{R_0}{R_i}\right)^n, \tag{8.39}$$

where n is the relaxation coefficient describing the law of stress decay with increasing distance (according to numerous works n is taken to be 2–3).

In the range of concentrations of inter-nodal atoms and vacancies typical for electrodeposited low- and high-melting metals, respectively, stress fields generated by the point defects overlap in space. The radius of the sphere of equal interaction potential R_{ia} depending on concentration of one or another type of point defects C is determined from the equation:

$$R_{ia} = \frac{R_0}{\sqrt{2}}\left(C^{-1/3} - 1\right). \tag{8.40}$$

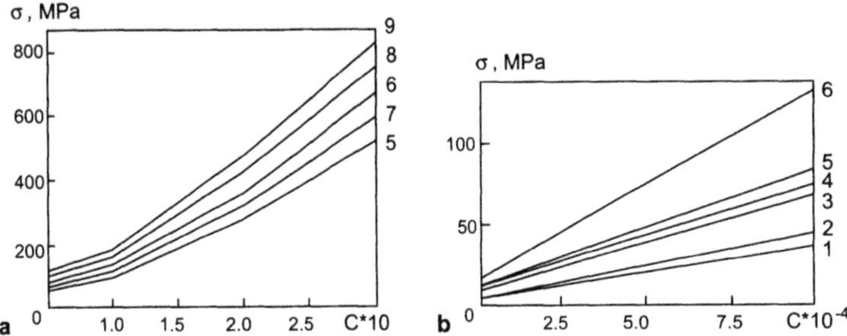

Fig. 8.18. Dependence of IS versus concentration of vacancies (**a**) and inter-nodal atoms (**b**) in electrodeposited antimony (*1*), bismuth (*2*), lead (*3*), indium (*4*), silver (*5*), copper (*6*), gold (*7*), iron (*8*) and nickel (*9*)

In view of equation (8.40) the values of stresses σ_{ia} at the points of equal interaction potential are calculated from the following equation:

$$\sigma_{ia} = \sigma_0 \left(\frac{R_0}{R_{ia}} \right)^n , \tag{8.41}$$

Fig. 8.18 shows internal stresses as functions of concentration of vacancies and inter-nodal atoms for several electrodeposited metals. As can be seen, when concentration of vacancies is higher than $0.5 \cdot 10^{-2}$ or concentration of inter-nodal atoms is higher than 10^{-4}, the stresses caused by these point defects are the prevailing contribution to the value of stretching IS and compression IS developing in the course of electrocrystallization of metals belonging to the first and the second groups, respectively.

For metals of the intermediate group depending on the conditions of electrodeposition the processes proceeding on the cathode and the resulting structural changes may be characteristic for metals either of the first or of the third group. Consequently, the metals of the second group tend to develop IS of both signs.

Thus, non-equlibrium point defects such as vacancies and inter-nodal atoms are the main driving force of IS formation in electroldeposited coatings. The prevailing type of defects in the crystal lattice of the coating is determined by the properties of metal and conditions of electrolysis and in turn determines the sign of IS. Apparently, by changing the proportions of defects in the coating or by reducing their concentration, control of IS and production of low-stressed galvanic coatings become possible.

Such an approach to the nature of IS provides explanation of many experimental facts and phenomena that are not always interpretable within the framework of the known models [8.18].

It should be noted that IS in galvanic coatings has extremely complicated nature determined by a large number of various factors. Nevertheless, the above discussion allows to consider the non-equlibrium point defects in galvanic coatings as the main cause for the development of IS. This does not rule out a possible additive effect of other factors that may introduce some characteristic features in the mechanism of IS formation, which still are consistent with the suggested model.

8.5.3 Methods of IS Measurement

Method of the Flexible Cathode

Experimental setup for IS determination by the flexible cathode method is a cuvette made of organic glass and fitted with a thermoregulating device (see Fig. 8.19a). The cathode in the form of a thin plate insulated with lacquer from the side opposite to the anode is fixed in the clamp with its lower end. The upper end of the cathode plate is 3–4 cm above the electrolyte level and is projected through the lens on the scale with the help of an illuminator fitted with a condenser. If the forming coating has stretching IS, the cathode plate with the coating is bent in the anode direction, while in the case of compression IS cathode bends in the opposite direction. The magnitude of deflection y_f is measured experimentally and is used in further IS calculations according to the following equation:

$$\sigma = \frac{E_c l_c y_f}{3S_c^2 l}, \tag{8.42}$$

where E_c is the cathode elasticity modulus; l_k and l are the thicknesses of the cathode and the coating, respectively; S_k is the length of the cathode; y_f is the deflection of the free end of the flexible cathode from its initial position.

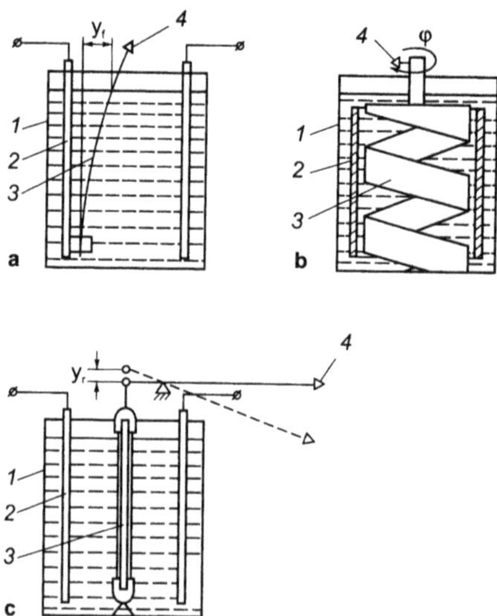

Fig. 8.19. Schemes for measurement of internal stresses by methods of flexible (**a**), spiral (**b**) and ribbon-type cathodes (**c**). *1* electrolyzer; *2* anode; *3* cathode; *4* indicator

Method of the Spiral Cathode

The main elements of a spiral contraction meter are a ribbon-type cathode folded into a spiral and placed inside a cylindrical anode (see Fig. 8.19b). Metal is deposited on the outer side of the spiral, while its inner side is protected with lacquer. Depending on the sign of IS, the spiral either twists or untwists turning the pointer by an angle φ. Angular deformation of the spiral measured by displacement of a light beam along the scale is used as input data for calculating IS using the following equation:

$$\sigma = \frac{E_c l_c^2 \varphi}{12\pi R n l}, \tag{8.43}$$

where E_c is the cathode elasticity modulus; l_c and l are the thicknesses of the cathode and the coating, respectively; φ is the spiral twisting angle; R is the radius of the spiral; n is the number of turns in the spiral.

In both these methods the swell of insulating lacquer results in parasitic deformation. An attempt to get rid of the lacquer in solution promoted the development of the method of IS determination by bending of a disk cathode [8.32]. In this method a layer of coating is deposited on a disk cathode that serves as a bottom of an electrolyzer. A gauge connected to the strain-measuring circuit is glued onto the center of the cathode from the outside. IS in the coating cause the cathode to bend thus inducing stresses in its outer layers which can be recorded using the gauge and the strain-measuring device. However, the disk cathode method produces comparatively large errors in the calculated values of IS, so it has not been widely accepted.

Method of Stretching-Compression of a Wire Cathode

The method of stretching-compression of a ribbon-type cathode turned out to be much more advanced [8.32]. Like the previous one, this method was "lacquerless" and, besides, for the first time it provided the possibility of direct IS determination through the deformation caused by the stresses, rather than using indirect indicators like bending, twisting, etc. In a setup with leverage-optical magnification (see Fig. 8.19c) the lower end of the ribbon-type cathode is fixed in the rod clamp, while the upper end is attached to a short lever arm through a coupling nut. The anodes are positioned on the walls of the cuvette parallel to the cathode. The uniformity of metal deposition is provided by the shape of the cuvette having a small gap between its walls and the anodes. Under the action of IS the ribbon cathode changes its length. The magnitude of this change denoted as y_r is measured by the leverage-optical system and serves as input data for calculating IS using the following equation:

$$\sigma = \frac{E_c l_c y_r}{2 S_c l}, \tag{8.44}$$

where E_c is the cathode elasticity modulus; l_c and l are the thicknesses of the cathode and the coating, respectively; S_c is the cathode ribbon length; y_r is the deformation of the ribbon-type cathode.

In electrocrystallization of certain metals (chromium, for instance) the circumstances like high temperatures, harsh medium and vigorous gas production at the cathode make automatic IS detection using the described techniques difficult. This problem can be solved by using a highly sensitive movable-electrode tube element combined with a wire cathode [8.41].

The installation (see Fig. 8.20) consists of an electrolysis cell, a movable-electrode tube for measuring the cathode deformation, and a device for recording the results of measurement. A tensioned wire cathode 0.5–1.0 mm in diameter is fixed in the cell and attached to the rod of the movable-electrode tube, which is set to zero with a micrometric screw and a spring. The installation provides continuous measurement of sign-variable cathode deformations with accuracy of at least 1 μm.

Laser Interferometry

All the described methods for IS determination have become classical. Reliable recommendations for preferable application of each method depending on its character and the research problem are currently available. Therefore today the techniques of IS determination are developing mainly in the direction of increasing the sensitivity of experimental setups. Application of resistor strain gauges, movable-electrode tubes and, finally, laser interferometers are the representatives examples of this tendency.

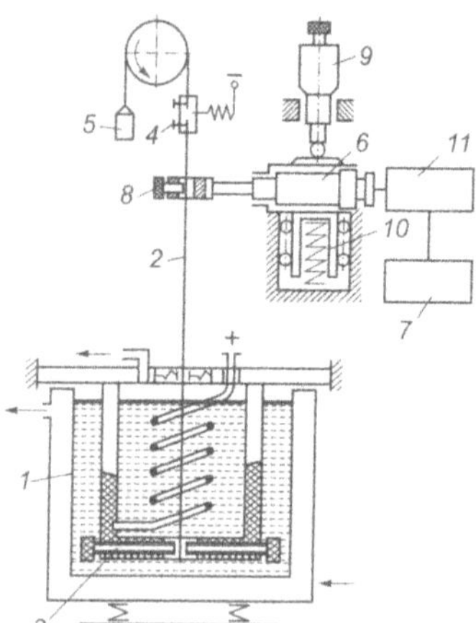

Fig. 8.20. Setup for measurement of internal stresses in electrolysis by the method of stretching-compression of a wire cathode. *1* electrolytic cell; *2* wire cathode; *3* screw; *4* clamp; *5* weight; *6* movable-electrode tube; *7* recording device; *8* screw; *9* micrometric screw; *10* spring; *11* matching unit

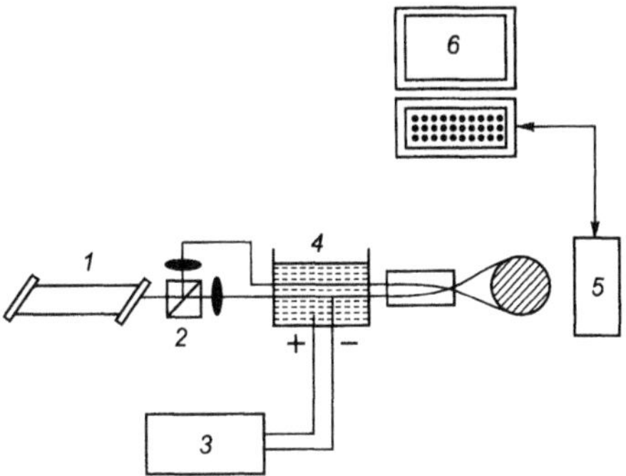

Fig. 8.21. Scheme of laser interferometer for determination of internal stresses by the method of stretching-compression of a glass light guide (by [8.6]). *1* He–Ne laser; *2* beam splitter; *3* power supply; *4* electrolysis cell; *5* light diode device; *6* microcomputer

The latter technique provides the highest attained sensitivity and accuracy, which makes laser interferometry the most promising way of registering IS. Butler and Ginley [8.6] have suggested the following scheme for application of the interference effect to determine IS in galvanic coatings (see Fig. 8.21).

The coating is deposited on the outer surface of an optical light guide in the form of a glass fiber 4 μm in diameter precoated with metal. Two light beams coming from the same source and passing through the main and auxilary light guides create the interference pattern. Changes in the length and the refractive index of the main light guide due to the action of IS in the coating shift the interference pattern, which allows quantitative determination of IS values.

However, an analysis shows that although the described scheme provides very high metrological characteristics, it still has some limitations. First, it is not universal and can be applied only to determination of IS using the wire cathode stretching-compression method and only for thin layers. And, second, because of the specific structure of the cathode the device only remotely approximates the real electrodeposition processes since the base material and the state of the surface have substantial influence on IS in the coating.

The authors of [8.49] have eliminated the mentioned shortcomings in a setup designed for IS detection and determination and based on the principle of measuring displacements with the help of laser interferometry.

Independently of the actual method of IS determination in all schemes linear or angular displacement of the cathode or some element connected to it is eventually detected. For the purpose of illustration let us consider a variant of IS measurement by the method of stretching-compression of a ribbon-type cathode.

The setup consists of an electrolysis cell and a laser interferometer (see Fig. 8.22). A half-transparent mirror splits the light beam coming from the laser into reference and measuring beams. The beams are reflected from the reference and main mirrors with outer reflecting coatings and combined again to produce an interference pattern at the registering device.

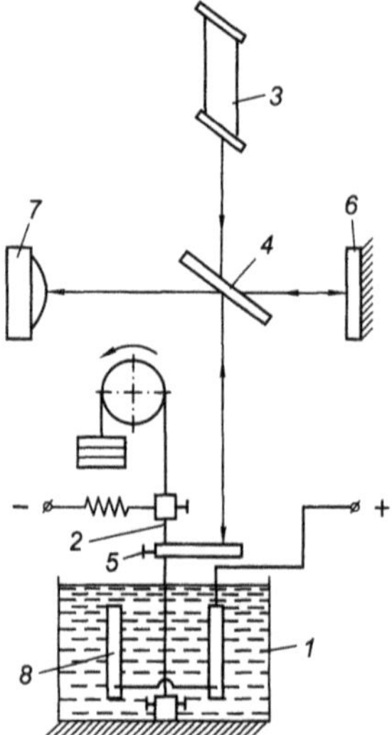

Fig. 8.22. Setup for detection of internal stresses by the method of stretching-compression of a ribbon-type cathode using laser interferometer. *1* electrolysis cell; *2* ribbon-type cathode; *3* He–Ne laser; *4* beam splitter; *5* main mirror; *6* reference mirror; *7* recorder; *8* anode

In the course of electrocrystallization the action of IS in the coating induces observable changes in the length of the ribbon-type cathode length. Deformation of the cathode causes a displacement of the main mirror connected to it from the position corresponding to zero value of IS in the coating. The displacement of the mirror creates a path-length difference between the main and the reference beams. A Michelson interferometer is tuned for operation with bands of equal width. Displacement of the working mirror by $\lambda/2$ leads to shift of the pattern of interference bands by one maximum $\Delta n = n\lambda$ ($n = 0,1,2\ldots$). Thus, observation of the order of interference bands allows determination of the cathode deformation and subsequent IS calculation using the known techniques with accuracy ±0.5 MPa.

Laser interferometry can be used with great efficiency in the studies of small deformations and stresses induced by various electrochemical processes including electrocrystallization, corrosion and hydrogen pickup by metals.

Interference Holography

Even more promising is application of interference holography. The principally new possibilities opened by interference holography are connected not only with enhanced accuracy and sensitivity of the method based on non-contact determination of stresses and deformations, but also with the ability of the interference pattern to visualize the character of deformation over the whole surface of the coating, thus revealing the IS distribution pattern.

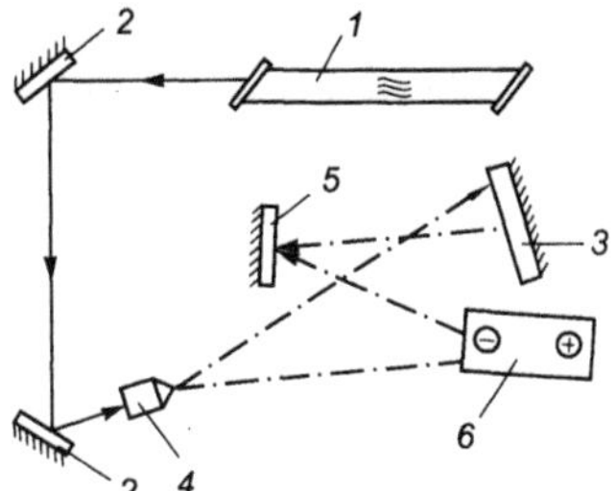

Fig. 8.23. Optical scheme of the setup for determination of internal stresses using interference holography. *1* He–Ne laser; *2* auxiliary mirror; *3* main mirror; *4* short-focus lens; *5* photographic plate; *6* cathode/specimen

An installation for IS determination using technique designed by the authors of [8.50] consists of an electrolysis cell with a flexible cathode and a holographic interferometer (see Fig. 8.23). The interferogram is recorded in the optical scheme of the off-axis holography.

A beam of light from a He–Ne laser is deflected by auxiliary mirrors with reflecting front surface and expanded by a microscope objective with a short-focus lens. The expanded beam simultaneously illuminates the specimen-cathode under test and the main mirror. The light reflected from the specimen (object beam) and the main mirror (reference beam) fall on the photographic plate producing an interferogram. Holographic interferorams are obtained by the method of double exposure. Two holograms of the specimen in different states, for example, before and after electrodeposition, are subsequently exposed on the same photographic plate. When the reference beam illuminates the developed plate, an interference band pattern carrying information about the changes in the object state that took place between the subsequent expositions appears on the background of the surface image.

Interpretation of the interference pattern is based on modeling the electrolysis cell cathode by a flat cantilever with one end rigidly fixed. IS in the coating exert flexure strain in the cathode in the direction perpendicular to the photographic plate, and the free end of the cantilever shifts along the W axis (see Fig. 8.24). Fig. 8.24b illustrates schematically an interference band pattern for this case. A light band adjacent to the stationary clamp is taken as zero ($n = 0$), and all subsequent bands are numbered in increasing order ($n = 1, 2, 3$, etc.). The magnitude of the specimen deflection W_d is related to the order (the number) of the band n by the following equation [8.47]:

$$W_d = n\lambda/(\cos\alpha + \cos\beta), \tag{8.45}$$

where λ is the wave length of the laser beam; α is the angle between the direction of the object beam and the normal to the photographic plate (observation angle); β is the angle between the direction of the reference beam and the normal to the photographic plate (illumination angle). In the given setup $\lambda = 0.6328$ μm, $\alpha = 18°$,

$\beta = 22°$. Thus, to determine deformation of the specimen in any point just the appropriate order of the interference band should be substituted into this equation. The value of IS can then be calculated with accuracy ~0.3 MPa using the known methods.

It should be mentioned that an experimental interferogram can be somewhat different from its theoretical counterpart, since a specimen subjected to the action of IS in the coating, besides pure deflection, undergoes other types of deformation. However, this circumstance does not introduce considerable errors in the results of measurements.

This method can be used to study the effect of such factors as electrodeposition regime, composition of electrolyte, thickness of the coating, state of the substrate, etc. on the sign and value of IS in coatings. The main reason preventing holographic interferometry from being widely used for solution of scientific and applied problems is the necessity to perform works in a specialized laboratory equipped with complicated stationary devices. To overcome this difficulty, an easy-to-install compact holographic installation has been designed to study IS in any research or industrial laboratory.

The installation shown in Fig. 8.25 is used for determination of IS by the method of flexible cathode, but can be easily adapted for any other technique. All constructional elements that are mainly standard units are positioned at two levels on the upper and the lower plate. The lower plate rests on pneumatic supports protecting the entire setup from vibrations. The plate accommodates a laser power supply, an electrochemical power supply (not shown), a hoist, a holographic camera and holographic camera controller.

The upper plate rests on supports coaxial with the pneumatic supports and accommodates a laser, a beam splitter with shutters, object and reference beam expanders and electrode holder. The beam expanders are connected through flexible light guides (not shown) to the outputs of the beam splitter. The components of the plate implement the optical scheme of the off-axis holography. The top part of the camera and the electrolyte bath mounted on the hoist are placed in the slots of the upper plate.

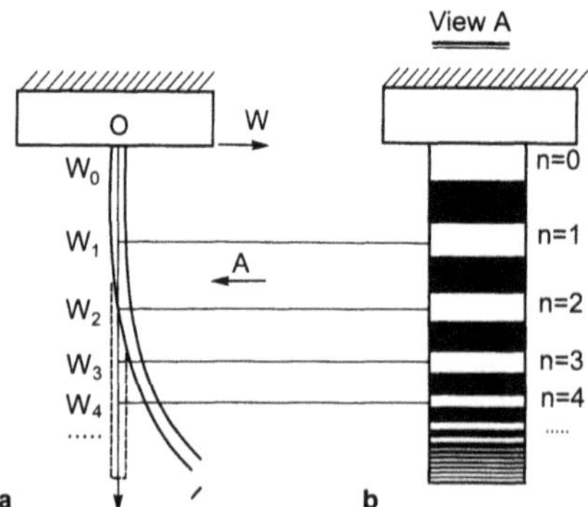

Fig. 8.24. Schematic representation of a specimen/cathode bending interpherogram

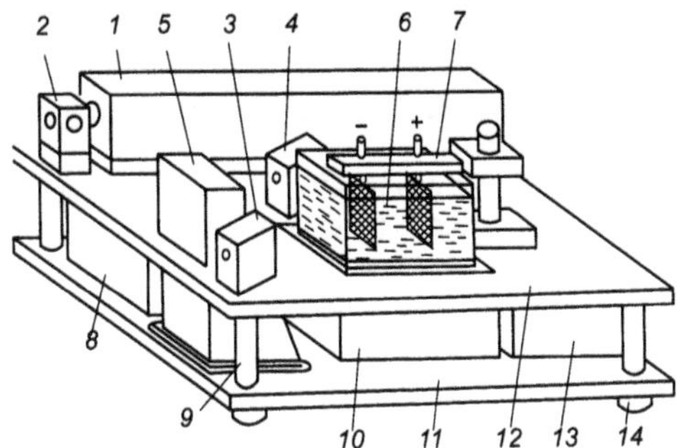

Fig. 8.25. Compact holographic setup for precise measurement of internal stresses in galvanic coatings (shown with removed light protective cap). *1* He–Ne laser; *2* beam splitter integrated with a shutter; *3* object beam expander; *4* reference beam expander; *5* holographic camera; *6* electrolyte bath; *7* electrode holder; *8* power supply; *9* upper plate support; *10* electrolyte bath hoist; *11* lower plate; *12* upper plate; *13* holographic camera controller; *14* vibration damping supports

Application of a holographic camera with recording medium on the basis of thermoplast considerably simplifies the procedure and reduces the time required for recording an interferogram as compared to conventional techniques using photographic emulsion. A special controller automatically controls recording according to a preset program.

In initial position the electrolyte bath is lowered down so that the electrodes are in open air. Once the controller is started up, the shutter combined with the beam splitter opens for the first exposure period thus allowing the camera to record the first cathode hologram. After the exposure is completed, the shutter closes, the hoist lifts the bath, and current in the electrochemical cell circuit is turned on for the pre-assigned period of time. After completion of the electrodeposition process the bath is lowered to the initial position. Thereafter the shutter opens again and the camera records the second cathode hologram on the same storage medium.

The installation has dimensions of only 900x540x440 mm^3 and is portable and easy-to-place.

References

8.1 Barvinok VA (1990) Stressed state control and properties of plasma-sprayed coatings (in Russian). Mashinostroenie, Moscow

8.2 Behnken H, Hauk V (1990) Influence of elastic and plastic strain of the stress of thin film composites. Thin Solid Films 193–194:333–341

8.3 Betsofen S Ya (1991) Specificity of residual stress measurements in TiN coatings. Vide, Couches Minces 47:153–157

8.4 Bijker MP, Van Drent WP, Lodder JC (1995) Stress measurements on magnetic multilayers. J Magn and Magn Mater 147:160–162

8.5 Broek D (1974) Elementary engineering fracture mechanics. Noordhoff, Leiden

8.6 Butler MA, Ginley DS (1987) In-site measurement of strain during electrodeposition. J of the Electrochemical Society 134:510–511

8.7 Chatterjee S, Edwards AG, Feigerle CS (1997) The nature of stresses in hot – filament chemical vapor-deposited diamond thin films on WC substrates. J Mater Sci 32:3355–3360

8.8 Cherepanov GP (1987) Modern problems of fracture mechanics (in Russian). Problemy Prochnosti 8:3–13

8.9 Friesen T, Haupt J, Gissler W, Вата A, Вата PB (1991) Ultrahard coatings from Ti–BN multilayers and by co-sputtering. Surface and Coating Technology 48:169–174

8.10 Harmsworth PD, Stevens R (1992) Microstructure of zirconia–yttria plasma-sprayed thermal barrier coatings. J Mater Sci 27:616–624

8.11 Huntz AM, Lebrun JL, Boumara A (1990) Relation between the residual stresses and the high-temperature oxidation resistance of superalloys protected by plasma-sprayed coatings. Oxide Metals 33:321–355

8.12 Ilinsky KL, Vasilchenko GS, Popov AA (1983) Effects of corrosion hard facing on failure toughness of steel 15H2NMFA (in Russian). Problemy Prochnosti 12:46–48

8.13 Ishida Tsuyoshi, Setoguchi Katsuya, Hiraki Kunihiro (1999) Thermal stress and residual stress control of thermally sprayed 80Ni20Cr coating. Int J A 42:119–125

8.14 Katayama Sakae, Hashimura Masayuki (1997) Effects of microcracks in CVD coating layers on cemented carbide and cermet substrates on residual stress and transverse rupture strength. Trans ASME J Manuf Sci and Eng 119:50–54

8.15 Khasui A (1975) Deposition techniques (in Russian). Mashinostroenie, Moscow

8.16 Kim JG, Yu Jin (1998) Comparative study of residual stresses measurement methods on CVD diamond films. Scr Mater 39:807–814

8.17 Knott JF (1973) Fundamentals and fracture mechanics. Butterworths, London

8.18 Kovensky IM, Povetkin VV (1989) On the nature of internal stresses in electrodeposited coatings (in Russian). J of Applied Chemistry 62:37–44

8.19 Kovensky IM, Povetkin VV (1999) Physical metallurgy of coatings (in Russian). Joint Venture “Internet Engineering”, Moscow

8.20 Krai MV, Davidson JL, Wert JJ (1993) Erosion resistance of diamond coatings. Wear 166:7–16

8.21 Krasovsky AYa, Veinstock VA, Kashtaljan YuA (1978) Application of linear and non-linear fracture mechanics to evaluation of resistance to crack development in structural steel 15H2NMFA (in Russian). Problemy Prochnosti 1:40–44

8.22 Kudinov VV, Ivanov VM (1981) Plasma deposition of high-melting coatings (in Russian). Mashinostroenie, Moscow

8.23 Kuroda S (1992) Fundamental phenomena in spray deposition of surface coatings. Res Activ:9–10

8.24 La Fontaine WR, Paszkiet CA, Korhonen MA, Li Che-Yu (1991) Residual stress measurements of thin aluminum metallizations by continuous indentation and X-ray stress measurement techniques. J Mater Res 6:2084–2090

8.25 Liebowitz H (1968) Fracture – An Advanced Treatise. Academic Press, New York

8.26 Marewski U, Stöver D, Hecker R (1991) Thermomechanical behaviour of thin oxide coatings. Surface and Coating Technology 46:47–63

8.27 McDonald B, Sih GC (1974) Fracture mechanics applied to engineering problems – strain energy density fracture criterion. Eng Fract Mech 6:361–386

8.28 Müller D, Fromm E (1995) Mechanical properties and adhesion strength of TiN and Al coatings on HSS, steel, aluminum and copper characterized by four testing methods. Thin Solid Films 270:411–416

8.29 Nisier SQ, Newaz GM (1998) Transient residual stress in thermal barrier coatings. Trans ASME J Appl Mech 65:346–353

8.30 Perry AJ, Sartwell BD, Valvoda V, Rafaja D, Williamson DL (1992) Residual stress and the effect of implanted argon in films of zirconium nitride made by physical vapor deposition. J Vac Sci and Technol A 10:1446–1452

8.31 Pokhmursky VI, Berezhnitsky LT, Gnyp IP (1984). Analytical determination of failure conditions of materials with protective coatings (in Russian). Zaschitnye Pokrytia na Metallah 18:25–27

8.32 Popereka MYa (1966) Internal stresses in electrodeposited metals (in Russian). West Siberian Press, Novosibirsk

8.33 Prost NE, Marsh KJ, Pook LP (1974) Metal fatigue. Clarendon Press, Oxford

8.34 Romaniv ON (1979) Fracture toughness of structural steels. Achievements of domestic physical metallurgy series (in Russian). Metallurgia, Moscow

8.35 Romaniv ON (1991) Determination of failure resistance (in Russian). In Bernstain ML, Rachstadt AG (eds) Physical metallurgy and thermal treatment of steel. Volume I, book 2. Metallurgia, Moscow, pp 324–350

8.36 Schutz HG, Globmann T, Stover D, Buchkremer HP, Jager D (1991) Manufacture and properties of plasma-sprayed Cr_2O_3. Mater and Manuf Processes 6:649–669

8.37 Shkoljnik LM (1978) Strategy of fatigue testing (in Russian). Metallurgia, Moscow

8.38 Shmyreva GP, Vorobiev GM (1976) Investigation into stresses of the first and second type in detonation-sprayed coatings (in Russian). Problemy Prochnosti 4:66–78

8.39 Shmyreva GP, Vorobiev GM (1983) Radiography determination of macrostresses in coatings (in Russian). Problemy Prochnosti 8:71–73

8.40 Sih GC (1973) Some basic problems in fracture mechanics and new concepts. Eng Fract Mech 5:365–377

8.41 Solovyova ZA, Agiev BU (1996) Internal stresses in chromium coatings measured during electrodeposition (in Russian). Zaschita Metallov 22:82–88

8.42 Sutta Yoshikasu, Toyoda Masao, Hirang Techio, Kusanagi Karamasa, Nakamura Sigeaki, Yamamoto Yochihiro, Mitani Hideki (1991) Analysis of residual stresses in evaporated and plating films. Trans Jap Weld Soc 22:16–22

8.43 Tawancy HM, Sridhar N, Abbas NM, Rickerby D (1998) Failure mechanism of a thermal barrier coating system on a nickel-base superalloy. J Mater Sci 33:681–686

8.44 Taylor KA, Emrick AJ (1992) Comparison of stress and structural composition of sputter deposited thick coatings of TiBz+Ni on polymeric composites. J Vac Sci and Technol A 10:1734–1739

8.45 Timoshenko SP, Goodier JN (1970) Theory of elasticity. McGraw-Hill, New York

8.46 Tushinsky LI, Plokhov AV, Stolbov AA, Sindeev VI (1996) Structural strength of the base metal–coating composition (in Russian). Nauka, Novosibirsk

8.47 Vest I (1982) Holographic interferometry (in Russian). Mir, Moscow

8.48 Vijgen RO, Dautrenberg JH (1995) Mechanical measurement of the residual stress in thin PVD films. Thin Solid Films 270:264–269

8.49 Voronov VS, Kovensky IM, Povetkin VV (1990) Determination of internal stresses in galvanic coatings by means of laser interferometry. Zaschita Metallov 26:318–320

8.50 Voronov VS, Kovensky IM, Povetkin VV (1992) Determination of internal stresses in galvanic coatings by means of holographic interferometry. Zaschita Metallov 28:695–698

8.51 Yamamoto Shuji, Ichimura Hiroshi (1992) Effects of intrinsic properties of TiN coatings on acoustic emission behaviour at scratch test. J Mater Res 7:2240–2247

8.52 Zhou Xianglin, Hu Hangi, Yu Jiahond (1997) Residual thermal stresses of laser cladding of intermetallic–ceramic composite coating. J Univ Sci and Technol Beijing 4:31–33

Conclusions

Last years have witnessed an explosive development of the technologies of protective and wear resistant coating deposition. The result is unquestionable success in increasing constructional strength of products by judicious deposition of coatings with state-of-the-art methods. However, reliability and durability of parts is enhanced not only by the technologies, but also by perfection of the techniques applied studying the structure and properties of coatings and coated materials.

The authors of this book sought to consider and critically analyze the possibilities of both classical and present-day techniques as applied to testing the base metal–coating composition.

Assessing the prospects for research development in this field, we can formulate a number of problems to be solved within the nearest years.

1. Analysis of the large scope of experimental data from various laboratories makes clear that the discrepancies and incomparability of the results produced by studies of coated materials can be attributed, first of all, to the lack of unified system of testing techniques. In the long run it is normative documents that ensure correctness and provide metrological basis of the experimental results. Whereas tribotechnical techniques of coating testing may be considered reasonably well standardized, such questions as consistency in evaluation of physical properties, bonding strength, fatigue characteristics and many others still remain open.
   The newly established standards should provide guidelines for sample selection and preparation of specimens, determine the required number and geometry of the specimens, describe the recommended equipment, regulate the procedure of testing and processing results.
2. A demand for design and commercial production of specialized instrumentation specifically to study the particular characteristics of coatings and coated materials (porosity, bonding strength, internal stresses, thermal conductivity, etc.) is clearly felt. These devices should be easy-to-use, sensitive, automated and inexpensive.
3. When designing new research techniques for the base metal–coating compositions, it should be taken into account that nowadays industry has considerably extended the range of products operating under extreme conditions including forced modes of dynamic, static and cyclic loading, abrasive wear imposing, aggressive medium attacks and so on. Therefore, along with conventional tests it is necessary to comprehensively employ such techniques as acoustic emission, quantitative analysis of wear products, continuous recording of structural changes in the metal-coating contact zone during operation in frictional pair

considering the effect of environment on the failure process. New disciplines, such as stereology, X-ray spectral microanalysis, nuclear gamma resonance, and radio spectroscopy should be more widely involved in studying the structure of the base metal-coating composition. The principles of fracture mechanics should be applied not only to evaluation of crack resistance, but also to calculation of the degree of wear in abrasive wear and to theoretical predictions of the base metal-coating bonding strength.

4. All works concerning studies of structure and properties of coatings, optimization of the deposition regimes and powder composition must be carried out with strict adherence to main principles of statistical processing of experimental data. Modern computers considerably accelerate research by relieving of routine computations, for instance, in evaluating the fatigue characteristics of coated specimens. Standard computer programs will provide improved calculation accuracy, correction for operational conditions and, as a result, reduction in specific quantity of metal of coated products while maintaining the desired level of constructional strength.
   Computer planning of experiments allows optimization of both strategy and tactics of designing and performing research interpretation of the obtained results. Use of mathematical statistics methods ensures the best outcomes with the least expenses when studying the base metal-coating composition.
5. Comprehensive analysis of the structure and properties of coated materials will assist in practical realization of combined hardening, when a coating, for instance, provides enhanced wear- and heat resistance, while bulk-hardened base metal has a sufficient margin of crack resistance. This is achieved through successful implementation of all basic structure-controlling dislocation mechanisms including creation of subgrains, polygons, cells and granular microstructural barriers to provide bulk hardening; separation of dispersed phases, introduction of dissolved substitution and insertion atoms and increase in dislocation density to form special surface properties. The resulting composite will satisfy the requirement of harmonious combination of reliability, durability and strength.

www.ingramcontent.com/pod-product-compliance
Ingram Content Group UK Ltd.
Pitfield, Milton Keynes, MK11 3LW, UK
UKHW021859190726
13853UKWH00003B/1346